Tom,

Thanks for your incredible support!

Eric [signature]

Ken Land [signature]

Sept. 14 2016

AF842518

The Springer Series on Demographic Methods and Population Analysis

Volume 40

In recent decades, there has been a rapid development of demographic models and methods and an explosive growth in the range of applications of population analysis. This series seeks to provide a publication outlet both for high-quality textual and expository books on modern techniques of demographic analysis and for works that present exemplary applications of such techniques to various aspects of population analysis.

Topics appropriate for the series include:

- General demographic methods
- Techniques of standardization
- Life table models and methods
- Multistate and multiregional life tables, analyses, and projections
- Demographic aspects of biostatistics and epidemiology
- Stable population theory and its extensions
- Methods of indirect estimation
- Stochastic population models
- Event history analysis, duration analysis, and hazard regression models
- Demographic projection methods and population forecasts
- Techniques of applied demographic analysis, regional and local population estimates and projections
- Methods of estimation and projection for business and health care applications
- Methods and estimates for unique populations such as schools and students

Volumes in the series are of interest to researchers, professionals, and students in demography, sociology, economics, statistics, geography and regional science, public health and health care management, epidemiology, biostatistics, actuarial science, business, and related fields

More information about this series at http://www.springer.com/series/6449

Anatoliy I. Yashin • Eric Stallard
Kenneth C. Land

Biodemography of Aging

Determinants of Healthy Life Span and Longevity

With contributions by:
Igor Akushevich
Liubov S. Arbeeva
Konstantin G. Arbeev
Irina Culminskaya
Mikhail Kovtun
Julia Kravchenko
Alexander M. Kulminski
Frank A. Sloan
Svetlana V. Ukraintseva
Deqing Wu

Springer

Anatoliy I. Yashin
Biodemography of Aging Research Unit, Center for Population Health and Aging
Duke Population Research Institute & Social Science Research Institute at Duke University
Durham, NC, USA

Eric Stallard
Biodemography of Aging Research Unit, Center for Population Health and Aging
Duke Population Research Institute & Social Science Research Institute at Duke University
Durham, NC, USA

Kenneth C. Land
Biodemography of Aging Research Unit, Center for Population Health and Aging
Duke Population Research Institute & Social Science Research Institute at Duke University
Durham, NC, USA

Department of Sociology
Duke University
Durham, NC, USA

ISSN 1389-6784 ISSN 2215-1990 (electronic)
The Springer Series on Demographic Methods and Population Analysis
ISBN 978-94-017-7585-4 ISBN 978-94-017-7587-8 (eBook)
DOI 10.1007/978-94-017-7587-8

Library of Congress Control Number: 2016934343

Printed on acid-free paper

This Springer imprint is published by Springer Nature
The registered company is Springer Science+Business Media B.V. Dordrecht

Preface

Some people live to age 100 years or more, others become sick and die prematurely. What factors determine human lifespan? Which biological mechanisms are involved in the regulation of health span and longevity? Which forces shape the age pattern and affect the time trend of human mortality? These questions constitute an emerging high priority research area because of the ever-increasing proportions of elderly individuals in most countries and the imperative to improve the health of these populations. Such improvements will require substantially better understanding of the regularities of individual aging-related changes in biological variables (biomarkers) that take place during the life course and their connections with aging-related declines in health and survival.

Better understanding can be reached by proper integration of relevant information accumulated in the field with newly collected data. The data for such studies are readily available today. Many informative datasets have been collected in longitudinal studies of aging, health, and longevity. The dimensions and scope of such information will continually increase over time—resulting in a supermassive "Big Data" problem. Dealing with such data requires efficient approaches that allow the analysis not only of subsets of available data (which practically all traditional methods do) but also the analysis of the entire dataset within the scope of a single comprehensive framework. The goal of this monograph is to show how questions about the connections between and among aging, health, and longevity can be addressed using the wealth of available accumulated knowledge in the field, the large volumes of genetic and non-genetic data collected in longitudinal studies, and advanced biodemographic models and analytic methods.

The distinguishing features of the aging-related declines in health and survival are the development of comorbidity and multimorbidity involving chronic diseases, medical conditions, frailties, and physical and cognitive impairments that are mutually dependent. The dynamic connections among trajectories of aging changes in biomarkers, risks of diseases, and mortality risks are important for evaluating long-term consequences of exposures to environmental factors, medical interventions, and other disturbances. This monograph visualizes aging-related changes in physiological variables and survival probabilities, describes methods, and summarizes the results of analyses of longitudinal data on aging, health, and longevity in humans performed by the group of researchers in the Biodemography of Aging Research Unit (BARU) at Duke University during the past decade.

The process of individual aging involves dynamic relationships among biological variables and their connections to health and survival outcomes. These relationships are captured in longitudinal data. Therefore, the focus of this monograph is studying dynamic relationships between aging, health, and longevity characteristics using longitudinal datasets. A substantial part of this research is based on the idea of using a specific stochastic process model (SPM) of longitudinal data. This model was developed in the 1970s by Max A. Woodbury who worked at the Center for Demographic Studies, Duke University, together with Kenneth G. Manton and Eric Stallard. This model was further developed by the current research team during the past decade to include responses of physiological variables and other biomarkers to persistent external disturbances, aging-related decline in resistance to stresses, adaptive capacity, connections between health and mortality, etc.

Another part of this research uses the Grade of Membership (GoM) model also developed in the 1970s at Duke University by Max A. Woodbury. To our knowledge this was one of the first attempts to address the "Big Data" problem. This model was also further developed by the current research team during the past two decades for application to longitudinal data on health and disability among the general elderly population, as well as for characterization of complex disease progression among Alzheimer's disease patients. Recent development has focused on the relaxation of the GoM model assumptions and a broadening of scope of application under the Linear Latent Structure (LLS) paradigm.

The research questions and related methods described above belong to the field of biodemography of aging—or, more narrowly, biomedical demography—because they link detailed biological and physiological information about aging-related changes in humans with population health and mortality characteristics. Demographers have always wanted to have a mortality model whose parameters could characterize the process of individual aging. Biodemographic concepts, models, and methods respond to this desire but they also make it clear that informative descriptions of the connections between and among aging, health, and longevity require longitudinal data.

In preparing this monograph, we were acutely aware of the needs and interests of the readers. The monograph is part of the Springer Series on Demographic Methods and Population Analysis; our focus on biodemography/biomedical demography meant that we needed to have an interdisciplinary and multidisciplinary biodemographic perspective spanning the fields of actuarial science, biology, economics, epidemiology, genetics, health services research, mathematics, probability, and statistics, among others. We achieved this broad-scaled perspective by enlisting as contributors the entire group of BARU researchers at Duke University whose combined areas of expertise spanned the disciplines listed above and who, most importantly, talked with each other on a daily basis and knew how to communicate their expertise to other collaborators with differing areas of expertise.

The author listings for each individual chapter give appropriate credit to the persons who were primarily responsible for drafting and editing that chapter. The overall integration of the various chapters into a coherent whole was the responsibility of the volume authors (Yashin, Stallard, and Land) who participated in the preparation of the final versions of all chapters and take joint responsibility for the contents of, and any errors in, the final version.

Our goal was a readable yet challenging presentation of the latest methods and results in our sub-field of the biodemography of aging. In addition to our strategy of enlisting multiple contributors, we also endeavored to help the reader by providing multiple sets of guideposts and way-stations, including Chaps. 1, 10, 11, 19, and 20—all of which were designed to either introduce the material coming up or to summarize what was previously presented. Thus, the interested reader can focus directly on the chapters/sections of greatest interest without having to read the entire volume. On the other hand, we presented some of the more innovative mathematical and statistical material in its natural form, with extensive verbal descriptive commentary, so that even the most sophisticated reader would find the material innovative and challenging. We did so, not because we particularly wanted to challenge the reader, but because the full resolution of the issues being considered required this level of treatment. Generally, we avoided derivations of the mathematical results; they are readily available in the published peer-reviewed literature. The one exception was the longitudinal GoM model in Chap. 17, which presented a fully self-contained derivation and explication of the model which previously did not exist in the literature.

The material in this monograph owes a special intellectual indebtedness to Max A. Woodbury who within a period spanning just several years during the 1970s made two seminal contributions that underlie the SPM and GoM modeling approaches developed and presented herein. Like all good ideas, these can be explained in just a few sentences:

1. In the SPM with a Gaussian covariate process and individually measured time-varying biomarkers, the average force of mortality among a cohort of survivors is represented in terms of variables that satisfy only ordinary differential equations and these completely describe the connection between individual-level and cohort-level survival processes.
2. The simple interchange of the order of summation and multiplication in the likelihood for the standard Latent Class Model yields a whole new powerful likelihood that forms the framework for GoM and LLS, both of which provide mechanisms for uncovering the hidden structure of large-scale high-dimensional highly correlated discretely coded measurement data.

The development of the many ideas, methods, and approaches in this monograph was motivated by challenging problems needed to be addressed and stimulated by the creative atmosphere which exists in the entire research field comprising the biodemography of aging, as well as by valuable comments and critical remarks obtained in numerous presentations and discussions of the results of analyses at seminars and local meetings at Duke University, at the semi-monthly meetings of the Long Life Family Study's Research and Development Committee, at the Population Association of America, at the Gerontological Society of America, and at other professional meetings. This development has benefitted immensely from the incredibly supportive and nurturing environment at Duke University, especially at the Social Science Research Institute, which has continued to enable our pioneering work in this research area to reach the point now represented in the present monograph. Financial support for the research presented in this monograph was primarily through NIA/NIH grants which are individually listed in the acknowledgements of each chapter. For all of this we are most grateful.

In closing, we acknowledge the work of Debra Fincham, our Program Coordinator, in assembling the chapters in a common format, making our corrections, and handling a myriad of other issues needed to complete this monograph; and we appreciate the support and patience of Evelien Bakker at Springer in the process leading to publication.

Durham, NC, USA Anatoliy I. Yashin
Durham, NC, USA Eric Stallard
Durham, NC, USA Kenneth C. Land

Contents

1 Introduction: The Biodemography of Complex Relationships Among Aging, Health, and Longevity 1
1.1 Introduction 1
1.1.1 Frailty Models 2
1.1.2 Biodemographic Ideas in Genetic Analyses of Human Longevity 3
1.1.3 Evolution of Aging, Health, and Mortality: Many Open Questions 4
1.1.4 Strehler and Mildvan's Model of Aging and Mortality 5
1.1.5 Historical Roots of the Stochastic Process Model 6
1.2 Information on Aging, Health, and Longevity from Available Data: Part I 7
1.3 Statistical Modeling and Other Advanced Methods of Analyzing Data on Aging, Health, and Longevity: Part II 10
1.4 Conclusion 13
References 14

Part I Information on Aging, Health, and Longevity from Available Data

2 Age Trajectories of Physiological Indices: Which Factors Influence Them? 21
2.1 Introduction 21
2.2 Data: The Framingham Heart Study (FHS) 23
2.3 Methods 25
2.4 Results 25
2.4.1 Average Age Trajectories of Physiological Variables 25

2.4.2 Age Trajectories of Standard Deviations (SD) of Physiological Variables 28
2.4.3 Age Patterns of Survival and Physiological Variables for Smokers and Non-smokers 29
2.4.4 Effects of Education on Survival and Average Age Trajectories of Physiological Indices 31
2.4.5 Age Trajectories of Long Lived (LL) and Short Lived (SL) Individuals 33
2.4.6 Effects of Disease on Dynamic Properties of Physiological Indices 37
2.4.7 Effects of Genetic Dose on Age Patterns of Physiological Indices 40
2.5 Conclusion 42
References 44

3 Health Effects and Medicare Trajectories: Population-Based Analysis of Morbidity and Mortality Patterns 47
3.1 Introduction 47
3.2 Data and Methods 48
3.2.1 Data: SEER-M and NLTCS-M 48
3.2.2 Definitions of Dates of Disease Onset and Dates of Recovery/Remission 50
3.3 Results 51
3.3.1 Age Patterns of Age-Associated Disease Incidence 52
3.3.2 Incidence Rates: Comparisons with Other Studies 54
3.3.3 Age-Adjusted Rates: Gender Disparities, Time Trends, and Sensitivity Analysis 63
3.3.4 Disability and Comorbidity Patterns of Incidence Rates 65
3.3.5 Mortality Age Patterns and Medicare Data 70
3.3.6 Recovery or Long-Term Remission 72
3.3.7 Risk Factors for Disease Incidence 74
3.3.8 Mutual Dependence in Disease Risks: Age-Patterns 78
3.3.9 Comorbidity and Multimorbidity 79
3.3.10 Predictive Population Models 81
3.4 Conclusion 86
References 89

4 Evidence for Dependence Among Diseases 95
4.1 Introduction 95
4.2 Data and Methods 96
4.3 Results 98

4.3.1 Empirical Analyses Reveal Negative Correlations among Major Causes of Death 98
4.3.2 A Dependent Competing Risk Model Capturing Negative Correlations Between Causes of Death 100
4.4 Discussion 102
4.4.1 Evidence of Trade-Offs Between Cancer and Aging 103
4.4.2 Trade-Offs Between Cancer and Other Diseases 104
4.4.3 Time Trends in Negative Correlations Between Cancer and Other Diseases 105
4.4.4 Cancer and Anti-aging Interventions 106
4.5 Conclusion 107
References 107

5 Factors That May Increase Vulnerability to Cancer and Longevity in Modern Human Populations 113
5.1 Introduction: Economic Prosperity, Longevity, and Cancer Risk 113
5.2 The Proportion of People Who Are More Susceptible to Cancer May Be Higher in the More Developed World 119
5.2.1 Improved Survival of Frail Individuals 120
5.2.2 Avoiding or Reducing Traditional Exposures 122
5.2.3 Burden of Novel and Nontraditional Exposures 123
5.3 Some of the Factors Associated with Economic Development and the Western Lifestyle May Antagonistically Influence Aging and Vulnerability to Cancer 128
5.3.1 Cancer and Aging: A Trade-Off? 128
5.3.2 Increased Exposure to Growth Factors 128
5.3.3 Later Menopause 129
5.3.4 Giving Birth at Later Age 130
5.4 Conclusion 131
References 132

6 Medical Cost Trajectories and Onset of Age-Associated Diseases 143
6.1 Introduction 143
6.2 Data and Methods 145
6.2.1 Data 145
6.2.2 Date of Disease Onset Definitions 146
6.2.3 Medical Cost Trajectories 147
6.3 Results 147
6.3.1 Medical Cost as Disease Severity 153
6.3.2 Forecasting Models 154
6.4 Discussion 156
References 160

7 Indices of Cumulative Deficits . . . 163
7.1 Introduction . . . 163
7.2 Conceptualization of the Deficits Index . . . 164
7.3 Cross-Sectional Age Patterns of the Deficits Index as Characteristics of Aging-Related Processes . . . 164
7.4 Deficits Indices and Age as Indicators of Aging-Related Processes . . . 165
7.4.1 Frequency Distributions . . . 166
7.4.2 Correlation of the DI and Age . . . 167
7.4.3 DI-Specific Age Patterns for Decedents and Survivors . . . 167
7.4.4 The DI and Age Patterns of Time to Death . . . 168
7.4.5 The DI and Age Specific Mortality Rates . . . 169
7.4.6 Relative Risks of Death . . . 170
7.5 Longitudinal Analyses: The DI as an Indicator and Predictor of Long Life . . . 172
7.5.1 Construction of Long- and Short-Life Phenotypes . . . 173
7.5.2 Longitudinal Changes of the Mean DI in the SL, LLD, and LLA Cohorts . . . 173
7.5.3 The DI as an Indicator of Frailty . . . 174
7.5.4 The Phenotypic Frailty Index (PFI) and the DI . . . 176
7.5.5 The PFI and DI as Predictors of Death . . . 177
7.5.6 Mid-to-Late Life DIs and Physiological Indices as Characteristics of Long-Term Survival . . . 180
7.5.7 The DI, Endophenotypes, and Long-Term Survival in the FHS . . . 181
7.6 Conclusion . . . 183
References . . . 183

8 Dynamic Characteristics of Aging-Related Changes as Predictors of Longevity and Healthy Lifespan . . . 187
8.1 Introduction . . . 187
8.2 Data and Methods . . . 190
8.2.1 Definitions and Evaluation of Dynamic Risk Factors . . . 190
8.2.2 Statistical Analyses . . . 196
8.3 Results . . . 201
8.3.1 Effects of Individual Dynamics of Physiological Indices at Ages 40–60 on Mortality Risk and Risk of Onset of "Unhealthy Life" at Ages 60+ . . . 201
8.3.2 Effects of Dynamic Characteristics of Physiological Indices with Non-monotonic Age Trajectories on Mortality Risk and Risk of Onset of "Unhealthy Life" . . . 201

8.3.3 Effects of Dichotomized Dynamic Characteristics of Physiological Indices with Non-monotonic Age Trajectories 202
8.3.4 Sensitivity Analyses 203
8.4 Discussion 204
8.5 Conclusion 207
References 208

9 The Complex Role of Genes in Diseases and Traits in Late Life: An Example of the Apolipoprotein E Polymorphism 211
9.1 Genes and Diseases in Late Life 211
9.2 The Antagonistic Role of the APOE Gene and Two Types of Sexually Dimorphic Tradeoffs: The Case of CVD and Cancer 213
9.2.1 The FHSO: Tradeoffs in the Effects of the APOE Polymorphism on the Ages at Onset of CVD and Cancer 213
9.2.2 The FHS: The Antagonistic Role of the APOE Polymorphism in CVD and Its Tradeoffs with Cancer 216
9.2.3 The FHS and the FHSO: Aging-Related Heterogeneity in a Changing Environment 218
9.3 Tradeoffs in the Effects of *APOE* on Risks of CVD and Cancer Influence Human Lifespan 222
9.3.1 The FHS and FHSO: Survival 222
9.4 Conclusion 228
References 228

10 Conclusions Regarding Empirical Patterns of Aging, Health, and Longevity 231

Part II Statistical Modeling of Aging, Health, and Longevity

11 Approaches to Statistical Analysis of Longitudinal Data on Aging, Health, and Longevity: Biodemographic Perspectives 241
11.1 Introduction 241
11.2 Statistical Approaches to Joint Analysis of Longitudinal and Time-to-Event Outcomes 244
11.2.1 Standard Joint Models and Their Extensions 244
11.2.2 The Use of Stochastic Processes to Capture Biological Variation and Heterogeneity in Longitudinal Patterns in Joint Models 250
11.3 Bringing Biology to Statistics: Biodemographic Models for Analysis of Longitudinal Data on Aging, Health, and Longevity 253
References 255

12 Stochastic Process Models of Mortality and Aging 263
12.1 Introduction 263
12.2 Models 266
12.2.1 General Description 266
12.2.2 Estimation Procedure 268
12.2.3 Simulation Studies 272
12.3 Discussion 275
12.3.1 To What Extent Can Mortality Rates Characterize Aging? 275
12.3.2 The Strehler and Mildvan Model 275
12.3.3 Comparing Two Versions of the Stochastic Process Model 276
12.3.4 Modeling Personalized Aging Changes 279
References 279

13 The Latent Class Stochastic Process Model for Evaluation of Hidden Heterogeneity in Longitudinal Data 285
13.1 Introduction 285
13.2 Approaches to the Incorporation of Hidden Heterogeneity in Analyses of Longitudinal and Time-to-Event Data 286
13.3 The Latent Class Stochastic Process Model 289
13.3.1 Specification of the Model 289
13.3.2 Likelihood Estimation Procedure 291
13.4 Simulation Studies 294
13.4.1 Simulation Study for Latent Class Stochastic Process Model 294
13.4.2 Simulation Study for Stochastic Process Model That Ignores Latent Classes 295
13.5 Discussion and Conclusion 298
References 300

14 How Biodemographic Approaches Can Improve Statistical Power in Genetic Analyses of Longitudinal Data on Aging, Health, and Longevity 303
14.1 Introduction 303
14.2 Simulation Studies of the Longitudinal Genetic-Demographic Model 306
14.3 Simulation Studies in Genetic Stochastic Process Model 310
14.4 Discussion 314
References 318

15 Integrative Mortality Models with Parameters That Have Biological Interpretations 321
15.1 Introduction 321
15.2 Conditional Risk of Death and Demographic Mortality Rate 323

15.3 Description of the Processes θ_t and Y_t and Their Connections to t 324
15.4 Evolution of the Conditional Distribution of θ_t and Y_t Among Those Who Survived to Age t 325
15.5 Gaussian Approximation 327
15.6 Conclusion 328
References 330

16 Integrative Mortality Models for the Study of Aging, Health, and Longevity: Benefits of Combining Data 331
16.1 Introduction 331
16.2 Observational Plans and Combining Data 332
16.2.1 Likelihood Function of Life Span Data 332
16.2.2 Longitudinal Data on Physiological Variables: Health Changes Are Not Observed: Observational Plan #1 333
16.2.3 Gaussian Approximation of the Model of Physiological Variables 335
16.2.4 Data on Health Transitions Without Measurements of the Physiological State: Observational Plan #2 337
16.2.5 The Likelihood of the Data on Health Transitions 338
16.2.6 Gaussian Approximation of the Model with Health Transitions 338
16.2.7 Discrete Time Observations of the Physiological State and Health Transitions: Observational Plan #3 339
16.2.8 Gaussian Approximation of the Model of Longitudinal Data on Physiological Variables and Health Transitions 341
16.3 A Simulation Study 343
16.3.1 The Model with Repeated Measurements of a Physiological Variable and Changes in Health State: Observational Plan #3 343
16.3.2 Combining Data with Observational Plans #1 and #2 347
16.4 Discussion and Conclusion 350
References 351

17 Analysis of the Natural History of Dementia Using Longitudinal Grade of Membership Models 353
17.1 Introduction 353
17.2 Methods 356
17.2.1 Model 356

17.2.2 Likelihood 361
17.2.3 Log-Likelihood 363
17.2.4 Derivatives of Log-Likelihood 363
17.2.5 Constrained Log-Likelihood 364
17.2.6 Kuhn-Tucker Conditions 364
17.2.7 Derivatives of Constrained Log-Likelihood 365
17.2.8 Constrained Newton-Raphson Procedures 367
17.2.9 Consistency and Asymptotic Normality 370
17.2.10 Model Testing 373
17.3 Data 375
17.3.1 National Long Term Care Survey 375
17.3.2 Sample Selection 377
17.4 Results 380
17.4.1 Model Selection 380
17.4.2 Model Description 383
17.4.3 Ancillary Analysis: Mortality 395
17.4.4 Ancillary Analysis: Acute and Long-Term Care 398
17.5 Discussion 407
Appendix 410
Synthesis of Known Results Regarding the Consistency of the General (Cross-Sectional) Empirical GoM Model 410
References 415

18 Linear Latent Structure Analysis: Modeling High-Dimensional Survey Data 419
18.1 Introduction 419
18.2 Linear Latent Structure Analysis 420
18.2.1 Structure of Datasets and Population Characteristics 420
18.2.2 LLS Task: Statistical, Geometrical, and Mixing Distribution Points of View 421
18.2.3 Moment Matrix and the Main System of Equations 423
18.3 Computational Algorithm for Estimating LLS Model 428
18.3.1 Moment Matrix Calculation 428
18.3.2 Computational Rank of the Frequency Matrix 429
18.3.3 Finding the Supporting Plane 429
18.3.4 Choice of a Basis 433
18.3.5 Calculation of Individual Conditional Expectations 433
18.3.6 Mixing Distribution 434
18.3.7 Properties of LLS Estimator 434

18.3.8 Clustering 435
18.3.9 Missing Data 436
18.4 Applications 436
18.4.1 Simulation Studies 436
18.4.2 LLS and Latent Class Models 437
18.4.3 LLS and Grade of Membership (GoM) Models 437
18.4.4 Application to the NLTCS Data 439
18.5 Discussion 441
References 443

19 Conclusions Regarding Statistical Modeling of Aging, Health, and Longevity 445

Part III Conclusions

20 Continuing the Search for Determinants of Healthy Life Span and Longevity 453
References 456

Chapter 1
Introduction: The Biodemography of Complex Relationships Among Aging, Health, and Longevity

Anatoliy I. Yashin, Eric Stallard, and Kenneth C. Land

1.1 Introduction

This monograph summarizes the results of selected studies of aging, health, and longevity recently conducted by the members of the Biodemography of Aging Research Unit at the Center for Population Health and Aging, Duke University. These studies deal with secondary analyses of available cross-sectional and longitudinal data. We endeavored to balance the topics of discussion to make them useful for researchers interested in better understanding the connections among aging, health, and longevity, innovative approaches to analyzing available data capable of integrating the body of knowledge accumulated in the field, as well as the results of analyses that demonstrate the efficiency of the proposed methods. The results of analyses of the members of the research group that are methodologically and historically linked to the topics discussed in this monograph but not included in corresponding chapters are reviewed in this introductory chapter. Since a comprehensive survey of the problems, methods, and research results in the entire field of biodemography of aging, health, and longevity was not the goal of this monograph, we acknowledge that many interesting ideas and research topics studied by other research groups are not reviewed in the chapters of this monograph.

Our interest in the biodemography of human aging was motivated by the desire to better understand factors and mechanisms responsible for age patterns and time trends in mortality rates and survival curves. The availability of human longitudinal and cross-sectional data on populations of study subjects made addressing these research questions possible and stimulated the development of methodological ideas on how these data could be used to better understand the forces and mechanisms shaping age patterns of human mortality rates. Substantial progress in manipulating rates of individual aging to increase active lifespan and longevity in populations of laboratory animals has encouraged researchers to search for genetic and non-genetic factors capable of extending life and improving population health in humans (Sierra et al. 2009). To understand whether and how such goals could be

A.I. Yashin et al., *Biodemography of Aging*, The Springer Series on Demographic Methods and Population Analysis 40, DOI 10.1007/978-94-017-7587-8_1

achieved, one has to better understand the biological mechanisms involved in the regulation of aging-related declines in health, wellbeing, and physiological functioning in human individuals, as well as connections among such declines and health and survival outcomes.

The biodemographic methods of studying human aging, health, and longevity allow for integration and efficient use of data and knowledge from relevant research fields including epidemiology, genetics, sociology, gerontology, environmental sciences, population genetics, etc. This chapter provides a selective account of important historical steps in the development of this research field in which members of the present research team have participated. It also illustrates how the integration of demographic and biological knowledge and data may contribute to progress in the field. Then it briefly describes the content and connections among the chapters of this monograph.

1.1.1 Frailty Models

Biodemographic thinking about mortality and survival started with attempts to explain the shape of the age trajectories of mortality rates by introducing additional variables affecting mortality risk. Survival experiments with large populations of laboratory animals provided evidence that age-specific mortality rates in a number of species increase exponentially over most of the adult age range (following Gompertz's curve) but, at the upper end of the adult age range, the rates decelerate, level off, or even decline. The results of these analyses were summarized in Vaupel et al. (1998). The downward deviation from the Gompertz curve has been explained by the presence of hidden (unobserved) heterogeneity in the chances of death. The demographic frailty model (Vaupel et al. 1979) employed an unobserved "fixed-frailty" variable to describe individual differences in susceptibility to death; the fixed frailty concept provided a convenient tool for analyzing mortality rates over the entire range of adult ages. The use of such models was a substantial step forward in understanding the need for going beyond pure demographic parametric descriptions of mortality curves to better understand the forces shaping the age patterns of cohort mortality rates and in elucidating the roles of compositional changes in such patterns resulting from processes of mortality selection in heterogeneous populations. Various versions of frailty and hidden heterogeneity models continue to be used in survival analyses (e.g., Erickson and Scheike 2015; de Castro et al. 2015; Sattar et al. 2015; Liu 2014).

Although the use of fixed frailty models created an opportunity for better description of late age mortality curves, it provided little information about the biological mechanisms influencing the mortality rates. Moreover, it was found that the class of fixed frailty models was not distinguishable from a class of models with randomly changing heterogeneity variables (Yashin et al. 1994). This observation indicated that additional information was needed to better understand the forces involved in regulation of the mortality rates. One way of adding such information was to introduce explanatory variables (observed covariates) into the fixed frailty models. Such models then treated hidden frailty as a random effect and corresponding studies were focused on estimation of effects of observed covariates on survival in the presence of hidden frailty. It was shown that the presence of hidden frailty modifies estimates of the effects of observed covariates on mortality risks. It was found that, when ignored, the presence of unobserved frailty attenuates the estimated effects.

An important class of extended frailty models used information on survival of related individuals (e.g., twins, siblings, and other relatives). These are models of shared and correlated frailty. For a number of such models (which differ in their frailty distributions), the multivariate survival function was described in a semi-parametric form. A remarkable property of such semi-parametric models was that one did not need to specify a parametric form for the baseline hazard rates. These hazard rates could be estimated semi-parametrically from bivariate or multivariate survival data. Multivariate survival models were used in analyses of the heritability of individual susceptibility to death (Yashin and Iachine 1995a, b), as well as the heritability of mortality by cause (Wienke 2010). The dependence among biologically related individuals is responsible for many interesting properties observed in studies of aging, health, and longevity. In particular, the fact that extreme longevity tends to run in families is a consequence of positive dependence among life spans of family members. This property makes multivariate survival models an efficient tool for studying factors and mechanisms affecting exceptional survival in families (Yashin and Iachine 1999a, b).

1.1.2 Biodemographic Ideas in Genetic Analyses of Human Longevity

The genetics of aging, longevity, and mortality has become the subject of intensive analyses, ranging from studies of candidate genes to genome-wide association studies (GWAS) (Nebel et al. 2011) that involve hundreds-of-thousands to millions of genetic variants (SNPs, i.e., single-nucleotide polymorphisms). However, most estimates of genetic effects on longevity in GWAS have not reached genome-wide statistical significance (after applying the Bonferroni correction for multiple testing) and many findings remain non-replicated. Possible reasons for slow progress in this field include the lack of a biologically-based conceptual framework that would drive development of statistical models and methods for genetic analyses of data, the presence of hidden genetic heterogeneity, the collective influence of many genetic factors (each with small effects), the effects of rare alleles, and epigenetic effects, as well as molecular biological mechanisms regulating cellular functions.

Another reason for slow progress in detecting genetic determinants of human aging and longevity could be the tendency to underestimate the role of demographic information in genetic analyses of these traits. The use of demographic data, models, and methods helped improve the accuracy of genetic estimates in genetic centenarian studies (Yashin et al. 1999). These analyses were able to detect non-monotonic (U-shaped) age trajectories of genetic frequencies corresponding to intersections of mortality rates for carriers and non-carriers of the respective genetic variants. The existence of such trajectories was later confirmed in Bergman et al. (2007). Alternative hypotheses describing possible mechanisms responsible for such age patterns of genetic frequencies (intersections of mortality rates) were discussed in Yashin et al. (2001b) and Bergman et al. (2007). The U-shaped patterns of genetic frequencies indicated that the effects of the corresponding genetic variants on mortality risk changed from harmful to beneficial during the life course. This property suggested that some genetic risk factors contributing to the mortality increase early in life are likely to be found in the genomes of long-lived individuals. This conclusion has been confirmed in a number of recent studies of human aging and longevity (Beekman et al. 2010).

The use of demographic information and models in analyses of data on genetically heterogeneous cohorts allowed researchers to compare the age patterns of mortality rates for carriers and non-carriers of candidate alleles and genotypes (Yashin et al. 2007b). Such comparisons were not possible using data on genetic frequencies alone. These methods were further extended in Arbeev et al. (2011). The improvement in the quality of genetic estimates from the joint analyses of genetic and demographic data was demonstrated in Yashin et al. (2013a). Recent genetic analyses of health-related traits revealed that genes affecting lifespan and healthspan do exhibit pleiotropic effects (Kulminski et al. 2011, 2013, 2015a, b; Yashin et al. 2012c, 2014, 2015, 2016). In comprehensive review Ukraintseva et al. (2016) summarized existing evidence and discussed possible mechanisms responsible for many such other puzzling effects of genetic risk factors.

1.1.3 Evolution of Aging, Health, and Mortality: Many Open Questions

The possibility of evolutionary origins of common aging-related diseases in situations where disease susceptibility alleles demonstrate deleterious or slightly deleterious effects was discussed in Pritchard (2001) and Reich and Lander (2001). Their results indicated that alleles affecting lifespan must show pleiotropic associations with healthspan, duration of unhealthy life, and mortality rates by cause. Under other population-genetics scenarios (Di Rienzo 2006), neutral genes can also be involved in the origin of common diseases. Since such genes have little or no effect on fitness, their associations with health-related traits are likely to be manifest in various trade-offs. Evidence for negative correlations between select diseases has been presented in Stallard (2002), Yashin et al. (2009), and Ukraintseva et al. (2010), among others. The existence of dependencies among diseases suggests the possibility of common genetic backgrounds for groups of health-related traits, and recent studies have confirmed the existence of pleiotropic associations of certain genes with health traits (Jeck et al. 2012). Thus, in addition to pleiotropic associations of genes between lifespan and health-related traits, such genetic associations can also be found among groups of health-related traits.

In parallel with demographic efforts focused on developing and implementing models of survival in heterogeneous populations, researchers in the field of aging have recognized the crucial role of aging-related changes in biomarkers in each individual that influence the chances of death. In contrast to the classical Gompertz mortality model specification of a rapid exponential decline in the "vitality function" with increasing age, reviews of physiological studies of aging carried out in the middle of the last century showed that physiological parameters characterizing many biological human capacities tended to decline almost linearly with age. To reconcile a linear decline in biological capacity with exponential increases in the rate of mortality, Strehler and Mildvan (1960) proposed a model of aging and mortality (the SM-model).

1.1.4 Strehler and Mildvan's Model of Aging and Mortality

In this model, mortality is viewed as a result of an interaction of aging-related decline in each organism's vitality function with a random process of energy demands. According to the SM-model, the death rate at a given age is proportional to "the frequency of stresses which surpass the ability of a subsystem to restore the initial conditions" at that age. The authors showed that, under such an assumption, the exponential increase in mortality (Gompertz's curve) results from the linear decline in vitality. The model allowed them to explain the negative correlation between the two parameters of the Gompertz curve (the SM-correlation – the correlation between the logarithm of the mortality rate of a population at the initial age of the range of adult ages studied in an analysis, and the slope of the logarithm of the mortality rates observed in empirical studies of the age patterns of mortality rates). The rectangularization pattern of survival improvement during the first half of the twentieth century in developed countries was in accordance with the SM correlation, as predicted by the SM-model. Recent cross-national SM-model analyses of mortality rates of both developed and developing countries for 1955–2003 found both heterogeneity among countries in the SM-correlation and increases over time of expected maximal survival ages (Zheng et al. 2011). These changes were linked to a decline in the average magnitudes of stresses experienced by successive population cohorts (Yashin et al. 2001a, 2002). Further developments of the SM model can be found in Li et al. (2013). Li and Anderson accommodated the later patterns of survival improvement that corresponded to almost parallel shift of survival curve to the right (Li and Anderson 2015). The importance of these studies lies in the fact that survival improvements can be explained by interactions between external challenges and internal aging-related changes. Several useful insights into the genetics of human longevity follow from these analyses.

The rectangularization pattern of survival improvement also is exhibited in groups of individual members of a cohort ordered by the number of longevity alleles that they carried (Yashin et al. 2012b). Since the environmental conditions in each such group remained the same, the SM model linked better survival with higher initial values of the vitality function, suggesting that exceptional survival was likely to have a genetic background. The more longevity alleles that were carried by study participants, the higher was the value of their initial stress resistance. In terms of genetic functions, the resistance to stresses is associated with repair capacity, redundancy, and other functions that increase an organism's resilience, or robustness. A useful illustration of how increased resilience may influence survival can be given by the model of saving lives (Vaupel and Yashin 1987). An increase in the "number of lives being saved" for individuals in the population can be interpreted as an increase in redundancy: for each lost life (premature death) saved, the saved individual has one or more "redundant" lives, depending on the relevant advances in medical technology (Yashin et al. 2012b). Further extensions of life saving model are discussed in Finkelstein (2013).

1.1.5 Historical Roots of the Stochastic Process Model

Strehler and Mildvan's theory (1960) stimulated further research in the direction of biodemographic analysis of mortality rates (Sacher and Trucco 1962). Woodbury and Manton (1977) introduced a multivariate stochastic process model of human mortality and aging. This model has been intensively used in analyses of longitudinal data on aging, health, and longevity (Manton and Stallard 1988; Manton et al. 1991, 1992, 1994, 1995; Woodbury and Manton 1983). This model was further extended in Yashin et al. (2008, 2011, 2012a) and Yashin and Manton (1997) to incorporate state of the art advances in aging research into the model structure and to link information on individual health histories with individual changes in physiological variables. In particular, the extended models described partly observed aging-related changes in physiological variables linked together with age-dependent unobserved variables that include resistance to stresses, adaptive capacities, physiological norms ("optimal" physiological states), stochasticity, allostatic adaptations, and allostatic load. Together, these extensions represent biological mechanisms of aging-related changes in humans that are consistent with existing biological knowledge about aging. The model describes how age-dependent unobserved variables interact with partly observed physiological indices and other factors, and links them with health and survival outcomes. This provides a convenient conceptual framework for comprehensive systemic analysis of aging-related changes in humans using longitudinal data and for linking these changes with genotypic profiles, and morbidity and mortality risks. The model has been used to develop unique efficient statistical methods for analyzing longitudinal data on aging, health, and longevity. The ideas and approaches briefly described above are further developed and discussed in detail in the chapters of this monograph. The monograph consists of three parts. **Part I** is motivational. Its goal is to inform readers about properties of aging-related changes represented by available data and to provide evidence of connections among diseases of the elderly and about possible determinants of health and survival outcomes. The chapters in this part discuss the results of analyses of various types of data using conventional statistical models. **Part II** contains chapters focusing on more sophisticated analyses using methods of advanced statistical modeling. **Part III** is a short chapter presenting conclusions.

1.2 Information on Aging, Health, and Longevity from Available Data: Part I

The process of aging involves changes in biological functioning during the life course. Data on such changes are collected in a number of longitudinal studies. That is why particular attention in this monograph is paid to the dynamic aspects of aging-related changes in biomarkers (e.g., physiological variables, composite indices, etc.) using data from human longitudinal studies. The individual age trajectories of such biomarkers show what is changing in human bodies when people get old, and how these changes develop during the life course. An important motivation for studying the aging process is to better understand its connections with health and survival outcomes. Such connections can be evaluated from longitudinal data.

Part I commences with Chap. 2, which presents the results of empirical analyses of data on age-related changes in physiological variables. These results show that average age trajectories of physiological variables follow remarkable regularities. Some of these trajectories are almost monotonic and others are non-monotonic. They depend on individuals' health status and gender, as well as on genetic and non-genetic factors. It is important to note that the shapes of the average age trajectories of these variables are formed by two major forces. One represents the biological mechanisms responsible for regulation of aging-related changes acting during an individual's life course. The other deals with compositional changes that take place in heterogeneous populations due to the process of mortality selection when individuals get older (Yashin et al. 2010). To understand the functioning of the biological machinery one has to separate the effects of these two forces. Such separation can be done using more sophisticated approaches to analyses of longitudinal data based on the statistical modeling described in **Part II**, Chaps. 11, 12, 13, 14, 15, and 16.

In addition to changes in biomarkers, the process of individual aging involves deterioration of health and developing chronic diseases. The fact that many elderly people suffer from several chronic conditions indicates that such diseases are likely to be dependent and have some common genetic or non-genetic risk factors. Chapter 3 describes age patterns of morbidity and comorbidity from observational data collected in the U.S. Medicare Files of Service Use (MFSU) for the entire

Medicare-eligible population of older U.S. adults. These data represent an example of Big Data analysis of current and historic health of older U.S. adults. The tremendous research potential of these data for evaluation of current and forecasting of future patterns of aging-related diseases among the older U.S. adults remains largely unexplored. This chapter presents results of epidemiologic and biodemographic analyses of these data. These results show how the age patterns of disease incidence, their time trends, recovery and long-term remission after disease onsets, interdependence of risks of multiple coexisting diseases, mortality at advanced ages, and multi-morbidity patterns can be evaluated. Empirical analyses, regression models, and methods of mathematical modeling are used to evaluate these health-related characteristics.

Chapter 4 provides additional evidence of dependence among diseases of the elderly. Traditional demographic calculations evaluating the contribution of disease to life expectancy reduction usually assume independence among causes of death. Such an assumption can be justified for some infectious diseases, but not for diseases of the elderly. The nature of these diseases differs from that of infectious diseases and deals with the complicated interplay among ontogenetic changes, senescence processes, and damages from exposures to hazardous environmental conditions. The determinants of such health disorders often contribute to the development of many health pathologies and their effects on disease risks may change with increasing age and time. The presence of such common risk factors makes diseases of the elderly mutually dependent. This chapter evaluates correlations among mortalities from cancer and other major health disorders, including heart disease, stroke, diabetes, Alzheimer's and Parkinson's diseases, and asthma using the Multiple Causes of Death (MCD) data. The analyses show significant negative correlations between cancer and some of the selected diseases. Possible mechanisms, including pleiotropic effects of genetic factors, are discussed.

Chapter 5 deals with factors increasing both longevity and cancer risk in human populations. Longevity and overall cancer incidence rates have increased over time in many countries in parallel with economic progress and the spread of the Western life style; the rates are also typically higher in more- than in less-developed countries. Could there be not merely an association but a causal connection here? This chapter investigates the possibility that some of the factors linked to high economic development and the Western life style may actually favour both longevity and vulnerability to cancer. The chapter provides a review of current evidence in support of this hypothesis and concludes that the higher overall cancer risk in affluent societies may in part be attributed to the higher proportion of individuals more *susceptible* to cancer, rather than to the higher burden of carcinogenic exposures in the respective populations. The proportions of susceptible people may have increased over time due to several key factors, including: (i) improved medical and living conditions that "relax" environmental selection and allow for survival of people with less efficient immune systems who may be more prone to cancer; (ii) novel/unusual exposures that are not carcinogenic themselves but may increase the body's vulnerability to established carcinogens (some new medicines and other agents will be discussed); (iii) several factors linked to the Western life style (such as delayed childbirth and food enriched with growth factors) that may postpone manifestations of physical aging and at the same time increase the body's susceptibility to cancer.

In response to aging-related health declines and other health-related challenges accompanying human lives, human society created, maintains, and continues to develop a health industry that aims to provide good health to its people. A major part of the economics of this industry is driven by medical costs. Such costs accompany each individual's health history. In Chap. 6, the trajectories of medical costs associated with the onset of twelve aging-related conditions are evaluated and analyzed. These conditions include acute coronary heart disease, stroke, ulcer, breast cancer, prostate cancer, melanoma, lung cancer, colon cancer, diabetes, asthma, Parkinson's disease, and Alzheimer's disease for older U.S. adults. The medical costs are associated with disease diagnosis and treatment. In the U.S., the prediction of future Medicare costs is crucial for health care planning, because almost all residents aged 65 years and older are enrolled in the Medicare system. The variables investigated in this chapter represent the sum of the medical costs associated with every person enrolled in the system. Individual costs deal with expenditures associated with disease onsets, their treatment, and subsequent costs of acute and chronic conditions. These trajectories were reconstructed using National Long Term Care Survey (NLTCS) data linked to the Medicare files of service use (NLTCS-M). A special procedure for selecting individuals with onset of each geriatric disease was developed and used for identification of the date of the disease onset. Among interesting research findings was the similarity of the time patterns of the individual medical cost trajectories for all studied diseases. This new approach yields a family of forecasting models with covariates. The dynamic relationships between Medicare expenditures and health indicators used in such models can lead to improved forecasting of Medicare costs.

Chapter 7 provides the rationale for construction, outlines the properties, and describes the applications of indices of cumulative deficits (DIs) in the analyses of data on aging, healthspan, and lifespan. The idea for such indices arose from the fact that observational studies typically measure not only major changes in health and well-being captured by well-defined risk factors (e.g., physiological measurements) but also various aging-related changes spread throughout hundreds of distinct variables that can be informative on longevity when accumulated in a single index. A DI is constructed as the proportion of failed (e.g., definitive deficits) or abnormal (e.g., doubtful deficits) health traits manifest by a given age – that is a summary measure of the average level of an organism's deterioration at a given age. A comparison of DI with clinical frailty was performed. The results suggest that integration of both approaches is highly promising for increasing the precision of the risk discrimination, especially among the most vulnerable part of the elderly population.

In Chap. 8 the dynamic properties of individual trajectories of aging-related changes in eight key physiological variables (body mass index, systolic and diastolic blood pressure, pulse pressure, pulse rate, blood glucose, hematocrit, and total cholesterol) in the Framingham Heart Study (FHS) participants are investigated, and connections between characteristics of these trajectories and human lifespan and healthspan estimated. It is well known from epidemiology that values of variables describing physiological states at a given age are associated with human morbidity and mortality risks. Much less well known are the facts that not only the values of these variables at a given age, but also characteristics of their dynamic behavior during the life course are also associated with health and survival outcomes. This chapter shows that, for monotonically changing variables, the value at age 40 (intercept), the rate of change (slope), and the variability of a physiological variable, at ages 40–60, significantly influence both healthspan and longevity after age 60. For non-monotonically changing variables, the age at maximum, the maximum value, the rate of decline after reaching the maximum (right slope), and the variability in the variable over the life course may influence healthspan and longevity. This indicates that such characteristics can be important targets for preventive measures aiming to postpone onsets of complex diseases and increase longevity.

Recently, participants of many longitudinal studies have been genotyped, so that large datasets of genetic information have become available for analyses together with phenotypic longitudinal data. Chapter 9 discusses the roles of genes in diseases in late life. Decades of studies of candidate genes show that they are not linked to aging-related traits in a straightforward fashion (Finch and Tanzi 1997; Martin 2007). Recent genome-wide association studies (GWAS) have supported this finding by showing that the traits in late life are likely controlled by a relatively large number of common genetic variants (e.g., Teslovich et al. (2010)). Further, GWAS often show that the detected associations are of tiny size (Stranger et al. 2011). This chapter considers several examples of complex modes of gene actions including genetic trade-offs, antagonistic genetic effects on the same traits at different ages, and variable genetic effects on lifespan. The analyses focus on the *APOE* common polymorphism.

1.3 Statistical Modeling and Other Advanced Methods of Analyzing Data on Aging, Health, and Longevity: Part II

Part II deals with more advanced methods of statistical analyses of data based on the idea of statistical modeling. As noted above, all of the empirical evidence indicates that aging is a multidimensional process that involves changes in many variables (biomarkers). Many biomarkers that play fundamental roles in aging-related changes remain unmeasured in longitudinal studies, so available longitudinal data are always incomplete. Information about some unobserved variables and their connections with observed biomarkers can be available from other aging studies. Mathematical and computer modeling allow for incorporating such information into a model's structure and estimating the model's parameters from the data. The use of methods of statistical modeling allows for evaluating age patterns of many hidden components of aging and their connections with components represented in longitudinal data and with morbidity/mortality risks. Several chapters of this monograph show how various dynamic models can be constructed and efficiently used in analyses of longitudinal data. Mathematical and computer modeling are used to represent available information about phenomena in a way that is informative for addressing research questions using available longitudinal or cross-sectional data. **Part II** starts with Chap. 11, which provides a brief review of approaches to statistical analyses of longitudinal data on aging that are relevant to the major topic of this monograph—the *Biodemography of Aging*. When relevant, it relates these approaches to the subsequent chapters in **Part II** of the book. Longitudinal data play a pivotal role in discovery related to aging, health, and longevity. There is a wide variety of statistical methods for analyzing longitudinal data; longitudinal data analysis is one of the most prolific areas of statistical science. This chapter presents the basics of the joint models of longitudinal and time-to-event outcomes and various extensions in the recent biostatistical literature and discusses them in the context of biodemographic applications.

Chapter 12 describes an approach to statistical analyses of longitudinal data based on the use of stochastic process models (SPMs) of human aging, health, and longevity. A better understanding of processes and mechanisms linking human aging with changes in health and longevity requires integrative statistical methods capable of taking into account relevant knowledge accumulated in the field when extracting useful information from new data. An important advantage of statistical analyses using SPMs compared to standard statistical methods of analyzing longitudinal data is the possibility of incorporating state of the art advances in aging research into the model structure and then using this model in statistical estimation procedures. Specifically, the proposed model incorporates variables characterizing resistance to stresses, adaptive capacity, and "optimal" (normal) physiological states. To capture the effects of exposure to persistent external disturbances, variables describing the effects of allostatic adaptation and allostatic load are also introduced into the model. These additional variables facilitate the description of the link between aging-related changes in physiological indices and morbidity and mortality risks. The approach provides researchers with a powerful conceptual framework for studying dynamic aspects of aging, and with an appropriate tool for analysis and systematization of information about aging and its connection with health and longevity.

Chapter 13 continues the model developments described in Chaps. 11 and 12. Various approaches that incorporate unobserved or hidden heterogeneity are ubiquitous in different scientific disciplines. Unobserved heterogeneity can arise because there may be some relevant risk factors affecting the outcome of interest that are either still unknown or just not measured in the data. The continuing interest in hidden heterogeneity can be seen in the recent books devoted to frailty models (Duchateau and Janssen 2008; Hanagal 2011; Wienke 2010), which have extensive (but not exclusively overlapping) lists of references. The modern era of revolutionary advances in genetics provides great opportunities and challenges for the field of biodemography and the need to integrate the principles of genetics and genomics into biodemography is apparent so that this field will continue to be on the forefront of demographic analyses (Carey 2008; Wachter 2008). The importance of "genetic biodemography" will continue to grow in the coming years because many studies that have collected data on biomarkers will include (or already have included) genetic information. The ongoing incorporation of genetic information into longitudinal studies is considered potentially "the most revolutionary element of the addition of biological data in large-scale surveys" (Suzman 2010) and such studies will "increasingly provide analyses of the interactions of genetic, biological, social, economic, and demographic characteristics" (Crimmins et al. 2010).

This chapter describes new approaches that model both time-to-event and longitudinal data. This excludes methods focusing on analyses of longitudinal data alone, where events are generally treated as nuisance factors to be adjusted for and approaches that do not include time-to-event information (e.g., onset of a disease), but include, for example, binary indicators such as prevalence of a disease. We also present a version of the stochastic process model (Yashin et al. 2007a) that accommodates such hidden heterogeneity, thus extending the earlier model (Yashin et al. 2008).

Chapter 14 presents results of simulation studies of a longitudinal genetic-demographic model illustrating that inclusion of biodemographic information in addition to follow-up data improves statistical power in analyses of genetic effects on mortality or morbidity risks. It also describes simulation studies of the genetic SPM, illustrating the increase in power of joint analyses of genotyped and non-genotyped participants of a longitudinal study compared to analyses of non-genotyped participants alone in different scenarios to test relevant biologically-based hypotheses. The results of these analyses and possibilities of further extensions of the approaches are discussed.

Chapter 15 describes an approach to integrative mortality modeling that represents mortality rates in terms of parameters describing aging-related changes in physiological variables and changes in health status with increasing age. In contrast to traditional demographic and actuarial models dealing with mortality data, such models can be used in statistical analyses of longitudinal data on aging, health, and longevity. The models use diffusion-type continuous-time stochastic processes for describing the evolution of physiological variables over the life course, and finite-state continuous-time processes for describing changes in health status during this period. The development of integrative mortality models involves integral-differential equations for conditional probabilities characterizing changes in physiological states and health status as individuals in a population under study get older. The approximation of changes in physiological states by a conditional Gaussian process, given current health state, simplifies the description and yields efficient methods of statistical modeling.

In Chap. 16, applications of integrated mortality models to the analysis of data from simulation experiments and from the Framingham Heart Study are described. The analyses show that model parameters can be evaluated from longitudinal data (Yashin et al. 2011). The application of these models to Framingham Heart Study data reveals important differences in physiological dynamics between healthy and sick individuals (Yashin et al. 2013b). The models can be successfully used in the joint analyses of data collected using different observational plans.

The need to efficiently analyze large-scale genetic data and longitudinal data on aging-related changes as well as data on human health and survival in selected populations of individuals exemplifies Big Data analysis. Analyses of such data would benefit from special approaches that take high dimensionality of the corresponding data into account.

Chapter 17 presents a new longitudinal form of the Grade of Membership (GoM) model for time-varying covariates. It provides a self-contained description of a new GoM estimation algorithm and its statistical properties, and illustrates its application with a substantively meaningful analysis of the progression of dementia among NLTCS respondents. The natural history of dementia is modeled as a complex irreversible multidimensional process governed by a latent three-dimensional bounded state-space process. Individual dementia cases were found to be initially widely dispersed in the latent state space. Over time, they moved to state-space locations associated with severe cognitive and physical impairment and dramatically increased need for care. The rate of progression of the disease was found to be highly variable over, but predictable from, the initial state-space location. This latter finding is currently being used to develop an improved prognostic model for individual Alzheimer's patients, their physicians, and caregiver teams.

In Chap. 18, the recently developed Linear Latent Structures (LLS) analysis model and its statistical properties are described. Applications of the LLS model to analyzing the NLTCS data are discussed. The results of the analyses are compared numerically and analytically to predictions of the Latent Class model (LCM) and the Grade of Membership (GoM) model. LLS analysis assumes that the mutual correlations observed in survey variables reflect a hidden property that can be described by a low-dimensional random vector. Applying the LLS model to the 1994 and 1999 NLTCS datasets (5,000+ individuals) with responses to over 200 questions on behavior factors, functional status, and comorbidities resulted in an identified population structure with a basis represented by "pure-type individuals," e.g., healthy, strongly disabled, having chronic diseases, etc. The estimated population structure and the score distributions are compared with predictions given by LCM and GoM analyses. The components of the vectors of individual LLS scores are used to make predictions of individual lifespans.

1.4 Conclusion

The evolutionary processes involved in forming the genetic structure of human populations as well as age patterns of human morbidity and mortality curves can be studied in the framework of Evolutionary Population Genetics. Many questions about the origin of human chronic diseases, aging-related changes, senescence, and connections among these traits can be addressed within this discipline.

The Biodemography of Aging deals with changes in genetically heterogeneous birth cohorts of individuals over age and time. Distributions of phenotypic traits within such cohorts change with increasing age. The genetic structure of the population cohort also experiences changes with increasing age. Biodemographic ideas, concepts, and methods facilitate the analysis of such changes and the assessment of their implications, leading to more efficient and informative analyses of demographic, longitudinal, and genetic data.

Acknowledgements The research reported in this chapter was supported by the National Institute on Aging grants R01AG027019, R01AG030612, R01AG030198, 1R01AG046860, and P01AG043352. The content is solely the responsibility of the authors and does not necessarily represent the official views of the National Institute on Aging or the National Institutes of Health.

References

Arbeev, K. G., Ukraintseva, S. V., Arbeeva, L. S., Akushevich, I., Kulminski, A. M., & Yashin, A. I. (2011). Evaluation of genotype-specific survival using joint analysis of genetic and non-genetic subsamples of longitudinal data. *Biogerontology, 12*(2), 157–166.

Beekman, M., Nederstigt, C., Suchiman, H. E. D., Kremer, D., van der Breggen, R., Lakenberg, N., Alemayehu, W. G., de Craen, A. J. M., Westendorp, R. G. J., Boomsma, D. I., de Geus, E. J. C., Houwing-Duistermaat, J. J., Heijmans, B. T., & Slagboom, P. E. (2010). Genome-wide association study (GWAS)-identified disease risk alleles do not compromise human longevity. *Proceedings of the National Academy of Sciences of the United States of America, 107*(42), 18046–18049.

Bergman, A., Atzmon, G., Ye, K., MacCarthy, T., & Barzilai, N. (2007). Buffering mechanisms in aging: A systems approach toward uncovering the genetic component of aging. *PLoS Computational Biology, 3*(8), e170.

Carey, J. R. (2008). Biodemography: Research prospects and directions. *Demographic Research, 19*, 1749–1757.

Crimmins, E., Kim, J. K., & Vasunilashorn, S. (2010). Biodemography: New approaches to understanding trends and differences in population health and mortality. *Demography, 47*(Supplement), S41–S64.

de Castro, M., Chen, M. H., & Zhang, Y. (2015). Bayesian path specific frailty models for multistate survival data with applications. *Biometrics, 71*(3, September), 760–771. doi:10.1111/biom.12298. PubMed PMID: 25762198, PubMed Central PMCID: PMC4567543, Epub 2015 Mar 11.

Di Rienzo, A. (2006). Population genetics models of common diseases. *Current Opinion in Genetics & Development, 16*(6), 630–636.

Duchateau, L., & Janssen, P. (2008). *The frailty model*. New York: Springer.

Eriksson, F., & Scheike, T. (2015). Additive gamma frailty models with applications to competing risks in related individuals. *Biometrics, 71*(3, September), 677–686. doi:10.1111/biom.12326. Epub 2015 Jun 1.

Finch, C. E., & Tanzi, R. E. (1997). Genetics of aging. *Science, 278*(5337), 407–411.

Finkelstein, M. (2013). Lifesaving, delayed deaths and cure in mortality modeling. *Theoretical Population Biology, 83*, 15, February–19. doi:10.1016/j.tpb.2012.10.005. Epub 2012 Oct 26.

Hanagal, D. D. (2011). *Modeling survival data using frailty models*. Boca Raton: Chapman & Hall/CRC.

Jeck, W. R., Siebold, A. P., & Sharpless, N. E. (2012). Review: A meta-analysis of GWAS and age-associated diseases. *Aging Cell, 11*(5), 727–731.

Kulminski, A. M., Culminskaya, I., Ukraintseva, S. V., Arbeev, K. G., Arbeeva, L., Wu, D., Akushevich, I., Land, K. C., & Yashin, A. I. (2011). Trade-off in the effects of the apolipoprotein E polymorphism on the ages at onset of CVD and cancer influences human lifespan. *Aging Cell, 10*(3), 533–541.

Kulminski, A. M., Culminskaya, I., Arbeev, K. G., Ukraintseva, S. V., Arbeeva, L., & Yashin, A. I. (2013). Trade-off in the effect of the APOE gene on the ages at onset of cardiocascular disease and cancer across ages, gender, and human generations. *Rejuvenation Research, 16*(1, February), 28–34. doi:10.1089/rej.2012.1362. PubMed PMID: 23094790, PubMed Central PMCID: PMC3582279.

Kulminski, A. M., Culminskaya, I., Arbeev, K. G., Arbeeva, L., Ukraintseva, S. V., Stallard, E., Wu, D., & Yashin, A. I. (2015a). Birth cohort, age, and sex strongly modulate effects of lipid risk alleles identified in genome-wide association studies. *PloS One, 10*(8, August 21), e0136319. doi:10.1371/journal.pone.0136319.eCollection2015. PubMed PMID: 26295473, PubMed Central PMCID: PMC4546650.

Kulminski, A. M., Arbeev, K. G., Culminskaya, I., Ukraintseva, S. V., Stallard, E., Province, M. A., & Yashin, A. I. (2015b). Trade-offs in the effects of the apolipoprotein E polymorphism on risks of diseases of the heart, cancer, and neurodegenerative disorders: Insights on mechanisms from the Long Life Family Study. *Rejuvenation Research, 18*(2, April), 128–135. doi:10.1089/rej.2014.1616. PubMed PMID: 25482294, PubMed Central PMCID: PMC4403014.

Li, T., & Anderson, J. J. (2015). The Strehler-Mildvan correlation from the perspective of a two-process vitality model. *Population Studies (Camb), 69*(1), 91–104. doi:10.1080/00324728.2014.992358. Epub 2015 Jan 30.

Li, T., Yang, Y. C., & Anderson, J. J. (2013). Mortality increase in late-middle and early-old age: Heterogeneity in death processes as a new explanation. *Demography, 50*(5, October), 1563–1591. doi:10.1007/s13524-013-0222-4. PubMed PMID: 23743628, PubMed Central PMCID: PMC4028711.

Liu, X. (2014). *Journal of Biometrics Biostatistics. 5*, pii: 1000191. PubMed PMID: 25525559; PubMed Central PMCID: PMC4267525.

Manton, K. G., & Stallard, E. (1988). *Chronic disease modelling: Measurement and evaluation of the risks of chronic disease processes*. New York: Oxford University Press.

Manton, K. G., Stallard, E., & Tolley, H. D. (1991). Limits to human life expectancy – Evidence, prospects, and implications. *Population and Development Review, 17*(4), 603–637.

Manton, K. G., Stallard, E., & Singer, B. (1992). Projecting the future size and health status of the United States elderly population. *International Journal of Forecasting, 8*(3), 433–458.

Manton, K. G., Stallard, E., Woodbury, M. A., & Dowd, J. E. (1994). Time-varying covariates in models of human mortality and aging: Multidimensional generalizations of the Gompertz. *Journals of Gerontology, 49*(4), B169–B190.

Manton, K. G., Woodbury, M. A., & Stallard, E. (1995). Sex differences in human mortality and aging at late ages: The effect of mortality selection and state dynamics. *Gerontologist, 35*(5), 597–608.

Martin, G. M. (2007). Modalities of gene action predicted by the classical evolutionary biological theory of aging. *Annals of the New York Academy of Sciences, 1100*, 14–20.

Nebel, A., Kleindorp, R., Caliebe, A., Nothnagel, M., Blanche, H., Junge, O., Wittig, M., Ellinghaus, D., Flachsbart, F., Wichmann, H. E., Meitinger, T., Nikolaus, S., Franke, A., Krawczak, M., Lathrop, M., & Schreiber, S. (2011). A genome-wide association study confirms APOE as the major gene influencing survival in long-lived individuals. *Mechanisms of Ageing and Development, 132*(6–7), 324–330.

Pritchard, J. K. (2001). Are rare variants responsible for susceptibility to complex diseases? *American Journal of Human Genetics, 69*(1), 124–137.

Reich, D. E., & Lander, E. S. (2001). On the allelic spectrum of human disease. *Trends in Genetics, 17*(9), 502–510.

Sacher, G. A., & Trucco, E. (1962). The stochastic theory of mortality. *Annals of the New York Academy of Sciences, 96*(4), 985–1007.

Sattar, A., Sinha, S. K., Wang, X. F., & Li, Y. (2015). Frailty models for pneumonia to death with a left-censored covariate. *Statistics in Medicine, 34*(14), 2266–2280. doi:10.1002/sim.6466. Epub 2015 Mar 2.

Sierra, F., Hadley, E., Suzman, R., & Hodes, R. (2009). Prospects for life span extension. *Annual Review of Medicine, 60*, 457–469.

Stallard, E. (2002). Underlying and multiple cause mortality at advanced ages: United States 1980–1998. *North American Actuarial Journal, 6*(3), 64–87.

Stranger, B. E., Stahl, E. A., & Raj, T. (2011). Progress and promise of genome-wide association studies for human complex trait genetics. *Genetics, 187*(2), 367–383.

Strehler, B. L., & Mildvan, A. S. (1960). General theory of mortality and aging. *Science, 132*(3418), 14–21.

Suzman, R. (2010). Prologue: Research on the demography and economics of aging. *Demography, 47*(Supplement), S1–S4.

Teslovich, T. M., Musunuru, K., Smith, A. V., Edmondson, A. C., Stylianou, I. M., Koseki, M., Pirruccello, J. P., Ripatti, S., Chasman, D. I., Willer, C. J., Johansen, C. T., Fouchier, S. W., Isaacs, A., Peloso, G. M., Barbalic, M., Ricketts, S. L., Bis, J. C., Aulchenko, Y. S., Thorleifsson, G., Feitosa, M. F., Chambers, J., Orho-Melander, M., Melander, O., Johnson, T., Li, X., Guo, X., Li, M., Shin Cho, Y., Jin Go, M., Jin Kim, Y., Lee, J. Y., Park, T., Kim, K., Sim, X., Twee-Hee Ong, R., Croteau-Chonka, D. C., Lange, L. A., Smith, J. D., Song, K., Hua Zhao, J., Yuan, X., Luan, J., Lamina, C., Ziegler, A., Zhang, W., Zee, R. Y., Wright, A. F., Witteman, J. C., Wilson, J. F., Willemsen, G., Wichmann, H. E., Whitfield, J. B., Waterworth, D. M., Wareham, N. J., Waeber, G., Vollenweider, P., Voight, B. F., Vitart, V., Uitterlinden, A. G., Uda, M., Tuomilehto, J., Thompson, J. R., Tanaka, T., Surakka, I., Stringham, H. M., Spector, T. D., Soranzo, N., Smit, J. H., Sinisalo, J., Silander, K., Sijbrands, E. J., Scuteri, A., Scott, J., Schlessinger, D., Sanna, S., Salomaa, V., Saharinen, J., Sabatti, C., Ruokonen, A., Rudan, I., Rose, L. M., Roberts, R., Rieder, M., Psaty, B. M., Pramstaller, P. P., Pichler, I., Perola, M., Penninx, B. W., Pedersen, N. L., Pattaro, C., Parker, A. N., Pare, G., Oostra, B. A., O'Donnell, C. J., Nieminen, M. S., Nickerson, D. A., Montgomery, G. W., Meitinger, T., McPherson, R., McCarthy, M. I., McArdle, W., Masson, D., Martin, N. G., Marroni, F., Mangino, M., Magnusson, P. K., Lucas, G., Luben, R., Loos, R. J., Lokki, M. L., Lettre, G., Langenberg, C., Launer, L. J., Lakatta, E. G., Laaksonen, R., Kyvik, K. O., Kronenberg, F., Konig, I. R., Khaw, K. T., Kaprio, J., Kaplan, L. M., Johansson, A., Jarvelin, M. R., Janssens, A. C., Ingelsson, E., Igl, W., Kees Hovingh, G., Hottenga, J. J., Hofman, A., Hicks, A. A., Hengstenberg, C., Heid, I. M., Hayward, C., Havulinna, A. S., Hastie, N. D., Harris, T. B., Haritunians, T., Hall, A. S., Gyllensten, U., Guiducci, C., Groop, L. C., Gonzalez, E., Gieger, C., Freimer, N. B., Ferrucci, L., Erdmann, J., Elliott, P., Ejebe, K. G., Doring, A., Dominiczak, A. F., Demissie, S., Deloukas, P., de Geus, E. J., de Faire, U., Crawford, G., Collins, F. S., Chen, Y. D., Caulfield, M. J., Campbell, H., Burtt, N. P., Bonnycastle, L. L., Boomsma, D. I., Boekholdt, S. M., Bergman, R. N., Barroso, I., Bandinelli, S., Ballantyne, C. M., Assimes, T. L., Quertermous, T., Altshuler, D., Seielstad, M., Wong, T. Y., Tai, E. S., Feranil, A. B., Kuzawa, C. W., Adair, L. S., Taylor, H. A., Jr., Borecki, I. B., Gabriel, S. B., Wilson, J. G., Holm, H., Thorsteinsdottir, U., Gudnason, V., Krauss, R. M., Mohlke, K. L., Ordovas, J. M., Munroe, P. B., Kooner, J. S., Tall, A. R., Hegele, R. A., Kastelein, J. J., Schadt, E. E., Rotter, J. I., Boerwinkle, E., Strachan, D. P., Mooser, V., Stefansson, K., Reilly, M. P., Samani, N. J., Schunkert, H., Cupples, L. A., Sandhu, M. S., Ridker, P. M., Rader, D. J., van Duijn, C. M., Peltonen, L., Abecasis, G. R., Boehnkeand, M., & Kathiresan, S. (2010). Biological, clinical and population relevance of 95 loci for blood lipids. *Nature, 466*(7307), 707–713.

Ukraintseva, S. V., Arbeev, K. G., Akushevich, I., Kulminski, A., Arbeeva, L., Culminskaya, I., Akushevich, L., & Yashin, A. I. (2010). Trade-offs between cancer and other diseases: Do they exist and influence longevity? *Rejuvenation Research, 13*(4), 387–396.

Ukraintseva, S., Yashin, A., Arbeev, K., Kulminski, A., Akushevich, I., Wu, D., Joshi, G., Land, K. C., Stallard, E. (2016). Puzzling role of genetic risk factors in human longevity: "risk alleles" as pro-longevity variants. *Biogerontology, 17*(1), 109–127.

Vaupel, J. W., & Yashin, A. I. (1987). Repeated resuscitation: How lifesaving alters life tables. *Demography, 24*(1), 123–135.

Vaupel, J. W., Manton, K. G., & Stallard, E. (1979). The impact of heterogeneity in individual frailty on the dynamics of mortality. *Demography, 16*(3), 439–454.

Vaupel, J. W., Carey, J. R., Christensen, K., Johnson, T. E., Yashin, A. I., Holm, N. V., Iachine, I. A., Kannisto, V., Khazaeli, A. A., Liedo, P., Longo, V. D., Zeng, Y., Manton, K. G., & Curtsinger, J. W. (1998). Biodemographic trajectories of longevity. *Science, 280*(5365), 855–860.

Wachter, K. W. (2008). Biodemography comes of age. *Demographic Research, 19*, 1501–1512.

Wienke, A. (2010). *Frailty models in survival analysis*. Boca Raton: Chapman & Hall/CRC.

Woodbury, M. A., & Manton, K. G. (1977). A random-walk model of human mortality and aging. *Theoretical Population Biology, 11*(1), 37–48.

Woodbury, M. A., & Manton, K. G. (1983). A theoretical model of the physiological dynamics of circulatory disease in human populations. *Human Biology, 55*(2), 417–441.

Yashin, A. I., & Iachine, I. (1995a). How long can humans live? Lower bound for biological limit of human longevity calculated from Danish twin data using correlated frailty model. *Mechanisms of Ageing and Development, 80*(3), 147–169.

Yashin, A. I., & Iachine, I. A. (1995b). Survival of related individuals: An extension of some fundamental results of heterogeneity analysis. *Mathematical Population Studies, 5*(4), 321–377.

Yashin, A. I., & Iachine, I. A. (1999a). Dependent hazards in multivariate survival problems. *Journal of Multivariate Analysis, 71*(2), 241–261.

Yashin, A. I., & Iachine, I. A. (1999b). What difference does the dependence between durations make? Insights for population studies of aging. *Lifetime Data Analysis, 5*(1), 5–22.

Yashin, A. I., & Manton, K. G. (1997). Effects of unobserved and partially observed covariate processes on system failure: A review of models and estimation strategies. *Statistical Science, 12*(1), 20–34.

Yashin, A. I., Vaupel, J. W., & Iachine, I. A. (1994). A duality in aging: The equivalence of mortality models based on radically different concepts. *Mechanisms of Ageing and Development, 74*(1–2), 1–14.

Yashin, A. I., De Benedictis, G., Vaupel, J. W., Tan, Q., Andreev, K. F., Iachine, I. A., Bonafe, M., DeLuca, M., Valensin, S., Carotenuto, L., & Franceschi, C. (1999). Genes, demography, and life span: The contribution of demographic data in genetic studies on aging and longevity. *American Journal of Human Genetics, 65*(4), 1178–1193.

Yashin, A. I., Begun, A. S., Boiko, S. I., Ukraintseva, S. V., & Oeppen, J. (2001a). The new trends in survival improvement require a revision of traditional gerontological concepts. *Experimental Gerontology, 37*(1), 157–167.

Yashin, A. I., Ukraintseva, S. V., De Benedictis, G., Anisimov, V. N., Butov, A. A., Arbeev, K., Jdanov, D. A., Boiko, S. I., Begun, A. S., Bonafe, M., & Franceschi, C. (2001b). Have the oldest old adults ever been frail in the past? A hypothesis that explains modern trends in survival. *Journals of Gerontology Series A: Biological Sciences and Medical Sciences, 56*(10), B432–B442.

Yashin, A. I., Begun, A. S., Boiko, S. I., Ukraintseva, S. V., & Oeppen, J. (2002). New age patterns of survival improvement in Sweden: Do they characterize changes in individual aging? *Mechanisms of Ageing and Development, 123*(6), 637–647.

Yashin, A. I., Arbeev, K. G., Akushevich, I., Kulminski, A., Akushevich, L., & Ukraintseva, S. V. (2007a). Stochastic model for analysis of longitudinal data on aging and mortality. *Mathematical Biosciences, 208*(2), 538–551.

Yashin, A. I., Arbeev, K. G., & Ukraintseva, S. V. (2007b). The accuracy of statistical estimates in genetic studies of aging can be significantly improved. *Biogerontology, 8*(3), 243–255.

Yashin, A. I., Arbeev, K. G., Akushevich, I., Kulminski, A., Akushevich, L., & Ukraintseva, S. V. (2008). Model of hidden heterogeneity in longitudinal data. *Theoretical Population Biology, 73*(1), 1–10.

Yashin, A. I., Ukraintseva, S. V., Akushevich, I. V., Arbeev, K. G., Kulminski, A., & Akushevich, L. (2009). Trade-off between cancer and aging: What role do other diseases play? Evidence from experimental and human population studies. *Mechanisms of Ageing and Development, 130*(1–2), 98–104.

Yashin, A. I., Arbeev, K. G., Akushevich, I., Ukraintseva, S. V., Kulminski, A., Arbeeva, L. S., & Culminskaya, I. (2010). Exceptional survivors have lower age trajectories of blood glucose: Lessons from longitudinal data. *Biogerontology, 11*(3), 257–265.

Yashin, A. I., Akushevich, I., Arbeev, K. G., Kulminski, A., & Ukraintseva, S. (2011). Joint analysis of health histories, physiological states, and survival. *Mathematical Population Studies, 18*(4), 207–233.

Yashin, A. I., Arbeev, K. G., Akushevich, I., Kulminski, A., Ukraintseva, S. V., Stallard, E., & Land, K. C. (2012a). The quadratic hazard model for analyzing longitudinal data on aging, health, and the life span. *Physics of Life Reviews, 9*(2), 177–188.

Yashin, A. I., Wu, D., Arbeev, K. G., Stallard, E., Land, K. C., & Ukraintseva, S. V. (2012b). How genes influence life span: The biodemography of human survival. *Rejuvenation Research, 15*(4), 374–380.

Yashin, A. I., Wu, D., Arbeev, K. G., & Ukraintseva, S. V. (2012c). Polygenic effects of common single-nucleotide polymorphisms on life span: When association meets causality. *Rejuvenation Research, 15*(4), 381–394.

Yashin, A. I., Arbeev, K. G., Wu, D., Arbeeva, L. S., Kulminski, A. M., Akushevich, I., Culminskaya, I., Stallard, E., & Ukraintseva, S. V. (2013a). How the quality of GWAS of human lifespan and health span can be improved. *Frontiers in Genetics, 4*, 125.

Yashin, A. I., Arbeev, K. G., Wu, D., Arbeeva, L. S., Kulminski, A. M., Akushevich, I., Culminskaya, I., Stallard, E., & Ukraintseva, S. V. (2013b, January 22). How lifespan associated genes modulate aging changes: Lessons from analysis of longitudinal data. *Frontiers in Genetics, 3*, 1–20.

Yashin, A. I., Wu, D., Arbeev, K. G., Arbeeva, L. S., Akushevich, I., Kulminski, A., Culminskaya, I., Stallard, E., Ukraintseva, S. V. (2014). Genetic structures of population cohorts change with increasing age: Implications for genetic analyses of human aging and life span. *Current Gerontology and Geriatrics Research. 1*(4), pii: 1020. PubMed PMID: 25893220; PubMed Central PMCID: PMC4398390.

Yashin, A. I., Wu, D., Arbeeva, L. S., Arbeev. K. G., Kulminski, A. M., Akushevich, I., Kovtun, M., Culminskaya, I., Stallard, E., Li, M., Ukraintseva, S. V. (2015) Genetics of aging, health, and survival: dynamic regulation of human longevity related traits. *Frontiers in Genetics, 6*(April 13), 122. doi:10.3389/fgene.2015.00122. eCollection 2015. PubMed PMID: 25918517; PubMed Central PMCID: PMC4394697.

Yashin, A. I., Arbeev, K. G., Arbeeva, L. S., Wu, D., Akushevich, I., Kovtun, M., Yashkin, A., Kulminski, A., Culminskaya, I., Stallard, E., Li, M., Ukraintseva, S. V. (2016). How the effects of aging and stresses of life are integrated in mortality rates: Insights for genetic studies of human health and longevity. *Biogerontology, 17*(1), 89–107.

Zheng, H., Yang, Y., & Land, K. C. (2011). Heterogeneity in the Strehler-Mildvan general theory of mortality and aging. *Demography, 48*(1), 267–290.

Part I
Information on Aging, Health, and Longevity from Available Data

Chapter 2
Age Trajectories of Physiological Indices: Which Factors Influence Them?

Anatoliy I. Yashin, Liubov S. Arbeeva, Konstantin G. Arbeev, Igor Akushevich, Alexander M. Kulminski, Eric Stallard, and Svetlana V. Ukraintseva

2.1 Introduction

Physiological variables and other biomarkers reflect the body's responses to numerous external and internal challenges and its ability to maintain reliable functioning in a changing environment. Some physiological variables are used for monitoring patients' health status, or for evaluating the efficiency of treatments in clinical trials. Many variables are measured in longitudinal studies of aging, health, and longevity. The data from these studies provide researchers with the unique opportunity to investigate how individual organisms change with increasing age, how these changes modulate risks of disease and death, and how they are influenced by genetic and non-genetic factors. The patterns of individual aging changes can be described in terms of characteristics of age trajectories of physiological variables and other biomarkers. Studying these age trajectories may yield insights into the nature of biological mechanisms involved in the regulation of aging. Important questions that can be addressed in this research include: How do the physiological age trajectories differ among individuals? Are their patterns gender-specific, and if so, then why? How do genetic and non-genetic factors associated with extreme longevity modulate the entire trajectories? Are there hidden (unobserved) biological mechanisms that regulate the dynamics of aging changes in physiological markers, and how could their contribution be estimated from available data?

To mediate the influence of internal or external factors on lifespan, physiological variables have to show associations with risks of disease and death at different age intervals, or directly with lifespan. For many physiological variables, such associations have been established in epidemiological studies. These include body mass index (BMI), diastolic blood pressure (DBP), systolic blood pressure (SBP), pulse pressure (PP), blood glucose (BG), serum cholesterol (SCH), hematocrit (H), and ventricular rate (VR).

For example, the effect of BMI on risks of disease and mortality was intensively studied in connection with metabolic syndrome. Freedman and colleagues (2006)

A.I. Yashin et al., *Biodemography of Aging*, The Springer Series on Demographic Methods and Population Analysis 40, DOI 10.1007/978-94-017-7587-8_2

showed that the connection between BMI and mortality risk is generally *J*-shaped for both genders and various age groups. The authors also found that this risk function changes with age. Relationships between mortality risk and BMI were also assessed in other studies (see Gelber et al. 2007; Gu et al. 2006; Klenk et al. 2009, among others; Zhou 2002).

The connection between diastolic blood pressure (DBP) and all-cause mortality risk has been investigated to better understand factors and mechanisms of cardiovascular diseases (CVD) (Cruickshank 1988, 2003; Staessen 1996). Special attention has been paid to the *J*-shape of the risk function (see Alderman 1996; Cruickshank 2003; Grassi et al. 2010, among others; Isles and Hole 1992; Messerli and Panjrath 2009). Franklin and colleagues (2001) studied changes in this risk function with age. Boshuizen and colleagues studied the connection between blood pressure and mortality risk in the elderly (Boshuizen et al. 1998). Questions of optimal blood pressure were discussed by Onrot (1993) and Townsend (2005), among others.

Anderson and colleagues (1987) evaluated the connection between serum cholesterol (SCH) and mortality using 30 years of follow-up data from the Framingham Heart Study. The authors found that each 1 % increase in total cholesterol produced a 2 % increase in coronary heart disease incidence among individuals between 60 and 70 years of age. Kronmal and colleagues (1993) found that the relationship between total cholesterol level and all-cause mortality was positive at age 40, negligible at ages 50–70, and negative at age 80. Other researchers (Manolio et al. 1992; Weverling-Rijnsburger et al. 2003) showed that CVD in old age was independent of total serum cholesterol levels. Weverling-Rijnsburger and colleagues (1997) proposed that this could be a result of selective mortality of those with the highest cholesterol levels in middle age. Weverling-Rijnsburger and colleagues (1997) and Schatz (Schatz et al. 2001) showed that low total serum cholesterol levels are associated with higher all-cause mortality in the oldest old. The relationship between serum cholesterol and all-cause mortality was also studied by Chyou and Eaker (2000) and Li et al. (2004a, b) among others.

The effects of resting heart rate (also called ventricular rate, VR) on cardiovascular mortality were described by Kannel and colleagues (1987). Mensink and Hoffmeister (1997) and Benetos and colleagues (1999) investigated the effects of resting heart rate on all-cause mortality. The connection between heart rate and mortality in the elderly also was investigated by Cacciatore and colleagues (2007). Kuzuya and colleagues (2008) found a J-shaped relationship between resting pulse rate and all-cause mortality in community dwelling older people with disabilities. Böhm and colleagues (2012) showed that resting heart rate in clinical conditions is associated with all-cause mortality, disability, and cognitive decline.

Taken altogether, the above results indicate that studying the age trajectories of physiological variables, as well as factors and mechanisms involved in their regulation, could substantially clarify complex connections among aging, health decline, and longevity, and provide useful insights into alternative strategies for the improvement of people's health (Kristjuhan 2012). None of the cited studies performed either systematic analyses of age-patterns of corresponding variables or of the roles of such patterns in mediating genetic influence on lifespan.

In this chapter, we use longitudinal data from the Framingham Heart Study (FHS) Original cohort to evaluate the average age trajectories of the eight physiological variables, including body mass index (BMI), diastolic blood pressure (DBP), systolic blood pressure (SBP), pulse pressure (PP), blood glucose (BG), serum cholesterol (SCH), hematocrit (H), and ventricular rate (VR). We show how these trajectories depend on various genetic and non-genetic factors affecting human lifespan.

2.2 Data: The Framingham Heart Study (FHS)

The longitudinal data on aging, health, and lifespan collected in the FHS include results of biennial measurements of physiological and health related variables during the life course of study participants, detailed data on their genetic background, and data on health and survival outcomes. The FHS (Original cohort) was launched in 1948 (Exam 1), with 5209 respondents (55 % females) aged 28–62 years at baseline and residing in Framingham, Massachusetts, who had not yet developed overt symptoms of cardiovascular disease (Dawber et al. 1951). The study has continued to the present with biennial examinations. To date, 31 exams of the Original Cohort have been conducted; data from exams 1–28 including detailed medical history, physical exams, laboratory tests, and ages 28–104, were used in this study.

Phenotypic Traits collected in the FHS cohorts over 60 years and relevant to our analyses include: life span, ages at onset of diseases (with the emphasis on cardiovascular disease (CVD), cancer, and diabetes mellitus), as well as indices characterizing physiological state. The occurrence of diseases (CVD and cancer) and death have been followed through continuous surveillance of hospital admissions, death registries, clinical exams, and other sources, so that all these events are included in the study.

To define the *age at onset of unhealthy life* in the present study, we used data on onsets of CVD, cancer (calculated from the follow-up data), and diabetes. CVD was defined as the first appearance of any one of the following codes for the variable EVENT in the Sequence of Cardiovascular Events (SOE) file provided by FHS: 1–19, 21–26, 30–49. These codes correspond to major CVD events (or death from such events) including myocardial infarction, angina pectoris, cerebrovascular accident (stroke, TIA), intermittent claudication, and congestive heart failure. The onset of diabetes was defined as the age at the first exam when an individual had a value of random BG exceeding 140 mg/dl, and/or took diabetes medication (oral hypoglycemic or insulin). The age of onset of unhealthy life was then defined as the minimum of ages at onset of these three diseases. If an individual did not contract any of these diseases during the observation period than then s/he was considered censored at the age of the last follow-up or death. Individuals who had any of these diseases before the first FHS exam were excluded from the analyses of "unhealthy life."

Data on eight physiological indices used in this study included: random *blood glucose* levels (BG, exams 1–4, 6, 8–10, 13–23, 26–28), *body mass index* (BMI, exams 1, 4, 5, 10–28), *diastolic blood pressure* (DBP, exams 1–28), *systolic blood pressure* (SBP, exams 1–28), *pulse pressure* (PP, exams 1–28), *ventricular rate* (VR, exams 4–28), *hematocrit* (H, exams 4–21), and *total cholesterol* (SCH, exams 1–11, 13–15, 20, 22–28). Some variables in the data were not measured or excluded for different reasons. For example, blood glucose was not measured at exams 5, 7, 11, 12, 24, and 25. Also, we used values of BG from the diabetes file and this file does not contain measurements for exam 28. BMI was calculated only for exams where both weight and height measurements were available. Hematocrit was measured at exams 4–21. However, the mean trajectories in exams 10 and 21 deviated substantially from the values recorded for the rest of the exams and the data from those exams accordingly were excluded from calculations. We created 13 age groups (<35, 35–39, . . ., 85–89, and 90+ years) and calculated empirical estimates of the mean values of the physiological indices in each group using pooled data on measurements from all exams.

The demographic characteristics of the Original FHS cohort are illustrated in Fig. 2.1. This figure shows a histogram of the age distribution of the participants of the Original FHS cohort at the first exam for males and females together with empirical estimates of survival functions for males, females, and males and females combined.

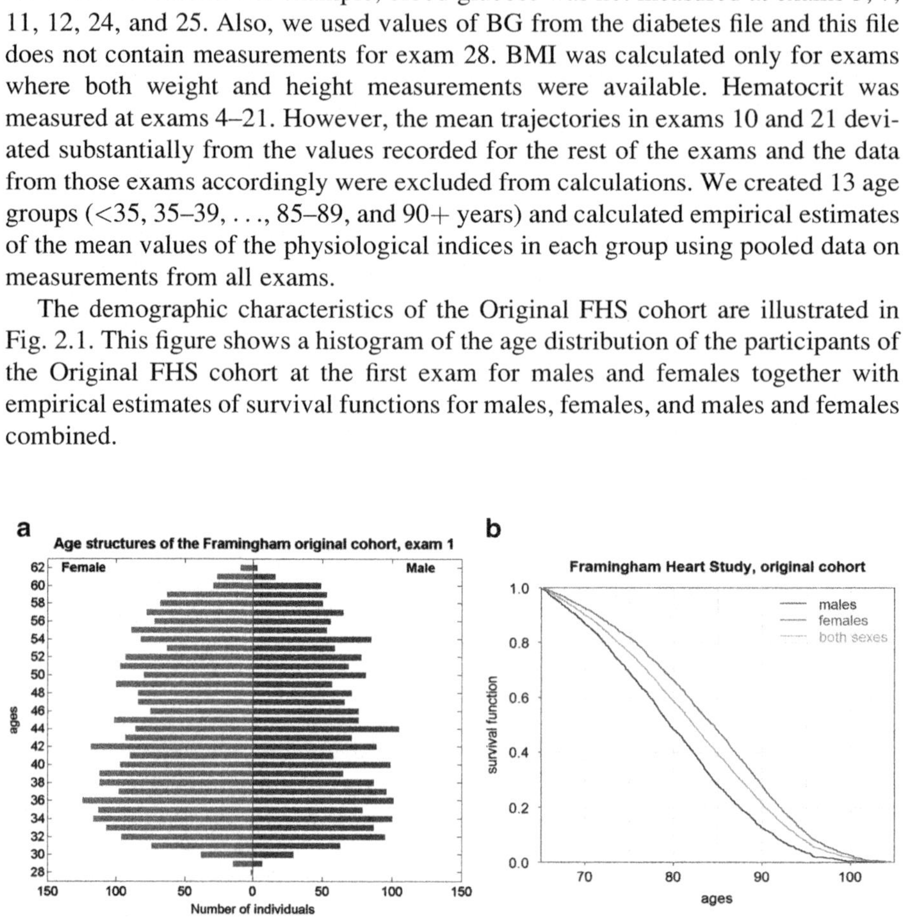

Fig. 2.1 (**a**) Age distribution of the participants of the Original Framingham cohort at the first exam for males and for females. (**b**) Kaplan-Meier estimates of survival functions for males, females, and males and females combined from the same cohort. In estimating the survival functions, it was assumed that the survival chances of individuals of the same gender from all sub-cohorts of the Original FHS cohort were the same. In this case, estimation of survival functions is reduced to constructing these functions from left truncated and right censored samples on life span data from population cohorts of males and females. The procedure for such a construction is readily available (Tsai et al. 1987)

Males and females have similar age ranges and comparable age distributions; the female survival from age 65 was better than that of males. This difference in survival was highly statistically significant.

2.3 Methods

We calculated empirical estimates of the mean values of the eight physiological indices (BMI, DBP, SCH, VR, SBP, PP, PR, and H) in age groups <35, 35–39, . . ., 85–89, and 90+ years for all participants of the Original FHS cohort (males and females) using pooled data on measurements from all exams. We selected groups of shorter-lived individuals (those dying at ages 75 or earlier; censored individuals were excluded from this group) and the 100 longest lived individuals (individuals with lifespans exceeding 92.7 years), and calculated average values of physiological indices in the same age groups of these individuals using pooled data on measurements from all exams. We also calculated the age patterns of physiological indices for individuals having different smoking statuses, as well as for individuals with different genetic backgrounds. These longitudinal data provide an opportunity to reveal some but not all the biological mechanisms involved in regulation of physiological aging changes. The existence of other mechanisms is supported by experimental animal studies and other research. For various reasons, all relevant components are not always measured in longitudinal human studies, including the FHS. The evaluation of such missing, or hidden, biomarkers is a challenging task; however, some progress in representing their effects can be achieved using methods of statistical modeling, as described in Chaps. 12 and 13.

2.4 Results

2.4.1 Average Age Trajectories of Physiological Variables

Empirical estimates of average age trajectories of physiological variables capture important regularities of aging-related changes and prepare our intuition for analyses of the dynamic properties of aging-related changes. These trajectories were generated by two mechanisms. One is responsible for biological aging-related changes and other involves the process of mortality selection in heterogeneous populations. These trajectories provide us with useful insights and ideas concerning linkages of these variables with lifespan and healthy lifespan. The average age trajectories of the eight physiological indices evaluated from the data on the Original FHS cohort are shown in Fig. 2.2a for males and females. Although all age patterns of physiological indices are non-monotonic functions of age, blood glucose (BG) and pulse pressure (PP) can be well approximated by monotonically increasing functions for both genders. The average age trajectories of hematocrit (H) for males and females have different shapes. The levels of H for males are higher than for females. For males, they tend to stay nearly constant until age 65 and then decline with age. For females, the average H values first increase to age 65 and then decline.

For both genders, the average values of body mass index (BMI) increase with age (up to age 55 for males and 65 for females), and then decline for both sexes. These values do not change much between ages 50 and 70 for males and between ages 60 and 70 for females. The male and female curves for systolic blood pressure (SBP) intersect around age 55. The female curve was initially lower than that of males. Both curves reach their maximum values at around age 75 and then decline. The values of diastolic blood pressure (DBP) for males were higher than those for females until age 55 and then became about the same. Both curves increase to age 55 and then decline for both sexes. For both genders, pulse pressure (PP) increases after age 40, reaches its peak at approximately 90 years of age and then shows a tendency to decline. The serum cholesterol (SCH) curve for males is initially higher

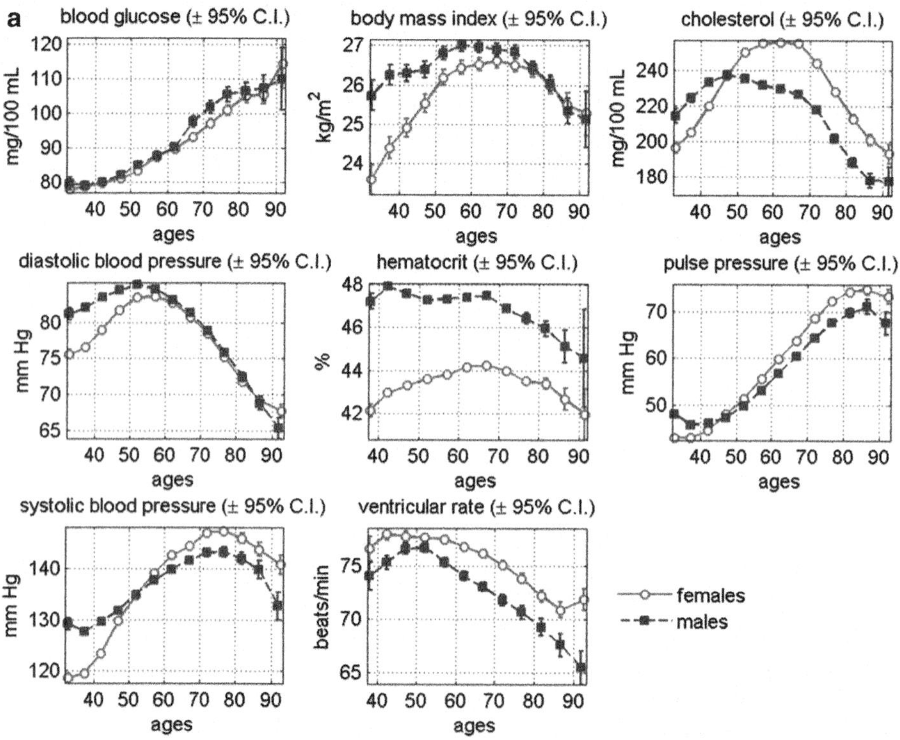

Fig. 2.2 (**a**) Average age trajectories of eight physiological variables for males and females evaluated from longitudinal data on the Original FHS cohort. (**b**) Average age trajectories of standard deviations of eight physiological variables for males and females evaluated from longitudinal data on the Original FHS cohort

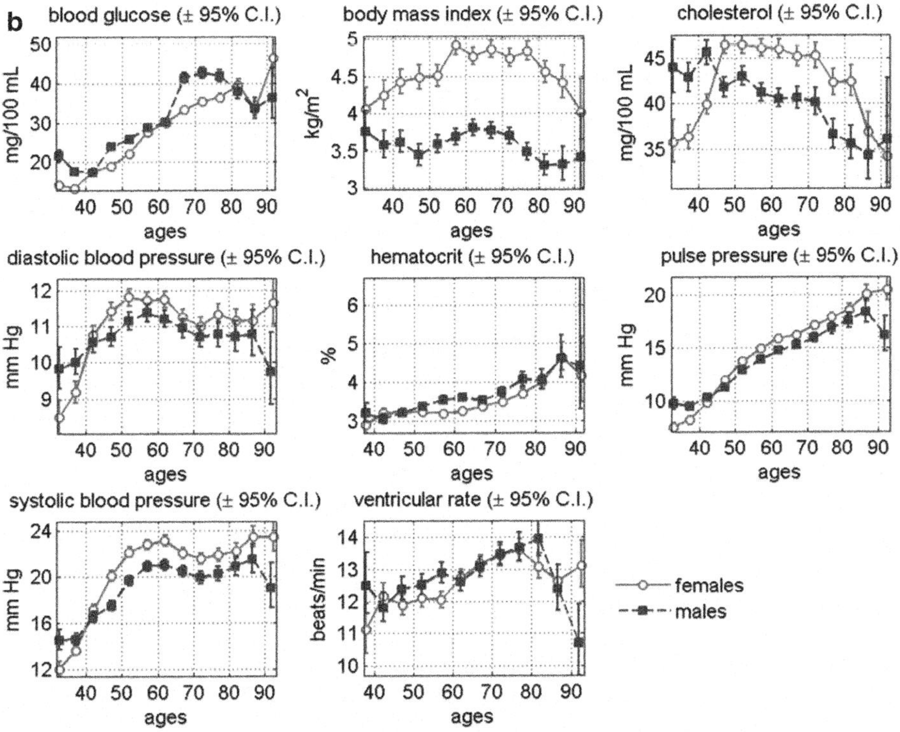

Fig. 2.2 (continued)

than that of females. The curves for males and females intersect around age 45. The female curve reaches its maximum value around age 60, stays at this level until age 70, and then declines. For males, a gradual rise in SCH levels ceases near age 50, then the level of SCH declines slowly to age 70. After age 70, the decline accelerates. The values of ventricular rate (VR) for males first increases, reaches a maximum around age 50 and then declines. For females, the average VR increases until age 45, and then shows a slight tendency to decline between ages 45 and 60. After age 60, it declines at a faster rate.

Except for blood glucose, all average age trajectories of physiological indices differ between males and females. Statistical analysis confirms the significance of these differences. In particular, after age 35 the female BMI increases faster than that of males. Both indices reach the same maximum value around age 60 and then show a similar pattern of decline after age 70. Diastolic blood pressure is higher among males until age 60, but increases more slowly. The maximum value of DBP occurs near ages 55–60 for both genders. The age pattern after age 60 is about the same for males and females. The average level of hematocrit is higher in males than in females during the entire age range. However, the rate of decline of this variable after age 70 was about the same for males and females. Average female pulse pressure tends to increase faster than that of males. Average female pulse (ventricular) rate always is higher than that of males. Between ages 35 and 50, the level of serum cholesterol is higher in males than in females. After age 50, the average values of male SCH are lower than those of females and remain lower for the remainder of the age range.

2.4.2 Age Trajectories of Standard Deviations (SD) of Physiological Variables

At each age, the values of physiological variables differ among individuals in the population. These differences were, however, within ranges that were compatible with life. Do these ranges change with age? The age trajectories of standard deviations of physiological indices yield useful insights into this question. These trajectories are likely to depend on initial differences in values of the physiological variable among individuals, on the variability acquired during the life course due to genetic differences in ontogenetic changes, and on random external disturbances affecting these variables differently in different individuals, as well as on forces reducing this variability due to mortality selection. The strength of the effects of mortality selection depends on how much the risk of death was influenced by the deviation of these variables from their normal values. Figure 2.2b shows how estimates of standard deviations (SD) of the eight physiological variables described above changed with ages for males and females.

The values of the standard deviations (SD) of BG for females are about the same between ages 35 and 40, then they increase monotonically between ages 40 and 80, decline until age 85, and then tend to increase again. For males, these values first decline until age 40, then increase until age 70, and then decline again. The values tend to be higher for males than for females until age 80. After age 80, the curves are about the same. The curves are also close to each other between ages 40 and 60. The age trajectories of the SD for BMI for the two genders are quite different. The female SD values are higher than those of males over the entire age interval from 30 to 95. They increase until age 55, remain about the same until age 75, and then decline. For males, these values tend to decline between ages 30 and 45, increase until age 65, decline until age 80, and then tend to increase again. The SD curves for DBP intersect for the two genders. The female's values of SD are lower than those of males until age 40 years. After that age, the females' values of SD are larger than those of males, without showing noticeable changes until the end of the age interval. The values of SD for PP remain small for most adult ages for both genders, with increases for both genders after age 80. The values of SD for SBP for males and females intersect at age 45. The female values are lower at the beginning and become higher afterwards. Both trajectories slightly decline between ages 60 and 70, and then slightly increase again. The SCH trajectories of SD for the two genders intersect at age 45. The female values are lower than those of males until age 45. Then they become higher and gradually decline until age 70. After age 70, the decline accelerates. For males, the values remain about the same until age 70, followed by a faster decline up to age 80. After this age, the SD values become statistically indistinguishable. The SD values for H increase with age for both genders, remaining about the same, with a tendency to be larger for males than for females with a decline after age 85. The VR SD increases with age for both genders until age 80 for males and until age 75 for females. Then the female SD declines until age 85 and shows a tendency to increase afterwards. The male SD continues to decline after age 80. Both curves were close to each other.

The differences in the shapes of male and female age trajectories may be related to differences in male and female survival distributions (Fig. 2.1c). Differences in average age trajectories of the physiological indices can be expected for groups of individuals of the same gender who have different exposures to deleterious risk factors such as cigarette smoking.

2.4.3 *Age Patterns of Survival and Physiological Variables for Smokers and Non-smokers*

That cigarette smoking is a risk factor for all-cause mortality is well known (Hubbard et al. 2009; Fenelon and Preston 2012; Preston et al. 2014). Figure 2.3a shows the age patterns of survival functions for smoking and non-smoking female participants of the Original FHS cohort. It can be seen that female smokers have worse survival than female non-smokers. Beyond overall survival, the question we now address is: Which of the eight physiological indices described above mediate the harmful influences of a smoking on survival? To address this question, Fig. 2.3b

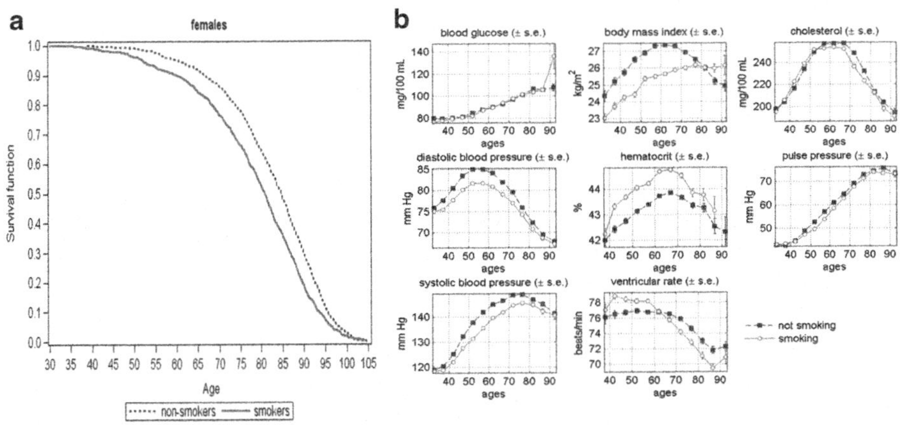

Fig. 2.3 (**a**) Age patterns of survival functions for smokers and non-smokers among female participants of the Original FHS cohort. (**b**) Average age trajectories of eight physiological variables for female smokers and non-smokers evaluated from longitudinal data on the Original FHS cohort

shows the average age patterns of the physiological indices for smoking and non-smoking females.

The average values of BG are increasing functions of age, and are about the same for female smokers and non-smokers until age 85. At age 90, the BG levels for smokers exceed those of non-smokers. However, the standard error at age 90 is much higher than at earlier ages, so the difference is not statistically significant.

The average values of BMI in female smokers are lower than those of non-smokers until age 80 at which point the BMI curves for smokers and non-smokers intersect. The smokers' BMI curve increases more slowly with age than that of nonsmokers. The maximum value of non-smokers' BMI is reached earlier (at age 60 versus age 75 for smokers) and the maximum value for non-smokers is higher that that of smokers. The smokers' curve does not change much after reaching its maximum. The average values of SCH tends to be slightly higher in female smokers until about age 55. Then these curves intersect and the values for smokers remains slightly lower than those of non-smokers until the end of the age range. The smokers' SCH reaches its maximum value slightly earlier than non-smokers (at age 50 versus age 55). Both curves decline after reaching their maximum values at about the same rate, and remain close to each other until the end of the age range. The average values of DBP are always lower for female smokers than for non-smokers. The maximum values of DBP are reached at about the same ages – around age 50. The largest difference between the two curves was at ages around the maximum point. The curves become closer to each other when age increases. The age trajectories for H are always higher for female smokers than for non-smokers, with maximum values around the same age, 65. The average values of PP are about the same until age 45. Then they become slightly lower for smokers and remain slightly lower until the end of the age range. The average values of SBP for female smokers are lower than those of non-smokers for the entire age range and have similar shapes. The average values of VR are higher for female smokers until age 65 when the curves intersect; after this age smokers have lower average values of VR than non-smokers.

The age patterns of average values of the eight physiological indices for smoking and non-smoking males are shown in Fig. 2.4.

After age 75 the BG values for male smokers become slightly higher than those for nonsmokers and remain higher until the end of age range. The average values of BMI are lower for male smokers until age 75. After this age, the BMI curves for smokers and non-smokers become indistinguishable. The average values of SCH tend to be slightly higher in male smokers until about age 55. Then these curves intersect; the curve for smokers becomes slightly lower than that of non-smokers until the end of the age range. The curves reach their maximum values at about the same age, 45 years. However, the decline in the SCH levels starts earlier among smokers (around age 45 versus 65 for non-smokers). The average values of DBP are slightly lower for male smokers than for non-smokers until about age 70. After age 85, the DBP curve for smokers becomes slightly higher than for non-smokers. The maximum values of DBP are reached at about the same ages. The average values of H are always higher for male smokers than for non-smokers, with maximum values

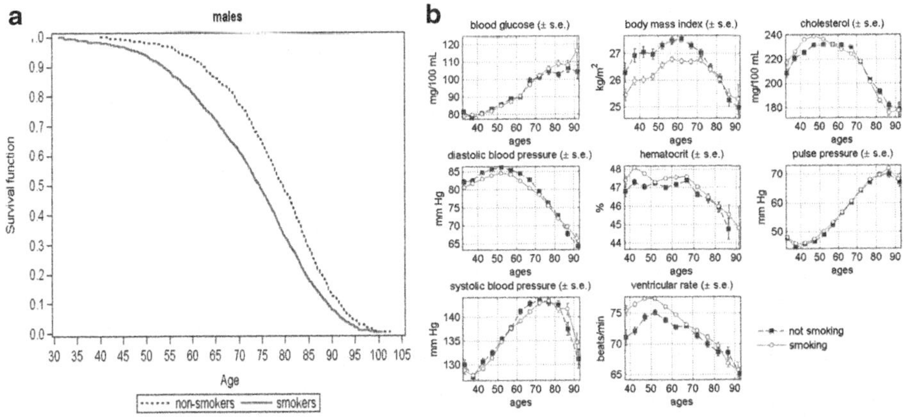

Fig. 2.4 (**a**) Age patterns of survival functions for smokers and non-smokers among male participants of the Original FHS cohort. (**b**) Average age trajectories of eight physiological variables for male smokers and non-smokers evaluated from longitudinal data on the Original FHS cohort

reached around the same ages. The average values of PP are about the same for male smokers and non-smokers for all ages. The average values of SBP for male smokers are lower than those of non-smokers until age 75. Then the curves intersect, with SBP for smokers thereafter being slightly higher than for non-smokers. The average values of VR are higher for male smokers until age 65, then they become closer to the VR curve for non-smokers until age 85. Then at age 85 the curves intersect and remain close to each other.

2.4.4 Effects of Education on Survival and Average Age Trajectories of Physiological Indices

Figure 2.5a shows average age patterns of eight physiological indices for the two groups of females from the Original FHS cohort that differ in their level of education (higher than 11th grade vs. less than or equal to 11th grade). For simplicity, we use notations LE and HE for the lower and higher educated groups respectively.

The average values of BG for both groups are about the same until age 45. Then the BG curve for the LE females becomes higher than that of the HE females until age 85 where the curves intersect. After this age, the BG curve for the HE group exceeds that of the LE group; however the differences between the curves are not statistically significant. The average values of BMI in the LE group are substantially higher than those among the HE group over the entire age interval. The maximum value for the LE curve is reached earlier (at age 50 vs. age 70 for the HE females). The average values of SCH are about the same for the two curves for all

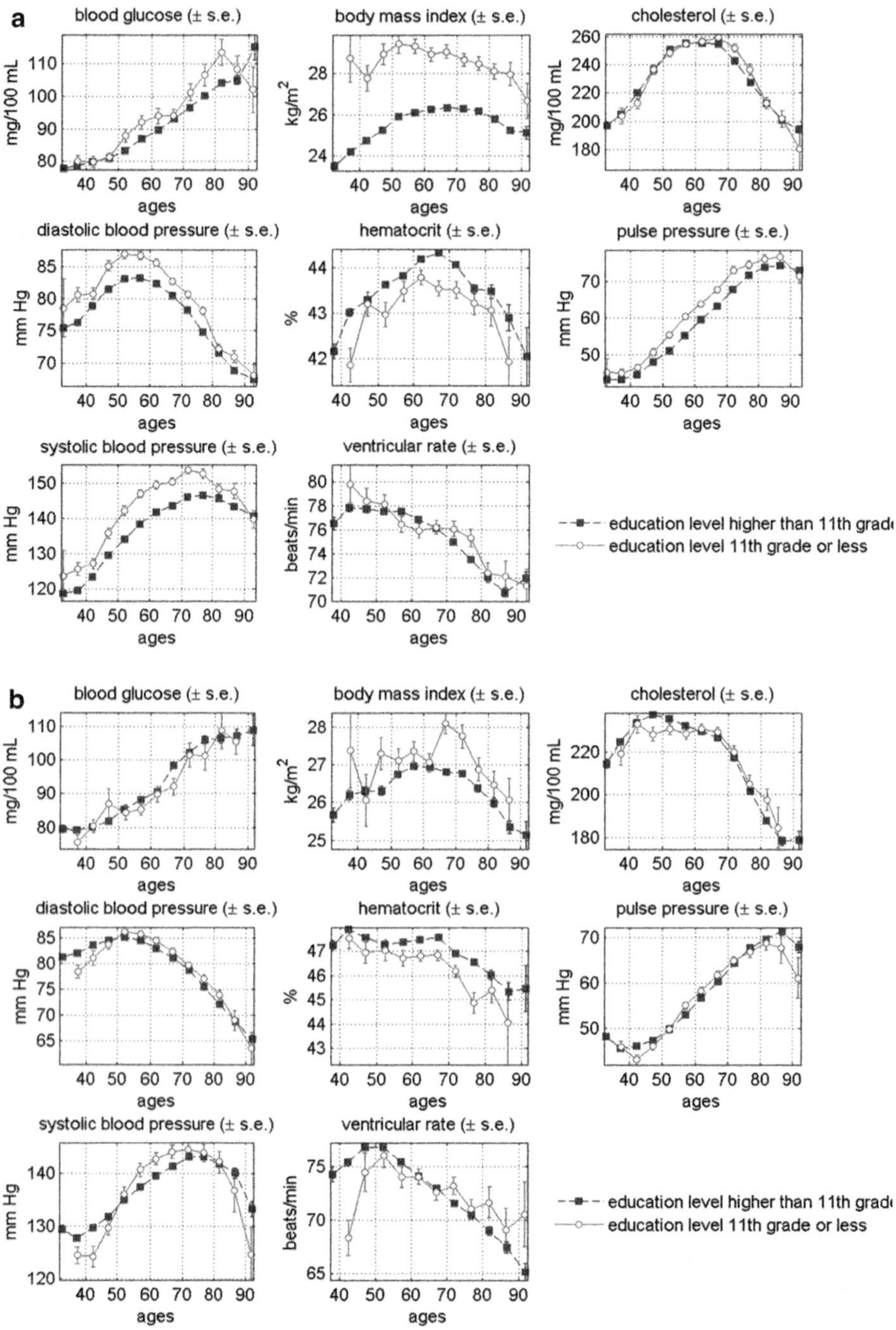

Fig. 2.5 (**a**) Average age patterns of eight physiological indices for female participants of the Original FHS cohort that differ in their level of education (higher than 11th grade vs. less than or equal to 11th grade). (**b**) Average age patterns of eight physiological indices for male participants of the Original FHS cohort that differ in their level of education (higher than 11th grade vs. less than or equal to 11th grade)

ages. The average values of DBP in the LE males are higher than those of the HE females for all ages. The largest difference between the two curves is around age 50. The average values of SBP in the LE females are higher than those of the HE females for all ages except above age 90 where the they are very similar. The average values of PP in the LE females are higher than those of the HE females for all ages except above age 90 where the they are very similar. The average values of H in the LE females are lower than those of the HE females for all ages. The largest difference between the two curves is around age 65. The average values of VR for the two curves are statistically indistinguishable over the entire age interval.

The age patterns of average values of eight physiological indices for the HE and LE males are shown in Fig. 2.5b.

The average values of BG for the HE and LE males are very similar over the entire age interval. The average values of BMI in the LE group tend to be higher than those of the LE group. However, the differences between groups are much smaller than for females. The average values of SCH are very similar for both groups over the entire age interval. The average values of DBP in the LE group are lower than those of the HE group until age 45. After this age, the two curves are very similar. The values of H for the LE group tend to be lower than for the HE group for all ages. The PP curves for the two groups are very similar over the entire age interval. The average values of SBP for the LE group are lower than those for the HE group until age 50. Then the curves intersect and the values of SBP in the LE males become higher than those of the HE males until age 80 where the curves intersect again and become statistically indistinguishable. The average values of VR are higher among the HE males until age 65 when the curves intersect and are similar for the next 10 years; however, above age 75 the HE curve is below the LE curve.

2.4.5 Age Trajectories of Long Lived (LL) and Short Lived (SL) Individuals

Figure 2.6a shows the average age trajectories of eight physiological indices for female participants of the Original FHS cohort for individuals having short lifespans (SL-group) (lifespan <75 years) and 100 individuals having the longest lifespans (LL-group). One can see that trajectories for the LL females are substantially different from those for the SL females in all eight indices.

Specifically, the average values of BG are higher and increase faster in the SL females. The entire age trajectory of BMI for the LL females is shifted to the right (towards an older age) compared to the SL females, and reaches its maximum about 10 years later. The values of BMI for SL females also starts to decline earlier (after about age 60). The average values of DBP among the SL females are higher than those of the LL females. The maximum value of the average DBP in the SL group is higher than that of the LL females and it is more distinct from adjacent values for

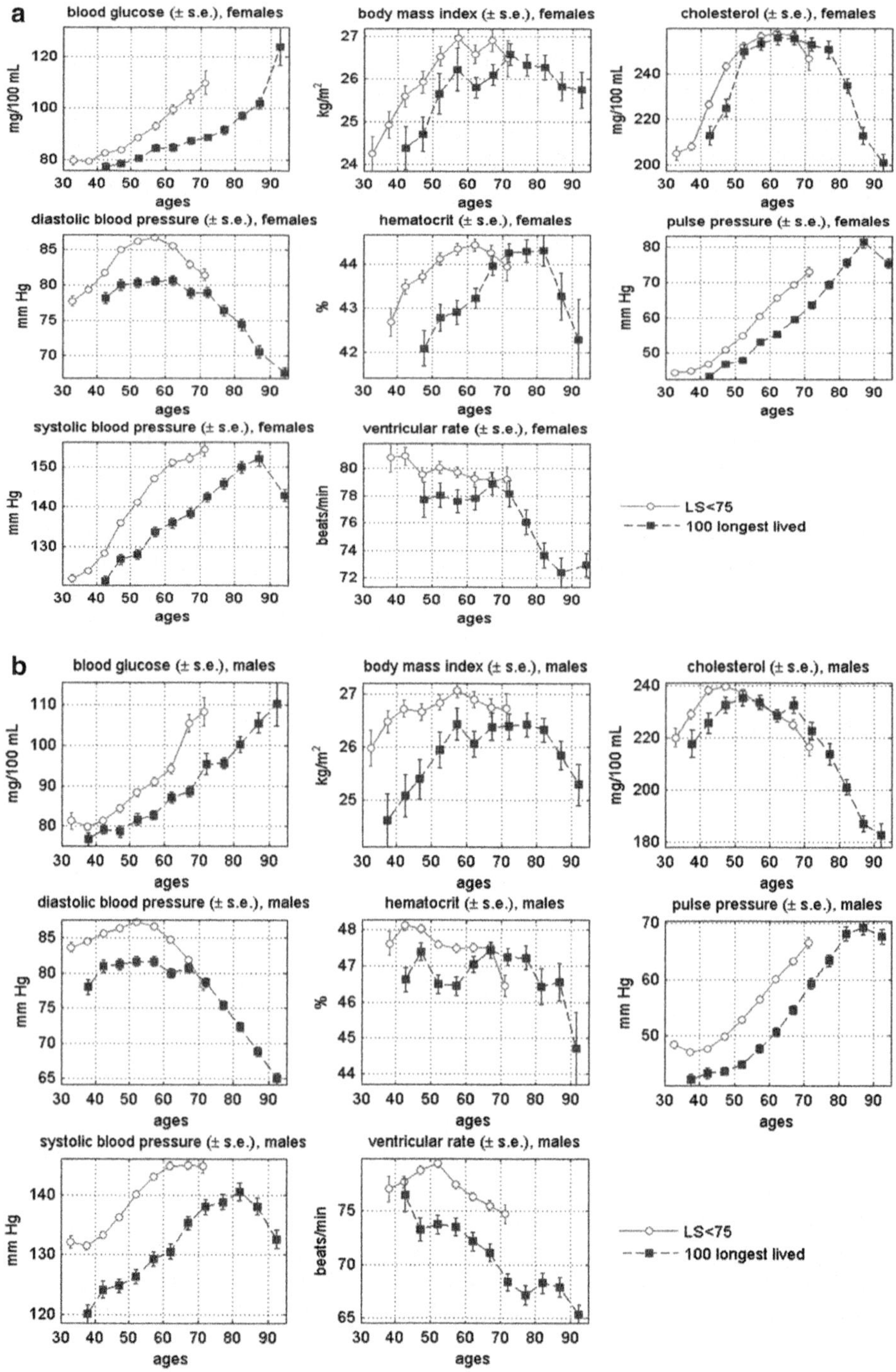

Fig. 2.6 (**a**) Average age trajectories of eight physiological indices for female participants of the Original FHS cohort having short lifespan (SL-group) (lifespan <75 years) and 100 individuals having the longest lifespans (LL-group). (**b**) Average age trajectories of eight physiological indices for male participants of the Original FHS cohort having short lifespan (SL-group) (lifespan <75 years) and 100 individuals having the longest lifespans (LL-group)

the SL than for the LL females. The average values of SBP are higher among SL females. They increase faster for SL than for LL females until age 60. The average values of PP for SL females are higher than those of LL females. Both curves increase at about the same rates. The average values of SCH are higher among the SL females up to age 65, where they reach their maximum value and then decline. At age 65, the average age trajectory of SCH for the SL females intersects that of the LL females. The average values of H are higher among SL females until age 65 when the curves intersect. The SL curve reaches its maximum value earlier (age 60 vs. age 75 for the LL females). The maximum values for the two curves are about the same. Overall, the H trend is similar to that for BMI. The values of VR for the SL females are higher until age 65 and then practically coincide with those of the LL females until age 75.

The average age trajectories of the eight physiological indices for SL and LL males are shown in Fig. 2.6b.

The average values of BG are higher and increase faster among the SL males than among the LL males. The average values of BMI are higher among the SL males. The entire age trajectory of BMI for the LL males is shifted to the right (towards an older age), compared to the SL, and reaches its maximum at a later age, similar to that in females. The values of BMI for SL males and females start to decline earlier (after about age 60 for the SL, while approx. 10 years later for the LL). The average values of DBP among the SL males are higher than those of the LL males until age 70. At ages 70–75, the values of DBP are practically indistinguishable between the two groups. The maximum value of the average DBP in the SL group is higher than that of the LL group and it is reached at about the same age, 55 years. The average values of SCH are higher among the SL males until age 50, reach their maximum value at about age 45, and then decline. The SCH curve for LL males reaches its maximum value at age 55. Between ages 55 and 65, the two curves are about the same. Then the SL curve becomes lower than the LL curve. The average values of H are higher among SL males until age 65 when the curves intersect. The SL curve has a clear maximum value at age 45 vs. two peaks years for the LL males (at ages 45 and 65). The maximum values for the two curves are close. The average values of PP for SL males are higher than those of LL males. Both curves increase at about the same rate. The average values of SBP are higher among SL males. They increase faster for SL than for LL males until age 60 after which the SL curve levels off. The values of VR for the SL males are higher than for the LL males.

Figure 2.7a shows the average age trajectories of eight physiological indices for female participants of the Original FHS cohort having values of healthy lifespan <75 years (short healthy lifespans (SHL)) and 100 females having the longest healthy lifespans (LHL). The average values of BG for SHL females are higher and increase faster than those of the LHL females. The BMI values for the SHL females are higher than for the LHL females for all ages. The decline after reaching the maximum is slower in the LHL females than in the SHL females, so the curves converge with increasing age up to 95 years. The values of SCH for the SHL females are slightly higher than for the LHL females until age 50. The curves are

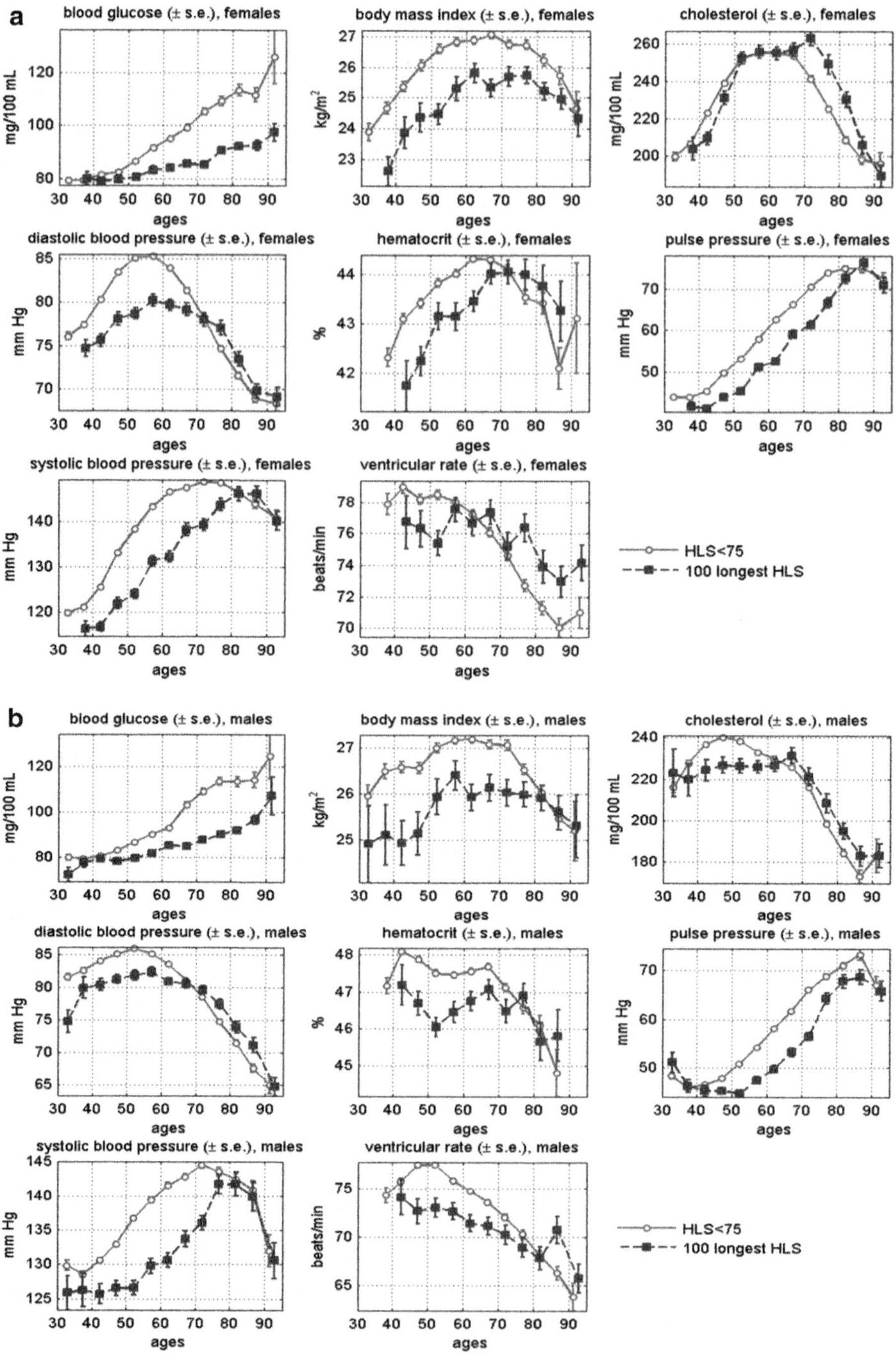

Fig. 2.7 (**a**) Average age trajectories of eight physiological indices for female participants of the Original FHS cohort having short healthy lifespans (HLS <75) and 100 individuals having the longest healthy lifespans (HLL-group). (**b**) Average age trajectories of eight physiological indices for male participants of the Original FHS cohort having short healthy lifespan (HLS <75) and 100 individuals having longest healthy lifespan (HLL-group)

about the same between 50 and 65 years of age. Then the SHL curve becomes lower than the LHL curve. The curves converge and become indistinguishable at about age 90. The values of DBP for the SHL females are higher and increase faster than for the LHL females until age 75. Then the curves cross over with lower SHL values thereafter. The H curves for the SHL females are higher until about age 70. Then the curves cross over with lower SHL values thereafter. The values of PP are higher among the SHL females until age 85. Then the curves converge. The values of SBP are higher among SHL females until age 85. Then the curves become statistically indistinguishable. The VR curves for the SHL females are higher until about age 55. Then the curves cross over with lower SHL values thereafter.

The SHL and LHL curves for males are shown in Fig. 2.7b. The average values of BG for SHL males are higher and increase faster than those of the LHL males; the pattern for males is very similar to the pattern for females. The BMI values for the SHL males are higher than for the LHL males up to age 80. The decline after reaching the maximum is much slower in the LHL males than in the SHL males. The SCH curves are about the same at the beginning for both male groups but then trend upwards for the SHL males with a peak at age 45 and a downward trend thereafter, with the two curves crossing over near age 60. After this age, the curve for the SHL males is lower than that of the LHL males through the end of age range. The values of DBP for the SHL males are higher than those of the HLL males until age 65 when the two curves intersect. After age 65, the DBP curve in the HLS group remains lower than that in the LHL group through the end of age range. The H curve for the SHL males is higher than for the LHL males until age 75. Then the curves become very similar. Both H curves for males start with higher values than for females. The PP curves for the SHL and LHL groups have similar patterns for males as for females. The values of SBP are higher among SHL males until age 75. Then the curves become very similar. The starting values for males are higher than for females. The VR curve for the SHL males is higher than for the LHL males until about age 80. Then the curves cross over with lower values thereafter for the SHL males.

2.4.6 Effects of Disease on Dynamic Properties of Physiological Indices

To better understand whether the presence of chronic disease affects the age dynamics of a physiological state, we distinguished between individuals having at least one of three major chronic diseases (cancer, CVD, and diabetes) and individuals free of such diseases.

Figure 2.8a shows the average age trajectories for eight physiological indices for healthy and unhealthy females. The average values of female BMI for the two groups are statistically indistinguishable until age 50. After this age, the BMI curves for healthy females are lower than for unhealthy females through the end of the age

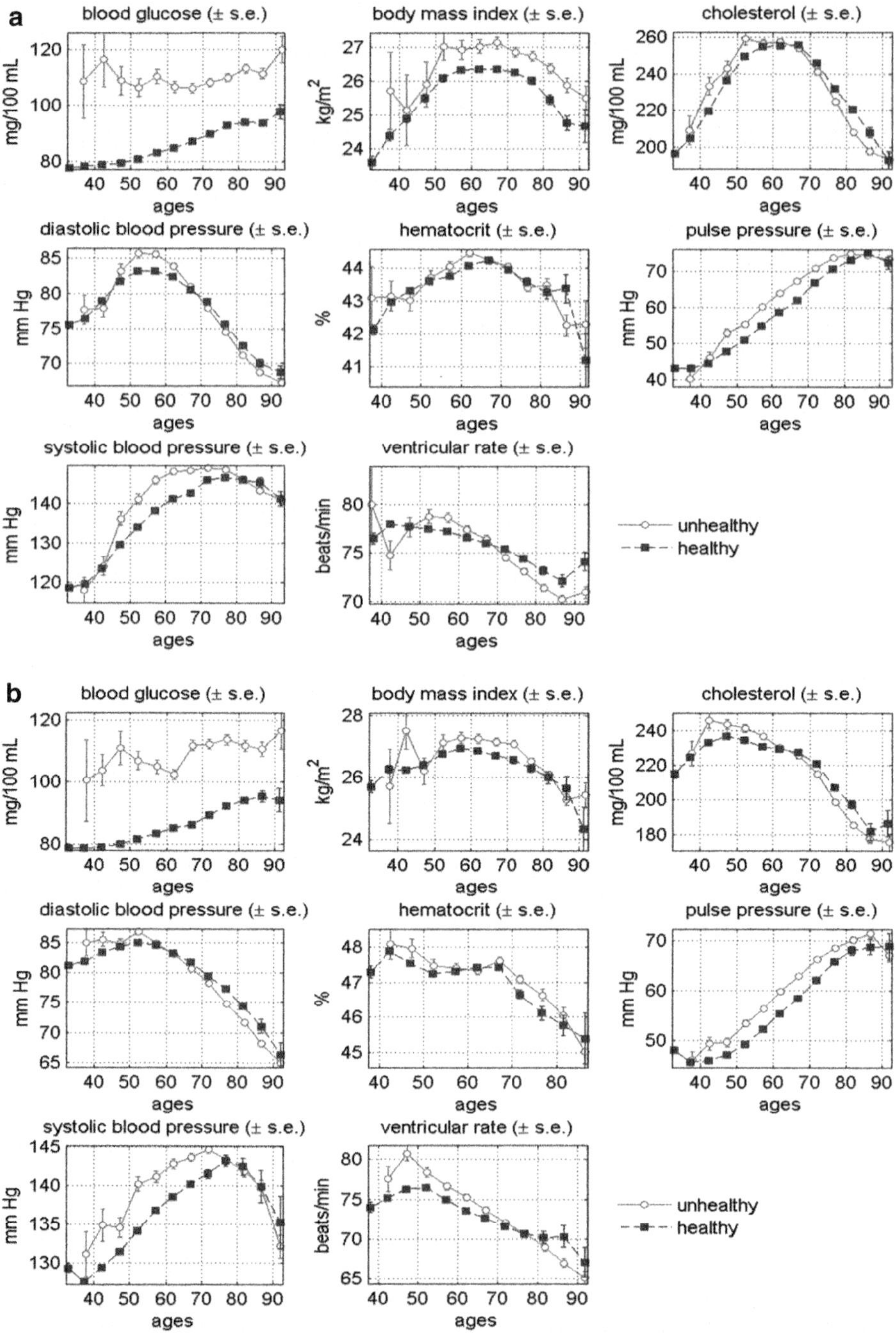

Fig. 2.8 (**a**) Average age trajectories of eight physiological indices for healthy and unhealthy female members of the Original FHS cohort. (**b**) Average age trajectories of eight physiological indices for healthy and unhealthy male members of the Original FHS cohort

range. Between ages 40 and 60, the average values of SCH for healthy females are lower than for unhealthy females. However, after this age, the values of this index for healthy females become higher than for unhealthy females. The SCH curve for unhealthy females appears to be characterized by a slight left-shift of the curve for healthy females. The average values of DBP for both female groups are about the same until age 45. Then between ages 45 and 65 the curve for unhealthy females becomes higher than that for healthy females. The curves intersect around age 70 after which the DBP values for unhealthy females becomes slightly lower than for healthy females. The average H values for the two groups are statistically indistinguishable at all ages except between 55 and 60 years, where the H values for unhealthy females are slightly higher than for healthy females. The average values of PP are about the same at ages 40 and 85, with higher PP values for unhealthy females in-between.

The SBP curve for unhealthy females is higher than that for healthy females between ages 45 and 80 with similar values outside this interval. The curve for unhealthy females increases faster and reaches its maximum earlier (about age 70) than that for healthy females (about age 75). The maximum value is slightly higher for unhealthy females. The VR curves are statistically indistinguishable at age 35; at age 40, the value for unhealthy females is lower than for healthy females. The curves intersect at age 45 and again at age 65, after which the VR curve for unhealthy females is lower than for healthy females.

Figure 2.8b shows average age trajectories for eight physiological indices for healthy and unhealthy males. The average values of BMI between ages 55 and 75 are slightly higher for unhealthy males than for healthy males. The curves are statistically indistinguishable above this age interval. The values of SCH in the two male groups are indistinguishable at age 40, but become higher for unhealthy males between ages 45 and 60. After age 65, the SCH curve for unhealthy males is lower than for healthy males through the end of the age range. The maximum value for the unhealthy males is slightly higher and reaches its maximum value earlier (about age 40) than for the healthy males (about age 45). The average values of DBP are indistinguishable until age 45, but become slightly higher for unhealthy males at age 50. At age 55, the curves intersect so that DBP for unhealthy males is slightly lower than for healthy males through the end of the age range. The average H values for the two groups are statistically indistinguishable from age 50 to 70. After these ages, the H curve for unhealthy males is slightly higher than for the healthy males until age 80, where the curves became indistinguishable again. The average values of PP are about the same at age 40 and at age 90 with higher values of PP for unhealthy males in-between. The SBP curve for unhealthy males is higher than for healthy males until age 75. After this age, the curves become indistinguishable. The curve for unhealthy males reaches its maximum value earlier (about age 70) than for healthy males (about age 75). The maximum value is slightly higher for unhealthy males. The VR curve for unhealthy males is higher than for healthy males until age 75. Then it becomes lower than for healthy males and remains so through the end of the age range.

Thus, the physiological states are dynamically connected with morbidity and mortality risks, and individual health influences the physiological dynamics and mortality risks. A particularly notable observation is the shift of the entire age trajectory of BMI for the LL males and females to the right (towards an older age), as compared with the SL group, and achieving its maximum at a later age. Such a pattern is markedly different from that for healthy and unhealthy individuals. The latter is mostly characterized by the higher values of BMI for the unhealthy people, while it has similar ages at maximum for both the healthy and unhealthy groups. This indicates that *health and extreme longevity can be mediated differently by the physiological variables*, such as BMI, and that longevity may be linked to a postponement of the physiological aging changes in BMI rather than to its "healthier" values.

Physiological aging changes usually develop in the presence of other factors affecting physiological dynamics and morbidity/mortality risks. Among these other factors are year of birth, gender, education, income, occupation, smoking, and alcohol use. An important limitation of most longitudinal studies is the lack of information regarding external disturbances affecting individuals in their day-to-day life.

2.4.7 Effects of Genetic Dose on Age Patterns of Physiological Indices

In a genome-wide association study (GWAS) of human life span using data from the Original FHS cohort (Yashin et al. 2012a), 27 longevity alleles having positive associations with life span were identified. These 27 alleles were used to construct polygenic score indices (Yashin et al. 2012b). One such index, called "the genetic dose index", was constructed for each genotyped individual by counting how many of the 27 longevity alleles were carried by that individual. The estimated influence of this index on life span was determined to be substantial and highly statistically significant.

Figure 2.9a shows how this index influences survival of female members of the Original FHS cohort. Females carrying fewer longevity alleles (<14) had worse survival than those carrying more longevity alleles (≥ 14).

Figure 2.9b shows how the average age trajectories for females of the eight physiological indices for carriers of fewer (<14) longevity alleles differs from carriers of more (≥ 14) longevity alleles. The average values of BG are about the same until age 75. After that age, the values of BG in the (<14)-group are lower than for the (≥ 14)-group. The average values of BMI for the (<14)-group are higher than for the (≥ 14) group until age 65. The curves are about the same until age 80, after which the values for the (<14)-group are lower than for the (≥ 14)-group. The average values of SCH are higher for the (<14)-group than for the (≥ 14)-group until age 85. Then the curves are about the same. The maximum value

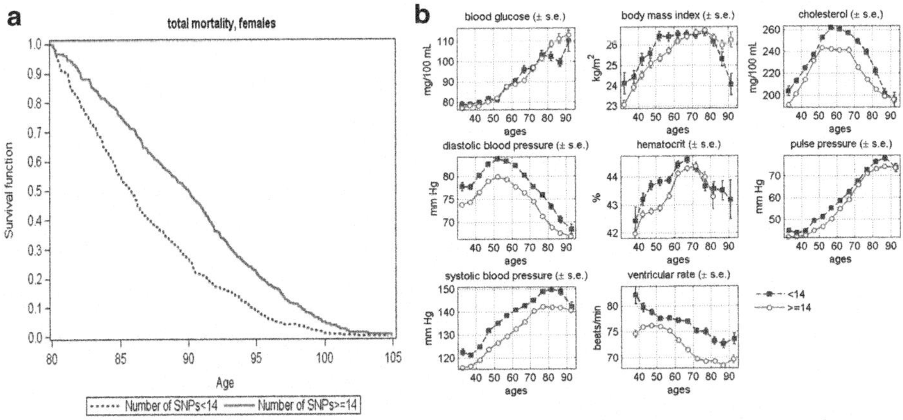

Fig. 2.9 (**a**) Kaplan-Meier estimates of survival functions for two groups of female members of the Original FHS cohort having different numbers of longevity alleles in their genomes. (**b**) Average age trajectories of eight physiological indices for genotyped female participants of the Original FHS cohort carrying different numbers of longevity alleles

of SCH is higher for the (<14)-group than for the (≥14)-group. The average values of DBP for the (<14)-group are higher than for the (≥14)-group for all ages. Both curves reach their maximum at age 50. The average values of H are higher for the (<14)-group until age 70. Between ages 70 and 80 the curves are about the same (the data after age 80 for the (≥14)-group are not available). The average values of PP for the (<14)-group are higher than for the (≥14)-group for all ages, except for the convergence at age 95. The average values of SBP for the (<14)-group are higher than for the (≥14)-group for all ages. The (≥14)-curve reaches its maximum value earlier than the (<14)-curve (age 75 vs. age 80). The average values of VR for the (<14)-group are higher than for the (≥14)-group for all ages.

Figure 2.10a shows how this index influences survival of male members of the Original FHS cohort. Males carrying fewer longevity alleles (<14) have worse survival than those carrying more longevity alleles (≥14).

The age patterns of the eight physiological indices for males carrying different numbers of longevity alleles are shown in Fig. 2.10b. This figure shows that the age patterns for average BG values for the two groups of males are similar to those of females, i.e., the two males groups are similar up to age 70 years after which the BG in the (<14)-group was lower than in the (≥14) group, except for the convergence at age 95.

The average values of BMI for the (<14)-group of males are slightly higher than those of (≥14)-group until age 45. Then the curves are about the same until age 55. After this age, the values of BMI for the (<14)-group are lower than for the (≥14)-group. The (<14)-curve reaches its maximum value earlier at age 55 and then declines. The maximum for the (≥14)-curve is at age 75. The average values of SCH are higher for the (<14) group than those for the (≥14)-group until age 50. The curves are about the same until age 70 after which the (≥14)-curve is lower. The maximum values of SCH for the two groups are about the same and reached at

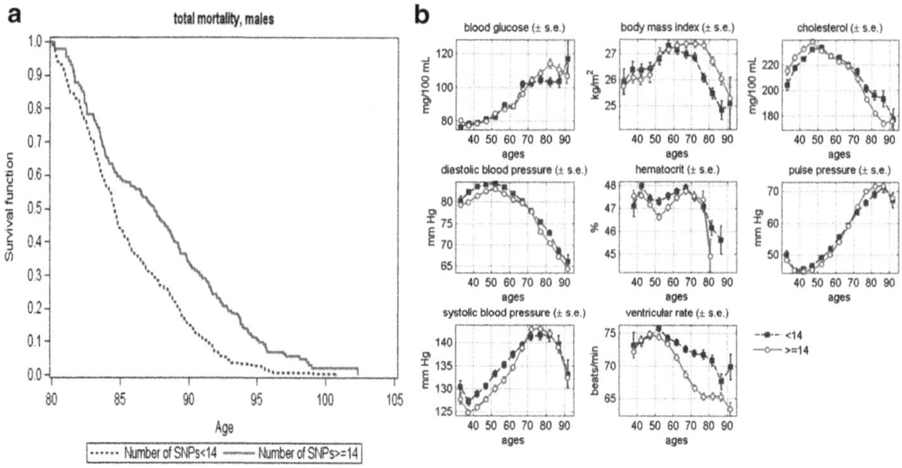

Fig. 2.10 (**a**) Kaplan-Meier estimates of survival functions for two groups of male members of the Original FHS cohort having different numbers of longevity alleles in their genomes. (**b**) Average age trajectories of eight physiological indices for genotyped male participants of the Original FHS cohort carrying different numbers of longevity alleles

about the same age, 45 years. The average values of DBP for the (<14)-group are slightly higher than for the ≥14)-group until age 65 after which they are about the same. After age 75, the (≥14)-curve is slightly lower again until the end of the age range. The average values of H are about the same at age 40; between ages 45 and 65, the values of H are higher for the (<14)-group. After age 65, the differences between curves are statistically insignificant. The average values of PP for the (≥14)-group are about the same between ages 35 and 45, slightly lower between ages 45 and 60, and slightly higher after age 70. The average values of SBP in the (<14)-group are higher until age 65, lower until age 80, similar thereafter. The values of VR for both groups are about the same between ages 35 and 50. After age 50, the values for the (≥14)-group decline much faster and farther than for the (<14)-group.

Comparisons of the age patterns in Figs. 2.9b and 2.10b allow us to conclude that genetic differences are small for average age patterns of BG and PP in both males and females, but are, larger and gender-specific for average age patterns of BMI, SCH, DBP, SBP, H, and VR. The factors responsible for these differences merit further study.

2.5 Conclusion

In this chapter, we evaluated the average age trajectories of eight physiological indices (BMI, DBP, SBP, PP, BG, SCH, H, and VR), using longitudinal human data from the Framingham Heart Study. We showed how these trajectories depend on various genetic and non-genetic factors affecting human lifespan. The empirical analysis illustrated the underused research potential of the FHS data and, hence, of other longitudinal panel data for investigating the dynamic aspects of aging, genetic and non-genetic factors affecting these dynamics, the role of aging in health deterioration, and the increasing chances of death with age.

Our results revealed that survival functions and average age trajectories of physiological indices are gender-specific and can be modulated by behavioral, social, economic, and genetic factors. The age-patterns of physiological changes differed between groups of short-lived and long-lived individuals, as well as between groups of individuals having shorter and longer healthspans. The physiological trajectories also depended on individuals' health status. A particularly notable finding was that health and extreme longevity can be mediated differently by the physiological variables, such as BMI, and that *longevity may be linked to a postponement of aging changes in physiological variables rather than to their "healthier" values.*

While our analysis captured dynamic behaviors of physiological variables and provided some insights into their connections with longevity and health, such as above, it still cannot fully explain: (1) the driving forces behind the patterns of the aging-related changes; and (2) the connections between age trajectories of physiological variables and health/survival outcomes. To further uncover such mechanisms and connections from analysis of the longitudinal human data, appropriate mathematical and computer models need to be developed and applied.

These new models have to link the observed changes in physiological variables with the anticipated biological mechanisms of aging in the presence of both harmful and beneficial external factors. This needs to be done in a way that can decompose the average age trajectories of physiological indices into at least two essential components: (1) a component that represents biological changes in an aging human body; and (2) a component that represents the compositional changes in the study cohort due to the process of differential mortality selection. These two components are mixed in the average age trajectories of physiological variables studied in this chapter, as well as in all existing studies of such trajectories. Additional components may surface when the age-related biological changes are further decomposed into sub-components dealing with (i) the senescence process per se, (ii) changes due to ontogenetic programming, (iii) changes occurring in response to persistent external disturbances, including the effects of compensatory adaptation to maintain the organism's functioning in the presence of disturbing and destructive forces. Several variations of these models will be discussed and applied to the analyses of longitudinal data in Chaps. 11, 12, 13, 14, 15, and 16.

Acknowledgements The research reported in this paper was supported by the National Institute on Aging grants R01AG027019, R01AG030612, R01AG030198, 1R01AG046860, and P01AG043352. The content is solely the responsibility of the authors and does not necessarily represent the official views of the National Institute on Aging or the National Institutes of Health. The Framingham Heart Study (FHS) is conducted and supported by the National Heart, Lung and Blood Institute (NHLBI) in collaboration with the FHS Investigators. This manuscript was prepared using a limited access dataset obtained from the NHLBI and does not necessarily reflect the opinions or views of the FHS or the NHLBI.

References

Alderman, M. H. (1996). Blood pressure J-curve: Is it cause or effect? *Current Opinion in Nephrology and Hypertension, 5*, 209–213.

Anderson, K. M., Castelli, W. P., & Levy, D. (1987). Cholesterol and mortality: 30 years of follow-up from the Framingham Study. *JAMA: Journal of the American Medical Association, 257*, 2176–2180.

Benetos, A., Rudnichi, A., Thomas, F., Safar, M., & Guize, L. (1999). Influence of heart rate on mortality in a French population – Role of age, gender, and blood pressure. *Hypertension, 33*, 44–52.

Bohm, M., Cotton, D., Foster, L., Custodis, F., Laufs, U., Sacco, R., Bath, P. M., Yusuf, S., & Diener, H. C. (2012). Impact of resting heart rate on mortality, disability and cognitive decline in patients after ischaemic stroke. *European Heart Journal, 33*, 2804–2812.

Boshuizen, H. C., Izaks, G. J., Van Buuren, S., & Ligthart, G. J. (1998). Blood pressure and mortality in elderly people aged 85 and older: Community based study. *British Medical Journal, 316*, 1780–1784.

Cacciatore, F., Mazzella, F., Abete, P., Viati, L., Galizia, G., D'ambrosio, D., Gargiulo, G., Russo, S., Visconti, C., Della Morte, D., Ferrara, N., & Rengo, F. (2007). Mortality and heart rate in the elderly: Role of cognitive impairment. *Experimental Aging Research, 33*, 127–144.

Chyou, P. H., & Eaker, E. D. (2000). Serum cholesterol concentrations and all-cause mortality in older people. *Age and Ageing, 29*, 69–74.

Cruickshank, J. M. (1988). Coronary flow reserve and the J-curve relation between diastolic blood pressure and myocardial infarction. *British Medical Journal, 297*, 1227–1230.

Cruickshank, J. (2003). The J-curve in hypertension. *Current Cardiology Reports, 5*, 441–452.

Dawber, T. R., Meadors, G. F., & Moore, F. E. (1951). Epidemiological approaches to heart disease: The Framingham Study. *American Journal of Public Health, 41*, 279–286.

Fenelon, A., & Preston, S. H. (2012). Estimating smoking-attributable mortality in the United States. *Demography, 49*, 797–818.

Franklin, S. S., Larson, M. C., Khan, S. A., Wong, N. D., Leip, E. P., Kannel, W. B., & Levy, D. (2001). Does the relation of blood pressure to coronary heart disease risk change with aging? The Framingham Heart Study. *Circulation, 103*, 1245–1249.

Freedman, D. M., Ron, E., Ballard-Barbash, R., Doody, M. M., & Linet, M. S. (2006). Body mass index and all-cause mortality in a nationwide U.S. Cohort. *International Journal of Obesity, 30*, 822–829.

Gelber, R. P., Kurth, T., Manson, J. E., Buring, J. E., & Gaziano, J. M. (2007). Body mass index and mortality in men: Evaluating the shape of the association. *International Journal of Obesity, 31*, 1240–1247.

Grassi, G., Quarti-Trevano, F., Dell'oro, R., & Mancia, G. (2010). The "J curve" problem revisited: Old and new findings. *Current Hypertension Reports, 12*, 290–295.

Gu, D. F., He, J., Duan, X. F., Reynolds, K., Wu, X. G., Chen, J., Huang, G. Y., Chen, C. S., & Whelton, P. K. (2006). Body weight and mortality among men and women in China. *JAMA: Journal of the American Medical Association, 295*, 776–783.

Hubbard, R. E., Searle, S. D., Mitnitski, A., & Rockwood, K. (2009). Effect of smoking on the accumulation of deficits, frailty and survival in older adults: A secondary analysis from the Canadian Study of Health and Aging. *The Journal of Nutrition, Health & Aging, 13*, 468–472.

Isles, C. G., & Hole, D. J. (1992). Is there a J-curve distribution for diastolic blood pressure. *Clinical and Experimental Hypertension. Part A Theory and Practice, 14*, 139–149.

Kannel, W. B., Kannel, C., Paffenbarger, R. S., & Cupples, L. A. (1987). Heart rate and cardiovascular mortality: The Framingham Study. *American Heart Journal, 113*, 1489–1494.

Klenk, J., Nagel, G., Ulmer, H., Strasak, A., Concin, H., Diem, G., Rapp, K., & VHM&PP Study Group. (2009). Body mass index and mortality: Results of a cohort of 184,697 adults in Austria. *European Journal of Epidemiology, 24*, 83–91.

Kristjuhan, U. (2012). Postponing aging and prolonging life expectancy with the knowledge-based economy. *Rejuvenation Research, 15*, 132–133.

Kronmal, R. A., Cain, K. C., Ye, Z., & Omenn, G. S. (1993). Total serum cholesterol levels and mortality risk as a function of age: A report based on the Framingham data. *Archives of Internal Medicine, 153*, 1065–1073.

Kuzuya, M., Enoki, H., Iwata, M., Hasegawa, J., & Hirakawa, Y. (2008). J-shaped relationship between resting pulse rate and all-cause mortality in community-dwelling older people with disabilities. *Journal of the American Geriatrics Society, 56*, 367–368.

Li, J. Z., Chen, M. L., Wang, S., Dong, J., Zeng, P., & Hou, L. W. (2004a). Apparent protective effect of high density lipoprotein against coronary heart disease in the elderly. *Chinese Medical Journal, 117*, 511–515.

Li, J. Z., Chen, M. L., Wang, S., Dong, J., Zeng, P., & Hou, L. W. (2004b). A long-term follow-up study of serum lipid levels and coronary heart disease in the elderly. *Chinese Medical Journal, 117*, 163–167.

Manolio, T. A., Pearson, T. A., Wenger, N. K., Barrett-Connor, E., Payne, G. H., & Harlan, W. R. (1992). Cholesterol and heart disease in older persons and women. Review of an NHLBI workshop. *Annals of Epidemiology, 2*, 161–176.

Mensink, G. B. M., & Hoffmeister, H. (1997). The relationship between resting heart rate and all-cause, cardiovascular and cancer mortality. *European Heart Journal, 18*, 1404–1410.

Messerli, F. H., & Panjrath, G. S. (2009). The J-curve between blood pressure and coronary artery disease or essential hypertension exactly how essential? *Journal of the American College of Cardiology, 54*, 1827–1834.

Onrot, J. (1993). Hypertension and the J-curve. How low should you go? *Canadian Family Physician, 39*, 1939–1943.

Preston, S. H., Stokes, A., Mehta, N. K., & Cao, B. (2014). Projecting the effect of changes in smoking and obesity on future life expectancy in the United States. *Demography, 51*, 27–49.

Schatz, I. J., Masaki, K., Yano, K., Chen, R., Rodriguez, B. L., & Curb, J. D. (2001). Cholesterol and all-cause mortality in elderly people from the Honolulu Heart Program: A Cohort Study. *Lancet, 358*, 351–355.

Staessen, J. A. (1996). Potential adverse effects of blood pressure lowering – J-curve revisited. *Lancet, 348*, 696–697.

Townsend, R. R. (2005). Can we justify goal blood pressure of <140/90 mm Hg in most hypertensives? *Current Hypertension Reports, 7*, 257–264.

Tsai, W. Y., Jewell, N. P., & Wang, M. C. (1987). A note on the product-limit estimator under right censoring and left truncation. *Biometrika, 74*, 883–886.

Weverling-Rijnsburger, A. W., Blauw, G. J., Lagaay, A. M., Knook, D. L., Meinders, A. E., & Westendorp, R. G. (1997). Total cholesterol and risk of mortality in the oldest old. *Lancet, 350*, 1119–1123.

Weverling-Rijnsburger, A. W. E., Jonkers, I., Van Exel, E., Gussekloo, J., & Westendorp, R. G. J. (2003). High-density vs low-density lipoprotein cholesterol as the risk factor for coronary artery disease and stroke in old age. *Archives of Internal Medicine, 163*, 1549–1554.

Yashin, A. I., Wu, D., Arbeev, K. G., Stallard, E., Land, K. C., & Ukraintseva, S. V. (2012a). How genes influence life span: The biodemography of human survival. *Rejuvenation Research, 15*, 374–380.

Yashin, A. I., Wu, D., Arbeev, K. G., & Ukraintseva, S. V. (2012b). Polygenic effects of common single-nucleotide polymorphisms on life span: When association meets causality. *Rejuvenation Research, 15*, 381–394.

Zhou, B. F. (2002). Effect of body mass index on all-cause mortality and incidence of cardiovascular diseases – Report for meta-analysis of prospective studies on optimal cut-off points of body mass index in Chinese adults. *Biomedical and Environmental Sciences, 15*, 245–252.

Chapter 3
Health Effects and Medicare Trajectories: Population-Based Analysis of Morbidity and Mortality Patterns

Igor Akushevich, Julia Kravchenko, Konstantin G. Arbeev, Svetlana V. Ukraintseva, Kenneth C. Land, and Anatoliy I. Yashin

3.1 Introduction

Determining national trends in health and vital status of the growing sector of older U.S. adults is a major public health concern and important issue for policymakers and governmental institutions. To better address the challenge of "healthy aging" and to reduce economic burdens of aging-related diseases, key factors driving the onset and progression of diseases in older adults must be identified and evaluated. An identification of disease-specific age patterns with sufficient precision requires large databases that include various age-specific population groups. Collections of such datasets are costly and require long periods of time. That is why few studies have investigated disease-specific age patterns among older U.S. adults and there is limited knowledge of factors impacting these patterns. Studies based on observational data could be a reasonable alternative. The assignment of subjects to a treatment group vs. a control group is not controlled by the investigator in observational studies; therefore, special attention has to be paid to the choice of statistical methodologies and the interpretation of the results obtained.

Information collected in U.S. Medicare Files of Service Use (MFSU) for the entire Medicare-eligible population of older U.S. adults can serve as an example of observational administrative data that can be used for analysis of disease-specific age patterns. These data, as any administrative health data, are generated through the routine administration of health care programs. Thus, the development of approaches to analyses of Medicare data and their application for discovery of substantive health-related results for the U.S. elderly population is well-timed and largely motivated by the lack of such comprehensive and representative analyses at a national level. Uncovering the properties of aging-related disease incidence, risk factors of disease onset and progression, co- and multimorbidities, and recovery/long-term remission among older adults can yield deep insights into theoretical aspects of the interrelations between disease incidence and senescence at advanced

A.I. Yashin et al., *Biodemography of Aging*, The Springer Series on Demographic Methods and Population Analysis 40, DOI 10.1007/978-94-017-7587-8_3

ages. These insights can aid in developing future health improvement strategies and forecasting the associated future Medicare expenditures.

In this chapter, we focus on a series of epidemiologic and biodemographic characteristics that can be studied using MFSU. The MFSU datasets provide an opportunity to investigate the demographic and epidemiologic properties of the general older U.S. population and cohorts of survivors with a wide spectrum of chronic and acute diseases.

3.2 Data and Methods

3.2.1 Data: SEER-M and NLTCS-M

Two datasets capable of generating national level estimates for older U.S. adults are the Surveillance, Epidemiology, and End Results (SEER) Registry data linked to MFSU (SEER-M) and the National Long Term Care Survey (NLTCS), also linked to MFSU (NLTCS-M). In the SEER-M and NLTCS-M datasets, the Medicare records are available for each institutional (inpatient, outpatient, hospice, skilled nursing facility, or home health agency) and non-institutional (carrier-physician-supplier and durable medical equipment providers) claim type. The so-called screener weights released with the NLTCS allow us to produce national population estimates (for a recent discussion, see Akushevich et al. 2013c). The extensive detail in these files allows for identification of disease incidence and recovery using computational algorithms designed to extract these events from administrative datasets. NLTCS-M represents a weighted random sample of the entire U.S. elderly population. SEER-M is a registry database and thus has much better statistical power but it represents the population of SEER areas only; therefore, SEER-M represents the U.S. general population only approximately. The age and sex distribution of the total SEER population is similar to non-SEER areas, though SEER areas have fewer whites, more urban residents, and fewer poor areas compared to non-SEER areas (Warren et al. 2002).

The SEER-M data are the primary dataset analyzed in this chapter. The expanded SEER registry covers approximately 26 % of the U.S. population. In total, the Medicare records for 2,154,598 individuals are available in SEER-M including individuals (i) with diagnosed carcinomas of breast (n = 353,285), colon (n = 222,659), lung (n = 342,961), and prostate (n = 448,410), and skin melanoma (n = 101,123); and (ii) from a random 5 % sample of Medicare beneficiaries residing in the SEER areas who had none of the above mentioned cancers. For the majority of persons, we have continuous records of Medicare services use from 1991 (or from the time the person reached age 65 after 1990) to his/her death. A small fraction of individuals (e.g., new patients who have been diagnosed with cancer in 2003–2005) has Medicare records beginning from 1998.

Table 3.1 Groups of diseases and the associated ICD-9 codes

Group of diseases	Disease with ICD-9 codes
Cardio- and cerebrovascular	Myocardial infarction (410.xx), angina pectoris (413.xx), stroke (431.xx, 433.x1, 434.x1, 436.xx), heart failure (428.xx)
Malignancies	Lung cancer (162.xx), colon cancer (153.xx), breast cancer (females) (174.xx), prostate cancer (185.xx), skin melanoma (172.xx), kidney cancer (189.xx), pancreatic cancer (157,xx)
Neurodegenerative	Parkinson's disease (332.xx), Alzheimer's disease (331.0)
Pulmonary	Chronic obstructive pulmonary disease (COPD) (490.xx, 491.xx, 492.xx, 493.xx, 494.xx, 495.xx, 496.xx), asthma (493.xx)
Bones/skeletal	Hip fracture (820.xx, 821.xx)
Endocrine and metabolic	Diabetes mellitus (250.xx), goiter (240.xx, 241.xx, 242.0x, 242.1x, 242.2x, 242.3x)
Miscellaneous	Chronic renal diseases with renal failure (403.xx, 404.xx, 581.xx, 582.xx, 583.xx, 585.xx, 586.xx, 587.xx, 588.xx, 250.4x, 249.4x), ulcer (531.xx, 532.xx, 533.xx, 534.xx), arthritis (714.0x, 714.1x, 714.2x, V82.1x)

The NLTCS-M data contain two of the six waves of the NLTCS: namely, the cohorts of years 1994 and 1999. These two waves were chosen primarily because high-quality Medicare follow-up data are available since 1991 and also because the complete 5-year follow-up after the NLTCS interview is accessible only for these two waves after 1991. In total, 34,077 individuals were followed-up between 1994 and 1999. These individuals were given the detailed NLTCS interview (those from the subcohorts of 1994 and 1999) which has information on risk factors. More than 200 variables were selected from the 1994 and 1999 surveys and were grouped as follows (a complete list of all variables used in the analysis is presented in Table 3.1 in the Electronic Supplemental Material in Akushevich et al. 2011a):

A. Demographic characteristics (four variables): sex, race, marital status, urban vs. rural residence.
B. Self-reported comorbidity (27 major medical conditions).
C. Daily living activities (22 variables): six activities of daily living (ADLs) with two severity levels, and ten instrumental activities of daily living (IADLs).
D. Range of motion (16 variables): reflecting ability to perform daily activities such as walking, using fingers to grasp and handle small objects, and climbing stairs.
E. Physical activity (29 variables, including 25 variables reflecting specific physical activities such as golf or tennis, measured in 1994 only).
F. Nutrition and social activities (30 variables, 24 of them representing a nutrition survey, measured in 1999 only).
G. Alcohol consumption and smoking (four variables): reflecting two severity levels.
H. Other functioning (28 variables): reflecting self-estimates of health, information about mood, habits, keeping in touch with friends and relatives, and satisfaction with individual's lives.

I. Characteristics of housing and neighborhood (23 variables): describing the area, housing and amenities where the individual lives, including information on whether the individual lives with other household members, and neighborhood characteristics.
J. Health insurance (six variables): containing information on coverage by Medicare, Health Maintenance Organization (HMO), Medicaid, etc.
K. Medical providers and prescription medicine (44 variables): providing information on the use of health care services and public and private expenditures for health care services.
L. Cognitive functioning (18 variables): describing cognitive status of individuals, including 10 variables measured in 1994 and 11 measured in 1999.
M. Income and assets (four variables): representing variables correlated with the socioeconomic status of individuals.
N. Body mass index (BMI) (five variables): representing BMI and dietary patterns.

3.2.2 Definitions of Dates of Disease Onset and Dates of Recovery/Remission

Disease incidence and recovery/remission rates were analyzed for aging-related conditions representing the major groups of diseases in the elderly, including: (i) circulatory (acute coronary heart disease (ACHD), myocardial infarction, angina pectoris, heart failure, and stroke), (ii) cancer (breast, prostate, lung, colon, and skin melanoma), (iii) neurodegenerative (Parkinson's and Alzheimer's diseases), (iv) endocrine and metabolic (diabetes mellitus and goiter), (v) pulmonary (Chronic Obstructive Pulmonary Disease (COPD), emphysema, and asthma), and (vi) other chronic conditions such as chronic renal disease, ulcer, and arthritis.

The majority of analyses presented in this chapter are based on identification of the date of the disease onset using information collected in the Medicare Claims files. The approach to the identification uses individual medical histories of the applicable disease reconstructed from Medicare files, combining all records with their respective ICD-9 codes. Several examples of individual medical histories reconstructed from MFSU are shown in Fig. 3.1. Also, Fig. 3.1 demonstrates an existence of periods of absence of MFSU records for certain diseases. Such periods could be associated with partial or complete long-term remission or even recovery from an applicable disease and are also subjects for investigation in this chapter.

Ages at onset of all diseases and recovery therefrom were reconstructed from MFSU using the following scheme. First, the individual medical histories history related to specific disease were reconstructed using ICD-9 codes (Table 3.1). Then a special procedure was applied for individuals with a history of the diseases to separate incident and prevalent cases and to identify the cases of disease onsets and disease recovery/remission. This procedure was based on two conditions applied to each medical history. The first condition allowed for identification of the first appearance of the disease code and the second was required for

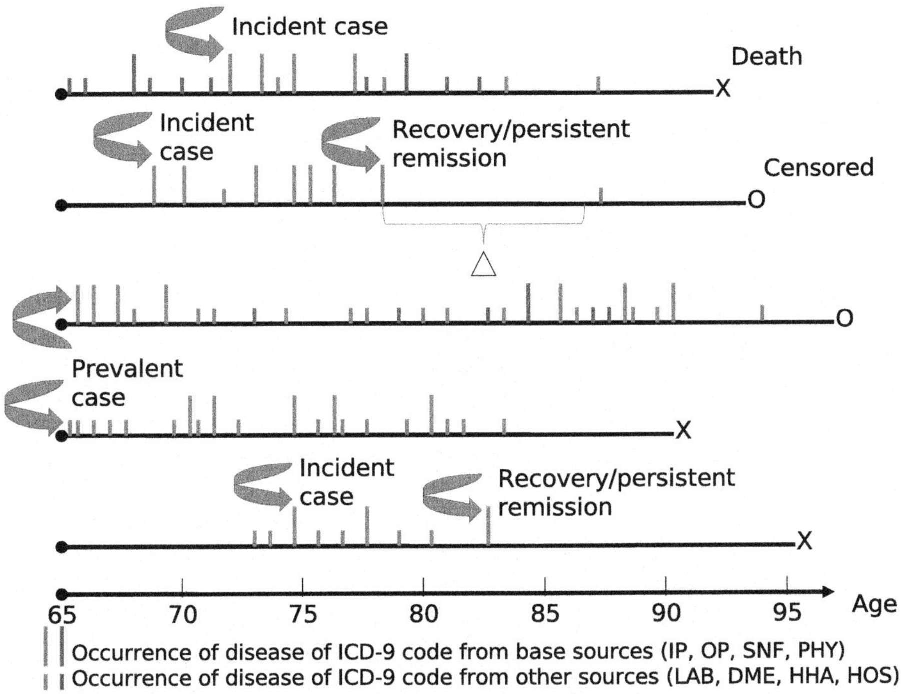

Fig. 3.1 Individual health trajectories. Four Medicare sources are considered base, i.e., inpatient (*IP*), outpatient (*OP*), carrier-physician-supplier (*PHY*), skilled nursing facilities (*SNF*). Other Medicare sources are hospice (*HOS*), home health agency (*HHA*), laboratory (*LAB*), and durable medical equipment (*DME*)

confirmation of disease presence. The individual Medicare history contains all records with a disease ICD-9 code; however, only records with a primary ICD-9 code and only from so-called base Medicare sources (inpatient care, outpatient care, physician services, and skilled nursing facilities) were used for the disease onset identification. This algorithm has been already used by our team to study recovery after stroke (Yashin et al. 2010), medical cost trajectories before and after age-related disease onset (Akushevich et al. 2011b), the wide spectrum of geriatric diseases incidence (Akushevich et al. 2012), and the role of behavior factors in cancer risk (Akushevich et al. 2011a). Further details about the applied algorithm are given in Chap. 6 of this monograph.

3.3 Results

Below we discuss the spectrum of results obtained using the MFSU and data linked with Medicare files.

First, we analyze morbidity and mortality patterns among older U.S. adults. We discuss the results of calculation of age patterns of disease incidence and comparisons with the results from other studies. Then we present the results of the calculation of age-adjusted disease incidence rates. This allows us to estimate time trends, comorbidity, and disability patterns of disease incidence with appropriate accuracy, as well as to perform detailed sensitivity analysis. Also, we include a discussion of uncertainties in calculations of mortality at advanced ages using MFSU.

Second, we investigate the phenomenon of recovery or long-term remission in patients with acute and chronic diseases. The main research question is whether patients who stopped visiting physicians are healthier (vs. those who continue visiting) and, therefore, could be considered recovered patients.

Third, we use multiple self-reported variables from the NLTCS interviews and individual follow-ups after these interviews up to the time of disease onset. A study of associations then allows us to identify disease-specific risk factors and describe high-risk groups based on self-reported measures.

Fourth, we investigate the phenomenon of multimorbidity in older U.S. adults. We construct a new multimorbidity index, compare its properties with the standard Charlson comorbidity index, and incorporate the new index into a model to project cohort-specific health status and mortality. We also describe a forecasting model that involves submodels of incidence, recovery, and mortality.

3.3.1 Age Patterns of Age-Associated Disease Incidence

Age patterns of incidence rates were assessed by stratifying the sample into relevant age categories (1 year, or several years). Empirical age-specific risks (λ_α) were calculated as a ratio of weighted numbers of cases to weighted person-years at risk: $\lambda_a = n_a/P_a$; where $n_a = \sum_n w_n, P_a = \sum_i w_i$, and w_i was the individual weight; n ran over all disease onsets detected in the age group, and i ran over all individuals at risk in α^{th} age group. The individual weights (NLTCS weights were calculated using U.S. Census data and released with the NLTCS data) were necessary to make the estimates representative of the entire U.S. elderly population, i.e., to take into account the effects of study design. The effects of study design also influence the calculation of standard errors (SEs) and confidence intervals of rate estimates. The formula used for calculating SEs is $\sigma_E = \sqrt{\lambda_a(1-\lambda_a)/P_{0\alpha}}$, where $P_{0\alpha}$ is the number of person-years estimated for unit weights. Thus, the standard errors were calculated based on the number of actually measured individuals. A generalization of the formula for SEs based on Wilson's approach (Brown et al. 2001) was used when $P_{0\alpha}$ or λ_α was small.

Age-patterns of acute and chronic disease incidence were evaluated using NLTCS-M and validated using SEER-M. The results are presented in Fig. 3.2 (Akushevich et al. 2012). As an additional control for the incidence rates, the

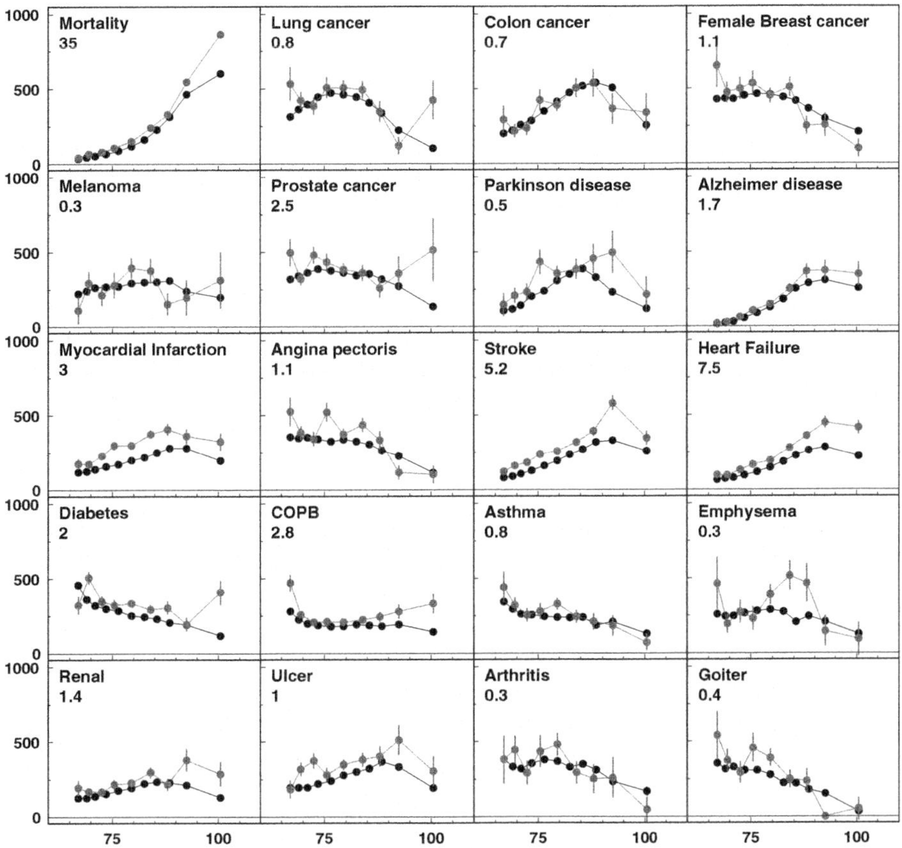

Fig. 3.2 Age-specific rates of total mortality and disease incidence calculated using NLTCS-Medicare (*larger standard errors*) and SEER-Medicare (*smaller standard errors*). Values on plots are rescale factors. Rates for different diseases are rescaled to use the same scale for all plots to facilitate comparisons of rates for different diseases: the original rates can be calculated by dividing the values obtained from the plots by the rescale factor (Color figure online)

total mortality rate was also estimated. Among studied diseases, incidence rates of Alzheimer's disease, stroke, and heart failure increased with age, while the rates of lung and breast cancers, angina pectoris, diabetes, asthma, emphysema, arthritis, and goiter became lower at advanced ages. The incidence rates of several non-cancer diseases (such as myocardial infarction, stroke, heart failure, diabetes, and ulcer) obtained from NLTCS-M were a little higher than the rates calculated using SEER-M. Similar differences were observed for total mortality rates at ages 85+. Both methodological and substantive aspects of these findings were discussed in (Akushevich et al. 2012).

Several types of age-patterns of disease incidence could be described. The first was a monotonic increase until age 85–95, with a subsequent slowing down, leveling off, and decline at age 100. This pattern was observed for myocardial infarction, stroke, heart failure, ulcer, and Alzheimer's disease. The second type had an earlier-age maximum and a more symmetric shape (i.e., an inverted U-shape) which was observed for lung and colon cancers, Parkinson's disease, and renal failure. The majority of diseases (e.g., prostate cancer, asthma, and diabetes mellitus among them) demonstrated a third shape: a monotonic decline with age or a decline after a short period of increased rates. Melanoma and emphysema can also be assigned to this pattern, yet their patterns could also be considered flat.

The occurrence of age-patterns with a maximum and, especially, with a monotonic decline contradicts the hypothesis that the risk of geriatric diseases correlates with an accumulation of adverse health events (such as genetic mutations, deterioration of vascular system, immunosenescence, etc.) during the life. Two processes could be operative in the generation of such shapes. First, they could be attributed to the effect of selection (Vaupel et al. 1998) when frail individuals do not survive to advanced ages. This approach is popular in cancer modeling and was successfully applied to SEER data (Kravchenko et al. 2011, 2012; Manton et al. 2009; Trussell and Richards 1985; Yashin et al. 2009). The second explanation could be related to the possibility of under-diagnosis of certain chronic diseases at advanced ages (due to both less pronounced disease symptoms and infrequent doctor's office visits); however, that possibility cannot be assessed with the available data (Enright et al. 1999; Solomon and Murphy 2005).

3.3.2 Incidence Rates: Comparisons with Other Studies

The agreement between the rates obtained from the two datasets (Fig. 3.2) was predictable, because the datasets have similar designs of data collection and the same computational approach was used for evaluation of diseases incidence. We next compared our results with those obtained in studies with different designs and, possibly, with different computational approaches. We focused on several major groups of diseases to highlight (1) the age-adjusted incidence rates, (2) the shapes of age patterns of incidence rates, and (3) sex differences in age patterns of disease incidence. In general, the datasets used for analysis of incidence rates in older populations were predominantly disease-specific (i.e., focused on a single rather than on multiple diseases). Also, they were not specifically oriented toward the elderly population but included a wide spectrum of age groups, among which the people aged 65+ years old (and especially 85+ years old) represented only a small fraction of the data. NLTCS-M data were used for this comparison because of its better correspondence to the U.S. general elderly population.

Four types of algorithms for identification of disease onset based on disease-specific medical histories (i.e., Algorithm A, B, C, and D) were considered. Algorithm A is the base algorithm: it is briefly described above in the subsection "Definitions of dates of disease onset and dates of recovery/remission" and more details are provided in Chap. 6. In Algorithm B, the confirmation by the second record is not required: i.e., only the first condition is valid. In Algorithm C, all codes (that are not necessary primary) are considered valid and the confirmation is also not required. In Algorithm D, a death event is not considered as the second, confirmation record.

3.3.2.1 Cancer

The most detailed U.S. data on cancer incidence come from the SEER Registry (Altekruse et al. 2009). Because age is widely-recognized as the most important risk factor for developing a cancer (Howlader et al. 2011) and because about 60 % of malignancies are diagnosed in persons aged 65+ years old (Hewitt and Simone 2000), datasets used for evaluating the age patterns of cancer incidence should focus on older populations. The exceptions to these patterns are breast and ovarian cancers, for which a higher proportion of diseases occur at ages younger than 65 years old (Parry et al. 2011). In the U.S., the estimated percent of cancer patients alive after being diagnosed with cancer (in 2008, by current age) was 13 % for those aged 65–69, 25 % for ages 70–79, and 22 % for ages 80+ years old (compared with 40 % of those aged younger than 65 years old) (Trask et al. 2008). In the previous section, we compared cancer rates obtained using Medicare records from NLTCS-M and SEER-M. The SEER-M incidence rates obtained using MFSU and SEER registry did not coincide; therefore, the estimates of the age-specific incidence rates obtained in the present study using the NLTCS-M data need to be compared with estimates obtained using the SEER registry. In most studies that compare age patterns of specific cancers with the patterns predicted from other data, the SEER Registry data were used as a "gold standard". In general, studies of algorithms of cancer incidence identification are rare. Among cancer sites, the most developed among others was the algorithm for the use of Medicare Claims data to identify women with incident breast cancer (Nattinger et al. 2004, 2006). In the patterns shown in Fig. 3.2, we used the basic algorithm (Algorithm A) that identifies cancer cases in the MFSU as "cancer" (i) when a record with the cancer code is confirmed by the second record/visit, or (ii) when the death occurred just after the first record/visit and only one record proving "cancer" existed (i.e., we assumed that in this case the cancer diagnosis was not confirmed by the second record/visit due to a forthcoming death). The fractions of confirmed/non-confirmed records in the NLTCS-M and SEER registry are different. For example, about 95 % of diagnosed cancers in the SEER registry are histologically confirmed, with less than 2 % of them coming from the death certificate or autopsy (Bleyer et al. 2006; Johnson and Adamo 2007; Ries et al. 2007). Therefore, we used the scheme for disease onset identification which excluded the second type of events, i.e., Algorithm D (see Fig. 3.3 for five cancer sites). All cancers, except melanoma, demonstrated good agreement of age-specific incidence rates. Rates for melanoma do not exceed two SEs from the SEER estimates. In summary, the results obtained in the present study demonstrated a good agreement of cancer incidence disease patterns with SEER registry data. Age patterns of incidence rates calculated using Algorithms A and D are presented in (Akushevich et al. 2013b).

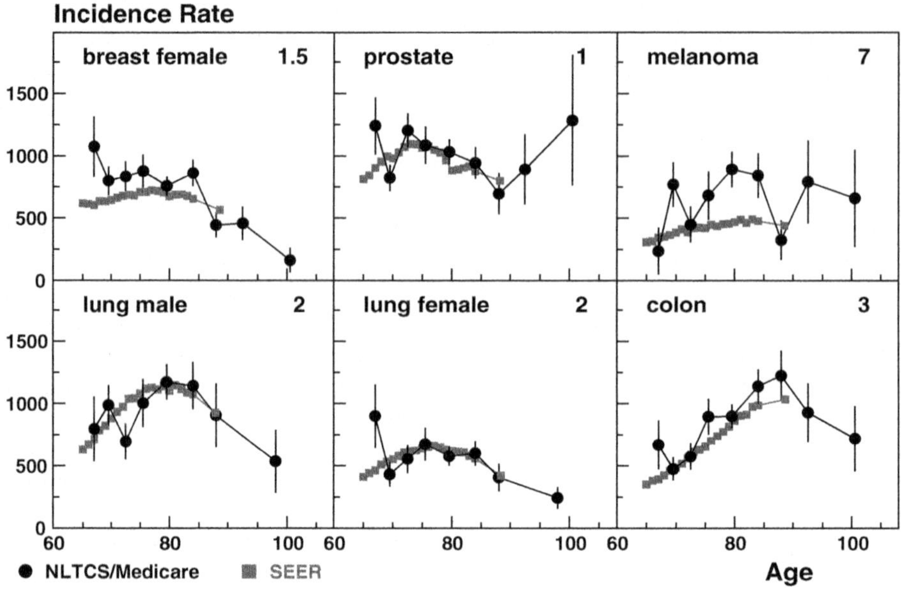

Fig. 3.3 Age specific cancer incidence rates: means and SEs of the NLTCS/Medicare and SEER data for 1994–2003. Rates for SEER above age 85 are shown at the mean age of cases above age 85. The number in the right upper corner is the renormalization factor: true incidence rates are obtained by dividing plotted rates on this factor

3.3.2.2 Heart Diseases and Stroke

There are a number of studies reporting increases in the prevalence of cardio- and cerebrovascular disease among older persons (Crimmins 2004), although more recent analyses have demonstrated that the increases were more significant among those approaching older age rather than among older adults (Freedman et al. 2007; Martin et al. 2009). Accordingly, updated information on this topic is very timely. Results on incidence rates of myocardial infarction, angina pectoris, stroke, and heart failure are presented in Figs. 3.4, 3.5, and 3.6: the results obtained in the present study are compared with the results obtained from several cohort studies (summarized in NIH/NHLBI 2006) such as the Atherosclerosis Risk in Communities (ARIC) study, the Cardiovascular Health Study (CHS), and the Framingham Heart Study (FHS). We have restricted our comparison of acute coronary heart disease (ACHD) patterns to myocardial infarction and angina pectoris, because ACHD is largely represented by these two diseases which are typically investigated in cohort studies. To obtain age patterns shown in Figs. 3.4, 3.5, and 3.6, we used the base strategy for disease onset identification (i.e., Algorithm A). However, the strategy without a requirement for another record confirming the primary diagnosis can be even more appropriate for acute diseases. Such a strategy was implemented in Algorithm B and was used for comparison of the ACHD patterns.

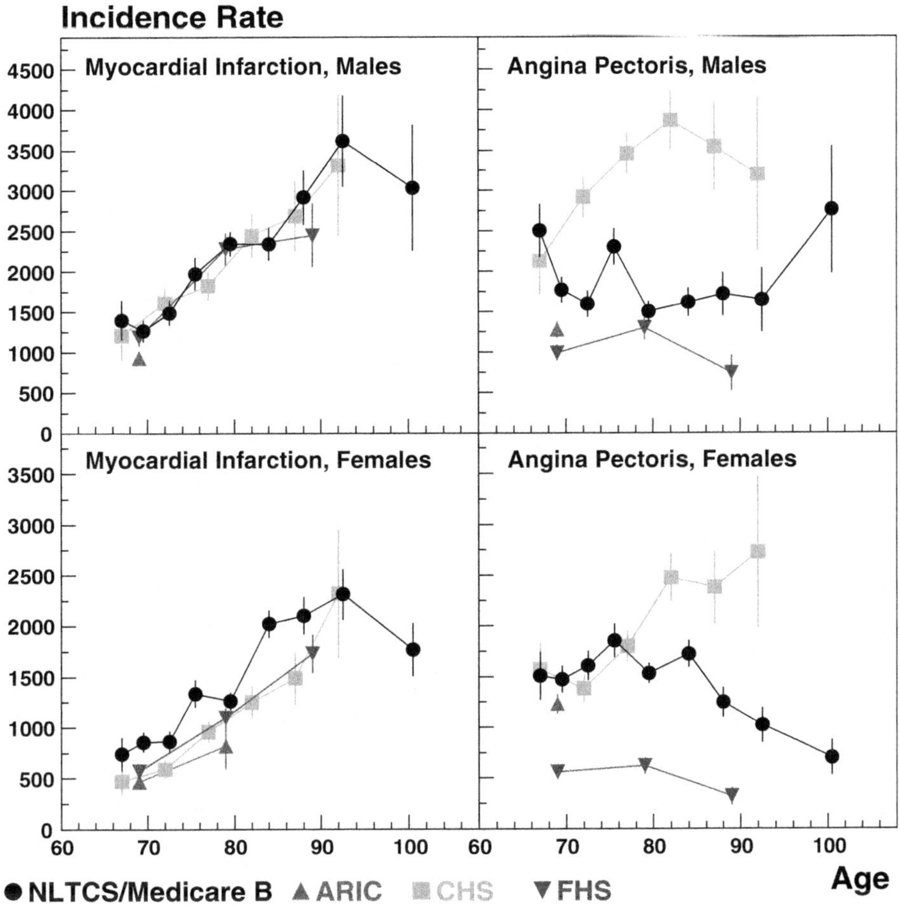

Fig. 3.4 Age-specific incidence rates for myocardial infarction and angina pectoris

The absolute values of incidence rates and the shapes of the age patterns of myocardial infarction are in a very good agreement between all cohort studies and also in agreement with our results. The rates of angina pectoris vary from study to study: e.g., the rates observed in the CHS were higher than the NLTCS rates by a factor of 2; the NLTCS rates were closer to those observed in the FHS. At least in part, that could be attributable to differences in definitions of angina pectoris incidence cases (as described in details in the Appendix of ref. NIH/NHLBI 2006), such as inclusion of the incidence events of angina pectoris diagnosed by a physician together with cases when patients received therapy with nitrates, beta-blockers, or calcium-channel blockers in the CHS. The estimated incidence rates for angina pectoris in the present study are between those found in these two prior studies.

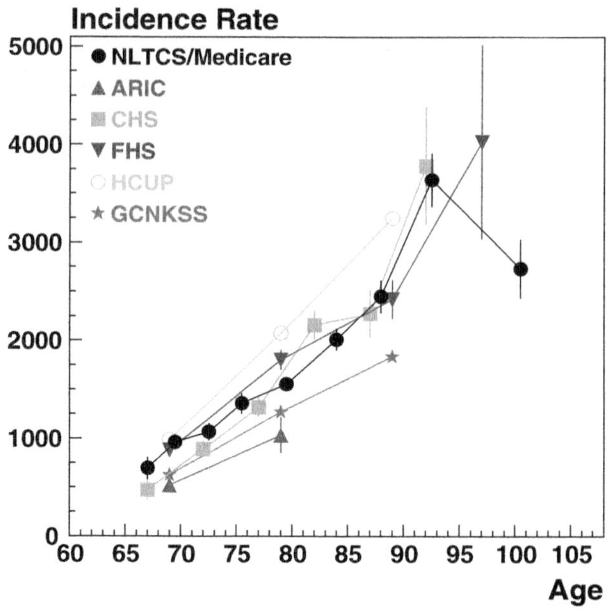

Fig. 3.5 Age-specific incidence rates for stroke

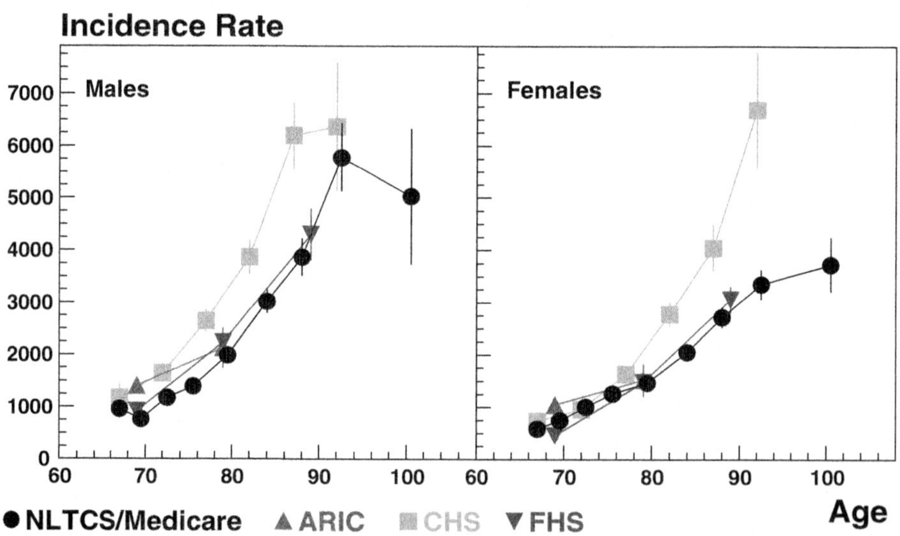

Fig. 3.6 Age-specific incidence rates for heart failure

For stroke, our results are in a good agreement with those obtained in the CHS and FHS cohort studies, as well as with the results of the Health Cost and Utilization Project (HCUP) study (Williams 2001) and the Greater Cincinnati-Northern Kentucky Stroke Study (GCNKSS) (Feigin et al. 2003). The estimated gender disparities in stroke incidence could be due to differences in the younger age group, as Fig. 3.1 shows that the rates at other ages do not differ. Therefore, the comparison study for stoke incidence rates was performed for the total population rather than for sex-specific subgroups. To calculate the rate for the total population, sex- or race-specific rates available in the HCUP and GCNKSS studies were combined with weights representing these population subgroups in the present study. Figure 3.5 shows that the results are in agreement between our estimated rates and those obtained in other studies.

Our estimates of age-specific incidence rates of heart failure (HF) also are in good agreement with the estimates obtained in the ARIC and FHS cohort studies and are significantly lower than those obtained from the CHS (see Fig. 3.6). That could be due to differences in disease incidence definitions used in each study (i.e., in terms of criteria used for diseases case selection/registration). For example, in the FHS cohort study, HF incidence was defined by combination of several major and minor criteria based on disease clinical symptoms; in the ARIC study, HF incidences were selected based on the hospital discharge ICD-9 codes 428 or 518.4; and in the CHS cohort study, HF incidence events were defined as being diagnosed by physician plus including the patients receiving specific medications (such as diuretic plus either digitalis, and vasodilator or angiotensin converting enzyme inhibitor) (NIH/NHLBI 2006). That can explain, at least in part, that incidence rates of HF in the CHS cohort study were higher than those obtained from the ARIC and FHS studies, as well as our estimates (see Fig. 3.6).

3.3.2.3 Diabetes

Diabetes affects about 21 % of the U.S. population aged 65+ years old (McDonald et al. 2009). However, while more is known about the prevalence of diabetes, the incidence of this disease among older adults is less studied. Our estimated incidence rates of diabetes mellitus (shown in Fig. 3.7) are in agreement with several studies performed on cohorts such as the Canadian Study of Health and Aging (Rockwood et al. 2000), the Zwolle Outpatient Diabetes project Integrating Available Care (ZODIAC-1, the Netherlands, Ubink-Veltmaat et al. 2003), and the U.K. Pooled Diabetes Study (Gatling et al. 2001). In these studies, the incidence rates of diabetes decreased with age for both males and females. In the present study, we find similar patterns, except for the first and the last points (Fig. 3.7), i.e., for ages 66 and 100 years. In the ZODIAC-1 study, diabetes type II incidence rates in 1998–2000 were slightly higher, and in the U.K. Poole Diabetes Study, the rates were slightly lower than our estimates.

Generally, the age trends and the absolute incidence rates in all of the studies considered correspond to our results. Age-specific predicted incidence rates of adult-onset diabetes were calculated in a population-based retrospective study using community-based medical records in Rochester, Minnesota (Leibson et al. 1997): in 1985, the incidence rate per 100,000 person-years was about 600 for ages 70–74, and about 500 for ages 80–84. These results are in agreement with our estimates for

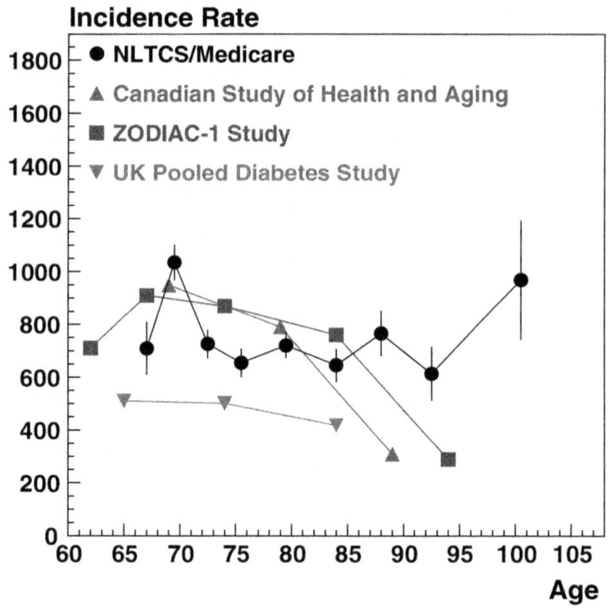

Fig. 3.7 Age-specific incidence rates for diabetes mellitus

these age groups. Another diabetes study (McBean et al. 2004) examined disease prevalence, incidence, and mortality from 1993 to 2001 among fee-for-service Medicare beneficiaries aged 67+ years old using a 5 % random sample of enrollees: a rate of 3000 per 100,000 was estimated. The reason for this disagreement could be different schemes for identification of diabetes onset (Hebert et al. 1999) from the Medicare data. In that study, it was required that a second record with a diabetes ICD code must be observed only if the first one was registered as an ambulatory claim (i.e., a physician/supplier or hospital outpatient claim). That could be the reason for the excess in incidence rates in McBean et al. (2004). We recalculated diabetes incidence rates using the approach described in (Hebert et al. 1999; McBean et al. 2004) and found an almost fivefold increase in the incidence rates compared to the estimates given by Algorithm A in the present study. The results for age-adjusted rates found by this approach were about 2700 for males and 2300 for females, which were close to those obtained by McBean et al. (2004).

3.3.2.4 Asthma

The prevalence of asthma among the U.S. population aged 65+ years old in the mid-2000s was as high as 7 % (Moorman 2007), with new cases occurring in older adults more frequently than is usually appreciated. However, older patients are more likely to be underdiagnosed, untreated, and hospitalized due to asthma than individuals younger than age 65 (Banerjee et al. 1987; Bellia et al. 2003; King and Hanania 2010). An inverse relationship between asthma prevalence in the older population and age was recently reported (Oraka et al. 2012). The inversion can be due to underdiagnosis of asthma when COPD or other comorbidities with similar nonspecific symptoms are present, or to undertreatment (also due to comorbidities) that can contribute to the death of the patients at younger ages which may appear as lower asthma prevalences at advanced ages. With such age patterns of asthma prevalence, it becomes important to know the exact age patterns of asthma incidence in this population group in order to understand whether the rates of newly diagnosed cases could contribute to the decreasing prevalence of this disease at advanced ages. However, while more is known about asthma prevalence among older adults, asthma incidence in this population is less studied. There are few studies of asthma onset among patients aged 65+ years old; moreover, the available studies are limited by a small number of patients and tend to group all patients older than 55 or 60 years old into a single category (Braman et al. 1991; Burr et al. 1979; Ford 1969; Lee and Stretton 1972). Similar to the results of the present study, asthma incidence rates have been shown to decrease with age in a population-based Rochester, Minnesota, study that analyzed age- and sex-specific incidence rates of definite and probable asthma in 1964–1983 (Ballard-Barbash et al. 2006). However, our results for absolute incidence rates were higher than in the Rochester study in which rates for males and females, respectively, at ages 65–74 were about 140 and 80 (per 100,000), 110 and 70 at ages 75–84, and about 60 and 50 at ages 85+. The higher rates in the present study could be due to an increasing trend of incidence of asthma in recent decades, as well as to the fact that the data from the Rochester study were obtained from the medical records retrospectively and predominantly involved Caucasians from a small Midwestern city. Also, slightly different male/female ratios in these two studies may also play a role.

Another study of asthma incidence that included data on older individuals was reported by ARIC (summarized by the NIH/NHLBI in ref. NIH/NHLBI 2006) for data collected in 1987–2001. Incidence rates (per 100,000) were estimated at 225 for ages 65–74 and 398 for ages 75–84 years old. These rates are in a better agreement with our results. Female-to-male ratios in age-adjusted rates were similar in the ARIC and NLTCS-M studies: i.e., 1.49 and 1.27, respectively. A study conducted in Moscow, Russia, covering a wide range of ages—from birth to 85 years old found that asthma risk declined steadily at ages 55+ in females and at ages 65+ in males, becoming very small among the oldest old (Ukraintseva and Sergeev 2000). This trend of declining asthma incidence with age is in agreement with our results.

3.3.2.5 Neurodegenerative Diseases (NDD)

The most common medical conditions reported in the NDD group are Alzheimer's (and other dementias) and Parkinson's diseases. Their incidence rates and age patterns in elderly populations have been estimated in several studies and meta-analyses. The prevalence and incidence of Alzheimer's disease increase exponentially with age, with the most notable rise occurring through the seventh and eight

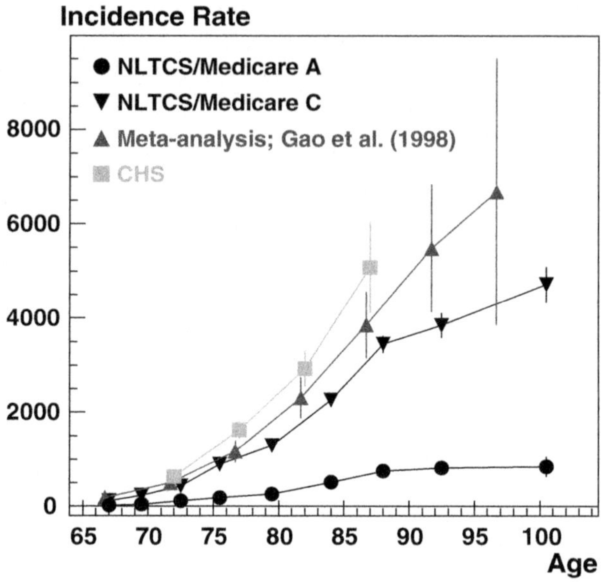

Fig. 3.8 Age-specific incidence rates for Alzheimer's disease

decades of life (Reitz et al. 2011). There are few variations in incidence rates obtained from the different studies for the population aged younger than 75 years old, whereas in the older age groups rates vary substantially. In part, methodological issues can account for observed variations. However, the variations in estimated rates might also reflect geographic differences associated with different prevalences of risk and protective factors across the U.S. We compared the results of our calculations of Alzheimer's disease incidence rates obtained from application of Algorithms A and C to the rates obtained in the meta-analysis (Gao et al. 1998) and in the CHS study (Fitzpatrick et al. 2004) (Fig. 3.8). The results obtained from Algorithm C were in agreement with these two studies, whereas the rates estimated from application of the base algorithm (Algorithm A) were lower than those from the above-mentioned studies. The findings of another study of neurodegenerative diseases among the older U.S. population—the Bronx Aging Study—indicate that whereas dementia incidence continues to increase beyond age 85, the rate of increase slows down (i.e., at ages 85+ vs. 65–84 years old). That suggests that dementia diagnosed at advanced ages might be related not to the aging process per se, but associated with age-related risk factors (de La Fuente-Fernández 2006; Hall et al. 2005). A similar pattern for Alzheimer's disease was observed in a study based on inpatient claims in the NLTCS-M data for 1984–2001: the decline of the risk was identified at ages 90+ years old (Ukraintseva et al. 2006). Based on the results of multiple epidemiological studies of incidence of Alzheimer's disease in Europe, North America, Asia, Africa, Australia, and South America, average annual age-specific incidence rates per 1000 person-years were found to increase across ages 65–95; i.e., incidence rates were increasing at these ages which is in agreement with our data, and the absolute rates were in general agreement with our Algorithm C data (de La Fuente-Fernández 2006).

Approximately 1–2 % of the population aged 65+ and up to 3–5 % aged 85+ years old suffer from Parkinson's disease (Fahn 2003). As for Alzheimer's disease, there also are substantial variations in reported incidence rates of Parkinson's disease, probably due to methodological differences between the studies, in particular, differences in case ascertainment and use of diagnostic criteria (Alves et al. 2008). After applying strict diagnostic criteria of Parkinson's disease, age-standardized incidence rates in population-based studies in the U.S. and Europe ranged from 8.6 to 19.0 per 100,000 population, while the surveys and studies based on broader inclusion criteria have yielded much higher incidence rates (Twelves et al. 2003; von Campenhausen et al. 2005). Studying incidence rates of Parkinson's disease is a challenging task: low incidence and prevalence of the disease, difficulties in establishing diagnosis, and the absence of population-based disease registries contribute to the lack of epidemiologic characteristics of this disorder (Van Den Eeden et al. 2003). There are few studies of Parkinson's disease incidence, especially in the oldest old, and its age patterns at advanced ages remain controversial (Mayeux et al. 1995; Morens et al. 1996). One incidence study of Parkinson's disease analyzed 1994–1995 data that came from the Kaiser Permanente Medical Care Program of Northern California: in this study, incidence rates per 100,000 were estimated at 38.8 in age group 60–69, 107.2 at ages 70–79, and 119.0 at ages 80–89 years old, with rates more than twice as high in males than in females aged 70+ years old (Van Den Eeden et al. 2003). Although our estimated rates for ages above 80 years old are higher by a factor of 1.5–2.0, since the statistical errors are large we can conclude that generally these results are in agreement with those observed in our study.

3.3.3 Age-Adjusted Rates: Gender Disparities, Time Trends, and Sensitivity Analysis

Age-adjusted rates (or directly standardized incidence rates) are weighted averages of the age-specific (crude) rates, where the weights are the proportions of persons in the corresponding age groups of a standard population. For the population aged 66+, they are calculated as $\lambda = \sum_{a=66}^{105+} \lambda_a P_a \left(\sum_{a'=66}^{105+} P_{a'} \right)^{-1}$. There are many ways to estimate SEs for age-adjusted rates (Breslow and Day 1987; Dobson et al. 1991; Fay and Feuer 1997). In this study, we used the simplest approach (in order to avoid dealing with uncertainties in SE estimation for low and even zero age-specific rates) based on the approximation suggested by Keyfitz (1966), in which SEs are estimated as $SE = \lambda / \sqrt{n_0}$, where n_0 is the unweighted sum of the cases.

Age-adjusted disease incidence rates are presented in Fig. 3.9 for all NLTCS-M cohorts (i.e., for cohorts of years 1994, 1999, and pooled for both time periods) and genders (i.e., males, females, and total population). The top panel in the figure shows the rates for circulatory diseases. ACHD (including myocardial infarction

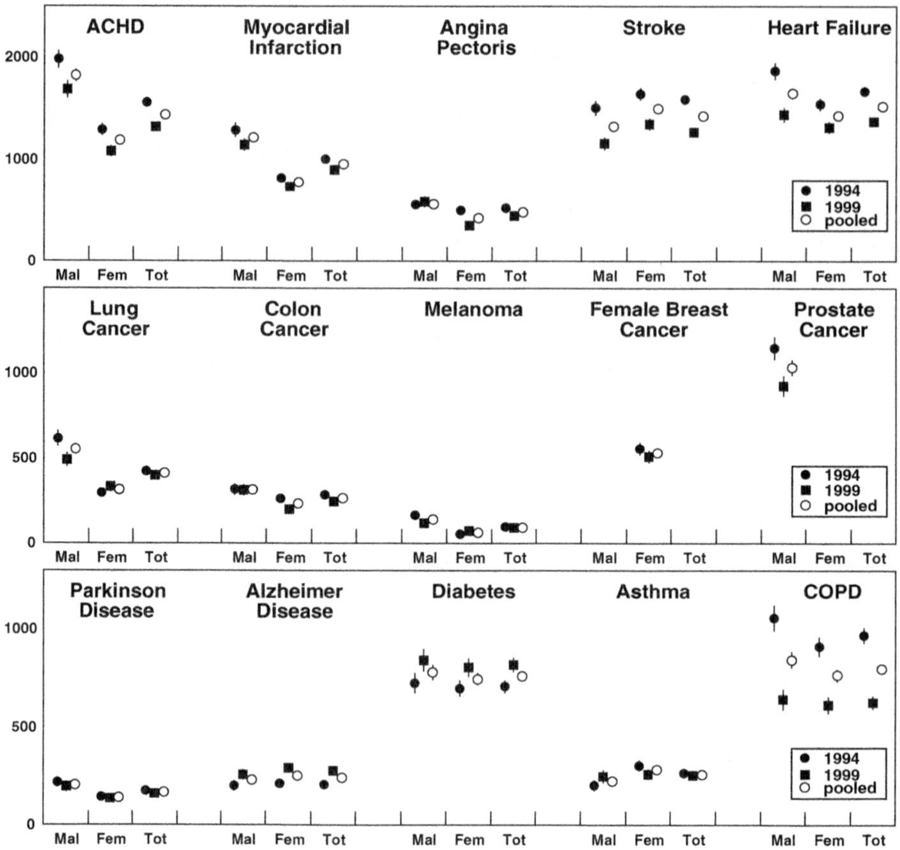

Fig. 3.9 Age-adjusted incidence rates per 100,000 of age-associated diseases with standard errors

and angina pectoris) and stoke had the highest rates for males and females, respectively among the analyzed diseases for cohorts of 1994, 1999, and for both cohorts pooled. The rate of heart failure is also high (comparable to the rate of ACHD) especially in males for the 1994 cohort and in the pooled analysis. Prostate and breast cancer are the cancers with the highest rates for males and females (the second panel in Fig. 3.9) followed by the lung cancer rates. Diabetes and COPD were other (i.e., non-circulatory and non-cancer) diseases with high rates. There were no significant male/female differences detected for them. Because sample weights were applied, the results for both the age-adjusted rates and standard errors are valid for the U.S. elderly population.

The results presented in Fig. 3.9 also provide information about time trends in the age-adjusted incidence rates: there was a significant 5-year decline in the incidence rates of circulatory diseases (including ACHD, stroke, and heart failure) and lung cancer for males, and an increase in rates of diabetes and Alzheimer's disease.

The time trends of disease incidence in older U.S. adults were also estimated from the SEER-M data and compared to those obtained from the NLTCS-M (Akushevich et al. 2013e). Dramatic increases of incidence rates of melanoma, goiter, chronic renal, and Alzheimer's disease were detected from 1992 to 2005. Besides specifying widely recognized time trends of age-associated diseases, new information was obtained for trends of asthma, ulcer, and goiter. The trends thus identified could be associated with changes in the socioeconomic status and demographic structure of the population, risk factors prevalence (e.g., smoking, obesity, etc.), as well as changes in prevention, screening, and diagnostic strategies.

3.3.3.1 Sensitivity Analysis

One disadvantage of large administrative databases is that certain factors can produce systematic over/underestimation of the number of diagnosed diseases or of identification of the age at disease onset. One reason for such uncertainties is an incorrect date of disease onset. Other sources are latent disenrollment and the effects of study design. To evaluate the effects of these uncertainties, we performed calculations with different definitions of the disease onset and used alternative censoring schemes to define individual observation periods. Table 3.2 presents the results of calculation of age-adjusted rates from NLTCS-M data when several alternative approaches were used. We can conclude that all calculated rates were relatively stable. Thus, columns V1–V3 represent calculations without age standardization, using as a standard the population of 1994, (V1) and without using NLTCS sample weights (V2). In the alternative censoring scheme (V3), the last day of observation is the latest day among (i) part B coverage; (ii) Medicare record in Part A or Part B; and (iii) response on interview in the next NLTCS wave (while in the basic calculation, the final date of observation is the earliest date among dates of disease onset or death, and the last date of cohort observation). Only minor changes in incidence rates obtained within V1–V3 strategies were detected. The results of calculations V4 and V5 reflect the effect of removing individuals from the cohort with different levels of additional coverage by a HMO (by different fractions of months covered by a HMO denoted by δ). Other calculations represent less (V6–V11) or more conservative (V12) approaches to the definition of the date at onset. Model V11 for the calculation of the diabetes age pattern is the same as that used by McBean et al. in their study (McBean et al. 2004). And model V12 is Algorithm D (as described in detail above in the subsection on age patterns of cancer incidence rates).

3.3.4 Disability and Comorbidity Patterns of Incidence Rates

Disability and comorbidity patterns were evaluated using the NLTCS-M data (Table 3.3). For this analysis, individuals were stratified by a disability index

Table 3.2 Age-adjusted incidence rates per 100,000 under alternative approaches to the definition of age at onset calculated for specific sex (S) and year (Y) of forming the cohort

	S	Y	V0	V1	V2	V3	V4	V5	V6	V7	V8	V9	V10	V11	V12
ACHD	*M*	*94*	1977	1954	2013	2027	1969	1739	1754	1814	2467	3553	5151	3933	1576
	M	*99*	1681	1684	1737	1769	1743	1446	1470	1540	2099	2915	4270	3054	1335
	F	*94*	1290	1274	1264	1266	1217	1113	1151	1225	1696	2560	3940	2935	995
	F	*99*	1076	1077	1097	1102	1098	944	950	997	1432	2238	3291	2055	842
Myocardial infarction	*M*	*94*	1287	1258	1309	1317	1242	1115	1144	1151	1493	1737	2446	1858	954
	M	*99*	1141	1141	1197	1215	1193	996	1013	1018	1227	1587	2080	1516	885
	F	*94*	816	791	782	783	756	695	723	738	947	1221	1728	1235	570
	F	*99*	729	731	764	767	746	648	662	662	868	1030	1403	985	544
Angina pectoris	*F*	*94*	552	541	566	569	581	510	545	558	980	1812	3147	1919	466
	F	*99*	579	581	560	570	579	475	480	526	855	1497	2493	1466	453
	M	*94*	499	496	500	501	480	433	466	499	870	1470	2517	1720	398
	M	*99*	349	349	351	353	369	306	309	343	646	1230	2024	1094	295
Stroke	*M*	*94*	1501	1441	1458	1468	1404	1251	1353	1314	1582	2332	3046	2077	1060
	M	*99*	1155	1156	1168	1185	1190	1003	1072	1047	1306	1983	2720	1583	854
	F	*94*	1640	1595	1583	1587	1536	1430	1525	1519	1770	2568	3311	2340	1239
	F	*99*	1344	1343	1333	1338	1307	1116	1196	1171	1447	2165	2922	1810	987
Heart failure	*M*	*94*	1864	1795	1869	1881	1784	1612	1706	1767	2713	3055	5345	4484	1389
	M	*99*	1437	1438	1465	1486	1501	1312	1367	1445	2292	2659	4670	3554	1215
	F	*94*	1540	1484	1511	1515	1476	1346	1458	1511	2348	3009	5084	4128	1189
	F	*99*	1313	1313	1323	1329	1349	1185	1274	1301	2042	2500	4397	3166	1084
Lung cancer	*M*	*94*	616	613	618	623	593	529	572	541	629	657	839	624	454
	M	*99*	491	490	488	495	484	432	511	462	496	612	711	532	373
	F	*94*	293	295	280	281	269	248	278	258	288	368	455	323	220
	F	*99*	331	330	332	333	322	289	342	302	324	393	454	330	264

Colon cancer	*M*	*94*	314	306	302	304	280	253	269	267	289	427	483	362	224
	M	*99*	311	311	321	326	322	296	308	300	356	440	526	377	270
	F	*94*	262	259	252	253	243	232	245	237	264	349	432	299	218
	F	*99*	196	195	194	195	192	177	200	187	200	282	361	220	170
Melanoma	*M*	*94*	163	159	161	162	149	131	131	137	142	225	277	147	110
	M	*99*	117	117	130	132	135	113	113	119	131	204	257	137	106
	F	*94*	53	51	49	49	45	42	44	46	57	139	166	73	36
	F	*99*	71	70	76	76	76	63	63	64	65	99	127	68	63
Breast cancer, Prostate cancer	*F*	*94*	555	559	572	573	541	498	508	511	544	802	906	680	485
	F	*99*	508	509	512	514	496	432	454	442	453	580	674	491	424
	M	*94*	1146	1148	1116	1125	1098	984	996	1033	1102	1649	2044	1419	944
	M	*99*	924	922	950	967	939	808	817	876	913	1305	1675	1131	792
Parkinson's disease	*M*	*94*	217	208	221	223	213	190	199	205	272	289	465	408	183
	M	*99*	195	195	186	189	179	156	159	171	313	282	531	416	150
	F	*94*	143	139	140	140	135	121	133	128	222	216	405	318	116
	F	*99*	134	134	127	127	130	117	130	130	216	209	413	283	107
Alzheimer's disease	*M*	*94*	199	187	200	201	192	174	179	194	396	343	737	607	151
	M	*99*	255	257	245	247	242	219	246	262	462	540	992	794	183
	F	*94*	210	197	192	192	189	183	225	221	532	439	1048	876	157
	F	*99*	290	291	287	288	296	264	301	323	717	670	1457	1114	239
Diabetes	*M*	*94*	724	721	718	723	725	681	715	772	1303	1787	3403	2529	646
	M	*99*	841	841	801	816	884	784	878	936	1478	2130	4158	2654	739
	F	*94*	697	696	702	704	707	650	678	728	1179	1665	3197	2219	611
	F	*99*	803	804	797	801	799	713	818	827	1260	1835	3636	2195	673

(continued)

Table 3.2 (continued)

	S	Y	V0	V1	V2	V3	V4	V5	V6	V7	V8	V9	V10	V11	V12
Asthma	*M*	*94*	201	200	202	204	198	174	198	211	395	744	1484	804	157
	M	*99*	247	248	263	267	260	228	260	255	478	723	1591	818	206
	F	*94*	303	307	309	310	301	287	329	316	550	934	1692	1045	278
	F	*99*	258	257	228	229	239	213	274	232	525	842	1679	901	200

The standard calculation (V0) of age-adjusted incidence rates (per 100,000) was performed on the screener NLTCS population, using the NLTCS weights, the four basic Medicare sources, only the primary diagnosis, at least two records (or death) in $\Delta = 0.3$ year, with a cut on frequency of HMO coverage $\delta = 0.005$, and age standardization using standard population of 1994. Other calculations are: (V1) no age standardization, i.e., age-specific rates are averaged using the population of each respective year, (V2) no NLTCS weights, V3) an alternative censoring strategy, (V4) $\delta = 0.5$, (V5) $\delta = 1$, (V6) all Medicare sources, (V7) $\Delta = 0.5$ years, (V8) no requirement for codes to be primary, (V9) no requirement of a second record (Algorithm B), (V10) apply both V8 and V9, (V11) alternative censoring strategy (McBean et al. 2004), and (V12) no death as a second event

Table 3.3 Disability and comorbidity patterns of incidence rates (per 100,000) of geriatric diseases

	Disability						Comorbidity			
	Non	IADL	1–2	3–4	5–6	Inst.	0	1	2	3+
ACHD	1442 (37)	1528 (183)	1596 (177)	1524 (244)	1997 (365)	947 (246)	1319 (43)	1384 (71)	1790 (126)	1930 (128)
Myocardial infarction	917 (28)	1197 (151)	1188 (150)	1179 (165)	1304 (219)	1149 (245)	741 (32)	970 (54)	1309 (97)	1386 (91)
Angina pectoris	475 (21)	485 (102)	592 (108)	793 (198)	643 (259)	224 (83)	422 (24)	492 (43)	500 (58)	740 (76)
Stroke	1294 (33)	2053 (220)	1885 (166)	3579 (468)	3621 (532)	2312 (339)	1182 (39)	1682 (78)	1557 (104)	2000 (104)
Heart failure	1395 (35)	1964 (212)	2228 (209)	2934 (331)	3168 (440)	2454 (343)	1289 (41)	1798 (81)	1760 (98)	1907 (108)
Lung cancer	413 (19)	395 (86)	594 (98)	417 (114)	223 (103)	313 (99)	306 (21)	578 (49)	451 (62)	518 (49)
Colon cancer	264 (15)	534 (98)	322 (73)	167 (58)	288 (106)	59 (25)	241 (18)	295 (33)	354 (50)	225 (29)
Melanoma	90 (8)	120 (58)	99 (43)	310 (103)	44 (29)	25 (18)	79 (10)	68 (14)	122 (24)	142 (25)
Breast cancer	554 (29)	480 (146)	404 (86)	416 (133)	108 (73)	272 (75)	532 (35)	657 (66)	464 (70)	363 (67)
Prostate cancer	1068 (49)	904 (270)	1128 (275)	74 (60)	206 (92)	506 (195)	1165 (63)	1085 (109)	839 (121)	686 (121)
Parkinson's disease	146 (11)	242 (66)	231 (52)	334 (101)	376 (117)	446 (108)	151 (14)	167 (25)	202 (30)	176 (26)
Alzheimer's disease	217 (13)	413 (94)	231 (47)	416 (110)	159 (50)	516 (88)	204 (16)	242 (26)	282 (34)	311 (33)
Diabetes	738 (26)	651 (139)	911 (140)	1214 (278)	1157 (343)	802 (172)	783 (32)	794 (60)	679 (70)	691 (75)
Asthma	242 (15)	211 (63)	599 (132)	413 (103)	426 (157)	145 (53)	175 (16)	370 (40)	352 (42)	382 (62)
COPD	761 (28)	711 (166)	1295 (188)	1337 (288)	1159 (274)	1590 (421)	858 (36)	758 (60)	724 (73)	691 (74)

Disability groups are nondisabled, IADL only, 1–2 ADLs, 3–4 ADLs, 5–6 ADLs, and institutional), and comorbidity group are in the units of the Charlson index (0, 1, 2, and 3 and more)

(with outcomes nondisabled, IADL only, 1–2 ADLs, 3–4 ADLs, 5–6 ADLs, and institutionalized) measured at the date of interview, i.e., at the beginning of the follow-up, and by the Charlson comorbidity index (according to the specifications described in Charlson et al. 1987; Quan et al. 2005) that was also measured at the date of interview (using the Medicare records for the period of a year prior to the date of interview). Comorbidity- and disability-specific rates are age-adjusted incidence rates calculated using the same equations applied for the stratified population.

The disability and comorbidity patterns were analyzed for selected diseases (see Table 3.3). For several diseases (e.g., myocardial infarction, stroke, heart failure, diabetes, asthma, and Parkinson's disease), the incidence rates were higher among individuals with severe disabilities, while for breast and prostate cancers the higher rates were registered among people with minor disabilities. Interestingly, for many diseases institutionalized individuals had lower incidence rates, and for several diseases (such as melanoma, lung cancer, colon cancer, and asthma) they had the lowest rates among all other disability groups, including non-disabled individuals. For all the diseases considered (except Alzheimer's disease), institutionalized individuals had lower disease rates. However, for neurodegenerative diseases such as Parkinson's (for females only) and Alzheimer's diseases, the rates among institutionalized individuals were the highest. Among individuals with high comorbidity indices (i.e., Charlson index), higher rates were observed for heart failure, melanoma, and Alzheimer's disease, while incidence rates of breast and prostate cancers, as well as diabetes, decreased with increasing comorbidity indices. More detailed analyses of comorbidity and disability patterns for circulatory diseases are presented in (Akushevich et al. 2013c).

3.3.5 Mortality Age Patterns and Medicare Data

The information available in Medicare data allows for more detailed analyses of uncertainties in estimates of mortality rates at advanced ages. We use all NLTCS data from 1982 to 2004/2005 and evaluate uncertainties in calculations of age-specific mortality rates using several scenarios. In this example, we follow Akushevich and Manton (2011).

The strategy for analysis of the mortality age patterns consists of the following steps. First, individuals were selected for analysis (27 deaths reported as occurring before 1982 were excluded). Second, a weight for each individual was assigned to project the results for the entire U.S. population and a time period when this weight was valid for an individual was defined. Several approaches using base weights and screener weights were used. Two ages were defined for each individual, namely, the age at enrollment and the age of death/censoring. The age of death was obtained from the Vital Statistics file for deceased persons. For individuals not marked as dead in the file, it is possible that some could be deceased, but information about their deaths was missing from the file. To resolve the issue, a set of approaches was developed to assign the censoring date (or the date of the end of follow-up) for these individuals using available information on these individuals from other sources. Specifically, we determined if one of three types of the event occurred: (i) service event in Medicare Part A (e.g., hospitalization); (ii) a payment to Medicare Part B (i.e., monthly premiums); and (iii) a payment of a monthly premium to HMO or other managed care plan.

We detected 90 persons paying Part B premiums that had no other (Part A) service use and who did not respond to the survey. The detailed analyses and comparison of mortality rates calculated for different approaches were important, because they can result in different age patterns of mortality rates. A specific analysis focused on the effects of non-responders: information on them was collected using the question "Reason for non-response in NLTCS", with the possible outcomes such as "deceased" and "gone abroad". Careful consideration of specific groups (e.g., 90 suspected non-responders who did not appear in Medicare Part A, but paid monthly Part B premiums) is important for several reasons. First, their contribution is significant, and therefore, if neglected it would significantly contribute to systematic uncertainty in mortality patterns at advanced ages. Second, such a small fraction of individuals is plausible because roughly 2 % of Part B enrolled persons do not participate in Part A benefits. Some small number of eligible individuals was not, however, covered by Medicare Part B (about 2 %). Additionally, a small proportion of persons would pay Part B premiums but would not be eligible for Part A (about 3 %). A third type of program eligibility would involve enrollment in managed care plans (e.g., HMOs) when the death was recorded. All these benefit categories involve positive actions to be recorded in the Medicare claims files. The final records are for persons who have responded to the survey which provides a direct confirmation of age reported on the Medicare claims data. All these persons should be enrolled in the Medicare program in some way, with enrollment necessary for being selected for NLTCS sample.

Figure 3.10a presents empirical estimates of mortality rates which suggest that there is a decline with age in hazard rates at extreme ages. Figure 3.10b presents various hazard functions estimated under the different alternative assumptions about data selection. The Basic Approach (marked by 0) is specified as having a censoring date as the date of the latest appearance in Medicare Part A (in records) or in Part B (in payments). Note that information about payments comes from the Denominator Files, and, therefore, can be less accurate. In our basic approach, the non-responses, gone abroad non-responders, and deceased non-responders were removed. Approaches 1–5 described below are defined as modifications of the basic approach:

- Approach marked by 1: no cut of the 90 suspected non-responses, i.e., those individuals who are (i) alive according to vital statistics, (ii) have no Medicare histories, and (iii) have premium payments till 2005.
- Approach marked by 2: censoring date is calculated according to Vital Statistics.
- Approach marked by 3: if a reason for non-responding (according to the NLTCS questionnaire) is "deceased" then this is a death case.
- Approach marked by 4: same as in #3, but this is censored, not a death case.
- Approach marked by 5: the moved abroad non-responders are not removed.

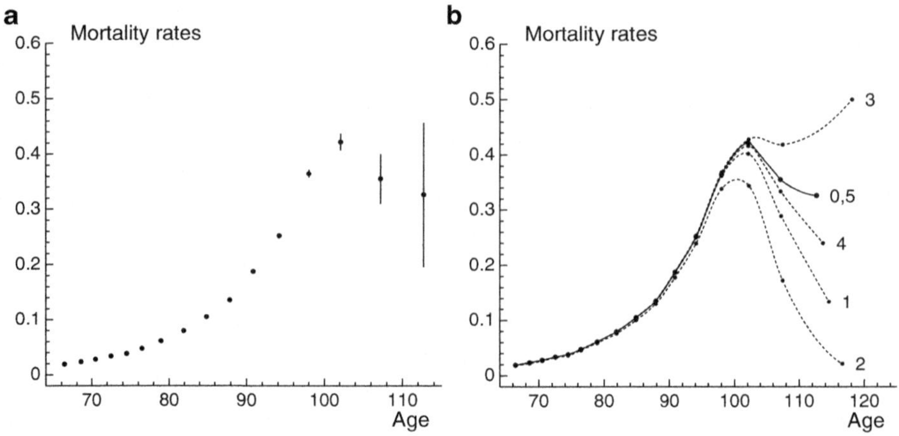

Fig. 3.10 (**a**) Results of base calculations (49,123 individuals were selected, i.e., 49,240 (total in the NLTCS) minus 27 with death before 1982 and minus 90 (nonresponders, not appearing in Medicare records, alive according to vital statistics, paid premiums until 2005) and (**b**) results of calculation of incidence rates using base (marked by 0) and five alternative approaches (see text for details)

Analyses of mortality patterns at old ages conducted on these data included empirical methods and applied demographic models of mortality. The empirical results showed that the increase in mortality rates with age stops at age 95 for males and at age 100 for females. Whether the decline in mortality rate above these ages is detected for the total population is still an open question, because there are multiple uncertainties accompanying estimates of mortality rates at advanced ages using Medicare data.

3.3.6 *Recovery or Long-Term Remission*

In this section, we use the two Medicare-linked datasets to investigate demographic and epidemiologic properties of the cohorts of survivors after certain chronic diseases were diagnosed. Analyzing individual trajectories, we found a subgroup of patients who had stopped using medical services after a certain period of time following the diagnosis. Who are these individuals? Are they a healthier or sicker subgroup of patients? If they are healthier, then they could be those who (i) have entered into a stable condition/long-term remission of chronic disease (in some cases such remission could be long enough so that a "recovery" term could be used); or (ii) have undergone a successful rehabilitation from acute diseases (e.g., myocardial infarction and stroke) without obvious complications affecting their quality of life. If they are a sicker group, then they could be the patients who (i) do not believe anymore in doctors' recommendations after medical treatment failed to improve their health and/or did not improve their quality of life (as substitution, they could rely on a treatment with naturalists, chiropractors, etc.); or (ii) were not able to pay the treatment expenses; or (iii) have moved to areas (e.g., rural) where they lacked transportation to reach the doctor's office for visits.

The formal definition of the recovery rate is as follows. An individual was considered to be recovered (or entered a sustained remission) at a given date if he/she did not have a Medicare record containing the respective ICD code during a period of time τ_d (e.g., 1, 2, or 3 years) after this date (Yashin et al. 2010). All Medicare records from the individual medical histories (i.e., records appearing in any Medicare sources and not necessary being primary) are used for the definition of recovery. The time periods τ_d are referred to as recovery times. An individual was censored at a date if that date plus the recovery time exceeded the date of the end of follow-up (i.e., in this case an individual does not have sufficient time for recovery). Note that since the identification of the date at onset requires confirmation by a record in another day, recovery on the day of diagnosis is not possible even if the length of service is 1 day. Another property of the recovery rate is that a recovery event within the 3-year strategy (i.e., for $\tau_d = 3$ years) implies a recovery event within a 1- or 2-year strategy, but not otherwise.

Kaplan-Meier estimates of not-yet-recovery probabilities for predetermined disease-specific recovery times are presented in Fig. 3.11. A comparison of curves for different time periods shows that the time trend was positive (remission chance increases) for the majority of acute and several chronic diseases, excluding cancers. To test the hypothesis about health status of recovered individuals (i.e., whether they are sicker or healthier), we used the Cox proportional hazards model with age at diagnosis and time after remission (equal to zero before remission). The estimates in Akushevich et al. (2013d) showed that the "recovered" patients (those who did not have medical care for 1, 2, or 3 years, depending on the corresponding scenario analyzed) had better survival. Therefore, they are (i) a "healthier" subgroup of elderly—they started feeling well enough to discontinue the use of medical services, or (ii) people who at the moment of diagnosis likely had a functional disorder rather than a serious disease, but similarity of disease symptoms led to their misinterpretation and "overdiagnosis". Also, the subcohort of "recovered" individuals can include patients who were disappointed in treatment results and stopped the therapy, turning instead to chiropractors, homeopaths, or herbal medicine, or patients who began experiencing difficulties with transportation to a physician's office. However, this subcohort is indistinguishable from the healthier (and much larger) fraction of recovered individuals.

An example of sensitivity analysis involving recovery rates was performed by Yashin et al. (2010) for time trends of recovery after stroke. The authors considered the following effects: (i) several different operational definitions of recovery and incidence rates; (ii) explicit representation of observed heterogeneity effects stratifying individuals by age, comorbidity, or disability; and (iii) other approaches to censoring strategies, selection of individuals, and study design effects. The results of these analyses indicated that positive trends in the recovery rate from stroke took place in all cases, independent of the definition of such rates.

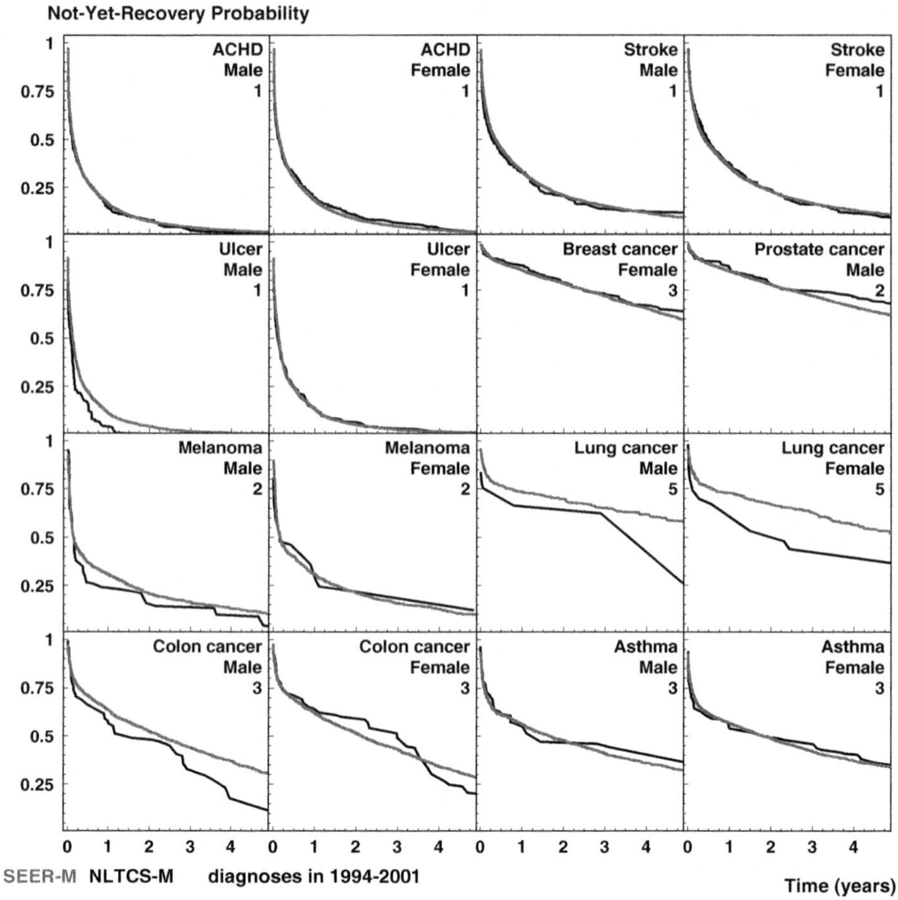

Fig. 3.11 "Not-yet-recovery" probability for geriatric diseases vs. time after diagnosis in years calculated using SEER-M (*thick lines*) and NLTCS-M (*thin lines*). Values on plots are "recovery" times, i.e., disease-specific time period without occurrence of respective ICD code in individual medical (Medicare) history (Color figure online)

Thus, Medicare data allow us to evaluate recovery rates from common acute and chronic diseases in older U.S. adults at the national level, using a new approach developed for quantitative analyses of individuals with recovery/long-term remission after onset of chronic diseases.

3.3.7 *Risk Factors for Disease Incidence*

Most chronic diseases are associated with multiple risk factors, and many of these factors are measurable and modifiable. If known, many of these risk factors are preventable: e.g., about 80 % of cancer risk factors, including behavioral/lifestyle-associated, are considered preventable. Combining information from several datasets, such as demographic surveys (e.g., NLTCS) and the MFSU, allows for testing of hypotheses about the contributions of specific risk factors to risks of aging-related chronic diseases. The primary purpose of the present analysis is to develop an approach to estimation of the contributions of measurable risk factors, including behavioral/lifestyle risk factors, to mortality and diseases incidence risks and to apply this approach to population data to find associations clarifying the role of behavioral/lifestyle risk factors to these risks in the U.S. elderly population.

Associations were investigated for each demographic, health, social, economic, and behavioral lifestyle variable (213 for the 1994 survey and 219 for the 1999 survey) for 26 types of outcomes (total mortality, seven non-cancer and five cancer incidences, for both the 1994 and 1999 NLTCS surveys). In total, 5616 associations were empirically estimated by comparison of age-adjusted incidence rates conditional on a specific outcome (i.e., a specific answer to a specific question). Relative risks were estimated as the ratios of the rates for alternative outcomes. Note that age-adjusted risks were calculated for subpopulations with different responses for certain questions/variables (e.g., current smokers and nonsmokers) using the same population weights for both outcomes. Therefore, the rates conditional on a specific outcome of each variable were adjusted for the total population age structure, thus taking into account a possible effect of age dependence of certain outcome prevalences. For example, lung cancer rates in smokers and non-smokers were adjusted for the age structure of the total population to include smokers, non-smokers, and individuals with missing information on smoking status.

Table 3.4 shows the results for hazard ratios of the variables that were most significantly associated with mortality. The selected variables included demographic characteristics, self-reported comorbidities, activities of daily living information, other activities including social factors, and medical care factors. The estimates for the 1994 and 1999 cohorts are in good agreement.

The following factors were found to be predictive of incidence of circulatory diseases:

- For ACHD: male (RR = 1.8); comorbidities such as diabetes (RR = 1.9), circulatory diseases (RR = 1.3–2.0), needs some additional devices in house (RR = 2.5), thinks that grocery and drug stores are not conveniently located (RR = 1.6), losing temper (RR = 1.6); for angina pectoris: low BMI (RR = 4.9), poor social contacts (RR = 2.2), financial or transportation problems (RR = 4.0), and for myocardial infarction: living in rural area (RR = 2.0) and smoking (RR = 2.7).
- For stroke: comorbidity such as diabetes (RR = 2.0); disability (ADL/IADL (RR = 1.6)) and other functionality, e.g., difficulties in washing hair (RR = 1.9) and/or lifting a 10-lb package like a bag of groceries and holding it for a few minutes (RR = 1.8); high BMI (RR = 2.4), not keeping in touch with relatives (RR = 2.0), disturbed memory (RR = 2.0); poverty characterized by food stamps receiving (RR = 2.2), and Medicaid coverage (RR = 2.0).

Table 3.4 Hazard ratios of lifestyle risk factors for mortality. Only variables with highly significant effects were selected ($p < 0.000002$)

Variable	Outcome	1994	1999	Variable	Outcome	1994	1999
Sex	Male	1.35	1.31	Healthy compared with others	Fair/ poor	1.93	2.14
Race	Nonwhite	1.3	1.23	Lose temper	Not at all	0.69	0.71
Marital status	Not marr.	1.44	1.33	Work on hobby	No	1.56	1.71
Parkinson's disease	No	0.43	0.47	Attend meeting of a club	No	1.68	1.64
Diabetes	No	0.69	0.65	Keep in touch with relatives	No	1.84	1.46
Cancer	No	0.57	0.45	Hospital overnight (last year)	No	0.59	0.56
Arteriosclerosis	No	0.77	0.67	Medical care in doctor's office	No	0.73	0.78
Heart attack (last year)	No	0.48	0.5	How many prescriptions	2+	1.45	1.55
Stroke (last year)	No	0.55	0.61	Medicine			
Emphysema (last year)	No	0.54	0.41	What is your street address	Not correct	2.04	3.05
Broken hip (last year)	No	0.46	0.54	What day of the week is this	Not correct	1.9	2.02
ADL eating	Can't	2.94	3.23	Walking for exercise	Yes	0.66	n/a
ADL getting in/out of bed	Can't	2.38	2.98	Gardening or yard work	Yes	0.6	n/a
ADL getting around inside	Can't	2.23	2.77	Vigorous activities	10+ min	0.64	0.54
ADL dressing	Can't	2.72	3.01	Avoid doing things because	Rarely	0.6	0.5
ADL bathing	Can't	2.38	2.72	Doesn't have enough energy			
ADL getting to bathroom	Can't	2.47	2.75	Satisfaction with life	Not satisfied	1.56	2.42
Currently smoke	No	0.63	0.57	Have a Medicaid card	No	0.72	0.65
How difficult to climb one flight of stairs	Can't	2.01	2.89	Processed meats such as frankfurters/luncheon meats?	Often	n/a	1.32

- For heart failure: being male (RR = 1.9), unmarried (RR = 2.2), feeling unhappy (RR = 2.5), having such comorbidities as diabetes (RR = 1.9) or cancer (RR = 1.9); disability (ADL/IADL (RR = 1.8)) and other functional disorders (RR = 1.9).

For neurodegenerative disease the identified factors were as follows:

- For Parkinson's disease: comorbidities such as cancer (RR = 4.3), frequent headaches (RR = 4.0), emphysema (RR = 5.5); needs some additional devices in house (RR = 6.0)), not satisfied with his/her life (RR = 5.4); having mental/ emotional problems (ever hospitalized (RR = 5.0), could not sleep like usual (RR = 4.0), forgets to do important things (RR = 4.0)); rarely drink coffee/tea (RR = 5.1).
- For dementia and Alzheimer's disease: comorbidities such as bronchitis (RR = 3.3); memory problems (forgets to do important things (RR = 3.2)); difficulties in making phone calls (RR = 4.0); difficulties with going outdoors (RR = 6.5); covered by any other public assistance program that pays for health care (RR = 9.8).

Several examples on other diseases:

- For ulcer: comorbidities such as permanent numbness (other besides paralysis and arthritis (RR = 3.5)), frequent severe headaches (RR = 5.0), bronchitis (RR = 3.0), trouble sleeping (RR = 4.2); depression (takes respective medicine (RR = 3.5)), being not satisfied with his/her life (RR = 4.9); having mental and emotional problems (such as losing temper (RR = 2.2)); disability (ADL/IADL (RR = 2.2)); using a hearing aid (RR = 3.2) and other devices (RR = 6.4).
- For diabetes: high BMI (RR = 3.2), self-reported overweight (2.0); poverty (Medicaid card (RR = 2.0), public assistance program that pays for health care (RR = 3.5)); disability (ADL/IADL (RR = 2.7)), cannot do everyday activities around the house (RR = 2.5), using hearing aid (RR = 2.5).
- For asthma: being female (RR = 3.7); self-reported obesity or overweight (RR = 2.7)); comorbidities such as heart attack (RR = 3.5), pneumonia (RR = 4.0), bronchitis (RR = 2.3); needs some additional devices in house (RR = 5.0), being not happy with his/her life (RR = 5.3); rarely eat fortified breakfast cereals (RR = 5.0).

The detected associations (i.e., relative risks) of selected factors with risks of breast, prostate, lung and colon cancers were discussed in Akushevich et al. (2011a). An overall view of the results of association analyses allowed the researchers to describe population groups of higher and lower risks of these cancers. For example, being a smoker was the main characteristic of elderly population group at higher risks of lung cancer, with comorbidity (e.g., emphysema), lower BMI, and poor functional status also each playing a role. Note that the most influential of potentially preventable risk factors can be detected with this approach using the NLTCS-Medicare linked dataset and for further deeper analyses employing other datasets with detailed risk factors.

The results of this analysis can be reformulated as a description of higher risk groups of major geriatric diseases in terms of variables measured in the NLTCS and the aggregated indices constructed from these variables.

3.3.8 *Mutual Dependence in Disease Risks: Age-Patterns*

Although multi-morbidity is common among older adults, for many aging-related diseases there is no information for the U.S. elderly population on how earlier-manifested diseases affect the risk of another disease manifested later during a patient's lifespan. Therefore, we investigated the phenomenon of multimorbidity in the U.S. elderly population by analyzing mutual dependence in disease risks, i.e., we calculated disease risks for individuals with specific pre-existing conditions (Akushevich et al. 2013a). In total, 420 pairs of diseases were analyzed. For each pair, we calculated age patterns of unconditional incidence rates of the diseases, conditional rates of the second (later manifested) disease for individuals after onset of the first (earlier manifested) disease, and the hazard ratio of development of the subsequent disease in the presence (or not) of the first disease. The most interesting (selected) results are presented in Fig. 3.12. Synergistic and antagonistic dependences in geriatric disease risks were observed among older U.S. adults confirming known and detecting new associations among a wide spectrum of age-associated diseases. More specifically, three groups of interrelations were identified: (i) diseases whose risk became much higher when patients had a certain pre-existing (earlier diagnosed) disease; (ii) diseases whose risk became lower than in the general population when patients had certain pre-existing conditions (a so called "trade-off" effect between earlier and later occurring diseases); and (iii) diseases for which "two-tail" effects were observed: i.e., when the effects are significant for both orders of disease precedence; both effects can be direct (either one of the diseases from a disease pair increases the risk of the other disease), inverse (either one of the diseases from a disease pair decreases the risk of the other disease), or controversial (one disease increases the risk of the other, but the other disease decreases the risk of the first disease from the disease pair). In general, the majority of disease pairs with increased risk of the later diagnosed disease in both orders of precedence were those in which both the pre-existing and later occurring diseases were cancers, and also when both diseases were of the same organ.

The existence of inverse associations for the later-in-life diagnosed disease risk may provide important insights into disease mechanisms and new opportunities for disease prevention and therapy that focuses on increases in healthy lifespan rather than concentrating efforts on reduction of risk for each particular disease alone.

Generally, the effect of dependence between risks of two diseases diminishes with advancing age. This could be because senescence itself becomes a leading risk factor of death in the oldest old, so that at the very old ages it matters less if a person is healthy or sick, because her/his vulnerability to death is high anyway due to a dramatic decline in the body's overall resistance to stresses attributed to aging.

Identifying mutual relationships in age-associated disease risks is extremely important since they indicate that development of seemingly very different cancer and non-cancer diseases may involve common biological mechanisms. This knowledge could help to develop disease prevention as an integrated field that targets an increase in healthy lifespan rather than a simple reduction in the risk of a particular health disorder. Moreover, a better understanding of the biological links between different diseases/groups of diseases can open new therapeutic horizons. The

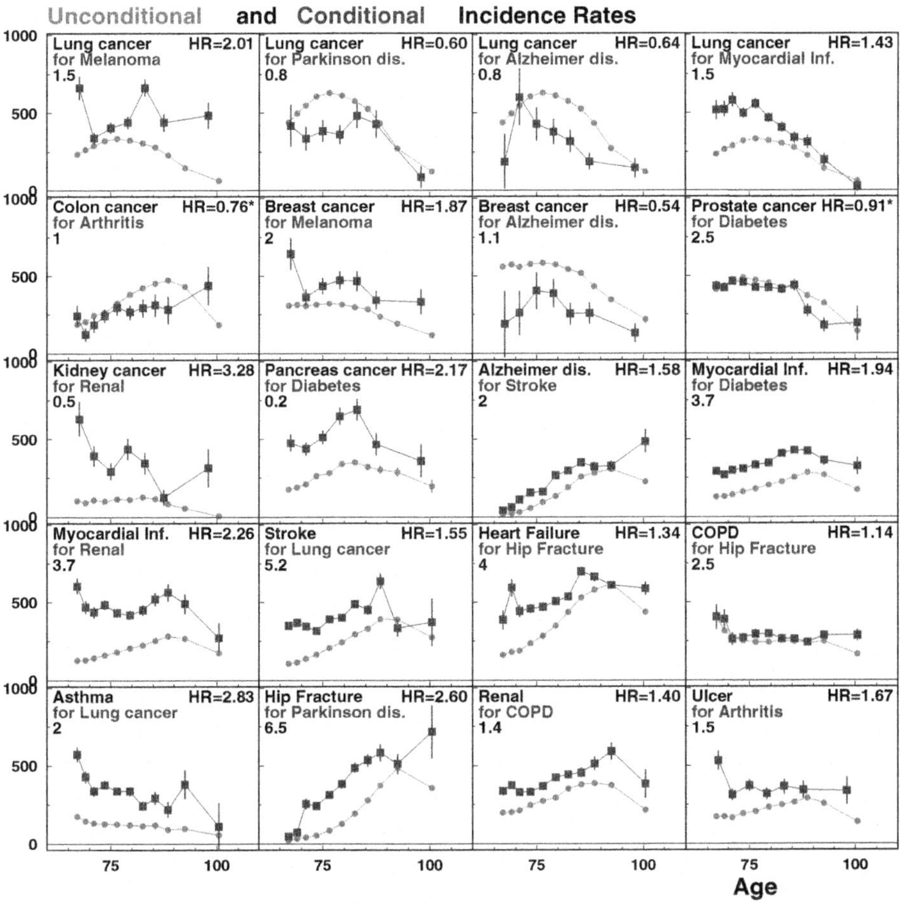

Fig. 3.12 Age-specific rates of disease incidence conditional on the onset of another disease calculated using SEER-Medicare (*squares*) and the corresponding unconditional rates (*dots*). Values on the plots are rescale factors. Rates for different diseases are rescaled to use the same scale on all plots to facilitate comparison of rates for different diseases: the original rate can be calculated by dividing the values obtained from plot by the rescale factor (Color figure online)

observed mutual dependence of cancer and non-cancer disease onset could be used for improvement of current models for forecasting future morbidity and mortality, for planning medical expenditures, and for optimization of screening and preventive strategies among older adults with multi-morbid conditions.

3.3.9 Comorbidity and Multimorbidity

The standard tool for measuring individual comorbidity is the *Charlson comorbidity index* (CCI) (Charlson et al. 1987; Diederichs et al. 2011). The CCI has been described as a "valid method of estimating risk of death from comorbid disease for use in longitudinal studies". It is the most frequently applied index among other health-related indices. However, the CCI has certain limitations with respect to the prediction of health state and mortality in an elderly population (Testa et al. 2009; Zekry et al. 2010). Specifically, the CCI does not take into account the severity of major diseases, overestimates heavily weighted conditions (such as AIDS) rarely encountered in the elderly, and underestimates some highly prevalent conditions (such as heart failure and Alzheimer's disease). In addition, since the mid-1990s, active screening strategies and new approaches to cancer treatment (e.g., prostate, breast, and cervical cancers) and changes in prevention and treatment of cardio and cerebrovascular diseases have changed the contributions of diseases to mortality in the U.S. elderly population. Therefore, the development of a high-precision tool for the prediction of health status and mortality of older persons is required. The MFSU data have the necessary information for developing such a tool. This new index is referred to as the *Adjusted for elderly population Multi-Morbidity Index* (AMMI). This index, like the CCI, is calculated by use of disease-specific weights, w_d, summed over all contributing diseases: $C(t) = \sum_d w_d I_d(t)$, where the term $I_d(t)$ indicates the presence or absence of the d condition in a patient, i.e., $I_d(t) = 1$ when a patient has the condition, and $I_d(t) = 0$ when he/she does not. The weights w_d are defined by considering the effect of individual diseases on the mortality rate. What is new compared to the CCI is the set of contributing diseases and disease-specific weights w_d that are made specific to the elderly population. We selected 48 disease conditions based on analysis of disease prevalence in the elderly population and causes of death using Multiple Cause of Death data. Then we used the Cox proportional hazard model with multivariate time-dependent predictions (i.e., individual disease prevalence estimated from Medicare data) to estimate the weights w_d. As in the majority of approaches for construction of co- and multimorbidity indices, disease-specific weights are simply rounded logarithms of hazard ratios of the respective conditions on mortality. This computation requires evaluated individual disease prevalence at all times of individual follow-up; therefore we used several additional specifications: (i) 1 year was used as the time period before the time point of interest to search for diagnosis codes, (ii) in the definition of the incidence rate calculated using Medicare histories four base sources (inpatient, outpatient, carrier-physician-supplier) were used, and (iii) both base and secondary diagnosis codes were used. Two basic properties of the index need to be verified: (i) variations across age, race, and gender strata; and (ii) associations between the shape of the index, i.e., slope or curvature in its age pattern, and mortality. The distributions of the indices for population subgroups are presented in Fig. 3.13. The CCI and AMMI were empirically evaluated for specific birth cohorts and their age patterns show that variation in the Charlson index over periods and cohorts are not reflected in the corresponding plots for total mortality (Fig. 3.14). In contrast, the AMMI fits much better.

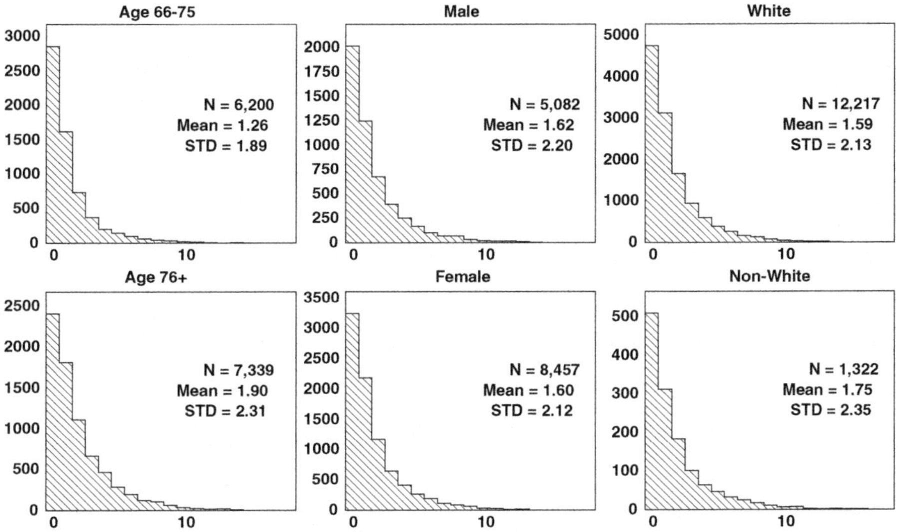

Fig. 3.13 Sex-, race-, and age-group-specific distributions of AMMI

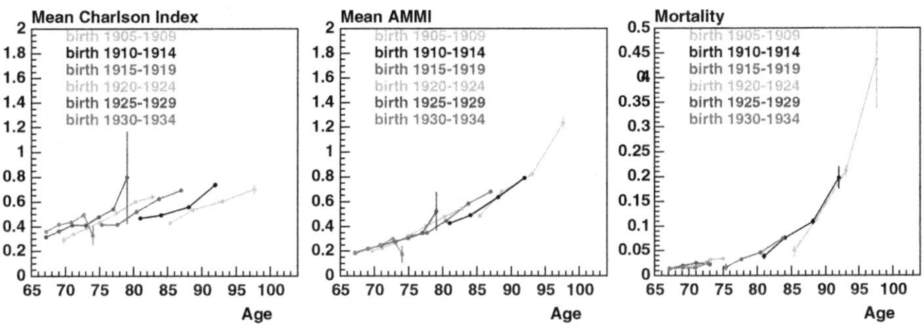

Fig. 3.14 Cohort specific patterns of CHI, AMMI, and mortality rates: NLTCS-M female population

3.3.10 *Predictive Population Models*

Standard approaches for probabilistic forecasting of the age pattern of human mortality, including the Lee-Carter method (Lee and Carter 1992) and its numerous generalizations and improvements (Booth 2006), are based on quite clear and simple assumptions and have been successfully applied in analyses of mortality data of different countries. However, these approaches do not take into account that, in population cohorts, trends in prevalence result from combinations of trends in incidence, population at risk, recovery, and patients' survival rates. Trends in the rates for one disease also may depend on trends in concurrent diseases, e.g., increasing survival from CHD contributes to an increase in the cancer incidence rate if the individuals who survived were initially susceptible to both diseases. An approach capable of resolving these limitations could focus on describing the dynamics of health status for individuals in sub-cohorts with hazard functions of morbidity and mortality dependent on health status. Below we consider two examples of such models. The first is a simple model in which the health state is conditioned on age and the multimorbidity index discussed in the previous section. The second describes health status in terms of disease prevalence.

The first approach has two components: the mortality model and the dynamic model for AMMI. The mortality model incorporates AMMI as a covariate:

$$\text{logit}(\text{Prob}(death = 1|C_0, C_3, Age, Y_b)) = u + \beta_0 \cdot C_0 + \beta_1 \cdot C_3 + \beta_{Age} \cdot (Age - 70) + \beta_b \cdot (Y_b - 1930)$$

In this model, C_0, the comorbidity index in the last month, is the major predictor of mortality. Its effect is linear in this logit regression model which yields an exponential relation of mortality to C_0. The exponential relation is expected because of the way C_0 is constructed with the HR evaluated using the Cox proportional hazard model. This was also confirmed by two separate models we analyzed: (i) with categorical AMMI (in which case the estimated pattern was close to linear) and (ii) with individual diseases as predictors (in which the estimated weights were close to those estimated using the Cox model). Second, C_3 is the multimorbidity index AMMI measured 3 months before the current month. Conditioning on previous values of the index is important, because it captures the effect of increasing comorbidity before death. Third, the effect of age is linear corresponding to a gamma-Gompertz model of mortality. Fourth, Y_b reflects the cohort effect. The linear form for this effect is justified by preliminary modeling analyses with a categorical cohort index that yielded an estimated pattern close to linear. The second model component, a dynamic model for AMMI, is of the form:

$$\Delta_{01} = C_0 - C_1 = \overline{u} + \overline{\beta}_1 \cdot C_1 + \beta_\Delta \cdot \Delta_{13} + \overline{\beta}_{Age} \cdot (Age - 70) + \overline{\beta}_b \cdot (Y_b - 1930)$$

with the same notation as in the mortality model. What is new here compared to the mortality model is: (i) the outcome variable difference between AMMIs measured in the current and last months; and (ii) one of the predictors difference between AMMIs measured in previous months ($\Delta_C = C_1 - C_3$). Parameter estimates for these two models using the NLTCS-M data are presented in Table 3.5. It can be seen that the estimated parameters for the multimorbidity index show an increasing time trend as indexed by birth cohort. One explanation of this is an increasing prevalence of diagnoses at earlier stages.

The model described above uses the multimorbidity index as an aggregated characteristic of individuals' health statuses. If the data are extensive (e.g., SEER-M data), the health status can be described more precisely, e.g., using the set of disease-specific indicator functions indicating the time-dependent prevalence of selected diseases. Three dynamic models are required to describe the changes in

Table 3.5 Parameter estimates for mortality and dynamic models involving AMMI

	NLTCS-medicare	
Parameter	Estimate	p-value
Mortality model		
Intercept	-7.31	<0.0001
Recent AMMI	0.437	<0.0001
3-month-prior AMMI	-0.154	<0.0001
Age	0.067	<0.0001
Birth cohort	-0.004	0.37
Dynamic model for AMMI		
Intercept	0.067	<0.0001
Recent AMMI	-0.042	<0.0001
Difference of recent AMMI	0.079	<0.0001
Age	0.0068	<0.0001
Birth cohort	0.002	<0.0001

individual health status during individual follow-up: models to predict mortality, incidence, and recovery (or long-term remission). They are given by the equations:

$$\operatorname{logit}\left(\operatorname{Prob}\left(death = 1 \middle| I^t, I^{t-1}a, Y_b\right)\right) = \widetilde{u} + \sum_{i=1}^{48}\left(\widetilde{\beta}_i I_i^t + \beta_i' I_i^{t-1}\right) + \sum_{i=1}^{48}\sum_{j>i}^{48}\widetilde{\beta}_{ij} I_i^t I_j^t + \widetilde{\beta}_a f(a) + \widetilde{\beta}_b f(Y_b)$$

$$\operatorname{logit}\left(\operatorname{Prob}\left(I_k^{t+1} = 1 \middle| I_k^t = 0, I_{i\neq k}^t, a, Y_b\right)\right) = \widetilde{u}^k + \sum_{\substack{i=1 \\ i\neq k}}^{48}\left(\widetilde{\beta}_i^k I_i^t + \beta_i'^k I_i^{t-1}\right) + \sum_{\substack{i=1 \\ i\neq k}}^{48}\sum_{\substack{j>i \\ j\neq k}}^{48}\widetilde{\beta}_{ij}^k I_i^t I_j^t + \widetilde{\beta}_a^k f_a^k(a) + \widetilde{\beta}_b^k f_b^k(Y_b)$$

$$\operatorname{logit}\left(\operatorname{Prob}\left(I_k^{t+1} = 0 \middle| I_k^t = 1, I_{i\neq k}^t, a, Y_b\right)\right) = \hat{u}^k + \sum_{\substack{i=1 \\ i\neq k}}^{48}\left(\hat{\beta}_i^k I_i^t + \hat{\beta}_i'^k I_i^{t-1}\right) + \sum_{\substack{i=1 \\ i\neq k}}^{48}\sum_{\substack{j>i \\ j\neq k}}^{48}\hat{\beta}_{ij}^k I_i^t I_j^t + \hat{\beta}_a^k \hat{f}_a^k(a) + \hat{\beta}_b^k \hat{f}_b^k(Y_b)$$

which, respectively, model the probability of death in the next month, the incidence of the k disease, and recovery/remission from the kth disease. Age and birth cohort effects can also be incorporated (possibly non-linearly) into these models.

This probability-of-death model was estimated using the NLTCS-M data. As expected, the majority of diseases had positive effects on mortality. For example, the odds ratio (OR) of heart failure was 2.33 (CI = 2.13–2.54), for myocardial infarction—3.07 (CI = 2.70–3.48), for cardiac arrhythmia—2.14 (CI = 1.98–2.32), for stroke—1.61 (CI = 1.47–1.77), for COPD—1.66 (CI = 1.52–1.81), for Alzheimer's disease—2.27 (CI = 1.99–2.6), and for metastatic cancer—6.86 (CI = 6.00–7.84). More than 100 significant interactions were found, which demonstrated the non-additive effects of co-occurrence of diseases, with some effects negative (antagonistic), e.g., diseases of peripheral veins and stroke, heart failure and MTS; and some effects positive (synergistic), e.g., inflammatory bowel disease and non-solid cancer, diabetes and alcohol abuse, cardiac arrhythmia and hypertension, emphysema and diabetes, and chronic liver disease and chronic peptic ulcer.

The effect of history was more important for acute than for chronic diseases, e.g., for myocardial infarction, estimates for each of the 3 months before death were: OR(t − 1) = 5.96 (CI = 4.64–7.65), OR(t − 2) = 0.59 (CI = 0.41–0.84), and OR(t − 3) = 0.69 (CI = 0.48–1.01), while for solid cancers with fast progression, estimates were, respectively: OR(t − 1) = 2.48 (CI = 1.98–3.10), OR(t − 2) = 1.32 (CI = 1.01–1.73), and OR(t − 3) = 1.32 (CI = 1.06–1.64). Age (in single years) also had a significant contribution: OR = 1.07 (CI = 1.065—1.075). The effect of time was nonlinear, but quadratic with a maximum at 1999–2000.

Estimation of the incidence and recovery models showed that comorbidity (i.e., the presence and absence of comorbid diseases) was the main predictor of incidence and recovery. For incidence, the ORs of the effects of comorbid diseases were in the 1.6–1.9 range. For recovery, the ORs of the effects of comorbid diseases were in the 0.6–0.8 range. Examples of specific effects of disease incidence and recovery are given in Table 3.6. Note that no strong predictors were found for neoplasms (non-strong predictors were low weight and anemia).

Table 3.6 Examples of the effects on disease incidence and recovery

Disease incidence	Effect (OR)
Myocardial infarction	Heart failure (1.81), Angina pectoris (1.70), age (1.03), time (0.97)
COPD	Asthma (3.7), emphysema (4.4), heart failure (1.87)
Alzheimer's disease	Time (1.09), dementia (9.4), Parkinson's (2.2), Cerebro (1.4)
Nephritis/nephrosis	Time (1.08), age (1.02), renal (6.6), diabetes (2.0)
Anemia	Age (1.04), time (1.04), low weight (1.77), cancer (2.0), IBD (2.7)
Osteoporosis/hip fracture	Metastatic cancer (1.8), time (1.12), age (1.04), RA (2.1)
Disease recovery	Effect (OR)
Cerebrovascular	Plegia (0.38), time (1.02), age (0.98), dementia (0.71), hypertension (0.89)
Chronic liver disease/cirrhosis	Alcohol abuse (0.59)
Anemia	Heart failure (0.88), low weight (0.81), nonsolid cancer (0.74), MTS (0.63)

To validate the three-equation model, we divided the NLTCS-M cohort into two subcohorts of equal size, thus yielding estimation and validation datasets. The validation procedure consisted of two steps. First, we estimated the model using the estimation dataset and then tested how well the fitted model predicts mortality outcomes in the validation dataset. This procedure can be formalized in terms of receiver operating characteristic (ROC) curves and the areas under the curves. The plots in Fig. 3.15 provide the ROC curves and area under curve (AUC) estimates for several models that differ by the set of predictors of mortality: (i) current disease prevalences (i.e., individual indicators of all diseases used in the analysis), (ii) disease prevalences and all paired interactions among them, (iii) current and last month prevalences (i.e., 1-level history), and (iv) current and two previous-month prevalences (two-level history). One can see that in all cases $AUC > 0.88$, therefore we can conclude that the quality of prediction is excellent. As a second step, we simulated individual trajectories using the estimated model using measures

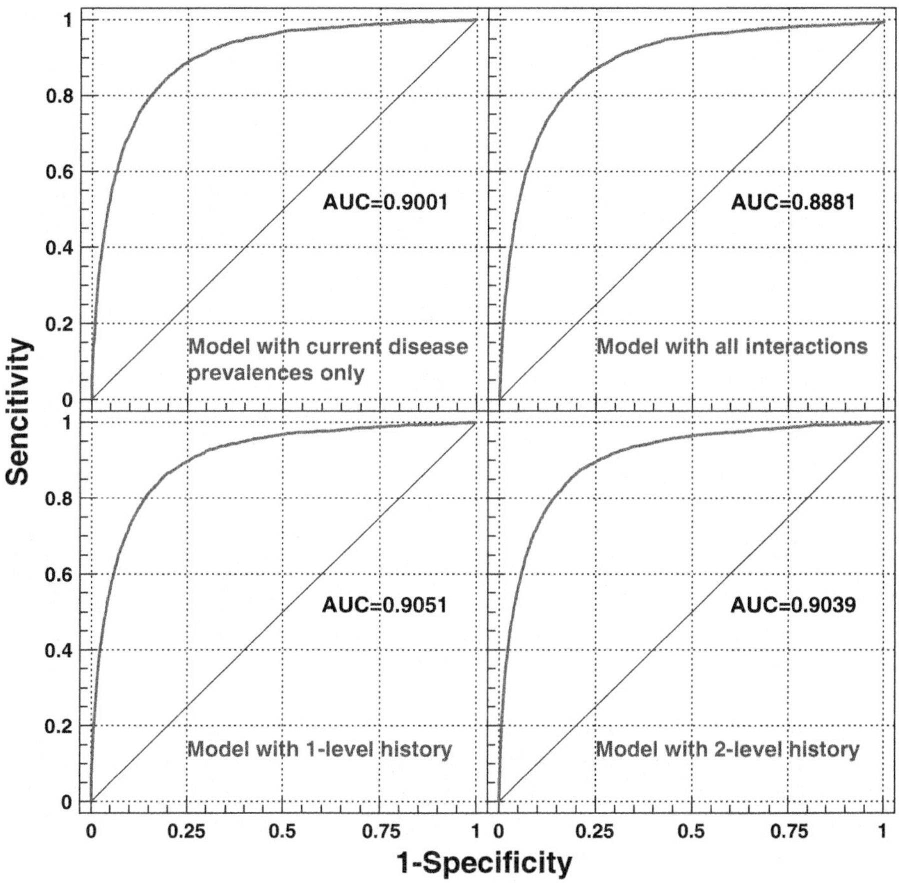

Fig. 3.15 ROC curves for the validation dataset

at the initial wave of the validation cohort. These simulations show excellent predictive performance for mortality outcomes in the first year (85–90 %). For longer time periods such as 5-years, the predictive accuracy deteriorates (68–75 %). This indicates that the incidence and recovery models could be improved.

3.4 Conclusion

In this chapter, we presented a spectrum of analyses of many important aspects in the biomedical demography of the U.S. population of older adults. All these analyses were based on Medicare data, a powerful and largely underexplored source of information about the health of the elderly population. Medicare data represent the collection of individual records of health-related information about established diagnoses and administrated medical procedures. In addition, Medicare data represent the U.S. population of older adults. Our analyses demonstrated how nonparametric and simple regression statistical methods allow researchers to analyze and evaluate many important health-related epidemiologic characteristics valid at the national level. Specifically, the topics discussed in this chapter included patterns of morbidity and mortality, recovery (or long-term remission), comorbidity and multimorbidity, risk factors of disease incidence and mortality, and projection modeling of health and mortality.

First, we analyzed the age-patterns of acute and chronic diseases in the elderly using the NLTCS-M and SEER-M data. The results of our comparative analyses of the age patterns evaluated from Medicare data with those obtained in other studies suggested that the national age-specific incidence patterns can be adequately evaluated from the Medicare data. We also discovered a series of new substantive findings regarding the shape of age-patterns of the diseases and their time trends. For example, males had higher rates of ACHD, heart failure, Parkinson's disease, skin melanoma, lung, and colon cancers, while females had higher rates of stroke and asthma. Another example is that a significant 5-year decline was observed for incidence rates of ACHD, stroke, heart failure, and prostate, lung (male) and colon (female) cancers, while the rates of diabetes, ulcer, and Alzheimer's disease increased. Note that the estimates of time trends become especially important in populations with increasing proportions of the elderly, for which maintaining good health at advanced ages is important. While mortality trends have been studied extensively, studies of morbidity trends are rare.

The dates of onset in these studies were identified using information collected in the MFSU with specific assumptions corresponding to specific calculation algorithms. Note that the date of onset of a certain chronic disease is a quantity which is not defined as precisely as mortality. This uncertainty makes difficult the construction of a unified definition of the date of onset appropriate for population studies. We compared several alternative definitions of the date of onset and identified the computational approach that most closely describes data collected in other studies on incidence of diseases. What is important is that this issue is not purely methodological. Different approaches to disease onsets can be used in clinical practice for different diseases. For example, the clinical criteria for circulatory disease onsets used in other studies are quite different which could result in different incidence rates. The diagnostic criteria of different heart studies are reviewed in the appendix of ref. (NIH/NHLBI 2006). We have briefly discussed this in our recent paper (Akushevich et al. 2012).

Analysis of the effects of comorbidity on disease onset showed that patients with higher comorbidity had higher rates of ACHD, stroke, heart failure, Alzheimer's disease, and melanoma. In this approach, comorbidity is represented in the form of the Charlson comorbidity index, a tool that allows the representation of comorbidity by a single variable combining disease indicators with weights evaluated from their effects on mortality. This representation of comorbidity is traditional, but it is only one possible approach. The richness of the Medicare data motivated us to search for more precise and detailed approaches for descriptions of co- and multimorbidity in U.S. elderly population.

A first approach to the analysis of multimorbidity in the U.S. elderly population developed in this chapter is based on the idea of dependent risks among geriatric diseases. We considered the risks of new disease onsets for population groups with a pre-existing condition. The risks of development of a new condition could be higher or lower than those in the general population for many reasons. For example, the effects of treatment of a prior disease can increase or decrease the risk of a later disease; shared behavioral risks and pleotropic effects of genes could also increase the risks of the later-developing diseases. A unified computational approach applied to all diseases considered within the same analysis allowed us to create a unified view of mutual interrelationships among the risks of cancer and non-cancer aging-associated diseases. Direct and inverse dependences in geriatric disease risks were observed among the U.S. elderly, confirming known and detecting new associations for a wide spectrum of diseases. A better understanding of biological links between different diseases (or between the groups of diseases—etiological or organ-specific) to which this research contributes can provide new therapeutic approaches for diseases with the shared pathological pathways.

Our second approach is based on the development of a new multimorbidity index for the U.S. older adult population and estimated using information from Medicare data. The stages of AMMI development included: (i) identification of multimorbidity patterns (or disease clusters) most often occurring for older adults; (ii) estimates of weights of specific diseases/disease clusters; and (iii) evaluating disease indicators during individual follow-up.

Two predictive models for mortality and dynamics of health status then were developed, estimated, and validated using Medicare data. The first model uses the AMMI as a characteristic of health status and predicts mortality and changes in AMMI in terms of their past values, age, and cohort. The second model includes three components: conditional models for mortality, incidence, and recovery. The models demonstrate excellent capabilities for predicting mortality rates. These models can be used for short-term predictions of the health and mortality of the

U.S. elderly population and for the analysis of "what-if" scenarios by consideration of specific interventions, e.g., the strategies of secondary prevention, new therapeutic approaches, and projected Medicare policy changes.

Medicare data also can be used for evaluating the epidemiologic characteristics of patient recovery via analysis of Medicare information during individual follow-up for the cohorts of patients with a specific disease onset. Specifically, we analyzed the rate of recovery of long-term remission by identifying patients who stopped visiting doctors during a 5-year follow-up after disease onset. We found that these patients (i.e., recovered individuals) had lower death rates than non-recovered patients; therefore, patients who stopped visiting doctors are a healthier subcohort. We also found that these patients had higher death rates than the general population for all diseases considered; thus implying that complete recovery does not occur. To our knowledge, this type of analysis has never been done before. The approach opens new opportunities for developing predictive models with time-dependent covariates representing health status. Such models could be further used to better quantify the contribution of age-related diseases to healthy life expectancy and to improve forecasts of health and mortality.

Medicare data are often linked to demographic surveys, thus allowing for joint analyses of survey and claims data. Using information about ADLs and IADLs in the NLTCS-M data, we identified cohorts of individuals with specific disabilities and found that, among individuals with severe disabilities, there were higher rates of stroke, heart failure (males), diabetes, asthma, and Parkinson's disease, while rates of breast and prostate cancers were higher for nondisabled or moderately disabled individuals. Another possible approach considers variables measured in a survey as potential risk factors of disease onsets (accessed from Medicare data) and mortality. These effects can be evaluated within association studies between the variables measured in the NLTCS-M and risks of all-cause mortality and morbidity extracted from the MFSU. Each of the variables (representing daily living activities, physical activities, smoking, alcohol consumption, social activities, self-reported comorbidity, health insurance, and medical providers and other groups of variables) was tested for association with the risks of all-cause mortality and morbidity extracted from the MFSU data. From this analysis, we identified the main factors predicting disease incidence and obtained a description of higher risk groups of major geriatric diseases in terms of variables representing distinct features of human aging. Groups of parameters for physical activity, tobacco consumption, comorbid conditions, demographic characteristics, health insurance, and medical care providers showed significant contributions to increasing or decreasing risk of incidence of the diseases considered. The most influential of potentially preventable lifestyle risk factors can be detected using this approach and applied to further deeper analyses, including other data sets with detailed risk factors. Potentially, the approaches developed, and results obtained, can be applied to developing more individualized forecasts and more individualized prevention strategies.

The utility of the Medicare data as demonstrated in our study is important because there are few data sources to study health effects at advanced ages in the national population. For example, heart disease and stroke account for more than

40 % of all deaths among persons aged 65–74 years and almost 60 % of those aged 85 years and older. However, there are no nationally representative data available on incidence, severity, or recurrence of acute coronary or stroke events in either the inpatient or outpatient settings. Therefore, Medicare-based datasets could be very useful for studying the epidemiology and biodemography of aging-related diseases and associated medical costs in the U.S. elderly population.

Acknowledgements The research reported in this chapter was supported by the National Institute on Aging grants R01AG027019, R01AG030612, R01AG030198, R01AG032319, R01AG046860, R21 AG045245, and P01AG043352.

References

Akushevich, I., & Manton, K. (2011, March 31–April 2). Mortality trajectories at the very advanced ages. Population Association of America 2011 Annual Meeting. Washington, DC

Akushevich, I., Kravchenko, J., Akushevich, L., Ukraintseva, S., Arbeev, K., & Yashin, A. I. (2011a). Cancer risk and behavioral factors, comorbidities, and functional status in the U.S. elderly population. *ISRN Oncology, 2011*, 415790.

Akushevich, I., Kravchenko, J., Akushevich, L., Ukraintseva, S., Arbeev, K., & Yashin, A. I. (2011b). Medical cost trajectories and onsets of cancer and non cancer diseases in US elderly population. *Computational and Mathematical Methods in Medicine, 2011*, 857892.

Akushevich, I., Kravchenko, J., Ukraintseva, S., Arbeev, K., & Yashin, A. I. (2012). Age patterns of incidence of geriatric disease in the U.S. elderly population: Medicare-based analysis. *Journal of American Geriatrics Society, 60*(2), 323–327.

Akushevich, I., Kravchenko, J., Ukraintseva, S., Arbeev, K., Kulminski, A., & Yashin, A. I. (2013a). Morbidity risks among older adults with pre-existing age-related diseases. *Experimental Gerontology, 48*(12), 1395–1401.

Akushevich, I., Kravchenko, J., Ukraintseva, S., Arbeev, K., & Yashin, A. (2013b). Population based analysis of incidence rates of cancer and non-cancer chronic diseases in the U.S. elderly using NLTCS/Medicare-linked database. *ISRN Geriatrics, 2013*, 943418.

Akushevich, I., Kravchenko, J., Ukraintseva, S., Arbeev, K., & Yashin, A. I. (2013c). Circulatory diseases in the U.S. elderly in the linked national long-term care survey-medicare database population-based analysis of incidence, comorbidity, and disability. *Research on Aging, 35*(4), 437–458.

Akushevich, I., Kravchenko, J., Ukraintseva, S., Arbeev, K., & Yashin, A. I. (2013d). Recovery and survival from aging-associated diseases. *Experimental Gerontology, 48*(8), 824–830.

Akushevich, I., Kravchenko, J., Ukraintseva, S., Arbeev, K., & Yashin, A. I. (2013e). Time trends of incidence of age-associated diseases in the U.S. elderly population: Medicare-based analysis. *Age and Ageing, 42*(4), 494–500.

Altekruse, S. F., Kosary, C. L., Krapcho, M., Neyman, N., Aminou, R., Waldron, W., Ruhl, J., Howlader, N., Tatalovich, Z., Cho, H., Mariotto, A., Eisner, M. P., Lewis, D. R., Cronin, K., Chen, H. S., Feuer, E. J., Stinchcomb, D. G., & Edwards, B. K. (2009). *SEER cancer statistics review, 1975–2007*. Bethesda: National Cancer Institute.

Alves, G., Forsaa, E. B., Pedersen, K. F., Gjerstad, M. D., & Larsen, J. P. (2008). Epidemiology of Parkinson's disease. *Journal of Neurology, 255*(5), 18–32.

Ballard-Barbash, R., Friedenreich, C., Slattery, M., & Thune, I. (2006). Obesity and body composition. In D. Schottenfeld & J. F. Fraumeni (Eds.), *Cancer epidemiology and prevention*. New York: Oxford University Press.

Banerjee, D., Lee, G., Malik, S., & Daly, S. (1987). Underdiagnosis of asthma in the elderly. *British Journal of Diseases of the Chest, 81*, 23–29.

Bellia, V., Battaglia, S., Catalano, F., Scichilone, N., Incalzi, R. A., Imperiale, C., & Rengo, F. (2003). Aging and disability affect misdiagnosis of COPD in elderly Asthmatics: The SARA study. *CHEST Journal, 123*(4), 1066–1072.

Bleyer, A. W., O'Leary, M., Barr, R. D., & Ries, L. A. G. (2006). *Cancer epidemiology in older adolescents and young adults 15 to 29 years of age, including SEER incidence and survival: 1975–2000*. Bethesda: National Cancer Institute.

Booth, H. (2006). Demographic forecasting: 1980 to 2005 in review. *International Journal of Forecasting, 22*(3), 547–581.

Braman, S. S., Kaemmerlen, J. T., & Davis, S. M. (1991). Asthma in the elderly – A comparison between patients with recently acquired and long-standing disease. *American Review of Respiratory Disease, 143*(2), 336–340.

Breslow, N. E., & Day, N. E. (1987). Statistical methods in cancer research. Volume II—The design and analysis of cohort studies. *IARC Science Publication*, (82), 1–406.

Brown, L. D., Cai, T. T., & DasGupta, A. (2001). Interval estimation for a binomial proportion. *Statistical Science, 16*, 101–117.

Burr, M., Charles, T., Roy, K., & Seaton, A. (1979). Asthma in the elderly: An epidemiological survey. *British Medical Journal, 1*(6170), 1041–1044.

Charlson, M. E., Pompei, P., Ales, K. L., & Mackenzie, C. R. (1987). A new method of classifying prognostic co-morbidity in longitudinal-studies – Development and validation. *Journal of Chronic Diseases, 40*(5), 373–383.

Crimmins, E. M. (2004). Trends in the health of the elderly. *Annual Review of Public Health, 25*, 79–98.

de La Fuente-Fernández, R. (2006). Impact of neuroprotection on incidence of Alzheimer's disease. *PLoS One*, 1, article 1. Available online at http://www.ncbi.nlm.nih.gov/pmc/articles/PMC1762379/

Diederichs, C., Berger, K., & Bartels, D. B. (2011). The measurement of multiple chronic diseases—A systematic review on existing multimorbidity indices. *The Journals of Gerontology Series A: Biological Sciences and Medical Sciences, 66*(3), 301–311.

Dobson, A. J., Kuulasmaa, K., Eberle, E., & Scherer, J. (1991). Confidence intervals for weighted sums of Poisson parameters. *Statistics in Medicine, 10*(3), 457–462.

Enright, P., McClelland, R., Newman, A., Gottlieb, D., & Lebowitz, M. (1999). Underdiagnosis and undertreatment of asthma in the elderly. *Chest, 116*(3), 603–613.

Fahn, S. (2003). Description of Parkinson's disease as a clinical syndrome. *Annals of the New York Academy of Sciences, 991*(1), 1–14.

Fay, M. P., & Feuer, E. J. (1997). Confidence intervals for directly standardized rates: A method based on the gamma distribution. *Statistics in Medicine, 16*(7), 791–801.

Feigin, V., Lawes, C., Bennett, D., & Anderson, C. (2003). Stroke epidemiology: A review of population-based studies of incidence, prevalence, and case-fatality in the late 20th century. *The Lancet Neurology, 2*(1), 43–53.

Fitzpatrick, A. L., Kuller, L. H., Ives, D. G., Lopez, O. L., Jagust, W., Breitner, J. C., Jones, B., Lyketsos, C., & Dulberg, C. (2004). Incidence and prevalence of dementia in the cardiovascular health study. *Journal of American Geriatrics Society, 52*(2), 195–204.

Ford, R. M. (1969). Aetiology of asthma – A review of 11,551 cases (1958 to 1968). *Medical Journal of Australia, 1*(12), 628–631.

Freedman, V. A., Schoeni, R. F., Martin, L. G., & Cornman, J. C. (2007). Chronic conditions and the decline in late-life disability. *Demography, 44*(3), 459–477.

Gao, S., Hendrie, H. C., Hall, K. S., & Hui, S. (1998). The relationships between age, sex, and the incidence of dementia and Alzheimer disease: A meta-analysis. *Archives of General Psychiatry, 55*(9), 809–815.

Gatling, W., Guzder, R. N., Turnbull, J. C., Budd, S., & Mullee, M. A. (2001). The Poole Diabetes Study: How many cases of Type 2 diabetes are diagnosed each year during normal health care in a defined community? *Diabetes Research and Clinical Practice, 53*(2), 107–112.

Hall, C., Verghese, J., Sliwinski, M., Chen, Z., Katz, M., Derby, C., & Lipton, R. (2005). Dementia incidence may increase more slowly after age 90: Results from the Bronx Aging Study. *Neurology, 65*(6), 882–886.

Hebert, P., Geiss, L., Tierney, E., Engelgau, M., Yawn, B., & McBean, A. (1999). Identifying persons with diabetes using Medicare claims data. *American Journal of Medical Quality, 14*(6), 270–277.

Hewitt, M., & Simone, J. (2000). *Enhancing data systems to improve the quality of cancer care*. Washington, DC: National Academy Prress.

Howlader, N., Noone A., Krapcho M., Neyman N., Aminou R., & Waldron, W. (2011). SEER cancer statistics review, 1975–2008, National Cancer Institute. Bethesda.http://seer.cancer.gov/csr/1975_2009_pops09/. Accessed October 25.

Johnson, C., & Adamo, M. (2007). *SEER program coding and staging manual 2007*. Bethesda: National Cancer Institute, National Institutes of Health, U.S. Department of Health and Human Services.

Keyfitz. (1966). Sampling variance of the standardized mortality rates. *Human Biology, 38*, 309–317.

King, M. J., & Hanania, N. A. (2010). Asthma in the elderly: Current knowledge and future directions. *Current Opinion in Pulmonary Medicine, 16*(1), 55–59.

Kravchenko, J., Akushevich, I., Seewaldt, V. L., Abernethy, A. P., & Lyerly, H. K. (2011). Breast cancer as heterogeneous disease: Contributing factors and carcinogenesis mechanisms. *Breast Cancer Research and Treatment, 128*(2), 483–493.

Kravchenko, J., Akushevich, I., Abernethy, A. P., & Lyerly, H. K. (2012). Evaluating the number of stages in development of squamous cell and adenocarcinomas across cancer sites using human population-based cancer modeling. *PLoS One, 7*(5), e37430.

Lee, R. D., & Carter, L. R. (1992). Modeling and forecasting U.S. mortality. *Journal of the American Statistical Association, 87*(419), 659–671.

Lee, H., & Stretton, T. (1972). Asthma in the elderly. *British Medical Journal, 4*(5832), 93–95.

Leibson, C. L., O'Bnen, P. C., Atkinson, E., Palumbo, P. J., & Melton, L. J., III. (1997). Relative contributions of incidence and survival to increasing prevalence of adult-onset diabetes mellitus: A population-based study. *American Journal of Epidemiology, 146*(1), 12–22.

Manton, K., Akushevich, I., & Kravchenko, J. (2009). *Cancer mortality and morbidity patterns in the U.S. Population: An interdisciplinary approach*. New York: Springer.

Martin, L. G., Freedman, V. A., Schoeni, R. F., & Andreski, P. M. (2009). Health and functioning among baby boomers approaching 60. *The Journals of Gerontology Series B: Psychological Sciences and Social Sciences, 64*(3), 369–377.

Mayeux, R., Marder, K., Cote, L., Denaro, J., Hemenegildo, N., Mejia, H., Tang, M., Lantigua, R., Wilder, D., & Gurland, B. (1995). The frequency of idiopathic Parkinson's disease by age, ethnic group, and sex in northern Manhattan, 1988–1993. *American Journal of Epidemiology, 142*(8), 820–827.

McBean, A. M., Li, S., Gilbertson, D. T., & Collins, A. J. (2004). Differences in diabetes prevalence, incidence, and mortality among the elderly of four racial/ethnic groups: Whites, blacks, hispanics, and asians. *Diabetes Care, 27*(10), 2317–2324.

McDonald, M., Hertz, R. P., Unger, A. N., & Lustik, M. B. (2009). Prevalence, awareness, and management of hypertension, dyslipidemia, and diabetes among United States adults aged 65 and older. *The Journals of Gerontology Series A: Biological Sciences and Medical Sciences, 64*(2), 256–263.

Moorman, J. E. (2007). *National surveillance for asthma—United States, 1980–2004*. Atlanta: Department of Health and Human Services, Centers for Disease Control and Prevention.

Morens, D., Davis, J., Grandinetti, A., Ross, G., Popper, J., & White, L. (1996). Epidemiologic observations on Parkinson's disease: Incidence and mortality in a prospective study of middle-aged men. *Neurology, 46*(4), 1044–1050.

Nattinger, A. B., Laud, P. W., Bajorunaite, R., Sparapani, R. A., & Freeman, J. L. (2004). An algorithm for the use of Medicare claims data to identify women with incident breast cancer. *Health Services Research, 39*(6), 1733–1749.

Nattinger, A. B., Laud, P. W., Bajorunaite, R., Sparapani, R. A., & Freeman, J. L. (2006). Clarification note to an algorithm for the use of medicare claims data to identify women with incident breast cancer (vol 39, pg 6, 2004). *Health Services Research, 41*(1), 302–302.

NIH/NHLBI. (2006). *Incidence and prevalence: 2006 chart book on cardiovascular and lung diseases*. Bethesda: National Institutes of Health, National Heart, Lung, and Blood Institute.

Oraka, E., Kim, H. J. E., King, M. E., & Callahan, D. B. (2012). Asthma prevalence among U.S. elderly by age groups: Age still matters. *Journal of Asthma, 49*(6), 593–599.

Parry, C., Kent, E. E., Mariotto, A. B., Alfano, C. M., & Rowland, J. H. (2011). Cancer survivors: A booming population. *Cancer Epidemiology, Biomarkers & Prevention, 20*(10), 1996–2005.

Quan, H. D., Sundararajan, V., Halfon, P., Fong, A., Burnand, B., Luthi, J. C., Saunders, L. D., Beck, C. A., Feasby, T. E., & Ghali, W. A. (2005). Coding algorithms for defining comorbidities in ICD-9-CM and ICD-10 administrative data. *Medical Care, 43*(11), 1130–1139.

Reitz, C., Brayne, C., & Mayeux, R. (2011). Epidemiology of Alzheimer disease. *Nature Reviews Neurology, 7*(3), 137–152.

Ries, L. A. G., Young, J. L., Keel, G. E., Eisner, M. P., Lin, Y. D., & Horner, M.-J. (2007). *SEER survival monograph: Cancer survival among adults: U.S. SEER program, 1988–2001, patient and tumor characteristics*. Bethesda: National Cancer Institute, SEER Program.

Rockwood, K., Awalt, E., MacKnight, C., & McDowell, I. (2000). Incidence and outcomes of diabetes mellitus in elderly people: Report from the Canadian Study of Health and Aging. *Canadian Medical Association Journal, 162*(6), 769–772.

Solomon, P. R., & Murphy, C. A. (2005). Should we screen for Alzheimer's disease? A review of the evidence for and against screening for Alzheimer's disease in primary care practice. *Geriatrics and Gerontology International, 60*(11), 26–31.

Testa, G., Cacciatore, F., Galizia, G., Della-Morte, D., Mazzella, F., Russo, S., Ferrara, N., Rengo, F., & Abete, P. (2009). Charlson comorbidity index does not predict long-term mortality in elderly subjects with chronic heart failure. *Age and Ageing, 38*(6), 734–740.

Trask, P. C., Blank, T. O., & Jacobsen, P. B. (2008). Future perspectives on the treatment issues associated with cancer and aging. *Cancer, 113*(S12), 3512–3518.

Trussell, J., & Richards, T. (1985). Correcting for unmeasured heterogeneity in hazard models using the Heckman-Singer procedure. *Sociological Methodology, 15*, 242–276.

Twelves, D., Perkins, K. S., & Counsell, C. (2003). Systematic review of incidence studies of Parkinson's disease. *Movement Disorders, 18*(1), 19–31.

Ubink-Veltmaat, L., Bilo, H., Groenier, K., Houweling, S., Rischen, R., & Meyboom-de Jong, B. (2003). Prevalence, incidence and mortality of type 2 diabetes mellitus revisited: A prospective population-based study in The Netherlands (ZODIAC-1). *European Journal of Epidemiology, 18*(8), 793–800.

Ukraintseva, S. V., & Sergeev, A. S. (2000). Analysis of genetic heterogeneity of bronchial asthma in relation with the age at the onset of disease. *Genetika, 36*(2), 201–205.

Ukraintseva, S., Sloan, F., Arbeev, K., & Yashin, A. (2006). Increasing rates of dementia at time of declining mortality from stroke. *Stroke, 37*(5), 1155–1159.

Van Den Eeden, S., Tanner, C., Bernstein, A., Fross, R., Leimpeter, A., Bloch, D., & Nelson, L. (2003). Incidence of Parkinson's disease: Variation by age, gender, and race/ethnicity. *American Journal of Epidemiology, 157*(11), 1015–1022.

Vaupel, J., Carey, J., Christensen, K., Johnson, T., Yashin, A., Holm, N., Iachine, I., Kannisto, V., Khazaeli, A., & Liedo, P. (1998). Biodemographic trajectories of longevity. *Science, 280* (5365), 855–860.

von Campenhausen, S., Bornschein, B., Wick, R., Bötzel, K., Sampaio, C., Poewe, W., Oertel, W., Siebert, U., Berger, K., & Dodel, R. (2005). Prevalence and incidence of Parkinson's disease in Europe. *European Neuropsychopharmacology, 15*(4), 473–490.

Warren, J. L., Klabunde, C. N., Schrag, D., Bach, P. B., & Riley, G. F. (2002). Overview of the SEER-Medicare data: Content, research applications, and generalizability to the United States elderly population. *Medical Care, 40*(8), IV-3–IV-18.

Williams, G. (2001). Incidence and characteristics of total stroke in the United States. *BMC Neurology, 1*, 2.

Yashin, A. I., Akushevich, I., Arbeev, K., Akushevich, L., Kulminski, A., & Ukraintseva, S. (2009). Studying health histories of cancer: A new model connecting cancer incidence and survival. *Mathematical Biosciences, 218*(2), 88–97.

Yashin, A., Akushevich, I., Ukraintseva, S., Akushevich, L., Arbeev, K., & Kulminski, A. (2010). Trends in survival and recovery from stroke: Evidence from the National Long-Term Care Survey/Medicare data. *Stroke, 41*(3), 563–565.

Zekry, D., Valle, B. H. L., Michel, J.-P., Esposito, F., Gold, G., Krause, K.-H., & Herrmann, F. R. (2010). Prospective comparison of six co-morbidity indices as predictors of 5 years post hospital discharge survival in the elderly. *Rejuvenation Research, 13*(6), 675–682.

Chapter 4
Evidence for Dependence Among Diseases

Anatoliy I. Yashin, Svetlana V. Ukraintseva, Igor Akushevich, Alexander M. Kulminski, Konstantin G. Arbeev, and Eric Stallard

4.1 Introduction

Understanding demographic and public health consequences of advances in medical technology and health care, climate change, industrial development, and other large-scale factors, is important for maintaining good population health. Interaction of these external factors with age-changes in the human body (e.g., due to ontogenetic programming or physical senescence) may affect susceptibility to complex diseases and generate dependence among them. Studying mechanisms of such dependence opens new opportunities for improving population health by developing adequate preventive measures and treatment strategies which could minimize the chances of harmful side effects. Indeed, factors associated with increased vulnerability to one disease may not always promote development of another disorder, but sometimes may be protective against it, or even favor overall survival and longevity, if the protective effect outweighs the detrimental one. If so, then a reasonable prevention strategy might include targeting the risk of death from all causes combined rather than the risks of separate diseases independently of each other, as implied in most today's preventive programs. A better understanding of the occurrence and consequences of trade-offs between major health disorders/causes of death may therefore have important health care implications.

To compare the effects of public health policies on a population's characteristics, researchers commonly estimate potential gains in life expectancy that would result from eradication or reduction of selected causes of death. For example, Keyfitz (1977) estimated that eradication of cancer would result in 2.265 years of increase in male life expectancy at birth (or by 3 % compared to its 1964 level). Lemaire (2005) found that the potential gain in the U.S. life expectancy from cancer eradication would not exceed 3 years for both genders. Conti et al. (1999) calculated that the potential gain in life expectancy from cancer eradication in Italy would be 3.84 years for males and 2.77 years for females.

A.I. Yashin et al., *Biodemography of Aging*, The Springer Series on Demographic Methods and Population Analysis 40, DOI 10.1007/978-94-017-7587-8_4

All these calculations assumed independence between cancer and other causes of death. The use of such an assumption would be completely justified more than a century ago when the leading causes of death were infectious diseases. However, for today's populations in developed countries, where deaths from chronic non-communicable diseases are in the lead, this assumption might no longer be valid. An important feature of such chronic diseases is that they often develop in clusters manifesting positive correlations with each other. The conventional view is that, in a case of such dependence, the effect of cancer eradication on life expectancy would be even smaller. As Keyfitz (1977) wrote: "since the most common kind of dependence must be a positive one, people saved from cancer would be more susceptible to heart and other diseases". The directions (positive or negative) of correlations among diseases can be empirically estimated using data on multiple causes of death.

The correlation between causes of death can be evaluated using the U.S. Data on Multiple Causes of Death (http://www.cdc.gov/nchs/products/elec_prods/subject/mortmcd.htm#1999-2002). The importance of such analyses was first demonstrated in Stallard (2002) wherein associations between diseases and their secular trends were evaluated by examining statistics on the joint distributions of causes of death for the years 1980, 1990, and 1998. Estimating ratios of the observed to the expected age-standardized joint frequencies of each pair of 15 selected conditions, Stallard found 57 associations or positive correlations of the disease indicator variables. He also demonstrated that Alzheimer's disease accompanies cancer deaths significantly less frequently (up to five times less) than expected. Stallard argued that any analysis of cause-specific mortality that does not account for the joint occurrence of multiple diseases among elderly decedents, as well as the difficulties inherent in selecting one of these diseases as the underlying cause, will be incomplete.

In this chapter, we investigate dependencies among major complex health disorders of the elderly using the Multiple Cause of Death (MCD) data, with emphasis on potential trade-offs between cancer and other diseases. We evaluate frequencies and associations among the specific diagnoses that appear most often in death certificates and are overall responsible for the majority of deaths in the U.S. to explore the magnitudes of correlations among causes of death, evaluate their temporal trends, and suggest plausible interpretations. Then we review experimental findings about connections between cancer and aging, as well as evidence of trade-off like relationships between cancer and longevity, and between cancer and other diseases in humans, to support our results with potential biological explanations.

4.2 Data and Methods

The Multiple Cause of Death (MCD) data files contain information about underlying and secondary causes of death in the U.S. during 1968–2010. In total, they include more than 65 million individual death certificate records. The information available in death certificates includes the date of death, geographic location (region, state, county, division) of death, place of residence (region, state, county, city, and population size), sex, race, age, marital status, state of birth, and origin of descent. In the present study, we used data for the period 1979–2004. The cause of death fields were coded using the ICD-9 for 1979–1998, and the ICD-10 for 1999 and later. The data were collected from death certificates filed in the vital statistics offices of each state and the District of Columbia.

The list of diseases of interest includes acute coronary heart disease (CHD), stroke (acute cerebrovascular accident, CVA), cancer (malignant neoplasm), diabetes mellitus, asthma, Parkinson's disease (PD) and Alzheimer's disease (AD). These are represented by the following ICD-9 and ICD-10 codes: Cancer, for all sites combined: ICD-9 (140–208) and ICD-10 (C00–C97); acute CHD: ICD-9 (410, 411, 413) and ICD-10 (I20–I24); Stroke (CVA): ICD-9 (431, 436) and ICD-10 (I61, I64); Diabetes mellitus (excluding gestational): ICD-9 (250, 648.0, V77.1) and ICD-10 (E10–E14); Asthma: ICD-9 (493) and ICD-10 (J45–J46); PD: ICD-9 (332) and ICD-10 (G20, F02.3); AD: ICD-9 (331.0 and 290.1) and ICD-10 (F00.x, G30.x).

Note that stroke (CVA) can also be represented by a broader number of ICD-9 and ICD-10 codes. For example, in the ICD-10, the whole group of I63.x codes refers to a cerebral infarction (WHO 2007; Kokotailo and Hill 2005). We, however, deliberately omitted this group to maintain better correspondence between the ICD-9 and the ICD-10 based on the available codes. That is, ICD-10 codes I63.x generally correspond to ICD-9-CM (an extended version of ICD-9) codes 433.x1 and 434.x1 (when the fifth digit is 1), both representing cerebral infarction. The fifth digits were, however, unavailable in the MCD data, so the use of the ICD-9 codes without the fifth digit allows only approximate correspondence between the two coding systems (e.g., using ICD-9 (431, 436, 433, 434) and ICD-10 (I61, I63, I64) for stroke).

The frequencies of diseases as well as joint disease frequencies are calculated as ratios of (1) the total numbers of the corresponding ICD codes or their selected combinations that appeared on death certificates of the selected population to (2) the total numbers of deaths. Since several diseases can appear on the same death certificate as underlying or contributory causes of death, the frequencies are summed to $n \times 100\,\%$ where n is the mean number of diseases per death certificate. Frequencies defined in this way take into account the contribution of secondary causes of death and illustrate the relative burden of each disease.

Bivariate correlations for the extent of co-occurrence/non-co-occurrence of two of the diseases listed above were calculated in terms of the above frequencies (denoted f_1 and f_2) and joint frequencies (f_{12}) which reflect patterns of co-occurrence/non-co-occurrence of each pair of diseases. The correlation coefficient was calculated using the formula

$$r = \frac{f_{12} - f_1 f_2}{\sqrt{f_1(1 - f_1) f_2(1 - f_2)}}.$$

This coefficient varies between −1 and +1. The border value $r=+1$ corresponds to a case in which the codes of the two conditions appear only together, then $f_{12}=f_1=f_2$ and $r=1$; The correlation is equal to −1 when in each death record there is a code from one (and only one) of two considered conditions, in which case $f_{12}=0$, $f_1=1-f_2$, and $r=-1$. The correlation is equal to 0 when codes of two groups appear independently. Then $f_{12}=f_1f_2$ and $r=0$.

4.3 Results

4.3.1 Empirical Analyses Reveal Negative Correlations among Major Causes of Death

Results of the analyses of the MCD data are shown in Figs. 4.1, 4.2, and 4.3. Figure 4.1 shows the temporal trends in the proportion of deaths from the nine ICD diseases identified above for the years 1979–2009.

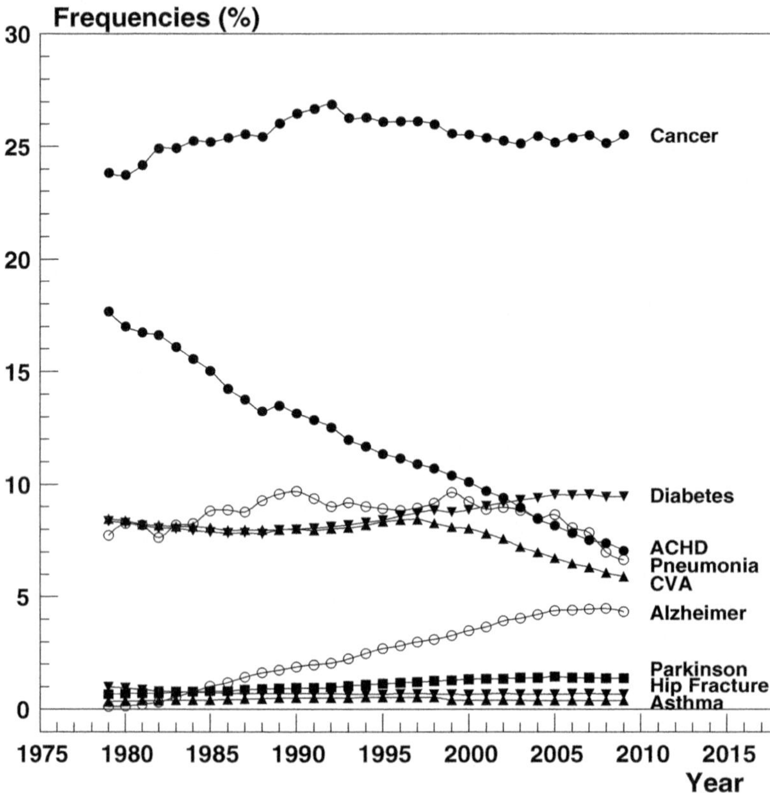

Fig. 4.1 Time trends in proportions of deaths from nine diseases from 1979 to 2009 (Note: Because the numbers are large, all standard errors are close to zero)

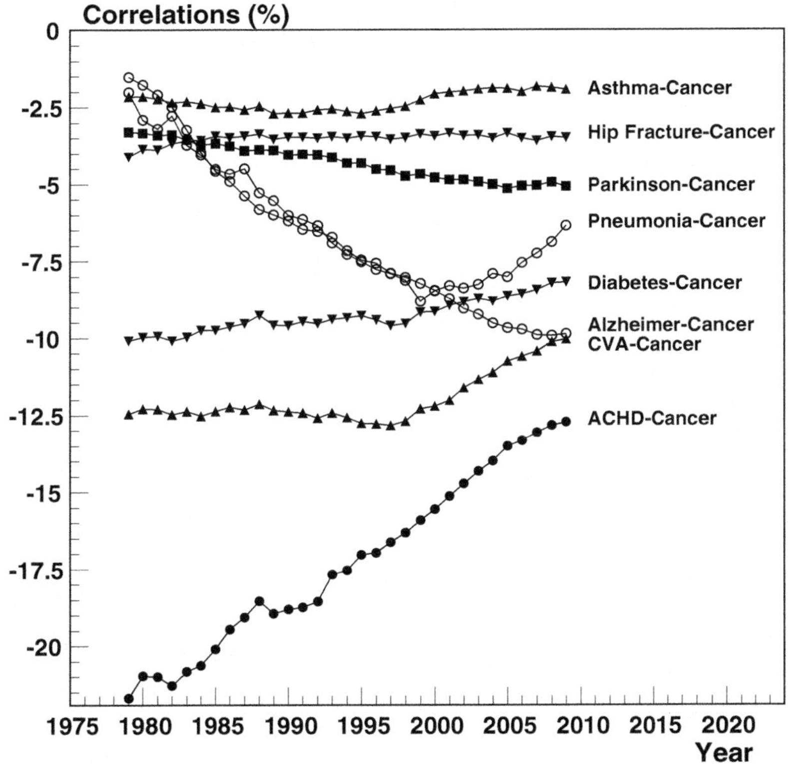

Fig. 4.2 Time trends in (negative) correlations between deaths from cancer and a number of other diseases between 1979 and 2009 (Note: Because the numbers are large, all standard errors are close to zero)

It can be seen from this figure that the frequency of deaths from acute CHD declined dramatically during the 31-year period, and the decline was relatively steady for the entire interval. The proportion of deaths from stroke (CVA) began to decline later, in the late-1990s. By contrast, the proportions of deaths from diabetes and AD increased over time, and those of cancer, asthma, and PD did not show substantial trends during this time period.

Figure 4.2 displays, for each year from 1979 to 2009, the estimated negative correlations between deaths from cancer and the co-occurrence/non-co-occurrence of selected diseases together with their time trends for the entire time period. It can be seen from this figure that there is a negative dependence for asthma, PD, AD, diabetes, CVA, and CHD. For asthma and diabetes, this correlation remained relatively constant during the entire time interval. For CVA, this correlation was relatively constant until 1998. Then its absolute value started to decline. For PD, its absolute value increased slightly but steadily during the entire period. A much faster increase in the absolute value of the correlation is evident for AD. The correlation between CHD and cancer has the highest absolute value. It also shows a substantial decline during the observational period.

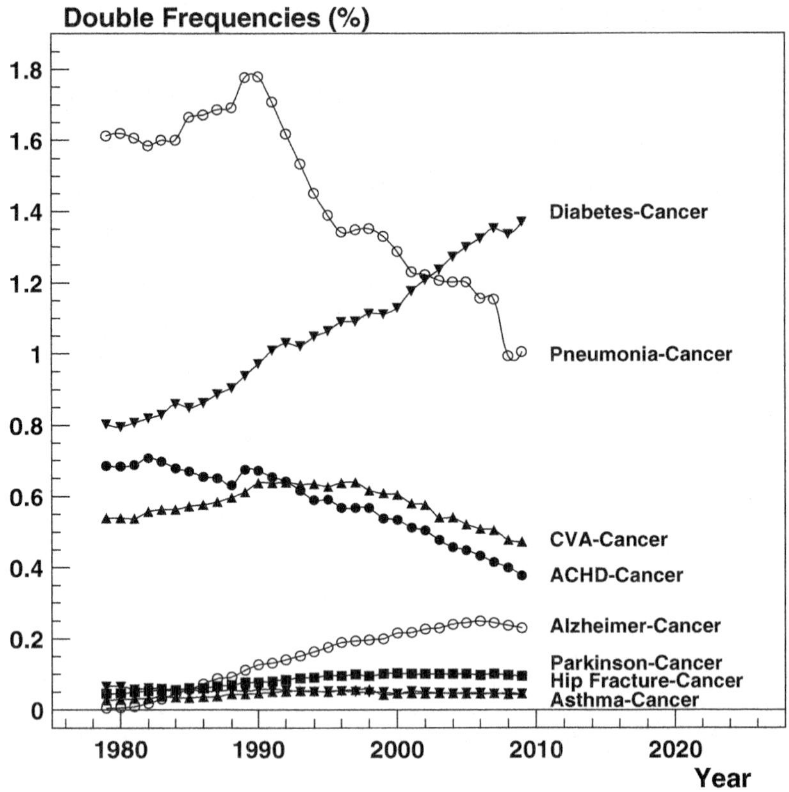

Fig. 4.3 Time trends in joint frequencies of deaths from cancer and selected diseases between 1979 and 2009 (Note: Because the numbers are large, all standard errors are close to zero)

Figure 4.3 shows estimates of joint frequencies of cancer and other diseases, as well as their time trends. One can see from this figure that among six diseases negatively correlated with cancer, diabetes most frequently appears together with cancer in the death certificates, and this frequency has a stable increasing time trend. The joint frequency of cancer and AD was close to zero in 1998. Then it increased steadily for the entire time interval, reaching 0.25 in 2004. The joint frequencies of PD/cancer and asthma/cancer show slight increases but remained small for the entire time interval. The frequency of CHD/cancer declined and CVA/cancer first increased and then declined.

4.3.2 A Dependent Competing Risk Model Capturing Negative Correlations Between Causes of Death

The simplest model describing negative correlations between competing risks is the multivariate lognormal frailty model. We illustrate the properties of such model for the bivariate case. The outline of this model is as follows.

Let $\mu_i(Z_i, x) = Z_i\mu_{0i}(x)$, $i = 1, 2$, be two random hazards in the dependent competing risk problem with two risks. Here $\mu_{0i}(x)$ are the baseline hazards at age x and Z_i, $i = 1, 2$, are frailties, which are correlated random variables having the bivariate lognormal distribution:

$$\binom{Z_1}{Z_2} \sim BVLogN\left(\begin{pmatrix} m_1 \\ m_2 \end{pmatrix}, \begin{pmatrix} \sigma_1^2 & \rho_z\sigma_1\sigma_2 \\ \rho_z\sigma_1\sigma_2 & \sigma_2^2 \end{pmatrix}\right).$$

Here m_1, m_2, σ_1^2, σ_2^2, and ρ_z are the means, variances, and correlation coefficient for the logarithms of the two frailties (i.e., they specify the associated bivariate normal distribution). The means, variances, and correlation coefficient for the frailties Z_i, $i = 1, 2$, are calculated from these parameters as follows:

$$m_{z_i} = EZ_i = e^{m_i + \frac{\sigma_i^2}{2}}, s_i^2 = Var(Z_i) = e^{2m_i + \sigma_i^2}\left(e^{\sigma_i^2} - 1\right), \ \rho z = corr(Z_1, Z_2)$$
$$= \frac{e^{\rho_z\sigma_1\sigma_2} - 1}{\sqrt{\left(e^{\sigma_1^2} - 1\right)\left(e^{\sigma_2^2} - 1\right)}}, i = 1, 2.$$

Traditionally, it is assumed in frailty models that the life spans are conditionally independent given the frailties and $mz = mz_1 = mz_2 = 1$. Simulation studies comparing different estimation strategies for such bivariate lognormal models are presented in Wienke et al. (2005). Bivariate lognormal frailty models, in contrast to the widely used gamma-frailty models (Yashin et al. 1995), allow for negative correlations between frailties as well as between life spans. Examples of positive and negative correlations between frailties and between life spans are presented in Fig. 4.4 (using Gompertz baseline hazard rates $\mu_{01}(x) = \mu_{02}(x) = \mu_0(x) = ae^{bx}$ with numerical values of the parameters a and b typical for human mortality data).

In this figure, when there is a positive correlation between frailties (graphs on the right side of the figure), the bivariate distributions of both frailty and life spans are spread along the main diagonal. However, when the correlation is negative (graphs on the left side of the figure), these distributions are spread in the directions opposite from the main diagonal. These effects are more pronounced in the frailty distributions than in the distributions of life span.

Even in this simple case there are several alternative ways to affect the marginal distribution of life span (which will further affect mortality from a specific cause). One alternative deals with a reduction of the baseline hazard. Another involves transformation of frailty distribution. These two strategies of reducing the mortality rate from one cause will produce different effects on the mortality rate from the other cause (Yashin and Iachine 1996). This model may be appropriate for illustration of the effects of dependence between competing risks. However, it is too simplified to be used for evaluation of consequences of disease prevention and treatment for which more sophisticated models of dependent competing risks that include the effects of changes in physiological and other variables affecting risks of diseases are necessary.

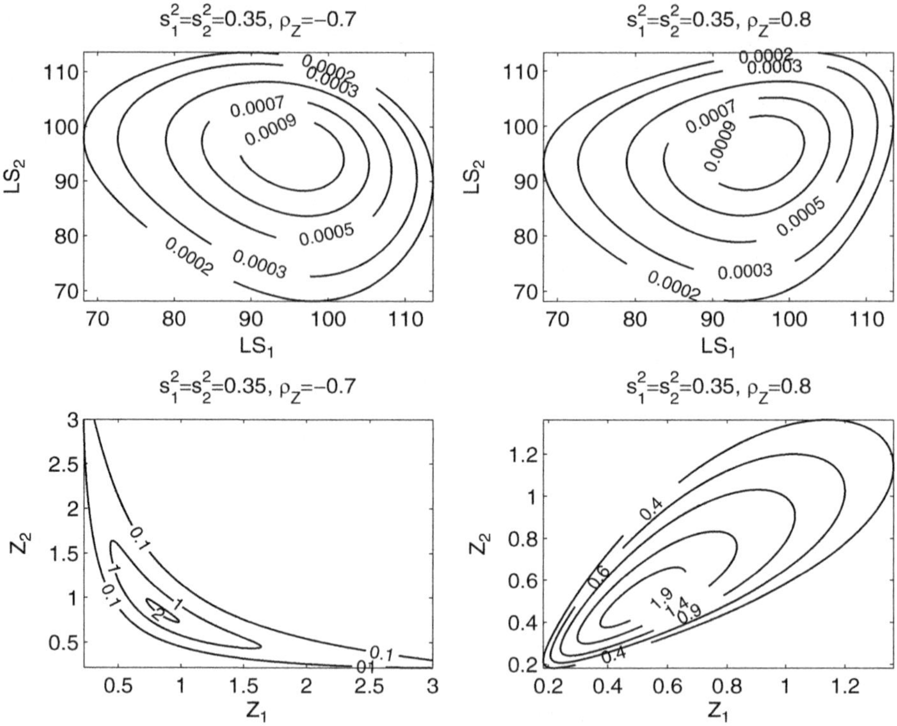

Fig. 4.4 Examples of positive and negative correlations between frailties (Z) and between life spans (LS) in the bivariate lognormal frailty model with different frailty distributions

4.4 Discussion

The twentieth century was characterized by persistent increases of survival rates in the populations of the developed world. These changes resulted from improvements in the quality of nutrition, living conditions, and progress in medical technology and health care. The effects of these improvements on population health were, however, complicated. For example, the mortality rate from CHD substantially declined (about threefold) during the second half of the last century. The overall cancer mortality rate, however, increased until the 1990s and only slightly declined afterwards in the U.S. (CDC 2007). The decline in the incidence rate for CHD first appeared in the 1950s. Only several decades later did a decline become visible for the incidence rate of cancer (Sytkowski et al. 1996; IARC 1965–2003).

Despite substantial progress in understanding factors and mechanisms responsible for such differences in trends among diseases, many important details remain unclear. The discordance of time trends in characteristics describing the two major human health disorders, cancer and CHD, was initially explained by the existence of common susceptibility factors. In particular, it was hypothesized that individuals whose lives were saved from cardiovascular death remained more susceptible to cancer than other individuals in the population. Although it has long been recognized that existing trends and dependence between other diseases and cancer can affect trends in cancer morbidity and mortality, no detailed analyses of cancer trends have been conducted that account for this possibility.

Studying these trends epidemiologically requires a dependent competing risk model. Using such a model, Rothenberg (1994) established that the contribution of the CHD mortality decline to the increase in cancer mortality has been small and does not account for the increasing age-specific risk of cancer among older persons. Llorca and Delgado-Rodriguez (2001) used a Markov chain model to analyze interrelations between CVD, CHD, and cancer in Spanish females. They found that declines in CVD and CHD mortality did not have an impact on cancer mortality. Although the absence of positive correlations between cancer and other selected diseases did not explain time trends in cancer rates for the old and oldest-old adults, it suggested a possible negative correlation between deaths from cancer and from other causes. The results of available studies of connections between cancer and aging, as well as between cancer and other diseases, suggest that such a negative correlation may have a strong biological basis.

4.4.1 Evidence of Trade-Offs Between Cancer and Aging

Studying the role of the *p53* gene in the connection between cancer and cellular aging, Campisi (2002, 2003) suggested that longevity may depend on a balance between tumor suppression and tissue renewal mechanisms. Tyner et al. (2002) and Donehower (2002) showed that mice carrying the *p53* mutation with a phenotypic effect analogous to the up-regulation of this gene have a lower risk of cancer development but their life span is reduced and accompanied by early tissue atrophy. Interestingly, the reduction of cancer mortality in super p53 mice was not accompanied by a decline in longevity in contrast to Donehower's p53 mutant mice (Garcia-Cao et al. 2002). This problem, typical of dependent competing risks, emphasizes the importance of studying various pathways of reduction of cancer mortality. Long-living mutant mice, $p66^{Shc-/-}$, have shown an impaired *p53* apoptotic response (Migliaccio et al. 1999). Introducing the null *p53* allele has been shown to protect $Ku80^{-/-}$ and $mTR^{-/-}$ mice from premature aging (Vogel et al. 1999; Chin et al. 1999), indicating that the senescence phenotypes were *p53*-dependent (Lim et al. 2000). Garcia-Cao et al. (2002) examined properties of "super *p53* mice". This type of mice was produced by transgenic introduction of one or two copies of *p53*. The mice were tumor resistant and did not exhibit any traits consistent with accelerated aging. Bauer et al. (2005) and Bauer and Helfand (2006) found that a reduction of *p53* activity in flies leads to lifespan extension. Although the mechanism by which *p53* regulates lifespan remains to be determined, these findings highlight the possibility that careful manipulation of *p53* activity during adult life may result in beneficial effects on healthy lifespan. Other tumor suppressor genes are also involved in regulation of longevity. Golubovski et al. (2006) examined survival in populations of *Drosopila* with a mutated *lgl* gene. Its ortolog, *Hugl-1*, is found in humans where it is mutated in 75 % of gut and prostate tumors. In addition to tumor suppression, the product of the *lgl* gene is an important part of the cytoskeleton and membranes. It participates in regulation of *cycline E* in the cell cycle, and plays an important part in the system transporting macromolecules. The study showed that animals heterozygous on the loss-of-function *lgl* tumor suppressor gene display a clear pre-adult viability advantage under stressful conditions (high 29 °C and low 16 °C temperatures). The survival and longevity advantage effect of the *lgl* loss-of-function is also observed in temperature stress conditions. One possible explanation of this stress-adaptive effect of reduced tumor suppressor dose might be a better resistance of *Drosophila* post-mitotic cells to a stress-associated apoptosis at old ages. The opposite manifestation of apoptotic and growth signaling pathways in cancer and aging was also reviewed in Ukraintseva and Yashin (2003a, b, 2004).

In humans, Dumont et al. (2003) demonstrated that a replacement of arginine (Arg) by proline (Pro) at position 72 of human p53 decreases its ability to initiate apoptosis, suggesting that these variants may differently affect longevity and vulnerability to cancer. Van Heemst et al. (2005) showed that individuals with the Pro/Pro genotype of p53 corresponding to reduced apoptosis in cells had significantly increased overall survival (by 41 %) despite a more than twofold increased proportion of cancer deaths at ages 85+, together with a decreased proportion of deaths from senescence related causes such as COPD, fractures, renal failure, dementia, and senility. It was suggested that human p53 may protect against cancer but at a cost of longevity. Orsted et al. (2007) examined survival among carriers Arg/Pro and Pro/Pro versus Arg/Arg genotypes of p53. The authors found an increase in median survival of 3 years for Pro/Pro versus Arg/Arg homozygotes which was not due to a decreased risk of cancer, but rather to increased survival after a diagnosis of cancer or other life-threatening disease, which may reflect a better ability to cope with stress in individuals with reduced apoptosis.

Other biological factors may also play opposing roles in cancer and aging and thus contribute to respective trade-offs (Ukraintseva et al. 2016). E.g., higher levels of IGF-1 were linked to both cancer and attenuation of phenotypes of physical senescence, such as frailty, sarcopenia, muscle atrophy, and heart failure, as well as to better muscle regeneration (Vasan et al. 2003; Renehan et al. 2004; Vinciguerra et al. 2010; Werner and Bruchim 2012; Sonntag et al. 2012; Ungvari and Csiszar 2012).

4.4.2 Trade-Offs Between Cancer and Other Diseases

The connection between cancer and longevity may potentially be mediated by trade-offs between cancer and other diseases which do not necessarily involve any basic mechanism of aging per se. In humans, it could result, for example, from trade-offs between vulnerabilities to cancer and AD, or to cancer and CVD (Ukraintseva et al. 2010; Tabares-Seisdedos et al. 2011; Kulminski et al 2011, 2013; Tabares-Seisdedos and Rubenstein 2013; Yashin et al. 2015; Ukraintseva et al. 2016). There may be several biological mechanisms underlying the negative correlation among cancer and these diseases. One can be related to the differential role of apoptosis in their development. For instance, in stroke, the number of dying neurons following brain ischemia (and thus probability of paralysis or death) may be less in the case of a downregulated apoptosis. As for cancer, the downregulated apoptosis may, conversely, mean a higher risk of the disease because more cells may survive damage associated with malignant transformation. It was shown that neurons die from apoptosis in oxygen-deprived brains, and a lower activity of the apoptotic signal leads to better survival of neurons after the stroke-induced ischemia (Barinaga 1998). Results of experimental studies suggest that medicated suppression of apoptosis may improve survival and recovery after stroke (Rosenberg et al. 2005; Harrison 2007; Kim et al. 2007; Fisher et al. 2006). On the other hand, the reduced apoptotic activity was shown to increase resistance of malignant tumors to anti-cancer therapy (Haffty and Glazer 2003). The trade-offs between cancer and various forms of CVD also may in part be related to the use of medications (Messerli et al. 2013). Also, the role of the apoptosis may be different or even opposite in the development of cancer and Alzheimer's disease (AD). Indeed, suppressed apoptosis is a hallmark of cancer, while increased apoptosis is a typical feature of AD (Lee et al. 2012; Hanahan and Weinberg 2011). If so, then chronically upregulated apoptosis (e.g., due to a genetic polymorphism) may potentially be protective against cancer, but be deleterious in relation to AD.

4.4.3 Time Trends in Negative Correlations Between Cancer and Other Diseases

The differential activity of apoptosis might be one factor contributing to the negative correlation between cancer and CHD or stroke. The observed weakening of this negative correlation over time could also be explained with the same concept. It is possible that those who survived stroke or MI are generally more resistant to apoptosis than those who didn't, and as such they may be relatively more vulnerable to cancer, as compared to those who died from MI or stroke. If so, then improvements in survival from stroke and MI may potentially contribute to a decline (in the absolute value) of the correlation between deaths from cancer and CHD over time shown in Fig. 4.2. It may happen as well that cancer treatment, which commonly induces apoptosis (e.g., with chemo or radiation therapy), increases the risk of stroke and CHD among cancer survivors and thus contributes to the weakening of the negative correlation between the diseases. The decline in absolute value of the negative correlation may also be the result of increased survival. This is because when individuals live longer, the chances of acquiring the additional co-morbidity increase.

Besides the pairs, cancer-CHD and cancer-CVA, the most prominent temporal trend in the analyses reported above was that for the correlation between cancer and AD. A strengthening of this negative correlation over time could occur, for example, if cancer treatment (e.g., with cytostatic drugs) is somehow protective against AD.

Additional analyses show that mortality from cancer may have a positive correlation with some diseases as Keyfitz (1977) expected. The presence of such a correlation may mask the effects of reducing mortality from cancer on total mortality and life expectancy for certain treatment strategies. However, as we mentioned above, the presence of a positive correlation between causes of death does not exclude the possibility of a treatment that will reduce mortality from both causes. The search for common susceptibility factors of external or internal origin could be the key for development of such a treatment strategy.

It is likely that, during the life course, the human organism undergoes the influence of factors capable of promoting both positive and negative correlation among diseases. Which way the co-morbidity pattern will develop probably depends on duration and levels of exposure to these factors during certain periods of an individual's life course. Studying these factors and the mechanisms of their action will help to better understand the regularities of aging-related declines in human health/well-being/survival status, and leading to more accurate evaluations of the consequences of interventions aiming to improve it.

4.4.4 Cancer and Anti-aging Interventions

Increased longevity can be associated not only with increased but also with decreased chances of cancer. Although the possibility that the rate of physiological aging can be significantly modified with some treatment is still being debated, several experimental studies demonstrated the possibility of simultaneous shifts of trajectories of cancer incidence and survival rates to the right under some interventions (Anisimov et al. 1987, 1998, 2000, 2001; Yashin et al. 2001). Treatment with L-DOPA (3,4-dihydroxy-L-phenylalanine) of female C3H/Sn mice increased their maximal life span together with a significant increase in tumor latency (Dilman and Anisimov 1980). The monoamine oxidase inhibitor Deprenyl (an anti-Parkinson's treatment) increased the life span of experimental mice, rats, and dogs (Ivy et al. 1994; Kitany et al. 1994; Piantanelli et al. 1994) and also inhibited the development of spontaneous and induced tumors (Kitani et al. 1994; ThyagaRajan et al. 1995; ThyagaRajan and Quadri 1999). Synthetic pineal tetrapeptide ALA-GLU-ASP-GLY (Epithalon) increased the life span of CBA mice and suppressed spontaneous neoplasm development (Anisimov et al. 2000). The most popular to-date "anti-aging" intervention, caloric restriction, often results in increased maximal life span along with reduced tumor incidence in laboratory rodents (Weindruch and Walford 1982; Blackwell et al. 1995). In Sheldon et al. (1995), an increase in lifespan of the food restricted animals was achieved primarily by a decrease in incidence and delay of onset of fatal tumors, of which lymphoma was the most prominent. Because the rate of apoptosis was significantly and consistently higher in food restricted mice regardless of age, James et al. (1998) suggested that caloric restriction may have a cancer-protective effect primarily due to the upregulated apoptosis in these mice.

4.5 Conclusion

Results of empirical analyses of MCD data revealed negative temporal correlations between deaths due to cancer and other complex diseases that are also major causes of death in the elderly. The negative correlations between deaths due to cancer and a number of other diseases indicate that individuals susceptible to cancer may be less susceptible to these diseases. These associations suggest the possibility of getting an additional (indirect) contribution to longevity increase from eradication or reduction of mortality from cancer. Indeed, those individuals whose lives were saved from cancer deaths could be more resistant to other diseases, which will ultimately result in additional longevity increases. The reality, however, is much more complicated, because the features of survivors after treatment depend on how the cancer treatment has been performed. The problem is well known to specialists in dependent competing risks: the effect of reduction or elimination of a selected risk on other risks depends on how such reduction or elimination was performed.

The connection between cancer and longevity may be modified by trade-offs between cancer and aging, or between cancer and other common diseases, such as AD, CHD, and stroke, among others. One potential biological mechanism underlying the negative correlation among cancer and other diseases could be related to the differential role of apoptosis in the development of these diseases.

Acknowledgements The research reported in this chapter was supported by the National Institute on Aging grants R01AG027019, R01AG030612, R01AG030198, 1R01AG046860, and P01AG043352. The content is solely the responsibility of the authors and does not necessarily represent the official views of the National Institute on Aging or the National Institutes of Health.

References

Anisimov, V. N. (1987). *Carcinogenesis and aging, vol. 1 and 2*. Boca Raton: CRC Press.

Anisimov, V. N. (1998). Physiological function of pineal gland (gerontological aspect). *Russian Physiology Journal, 83*, 1–10.

Anisimov, V. N., Khavinson, V. K., Zavarzina, N. Y., Zabezhinski, M. A., Zimina, O. A., Popovich, I. G., Shtylik, A. V., Malinin, V. V., Morozov, V. G., Arutjunyan, A. V., Oparina,

T. I., & Prokopenko, V. M. (2000). Effect of peptide bioregulators and melatonin on biomarkers of aging, life span and tumor development in mice. *Advances in Gerontology, 4*, 88–96.

Anisimov, V. N., Khavinson, V. K., Mikhalski, A. I., & Yashin, A. I. (2001). Effect of synthetic thymic and pineal peptides on biomarkers of ageing, survival and spontaneous tumour incidence in female CBA mice. *Mechanisms of Ageing and Development, 122*(1), 41–68.

Barinaga, M. (1998). Stroke-damaged neurons may commit cellular suicide. *Science, 281*(5381), 1302–1303.

Bauer, J. H., & Helfand, S. L. (2006). New tricks of an old molecule: Lifespan regulation by p53. *Aging Cell, 5*(5), 437–440.

Bauer, J. H., Poon, P. C., Glatt-Deeley, H., Abrams, J. M., & Helfand, S. L. (2005). Neuronal expression of p53 dominant-negative proteins in adult Drosophila melanogaster extends lifespan. *Current Biology, 15*, 2063–2068.

Blackwell, B. N., Bucci, T. J., Hart, R. W., & Turturro, A. (1995). Longevity, body weight, and neoplasia in ad libitum-fed and diet-restricted C57BL6 mice fed NIH-31 open formula diet. *Toxicologic Pathology, 23*(5), 570–582.

Campisi, J. (2002). Cancer and aging: Yin, yang, and p53. *Science of Aging Knowledge and Environment, 1*, pe1.

Campisi, J. (2003). Cancer and ageing: Rival demons? *Nature Reviews Cancer, 3*(5), 339–349.

CDC. (2007). Trends in health and aging. *Mortality*. http://209.217.72.34/aging/ReportFolders/ReportFolders.aspx. Accessed on Jan 2008.

Chin, L., Artandi, S. E., Shen, Q., Tam, A., Lee, S. L., Gottlieb, G. J., Greider, C. W., & DePinho, R. A. (1999). p53 deficiency rescues the adverse effects of telomere loss and cooperates with telomere dysfunction to accelerate carcinogenesis. *Cell, 97*, 527–538.

Conti, S., Farchi, G., Masocco, M., Toccaceli, V., & Vichi, M. (1999). The impact of the major causes of death on life expectancy in Italy. *International Journal of Epidemiology, 28*(5), 905–910.

Dilman, V. M., & Anisimov, V. N. (1980). Effect of treatment with phenofromin, dyphenylhydantoin or L-DOPA on life span and tumor incidence in C3H/Sn mice. *Gerontology, 26*, 241–245.

Donehower, L. (2002). Does p53 affect organismal aging? *Journal Cellular Physiology, 192*, 23–33.

Dumont, P., Leu, J. I., Della Pietra, A. C., III, George, D. L., & Murphy, M. (2003). The codon 72 polymorphic variants of p53 have markedly different apoptotic potential. *Nature Genetics, 33*, 357–365.

Fisher, M., Dávalos, A., Rogalewski, A., Schneider, R., Ringelstein, E. B., & Schäbitz, W.-R. (2006). Toward a multimodal neuroprotective treatment of stroke. *Stroke, 37*(4), 1129–1136.

García-Cao, I., García-Cao, M., Martín-Caballero, J., Criado, L. M., Klatt, P., Flores, J. M., Weill, J. C., Blasco, M. A., & Serrano, M. (2002). "Super p53" mice exhibit enhanced DNA damage response, are tumor resistant and age normally. *EMBO Journal, 21*(22), 6225–6235.

Golubovski, M. D., Weisman, N. Y., Arbeev, K. G., Ukraintseva, S. V., & Yashin, A. I. (2006). *Experimental Gerontology, 41*(9), 819–827.

Haffty, B. G., & Glazer, P. M. (2003). Molecular markers in clinical radiation oncology. *Oncogene, 22*, 5915–5925.

Hanahan, D., & Weinberg, R. A. (2011). Hallmarks of cancer: The next generation. *Cell, 144*(5), 646–674. doi:10.1016/j.cell.2011.02.013. Review.

Harrison, C. (2007). Neurological diseases: New avenues for stroke treatment. *Nature Reviews Drug Discovery, 6*, 520.

IARC (The International Agency for Research on Cancer [France]). (1965–2003). *Cancer incidence in five continents* (Vol. I–VIII). IARC Science Publication IARC, Lyon.

Ivy, G. O., Rick, J. T., & Murphy, M. P. (1994). Effects of L-deprenyl on manifestations of aging in the rat and dog. *Annals of the New York Academy of Sciences, 717*, 45–59.

James, S. J., Muskhelishvili, L., Gaylor, D. W., Turturro, A., & Hart, R. (1998). Upregulation of apoptosis with dietary restriction: Implications for carcinogenesis and aging. *Environmental Health Perspectives, 106*(Suppl.1), 307–312.

Keyfitz, N. (1977). What difference would it make if cancer were eradicated? An examination of the Taeuber paradox. *Demography, 14*(4), 411–418.

Kim, H. J., Rowe, M., Ren, M., Hong, J. S., Chen, P. S., & Chuang, D. M. (2007). Histone deacetylase inhibitors exhibit anti-inflammatory and neuroprotective effects in a rat permanent ischemic model of stroke: Multiple mechanisms of action. *Journal of Pharmacol and Experimental Therapeutics, 321*, 892–901.

Kitani, K., Kanai, S., Carrillo, M. C., & Ivy, G. H. (1994). (-)Deprenyl increases the life span as well as activities of superoxide dismutase and catalase but not of glutathione peroxidase in selective brain regions in Fisher rats. *Annals of the New York Academy of Sciences, 717*, 60–71.

Kokotailo, R. A., & Hill, M. D. (2005). Coding of stroke and stroke risk factors using international classification of diseases, revisions 9 and 10. *Stroke, 36*, 1776–1781.

Kulminski, A. M., et al. (2011). Trade-off in the effects of the apolipoprotein E polymorphism on the ages at onset of CVD and cancer influences human lifespan. *Aging Cell, 10*, 533–541. doi:10.1111/j.1474-9726.2011.00689.x.

Kulminski, A. M., Culminskaya, I., Arbeev, K. G., Ukraintseva, S. V., Arbeeva, L., & Yashin, A. I. (2013). Trade-off in the effect of the APOE gene on the ages at onset of cardiovascular disease and cancer across ages, gender, and human generations. *Rejuvenation Research, 16*, 28–34. doi:10.1089/rej.2012.1362.

Lee, J. H., Cheon, Y. H., Woo, R. S., Song, D. Y., Moon, C., & Baik, T. K. (2012). Evidence of early involvement of apoptosis inducing factor-induced neuronal death in Alzheimer brain. *Anatomy Cell Biology, 45*(1), 26–37. doi:10.5115/acb.2012.45.1.26. PubMed PMID: 22536549, PubMed Central PMCID: PMC3328738, Epub 2012 Mar 31.

Lemaire, J. (2005). The cost of firearm deaths in the United States: Reduced life expectancied and increased insurance costs. *The Journal of Risk and Insurance, 72*(4), 679–683.

Lim, D. S., Vogel, H., Willerford, D. M., Sands, A. T., Platt, K. A., & Hasty, P. (2000). Analysis of ku80-mutant mice and cells with deficient levels of p53. *Molecular Cell. Biology, 20*, 3772–3780.

Llorca, J., & Delgado-Rodriguez, M. (2001). Competing risks analysis using Markow chains: Impact of cerebrovascular and ischaemic heart disease in cancer mortality. *International Journal of Epidemiology, 30*, 99–101.

Messerli, F. H., Bangalore, S., Torp-Pedersen, C., Staessen, J. A., & Kostis, J. B. (2013). Cardiovascular drugs and cancer: Of competing risk, smallpox, Bernoulli, and d'Alembert. *European Heart Journal, 34*(15), 1095–1098. doi:10.1093/eurheartj/ehs158. Epub 2012 Jul 19.

Migliaccio, E., Giorgio, M., Mele, S., Pelicci, G., Reboldi, P., Pandolfi, P. P., Lanfrancone, L., & Pelicci, P. G. (1999). The p66shc adaptor protein controls oxidative stress response and life span in mammals. *Nature, 402*(6759), 309–313.

Ørsted, D. D., Bojesen, S. E., Tybjaerg-Hansen, A., & Nordestgaard, B. G. (2007). Tumor suppressor p53 Arg72Pro polymorphism and longevity, cancer survival, and risk of cancer in the general population. *Journal of Experimental Medicine, 204*(6), 1295–1301.

Piantanelli, L., Zaia, A., & Rossolini, G. (1994). Influence of L-deprenyl treatment on mouse survival kinetics. *Annals of the New York Academy of Sciences, 717*, 72–78.

Renehan, A. G., Zwahlen, M., Minder, C., O'Dwyer, S. T., Shalet, S. M., & Egger, M. (2004). Insulin-like growth factor (IGF)-I, IGF binding protein-3, and cancer risk: Systematic review and meta-regression analysis. *Lancet, 363*, 1346–1353. doi:10.1016/s0140-6736(04)16044-3.

Rosenberg, G., Angel, I., & Kozak, A. (2005). Clinical pharmacology of DP-b99 in healthy volunteers: First administration to humans. *British Journal of Clinical Pharmacology, 60*(1), 7–16.

Rothenberg, R. B. (1994). Competing mortality and progress against cancer. *Epidemiology, 5*(2), 197–203.

Sheldon, W. G., Bucci, T. J., Hart, R. W., & Turturro, A. (1995). Age-related neoplasia in a lifetime study of ad libitum-fed and food-restricted B6C3F1 mice. *Toxicologic Pathology, 23*(4), 458–476.

Sonntag, W. E., Csiszar, A., de Cabo, R., Ferrucci, L., & Ungvari, Z. (2012). Diverse roles of growth hormone and insulin-like growth factor-1 in mammalian aging: Progress and controversies. *Journals of Gerontology Series A-Biological Sciences and Medical Sciences, 67*, 587–598. doi:10.1093/gerona/gls115.

Stallard, E. (2002). Underlying and multiple cause mortality at advanced ages: United States 1980–1998. *North American Actuarial Journal, 6*(3), 64–87.

Sytkowski, P. A., D'Agostino, R. B., Belanger, A., & Kannel, W. B. (1996). Sex and time trends in cardiovascular disease incidence and mortality: The Framingham Heart study, 1950–1989. *American Journal of Epidemiology, 143*(4), 338–350.

Tabares-Seisdedos, R., & Rubenstein, J. L. (2013). Inverse cancer comorbidity: A serendipitous opportunity to gain insight into CNS disorders. *Nature Reviews Neuroscience, 14*, 293–304. doi:10.1038/nrn3464.

Tabares-Seisdedos, R., et al. (2011). No paradox, no progress: Inverse cancer comorbidity in people with other complex diseases. *Lancet Oncology, 12*, 604–608. doi:10.1016/s1470-2045 (11)70041-9.

ThyagaRajan, S., & Quadri, S. K. (1999). L-deprenyl inhibits tumor growth, reduces serum prolactin, and suppresses brain monoamine metabolism in rats with carcinogen-induced mammary tumors. *Endocrine, 10*, 225–232.

ThyagaRajan, S., Meites, J., & Quadri, S. K. (1995). Deprenyl reinitiated estrous cycles, reduces serum prolactin, and decreases the incidence of mammary and pituitary tumors in old acyclic rats. *Endocrinology, 136*, 1103–1110.

Tyner, S. D., Venkatachalam, S., Choi, J., Jones, S., Ghebranious, N., Igelmann, H., Lu, X., Soron, G., Cooper, B., Brayton, C., Hee Park, S., Thompson, T., Karsenty, G., Bradley, A., & Donehower, L. A. (2002). p53 mutant mice that display early ageing-associated phenotypes. *Nature, 415*, 45–53.

Ukraintseva, S. V., & Yashin, A. I., (2003a). Individual aging and cancer risk: How are they related? *Demographic Research*, *9*(8), 163–196. http://www.demographic-research.org/

Ukraintseva, S. V., & Yashin, A. I., (2003b). Opposite phenotypes of cancer and aging arise from alternative regulation of common signaling pathways. Annals of the New York Academy of Sciences, *1010*, 489–492.

Ukraintseva, S. V., & Yashin, A. I. (2004). Cancer as "Rejuvenescence". *Annals of the New York Academy of Sciences, 1019*, 200–205.

Ukraintseva, S. V., et al. (2010). Trade-offs between cancer and other diseases: Do they exist and influence longevity? *Rejuvenation Research, 13*, 387–396. doi:10.1089/rej.2009.0941.

Ukraintseva, S., Yashin, A., Arbeev, K., Kulminski, A., Akushevich, I., Wu, D., Joshi, G., Land, K. C., & Stallard, E. (2016). Puzzling role of genetic risk factors in human longevity: "risk alleles" as pro-longevity variants. *Biogerontology*, *17*(1), 109–127.

Ungvari, Z., & Csiszar, A. (2012). The emerging role of IGF-1 deficiency in cardiovascular aging: Recent advances. *Journals of Gerontology. Series A, Biological Sciences and Medical Sciences, 67*(6), 599–610. doi:10.1093/gerona/gls072. PubMed PMID: 22451468, PubMed Central PMCID: PMC3348495, Epub 2012 Mar 26. Review.

van Heemst, D., Mooijaart, S. P., Beekman, M., Schreuder, J., de Craen, A. J., Brandt, B. W., Slagboom, P. E., & Westendorp, R. G. (2005). Long Life study group. Variation in the human TP53 gene affects old age survival and cancer mortality. *Experimental Gerontology, 40*(1–2), 11–15. Review.

Vasan, R. S., et al. (2003). Serum insulin-like growth factor I and risk for heart failure in elderly individuals without a previous myocardial infarction: The Framingham Heart Study. *Annals of Internal Medicine, 139*, 642–648.

Vinciguerra, M., Musaro, M., & Rosental, N. (2010). Regulation of muscle atrophy in muscle ageing and disease. In N. Tavernarakis (Ed.), *Protein metabolism and homeostasis in aging* (pp. 211–233). Austin: Springer.

Vogel, H., Lim, D. S., Karsenty, G., Finegold, M., & Hasty, P. (1999). Deletion of Ku86 causes early onset of senescence in mice. *Proceedings of the National Academy of Sciences of the United States of America, 96*, 10770–10775.

Weindruch, R., & Walford, R. (1982). Dietary restriction in mice beginning at 1 year of age: Effect on life-span and spontaneous cancer incidence. *Science, 215*(4538), 1415–1418.

Werner, H., & Bruchim, I. (2012). IGF-1 and BRCA1 signalling pathways in familial cancer. *The Lancet Oncology, 13*, e537–e544. doi:10.1016/s1470-2045(12)70362-5.

WHO. (2007). ICD-10 on-line, version 2007. Cerebrovascular diseases (I60–I69) http://www.who.int/classifications/apps/icd/icd10online/?gi60.htm+i64

Wienke, A., Arbeev, K. G., Locatelli, I., & Yashin, A. I. (2005). A comparison of different bivariate correlated frailty models and estimation strategies. *Mathematical Biosciences, 198*, 1–13.

Yashin, A. I., & Iachine, I. A. (1996). *Surprising dynamics of hazards in the dependent competing risks problem: The case of correlated frailty* (Population studies of aging #18). Odense: Odense University.

Yashin, A. I., Vaupel, J. W., & Iachine, I. A. (1995). Correlated individual frailty: An advantageous approach to survival analysis of bivariate data. *Mathematical Population Studies, 5*(2), 145–159.

Yashin, A. I., Ukraintseva, S. V., De Benedictis, G., Anisimov, V. N., Butov, A. A., Arbeev, K., Jdanov, D. A., Boiko, S. I., Begun, A. Z., Bonafe, M., & Franceschi, C. (2001). Have the oldest old adults ever been frail in the past? A hypothesis that explains modern trends in survival. *Journal of Gerontology: Biological Sciences, 56*(10), B432–B442.

Yashin, A. I., Wu, D., Arbeeva, L. S., Arbeev, K. G., Kulminski, A. M., Akushevich, I., Kovtun, M., Culminskaya, I., Stallard, E., Li, M., & Ukraintseva, S. V. (2015). Genetics of aging, health, and survival: Dynamic regulation of human longevity related traits. *Frontiers in Genetics, 6*, 122. doi:10.3389/fgene.2015.00122.eCollection2015. PubMed PMID: 25918517, PubMed Central PMCID: PMC4394697.

Chapter 5
Factors That May Increase Vulnerability to Cancer and Longevity in Modern Human Populations

Svetlana V. Ukraintseva, Konstantin G. Arbeev, Igor Akushevich, Alexander M. Kulminski, Eric Stallard, and Anatoliy I. Yashin

5.1 Introduction: Economic Prosperity, Longevity, and Cancer Risk

According to the IARC (International Agency for Research on Cancer, WHO) data, SEER, and other epidemiological sources, the overall cancer incidence rate is generally higher in the more developed regions of the world (Figs. 5.1a, 5.1b, 5.1c, 5.2a, 5.2b and 5.3).[1] It has also increased during the second half of the twentieth century around the globe in association with economic progress and the spread of the Western lifestyle (Figs. 5.2 and 5.4) (CI5 1966–2013; Ferlay et al. 2013; Howlader et al. 2014; Ries et al. 2000; Jemal et al. 2008, 2011; Ukraintseva et al. 2008). The higher cancer risk in more developed countries is largely attributed to the higher incidence rates of many common cancer sites (especially, lung, male prostate, female breast, colon, melanoma, kidney, pancreas,

[1] By "cancer risk" in this chapter, we refer to the risk for all cancers combined, if not stated otherwise. In this regard, one needs first to explain why we believe that it is appropriate to discuss common risk factors for overall cancer, considering that "cancer" is the generic term for more than 100 diseases, each characterized by specific etiology, pathogenesis, and tissue localization. The development of cancer has multiple causes, including genetic predisposition, infectious agents, and exposure to chemical or physical carcinogens. If so, then how could we discuss the risk factors for overall cancer? As far as cancer is concerned, this is justified because most cancers share common key features or hallmarks. They include uncontrolled abnormal growth of cells, their potential immortality due to evasion of apoptosis, de-differentiation, and capacity for invasion and metastasis (Ukraintseva and Yashin 2003b; Hanahan and Weinberg 2000, 2011). These common features suggest that there may exist common risk factors for the different cancers. For example, chronic inflammation might be one such factor, because it facilitates almost all cancer features described above (Coussens and Werb 2002; Coussens et al. 2013). In this chapter, we mainly discuss common risk factors for cancer, especially those linked to economic prosperity and the Western lifestyle and those that may influence both individual vulnerability to cancer and aging/ longevity in humans.

A.I. Yashin et al., *Biodemography of Aging*, The Springer Series on Demographic Methods and Population Analysis 40, DOI 10.1007/978-94-017-7587-8_5

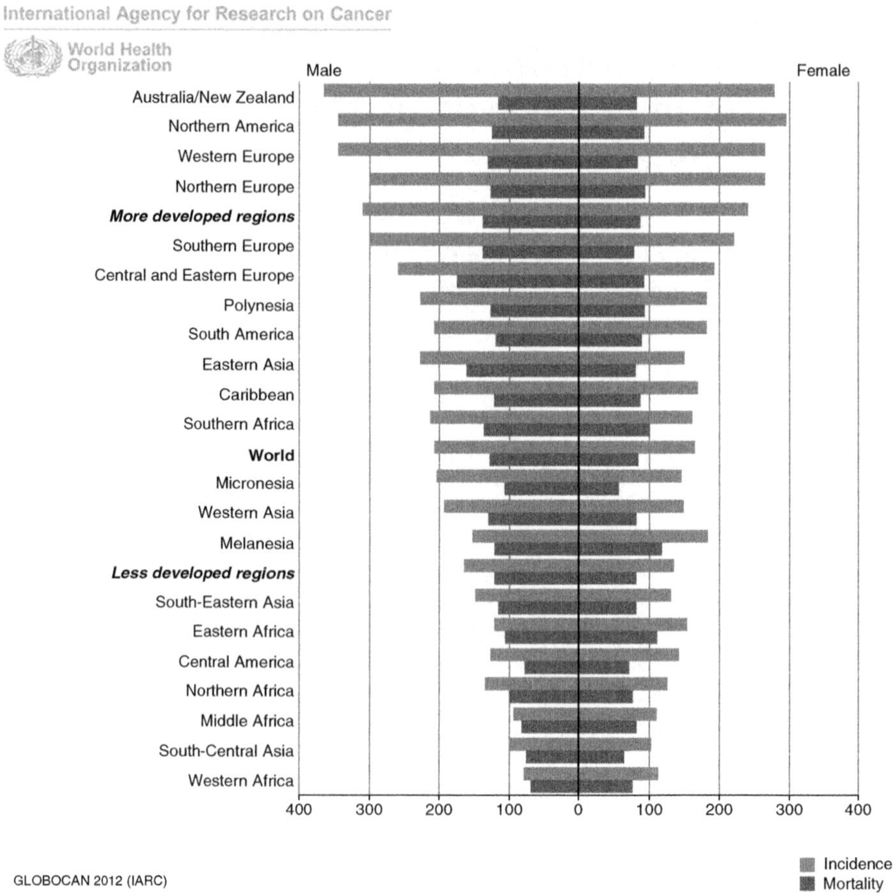

Fig. 5.1a Age-standardized cancer rates (per 100,000): all cancers but skin. More vs. less developed regions GLOBOCAN 2012 (Ferlay et al. 2013), http://globocan.iarc.fr, Section of Cancer Surveillance (accessed 9/15/2014)

leukemia, non-Hodgkin lymphoma (NHL), male bladder, and female thyroid and uterus) in these countries (Fig. 5.1) (CI5 1966–2013; GLOBOCAN 2012).

After a long-term increase, the incidence rates for all cancer sites combined showed a deceleration or a decline starting in the 1990s in some developed countries, especially in the U.S., and mostly in males (CI5 1966–2013; Ries et al. 2000; Ferlay et al. 2013; Howlader et al. 2014; Edwards et al. 2014) (Fig. 5.2a and Fig. 5.2b). In the U.S., the decline was largely due to decreasing rates of some of the common cancer sites (male lung and prostate, female breast and cervix, and colon and stomach in both sexes) (Howlader et al. 2014; Edwards et al. 2014; Jemal et al. 2008, 2013). The reference is usually made to declining exposure to tobacco smoking for lung cancer, use of screening with removal of precancerous polyps for colorectal cancer, controlling the *H. pylori* for stomach

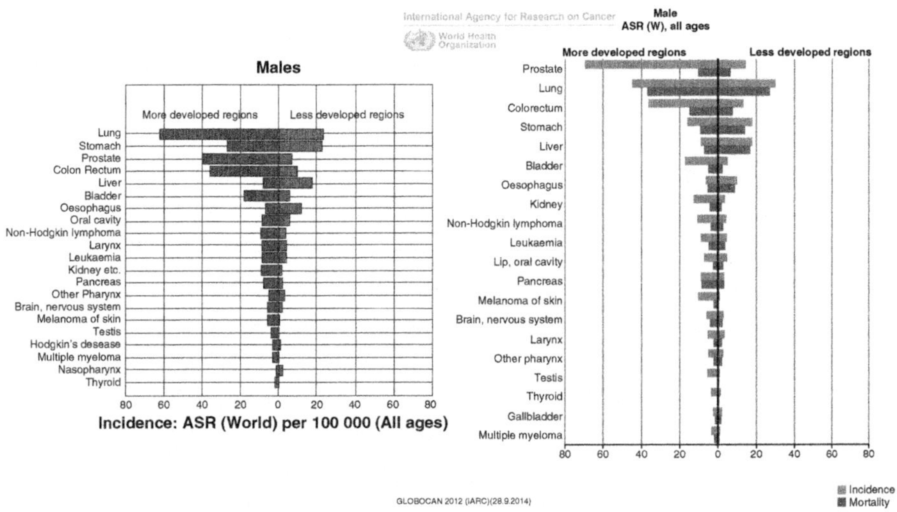

Fig. 5.1b Age-standardized cancer rates (per 100,000): Top 20 cancers. More vs. less developed regions. Males. GLOBOCAN 2012 (Ferlay et al. 2013), http://globocan.iarc.fr (accessed 9/28/2014). Shown in comparison with the past chart from GLOBOCAN 2000 representing 1990s (smaller figure on the *left*).

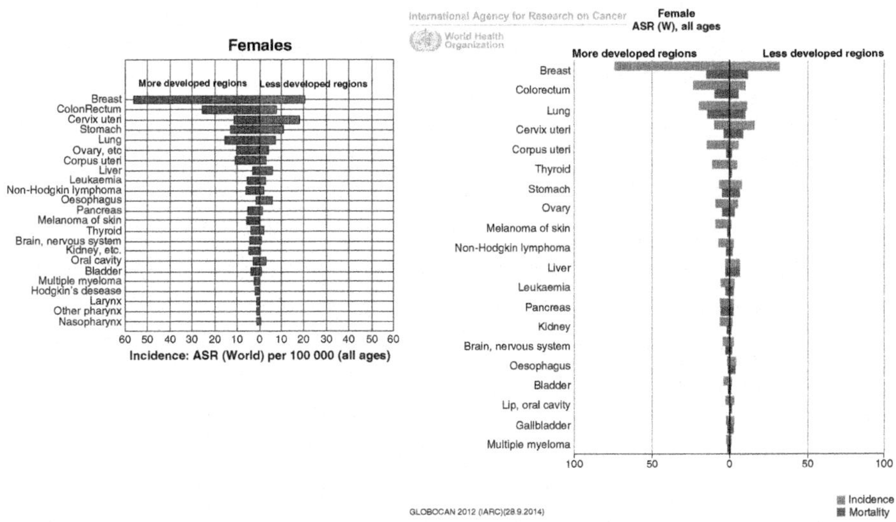

Fig. 5.1c Age-standardized cancer rates (per 100,000): Top 20 cancers. More vs. less developed regions. Females. GLOBOCAN 2012 (Ferlay et al. 2013), http://globocan.iarc.fr (accessed 9/28/2014). Shown in comparison with the past chart from GLOBOCAN 2000 representing 1990s (smaller figure on the *left*)

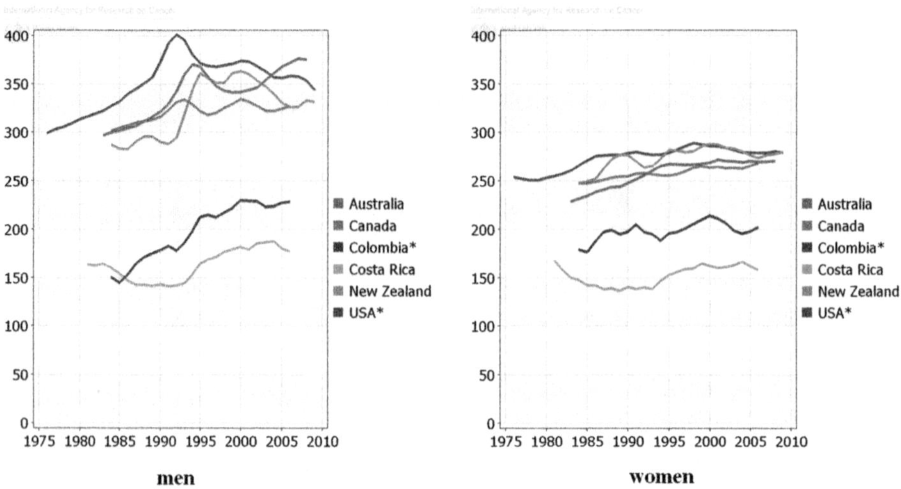

Fig. 5.2a Trends in all-cancer incidence rates in selected countries: age-standardized rate (world) per 100,000: (*a*) men, (*b*) women. GLOBOCAN 2012 (Ferlay et al. 2013), http://globocan.iarc.fr, Section of Cancer Surveillance (accessed 9/15/2014)

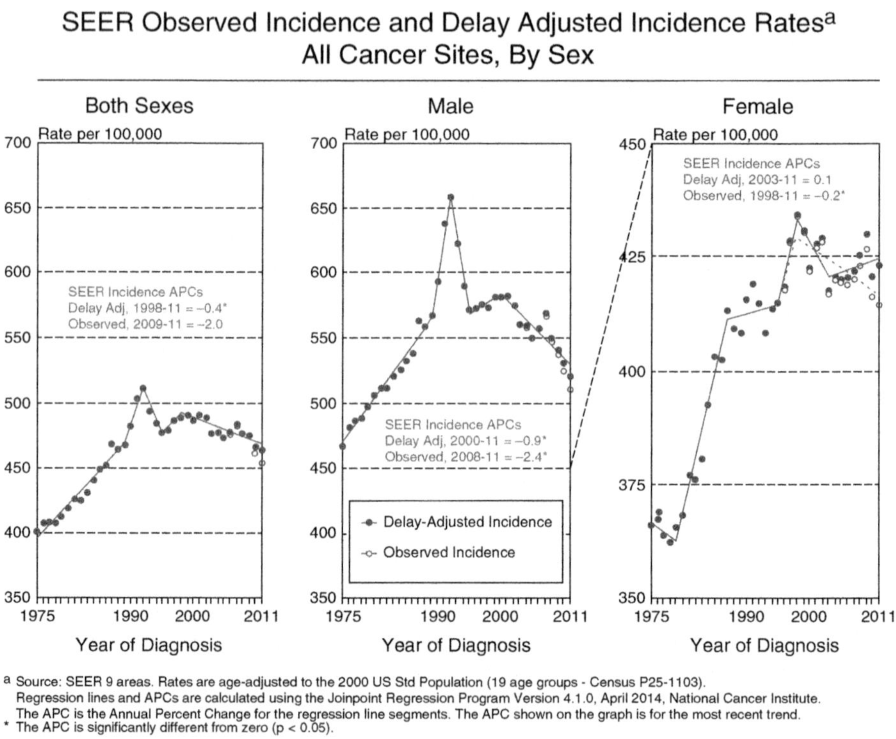

Fig. 5.2b All cancer age-adjusted incidence rates, USA 1975–2011 (SEER: http://seer.cancer.gov/statistics/summaries.html; accessed 9/15/2014)

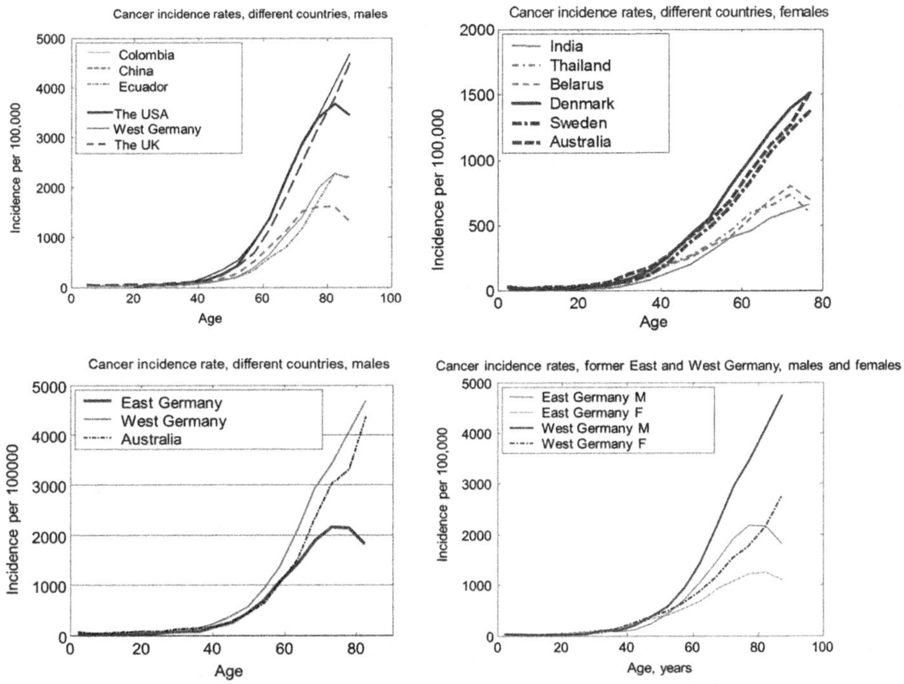

Fig. 5.3 Age-patterns of cancer incidence rate (all sites, but skin), males and females, 1988–1992, average annual (CI5 1966–2013). More developed regions in comparison with less developed ones

cancer and papilloma virus infection for cervix cancer, reduced HRT use for breast cancer, and decreased detection due to recent leveling off of the screening for breast and prostate cancers. However, for the majority of other common cancer sites the incidence rates continued to increase in the U.S., for which explanations have not been fully elucidated (CI5 1966–2013; Jemal et al. 2008, 2013; Ukraintseva et al. 2008; Edwards et al. 2014).

The overall cancer risk has continued to increase in most countries, especially in quickly developing ones, and in countries with a relatively recent history of rapid economic growth and adoption of the Western lifestyle (e.g., Japan, Singapore, and some East European Countries). This increase involved multiple cancer sites, such as thyroid, melanoma, kidney, pancreas, leukemia, liver, myeloma, male NHL, female uterus, and childhood cancer, among others (CI5 1966–2013; Ukraintseva et al. 2008; Jemal et al. 2011; Ferlay et al. 2013; Edwards et al. 2014).

Currently, the overall cancer incidence rate (age-adjusted) in the less developed world is roughly half that seen in the more developed world (Fig. 5.1a, 5.1b, 5.1c) (Jemal et al. 2011; Ferlay et al. 2013). The age curves of the cancer incidence rates displayed in Fig. 5.3 suggest that factors linked to economic prosperity may be more important contributors to the differences in cancer risk between more and less developed regions than ethnic, geographic, and climate related factors.

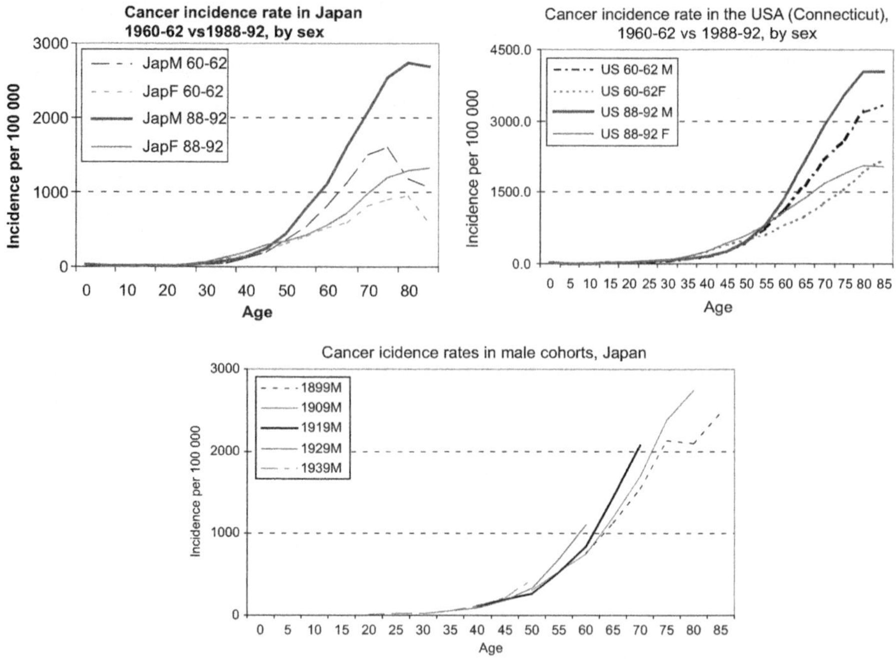

Fig. 5.4 Age-patterns of cancer incidence rates (all sites combined, average annual) in the same country in different time periods or different cohorts (CI5 1966–2013)

For countries with similar levels of economic development but different climate and ethnic characteristics (e.g., West Germany vs. Australia), the cancer rate patterns look much more similar than for the countries that share the same geographic location, climate, and ethnic distribution, but differ in the level of economic development (e.g., East vs. West Germany before reunification). This suggests that different countries may share common factors linked to economic prosperity that could be primarily responsible for the modern increases in overall cancer risk. What are these factors?

Traditional explanations of the higher overall cancer incidence rates in the more developed world involve population aging, improved cancer diagnostics, and elevated exposure to carcinogens. Population aging (increases in the proportion of older people) may indeed partly explain the rise in the global cancer burden (Jemal et al. 2011); however, it cannot explain increases in age-specific cancer incidence rates over time (Fig. 5.4). Improved diagnostics and elevated exposures to carcinogens may explain increases in rates for selected cancer sites, but they cannot fully explain the increase in the overall cancer risk, nor incidence rate trends for most individual cancers (Jemal et al. 2008, 2013).

Could life in affluent societies make people more *susceptible* to cancer, so that the increased overall cancer risk there would be a result of, on average, higher individual vulnerability to cancer rather than merely the result of improved diagnostics and a higher carcinogenic burden?

Human longevity (measured both by increases in life expectancy and increases in proportions of the longest lived people) also dramatically increased during the second half of the twentieth century, along with economic progress and the spread of the Western lifestyle, with a dominance of adult and oldest-old mortality reduction (Vaupel et al. 1998; Canudas-Romo 2010). Typical explanations of the modern rise in human longevity and in the proportion of centenarians, especially in developed countries, include saving lives due to better medical and living conditions (Finch et al. 2014), as well as a possible increase in the fraction of people who biologically age slower (Yashin et al. 2001).

Longevity and the overall cancer risk are thus both higher in affluent societies. Could it be that the same factors linked to economic prosperity and Westernization actually promote both? And could some of these factors also intervene in physiological aging processes in humans? Answering these questions is vital for understanding the mechanisms of both aging and cancer development.

Here we propose that the association between the overall cancer risk and the economic progress and spread of the Western lifestyle could in part be explained by the *higher proportion of individuals more susceptible to cancer* in the populations of developed countries, and discuss several mechanisms of such an increase in the proportion of the vulnerable. We also hypothesize that some of the factors that may enhance susceptibility to cancer in affluent societies may also favor longevity, possibly through beneficial effects on physical and reproductive aging. Below we discuss current evidence in support of this view.

5.2 The Proportion of People Who Are More Susceptible to Cancer May Be Higher in the More Developed World

Improved diagnostics and increased exposure to carcinogenic factors do not appear to fully explain the observed association between cancer risk and economic prosperity. An alternative explanation could be that people in more developed countries may be on average more susceptible to cancer, so that at the same level of a carcinogenic exposure, the more susceptible individuals would end up with a higher risk of cancer than the less susceptible ones. There are epidemiological, demographic, and biological indicators of the possibility of such a scenario, and we will discuss relevant examples in this section. We will combine these examples into several categories according to potential mechanisms connecting economic progress/Westernization with the increase in the proportion more susceptible to cancer in the respective populations. These mechanisms include but are not limited to:

(i) *Improved survival of frail individuals.* Better medical and living conditions in developed countries contribute to "relaxation" of environmental selection and allow for survival of individuals with less efficient immune systems, who would otherwise have died in the past. The less efficient immune systems may in turn be less capable of controlling cancer, making these individuals more vulnerable to it.

(ii) *Avoiding or reducing traditional exposures.* Excessive disinfection and hygiene typical of the developed world can diminish exposure to some factors that were abundant in the past, such as dirt, unsanitary conditions, and diverse microbial communities. Such exposures can be essential for proper training of the immune system, especially in youth, and for forming adequate immune responses later in life. Insufficiently or improperly trained immune systems may be less capable of resisting cancer.

(iii) *Burden of novel exposures.* Some new medicines, cleaning agents, foods, etc., that are not carcinogenic themselves may still affect the natural ways of processing carcinogens in the body, and through this increase a person's susceptibility to established carcinogens. Also, organismal resources are not unlimited, so that the increased burden of novel, even individually harmless, exposures on the xenobiotic processing system may reduce its capacity to address real threats and thus increase the body's vulnerability to cancer.

(iv) *Some of the factors* linked to economic prosperity and the Western lifestyle (e.g., delayed childbirth and food enriched with growth factors) *may antagonistically influence aging and cancer risk.* That is, such factors may attenuate some phenotypes of physical and reproductive aging, and, at the same time, increase the body's vulnerability to cancer. The latter suggests a trade-off between cancer and aging that may contribute to concurrent increases in cancer risk and longevity in modern populations.

5.2.1 Improved Survival of Frail Individuals

More developed countries have higher living standards and quality of medical care. These achievements, however, may lead to a "relaxation" of environmental selection, thereby facilitating the survival of individuals with various genetic and immune deficiencies, who would likely have died in the past. These survivors may contribute to a higher proportion of people who are more vulnerable to diseases (including cancer) in the populations of developed countries. Below are several epidemiological indicators of the possibility of such a scenario.

Improved Survival During Childhood There was a dramatic decline in infant and childhood mortality in developed countries during the last century. For example, the infant mortality rate in the United States was about 6 % of live births in 1935, 3 % in 1950, 1.3 % in 1980, and 0.6 % in 2010. That is, it declined tenfold over the course of 75 years (Singh and van Dyck 2010; Health U.S. 2013). Most newborns in developed countries now reach reproductive age. This decline in mortality was largely due to radical improvements in the survival of infants with birth defects and infectious diseases, particularly with severe respiratory and intestinal infections (Singh and van Dyck 2010). Contrarily, childhood mortality (up to 5 years of age) in some regions of the less developed countries such as India was until recently nearly 20 %. This indicates that the pressure of environmental selection could be much higher in the least developed countries compared to the most developed ones. However, the better survival in the more developed world has its shadowy side. Because almost all children (including those with immunity deficiencies) survive, the proportion of the children who are inherently more vulnerable could be higher in the more developed countries. This is consistent with a typically higher proportion of children with *chronic* inflammatory immune disorders such as asthma and allergy in the populations of developed countries compared to less developed ones (Pearce and Douwes 2006). People with such disorders may be more susceptible to some cancers (Ukraintseva et al. 2010; Josephs et al. 2013).

Cancer Incidence in Countries with Shorter and Longer Histories of Economic Growth If improved living conditions do facilitate survival of people who are more susceptible to cancer, then developed countries with shorter histories of economic prosperity should have lower overall cancer incidence rates than countries with a longer history of economic growth, particularly at old ages. This is because the older individuals in the rapidly developing countries have experienced an improved quality of life only recently, whereas they faced more difficult living conditions earlier in their life. In such circumstances, robust individuals were more likely to survive the environmental selection and reach old age. This should result in a lower proportion of individuals who are susceptible to cancer among the elderly in recently developed countries compared to countries with longer histories of economic growth. Figure 5.5 supports this prediction. It shows that the

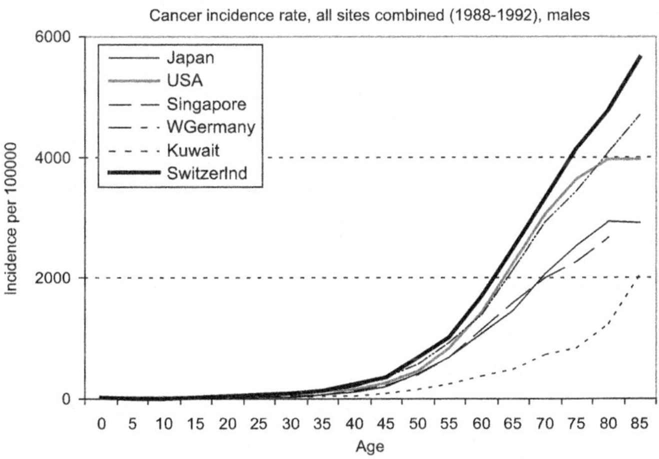

Fig. 5.5 Comparison of age trajectories for the overall cancer incidence rate among developed countries with different histories of economic development (CI5 1966–2013). One can see that countries with more recently developed economies (Japan, Singapore, and Kuwait) have lower cancer incidence rates than countries with the longer history of economic growth, such as the USA, Switzerland, and West Germany, especially at old ages

age-specific cancer incidence rates in Japan, Singapore, and Kuwait (the "younger" developed countries) are lower than in the United States, the United Kingdom, and Switzerland (the "older" developed countries), especially at older ages, despite the similar quality of cancer diagnostics in these countries nowadays.

5.2.2 Avoiding or Reducing Traditional Exposures

The better living conditions in developed countries have a downside in excessive hygiene and body cleansing. Excessive disinfection and hygiene may prevent or diminish some exposures that were abundant in the past, such as dirt, unsanitary conditions, or diverse microbial communities, among others. Until recently, these exposures were an inherent part of human living and our immune system learned to develop by interacting with them (Sing and Sing 2010). Over-reduction of such traditional exposures may result in an insufficiently/improperly trained immune system early in life, which could make it less able to resist diseases, including cancer later in life, thus contributing to the increased proportion of vulnerable individuals in adult populations of developed countries. There is accumulating evidence of the important role of these effects in cancer risk.

An earlier study by a National Cancer Institute team suggested that improved public hygiene conditions, as measured by a decreased prevalence of hepatitis A virus infection, were also associated with higher incidence rates of acute lymphoblastic leukemia (ALL) in children (Smith et al. 1998). More recently, it was shown that long-term exposure to microbial endotoxins can stimulate an anti-cancer immune response and reduce the risk of lung cancer by 40 % (with a 20-year lag time), while the lack of such exposure increases the risk (Mastrangelo et al. 2005; Astrakianakis et al. 2007). Excessive body cleansing during childhood also decreases exposure to helminths (traditional in the past) that is important for the proper development of immunoregulatory mechanisms (Oikonomopoulou et al. 2013). The excessive cleansing and avoiding contact with dirt and dust may lead not only to reduced exposure to particular microorganisms, but also to reduced exposure to a *diversity* of molecules found in dirt and dust, some of which may play a role in immune priming and immune system development in the long term (Sing and Sing 2010).

A number of studies have connected excessive disinfection and lack of antigenic stimulation (especially in childhood) of the immune system in Westernized communities with increased risks of both chronic inflammatory diseases and cancer (Krämer et al. 1999; Lange et al. 2003; Hajdarbegovic et al. 2012; Oikonomopoulou et al. 2013; Francescone et al. 2014; Sheflin et al. 2014). For example, it was shown that some changes in traditional exposures may lead to microbial dysbiosis in the human body, which in turn may promote chronic inflammation (e.g., in the gut) and favor cancer development (Sheflin et al. 2014). Since chronic inflammation plays an important role in cancer development (Coussens and Werb 2002; Coussens et al. 2013), these studies warrant deeper

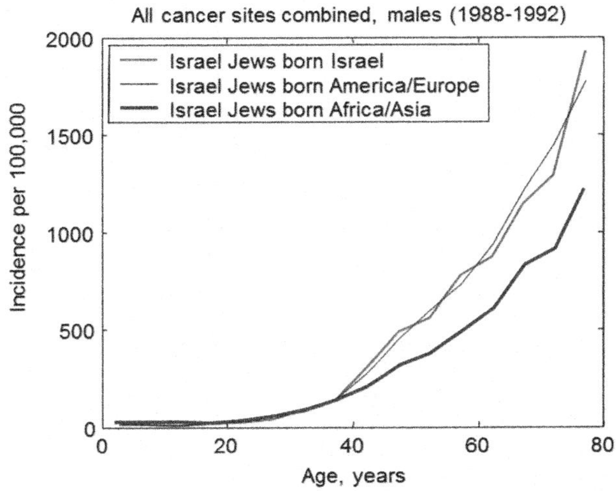

Fig. 5.6 Comparison of age trajectories of the overall cancer incidence rate among Jews born in more and less developed regions of the world (CI5 1966–2013)

research towards understanding the impact of reduced traditional exposures on cancer development through inflammatory mechanisms.

Differences in cancer rates among migrants to the same country provide additional epidemiological indicators of the importance of early environmental exposures in shaping susceptibility to cancer later in life. The IARC data on migrants to Israel (CI5I 1966–2013) allow for comparison of the age trajectories of cancer incidence rates between adult Jews who live in Israel but were born in other countries (Fig. 5.6).

The age curves of cancer incidence in Fig. 5.6 show that Jews born in less developed regions (Africa and Asia) have overall lower cancer risk than those born in the more developed regions (Europe and America). The discrepancy is unlikely to be due to differences in cancer diagnostics because at the moment of diagnosis all these people were citizens of the same country with the same standard of medical care. These results suggest that surviving childhood and growing up in a less developed country with diverse environmental exposures might help form resistance to cancer that lasts even after moving to a high risk country.

5.2.3 Burden of Novel and Nontraditional Exposures

Many behavioral and dietary habits, new medicines, foods, and chemicals, are typical of economically prosperous countries but not common in the less developed ones. Some of the novel and non-traditional exposures associated with the economic prosperity and Western lifestyle, not being formally carcinogenic, might affect the natural ways of processing carcinogens in the body, or favor chronic inflammation, and through this increase a person's susceptibility to cancer. Other factors that are not carcinogenic when considered individually, together might create an excessive burden on the body's xenobiotic processing system. This burden may reduce the capacity of this system to address real threats, and thus increase the body's susceptibility to cancer. Here we present some recent evidence in support of this view. Many of the relevant examples are provided by the IARC Monographs on the Evaluation of Carcinogenic Risks to Humans (IARC Monographs 1972–2014), a valuable WHO resource.

Western Pattern of Food Consumption The Western pattern of food consumption is characterized by a high content of animal protein (from meat, eggs, milk, cheese, etc.), fat, and purified sugar, as well as a low content of crude plants and grains in the everyday diet. A number of human and animal studies suggest a causal connection between the spread of Western dietary habits and changes in vulnerability to some cancers (Watson and Collins 2011; Mosby et al. 2012). For example, individuals with relatively high consumption of animal protein may face a significantly increased risk of colon cancer (Willett 1989; Ananthakrishnan et al. 2015; Carr et al. 2015). Kagawa et al. (1978) described the traditional pattern of food consumption in Japan in the past as characterized by a high proportion of crude grain (barley) and a low proportion of any kind of animal protein. Since the 1950s, this diet has been gradually replaced by one that includes a high proportion of protein and a low proportion of barley. Other components of the Japanese diet (e.g., vegetables) did not change much during the same period. The colon cancer incidence rate has been increasing in Japan since then, while the rate of stomach cancer has been decreasing (CI5 1966–2013). The mechanism could involve a trade-off between reduced damage to the stomach and increased microbial dysbiosis. That is, on the one hand, an excessive amount of crude fibers in food can harm the stomach's mucous membrane and promote inflammation, thus potentially increasing the stomach's vulnerability to cancer. On the other hand, a decreased fiber intake may diminish this particular harm, but at the same time it may suppress intestinal motility and thus elevate the risk of colon cancer. This is because the motility prevents the intestine from festering, and the festering creates an environment conducive to the development of microbial dysbiosis in the intestine. The dysbiosis, in turn, may favor carcinogenic production by the colon bacteria, such as *E.coli* (Falk et al. 1998; Parsonnet 1999). An increased intake of meat promotes food festering in the intestine accompanied by the bacterial imbalance and respective increase in internal carcinogenic exposure (Parsonnet 1999). A number of recent studies strongly support the idea that changes in the traditional pattern of food consumption may lead to microbial dysbiosis (both on the skin and in the gut), which in turn may favor cancer development through inflammatory and other mechanisms (Francescone et al. 2014; Sheflin et al. 2014).

New Medicines and Other Chemicals Some pharmaceuticals and dietary supplements that are prevalent in developed but not in developing countries may influence vulnerability to cancer, although they have not been individually shown to be carcinogenic. The following illustration shows how it could happen. A well-studied chemical, benzpyrene, is a non-direct carcinogen. It needs to be metabolically processed in the body before it can become harmful. First, oxidative enzymes (e.g., cytochrome P450) decompose benzpyrene into intermediate products of the metabolism. Several of these products (e.g., phenol) are already carcinogenic. Other substances (e.g., glutathione) bind these carcinogenic products to deactivate them and take them out of the metabolism. If there is the right balance between cytochrome P450 and glutathione in a cell, then the carcinogen is quickly deactivated without harming the organism. However, if the amount of glutathione is not in the right balance with the amount of cytochrome P450, the carcinogen deactivation process is delayed or incomplete. As a result, carcinogenic metabolites may accumulate in the body and increase the chances of developing cancer. The right balance of chemical players in the processing of carcinogens could, therefore, be an important factor in vulnerability to cancer (Diggs et al. 2013). The high burden of new substances on the body's systems processing xenobiotics may disturb the delicate balance of events and processes leading to neutralization of carcinogens and through this to increased vulnerability to cancer. Research on the simultaneous exposures to carcinogenic and non-carcinogenic compounds provides support for this mechanism.

Paracetamol (Acetaminophen, Tylenol, Contac) is a non-prescription antipyretic which has been used extensively in developed countries since 1946. The drug is not classifiable by the IARC for carcinogenicity to humans. However, animal experiments have shown that paracetamol increases the incidence of renal adenomas induced by an established carcinogen, N-nitrosoethyl-N-hydroxyethylamine (IARC Monographs, Vol. 50). That is, paracetamol, being not harmful on its own, may enhance the body's susceptibility to an established carcinogen, so that a lower level of carcinogenic exposure may be required to induce tumor growth in the more susceptible individual.

Antibiotics People who are frequently treated with antibiotics may have decreased diversity of microbial community, especially in the gut and on the skin. The decreased microbial diversity itself was shown to be associated with increased risks of several cancers, most notably with colon cancer (Modi et al. 2014; Ahn et al. 2013). The mechanism could be that the decreased microbial diversity after treatment with antibiotics creates the conditions for bacterial imbalance in the colon. As mentioned above, this imbalance (dysbiosis) may result in the suppression of the bifidobacteria and the promotion of *E.coli* bacteria (Falk et al. 1998). The latter have a propensity to transform normal metabolic products (e.g., bile acids) into internal carcinogens, thereby increasing the risk of colon cancer (Parsonnet, 1999). Antibiotics may also influence the metabolism of external carcinogens. Metronidazole, an antibiotic which can destroy *H. pylori* and decrease the risk of stomach cancer, also increases the incidence of colon cancer in rats, induced by the administration of an established carcinogen (Sloan et al. 1983; IARC Monographs, Suppl. 7). Chloramphenicol, an antibiotic broadly used since the 1950s, increased the incidence of lymphomas induced by an established carcinogen in mice, while the drug alone did not show clear carcinogenic effect (IARC Monographs, Vol. 50).

On the other hand, antibiotics may sometimes reduce cancer risks. This may be particularly relevant to stomach cancer, whose risk is typically lower in the more developed world. A number of studies discussed in the IARC Monographs on the Evaluation of Carcinogenic Risks to Humans (1972–2014) established an association between seropositivity for the *H. pylori* bacteria and stomach cancer. An estimate of the relative risk was about four times the natural risk of this cancer (IARC Monographs, Vol. 61). A possible mechanism involves the cancer promoting effects of chronic inflammation which accompanies the infection. The prevalence of *H. pylori* infection is substantially lower in developed countries than in developing ones. In both, the prevalence is higher in the lower socioeconomic classes. A progressive reduction in the rate of this infection in successive birth cohorts in the developed countries (IARC Monographs, Vol. 61) is held to be the result of improved hygiene and the spread of antibiotics which can destroy *H. pylori* bacteria. Antibiotic treatment may therefore decrease the risk of stomach cancer. So, the increase in colon cancer and the decrease in stomach cancer risks, which occurred along with economic progress, could in part be explained not only by the change in food patterns, but also by the population-wide exposure to antibiotics.

Hormone Replacement Therapy (HRT) Menopausal and early postmenopausal HRT with various combinations of estrogen and progestin is more common in developed than in developing countries. At the peak of its use in 1999, approximately 20 million women in the developed world used HRT, including about half of all women aged 50–65 years in the U.S. HRT is thought to be in part responsible for differences in incidence rates of female hormone-dependent cancers between more and less developed countries (IARC Monographs, Vol. 72, Vol. 91). Postmenopausal HRT use in the U.S. has dropped since 2002, particularly for continuous HRT, following the report of adverse effects by the Women's Health Initiative's estrogen plus progestin trial. Prescriptions for HRT declined from 61 million prescriptions in 2001 to 21 million in 2004. This was followed by a decline in the risk of estrogen-receptor-positive breast cancer. For example, the age-adjusted incidence rate of breast cancer in women who were 50 years of age or older fell 6.7 % in the United States in 2003 (IARC Monographs, Vol. 91; Ravdin et al. 2007; Ukraintseva et al. 2008).

Oral Contraceptives Another female hormonal treatment, oral contraception for pregnancy prevention, is even more prevalent than postmenopausal therapy. Oral contraceptives usually include both estrogen and progesterone. This treatment has been popular in developed countries since the 1960s. Today, worldwide, more than 100 million women, an estimated 10 % of all women of reproductive age, use combined hormonal contraceptives. Current use of these drugs is greatest in developed countries (16 %) and is lower in developing ones (6 %). They have been shown to increase the risk of breast and liver cancer, while being protective against endometrial and ovarian cancers in women (IARC monographs, Vol. 91). Starting in the 1970s, the decline in female endometrial and ovarian cancer incidence rates, as well as the increase in incidence rates of breast and liver cancers in the United States, could in part be related to the nationwide exposure of American women to hormonal contraceptives.

The mechanism of the effect of estrogen, alone and in combination with progestin, on the female organism is very complex and depends on age, stage of ontogeny, and target tissue. Cancer promoting properties could in part be related to the estradiol-associated stimulation of growth hormone release (Veldhuis et al. 2004), which could potentially lead to excessive cell proliferation, while cancer protective properties could be linked to increased regenerative potential and competitive ability of normal host cells surrounding a transformed cell or latent tumor (Ukraintseva and Yashin 2003b, 2005) or to other mechanisms. What is clear is that HRT and the estrogen-progestogen contraceptives are not simply carcinogens but may differentially influence susceptibility to cancer in different tissues and periods of life.

Household Chemicals Chemicals used at home or in small businesses, such as components of plasticware, cleaning agents, flame retardants, and others, are normally tested for carcinogenic properties before market introduction, and therefore are unlikely to be directly carcinogenic. However, the pre-market testing usually does not take into account that the products may occasionally be consumed with drink or food, or through the skin. The problem is that the clearance of such chemicals from surfaces (e.g., glasses, plates, or clothes from washing detergents) can be poor, and their residuals may enter the body and potentially accumulate in amounts sufficient to harm it. Research on this important topic is emerging though still rather limited. It is increasingly recognized that many household chemicals that are common in developed countries may have "*endocrine-disrupting*" effects, meaning that they may interfere with hormonal processes in the body and consequently increase an organism's susceptibility to various health disorders, including cancer, especially in people who lack detoxifying enzymes (e.g., De Coster and van Larebeke 2012; The 2013 Berlaymont Declaration on Endocrine Disrupters). Examples of relevant chemicals that are currently discussed as having endocrine-disrupting properties with possible health consequences, include (but are not limited to): Bisphenol A (BPA), some flame retardants, phthalates, pesticides, solvents, household cleaning products, air fresheners, hair dyes, cosmetics, and sunscreens (e.g., Travier et al. 2002; Zota et al. 2010; De Coster and van Larebeke 2012; The 2013 Berlaymont Declaration on Endocrine Disrupters).

Quantity Versus Quality There may be several biological mechanisms by which exposure to large numbers of new, individually harmless chemicals may increase susceptibility to cancer. Some may involve cumulative effects, "cocktail effects", and synergistic interactions of chemicals (Cedergreen 2014). For example, BPA and estrogen may both interact with estrogen receptors (Gao et al. 2015), so their cumulative effect on these receptors may be more "endocrine-disrupting" and influential to cell proliferation and migration than when they act alone. There may also be less specific mechanisms, in which variety and quantity of the chemicals rather than their qualities play a major role. Indeed, in developed countries, people are typically exposed to *many* new chemicals at once. When these chemicals end up in the body all together, they may create an excess burden on the body's xenobiotic processing system. We speculate that the unusually large numbers of unfamiliar compounds entering the xenobiotic processing system may reduce its efficiency to respond to real threats, and thus potentially increase the body's vulnerability to cancer, even in absence of particular oncogenic effects of individual compounds. This potential mechanism deserves further investigation.

5.3 Some of the Factors Associated with Economic Development and the Western Lifestyle May Antagonistically Influence Aging and Vulnerability to Cancer

5.3.1 Cancer and Aging: A Trade-Off?

One paradoxical feature of economic progress is the concurrent increase in longevity and overall cancer risk in affluent societies. As discussed above, some factors linked to economic development and the Western lifestyle may contribute to increased vulnerability to cancer. Here we show that some of such factors may influence both cancer and aging/ontogeny related traits (e.g., growth, reproductive period, and physical senescence), sometimes antagonistically. Those factors that increase vulnerability to cancer but also attenuate some phenotypes of physical and reproductive aging might contribute to increases in both cancer risk and longevity in modern human populations. Below we provide evidence from human and animal studies supporting such a possibility.

5.3.2 Increased Exposure to Growth Factors

People in developed countries have virtually unlimited access to dense nutritious food, such as meat, fat, and sweets, which may promote growth (increase in height and weight) and affect the metabolism of internal growth factors (Kaklamani et al. 1999; Giovannucci et al. 2004; Larsson et al. 2005; Bujnowski et al. 2011; TeMorenga et al. 2012). Increased height, weight, and levels of internal growth factors are in turn considered to play an important role in cancer development and have been associated with risks of several common cancers (Giovannucci et al. 2004; Renehan et al. 2004; Pollak2008; Batty et al. 2009; Moore et al. 2009; Key et al. 2010; Yoshimoto et al. 2011; Kabat et al. 2013; Mellemkjær et al. 2012; Davis et al. 2011). Potential mechanisms may involve enhanced cell growth and proliferation, and the anti-apoptotic effects of growth factors favoring survival of transformed cells and latent tumors (van der Veeken et al. 2009; Bruchim et al. 2009; Arnaldez and Helman, 2012).

As for aging, the higher levels of internal growth factors, such as IGF-1, estrogens, and some others, have been linked to attenuation of phenotypes of physical senescence, including elderly frailty, sarcopenia, muscle atrophy, heart failure, hip fractures, as well as to better muscle regeneration and healing (e.g., Ruiz-Torres and Soares de Melo Kirzner 2002; Vasan et al. 2003; Roubenoff et al. 2003; Vinciguerra et al. 2010; Conti et al. 2011; Musaro 2012; Thornton 2013; Yeap et al. 2013; Locatelli and Bianchi 2014; Levine et al. 2014).

In rodent models, a *reduction* in growth factors/IGF-I signaling is often correlated with increased longevity (e.g., Bartke et al. 2003). This increase in longevity is largely attributed to reduced cancer risk in the laboratory animals (Ikeno et al. 2009). At the same time, the overexpression of IGF-1 was shown to attenuate the aging-associated cardiac, cerebrovascular, and cognitive decline in older animals (Torella et al. 2004; Trejo et al. 2007; Sonntag et al. 2013). Mouse studies have shown that high protein intake and upregulated GHR-IGF-1 signaling favor the incidence and progression of several cancers (e.g., breast and melanoma tumors); however, a low protein diet had detrimental effects in the very old (Levine et al. 2014).

Overall, data support the possibility of trade-offs between the effects of growth factors on certain cancer and aging-related phenotypes. The pro-cancer properties of growth factors could be related to upregulated growth and proliferation, and anti-apoptotic effects. Anti-aging properties could be related to decelerated muscle loss, better tissue regeneration, and cell survival. Growth factors may especially contribute to cancer risk and mortality *before* the oldest old age, when the incidence rate reaches its peak in the population for most cancers (Ukraintseva and Yashin 2003a; Ukraintseva et al. 2008; Akushevich et al. 2012). They may also contribute to extreme longevity because higher levels of growth factors can be particularly beneficial for survival at very old ages (90+), when physical senescence and related disorders (e.g., heart failure due to muscle atrophy) become leading contributors to mortality risk. Also, several major diseases that could potentially benefit from higher levels of growth factors (such as stroke and AD) reach maximal incidence risk at oldest old ages (90+), when cancer risk is already declining (Ukraintseva and Yashin 2003a; Johnsen et al. 2005; Ukraintseva et al. 2008, 2010; Akushevich et al. 2012; Duron et al. 2012; Dong et al. 2014). This suggests that the timing of exposure may be important for the pro-cancer or pro-longevity effects of growth factors.

5.3.3 Later Menopause

The median age of menopause is generally higher in more developed countries. This age typically varies from 44–49 years in less developed regions (e.g., Mexica, India, Africa) to 50–54 years in more developed ones (e.g., the UK, the

United States) (MacMahon and Worcester 1966; McKinaly et al. 1972; Garrido-Latorre et al. 1996; Kriplani and Banerjee 2005; Dratva et al. 2009). It also tends to be higher in upper socio-economic groups within the same country (e.g., Hardy and Kuh 2005).

Later menopause has been associated with elevated risks of female hormone-related cancers in postmenopausal women, especially with breast, ovarian, and endometrial tumors (Franceschi et al. 1991; Ossewaarde et al. 2005; Mondul et al. 2005; Collaborative group, 2012).

At the same time, later menopause was linked to increased overall survival and longevity in several large studies. For example, the natural menopause that occurred at ages 50–54 vs. 40–44 years was associated with longer survival in large cohorts of Dutch and American women (Ossewaarde et al. 2005; Mondul et al. 2005). This longer survival was accompanied by significantly increased mortality from breast, endometrial, and ovarian cancers, as well as reduced mortality from pneumonia, influenza, and falls (which are common causes of death in the very old). Later menopause was linked to a lower total mortality risk, and to reductions in cardiovascular deaths and heart failure, in several other studies (Snowdon et al. 1989; Jansen et al. 2002; de Kleijn et al. 2002; Rahman et al. 2015). The postponed menopause can also be accompanied by signs of slower physiological aging. For example, it was significantly associated with slower cognitive aging, especially with better memory in naturally postmenopausal elderly women (McLay et al. 2003; Tierney et al. 2013).

5.3.4 *Giving Birth at Later Age*

Age at childbirth has increased across world populations along with economic progress and adoption of the Western lifestyle. This age is typically higher in younger compared to older generations in developed countries (Morabia and Costanza 1999; Savage et al. 2013; Baghurst et al. 2014).

Older age at childbirth (first, last, and average) has been associated with elevated risks of several human cancers in mothers, especially with breast cancer and melanoma (Ewertz et al. 1990; Wohlfahrt and Melbye 2001; Li et al. 2014). For example, women who gave their first birth after age 35 had a risk increase for breast cancer by 40 % compared to mothers who experienced their first birth before age 20. And the relative risk for melanoma was 1.47 in women of the oldest versus youngest age at first birth in a meta-analysis (Ewertz et al. 1990; Li et al. 2014).

On the other hand, an older age at birth, especially that of the last child, shows a positive association with mother's survival toward very old age (Helle et al. 2005; McArdle et al. 2006; Sun et al. 2015). For example, women who had their last child after age 33 had twice the odds for survival to the top 5th percentile of survival for their birth cohorts compared with women who had their last child by age 29 years (Sun et al. 2015). Our unpublished study of 2,401 Danish Twins aged 75+ (LSADT 1995–2000) also revealed that women who were in their 40s at the time of birth of their last child had about 22 % lower death rates after age 75 compared to women who had their last child before age 40. In line with these data, experimental animal studies found that selection for late reproductive ability results in increased longevity of Drosophila flies after just a few generations of such selection (e.g., Rose and Charlesworth 1981).

Older parental age (both maternal and paternal) was also linked to increased cancer risks in the *offspring*, especially of leukemia, brain tumors, and breast and prostate cancers (Hemminki et al. 1999; Zhang et al. 1999; Hodgson et al. 2004). Increasing parental age in developed countries may therefore also contribute to a higher vulnerability to cancer in these countries. Notably, in laboratory rodents, a higher father's age was associated with increased susceptibility to an established carcinogen in mice offspring (Anisimov and Gvardina 1995). It is not clear so far if having an older parent carries anti-aging/pro-longevity benefits for the offspring. One potential mechanism of how the late birth may do both (i.e., increase vulnerability to cancer and attenuate physical aging in offspring) might involve trade-off-like effects of internal growth factors, such as estrogens and IGF-1, on the cancer and aging phenotypes (Yang et al. 2005; Levine et al. 2014). It was shown that children of mothers who were 30–35 years of age at childbirth were taller and displayed a 19 % increase in IGF-I concentrations compared to offspring of mothers who gave birth prior to age 30 (Savage et al. 2013). An association has also been found between older paternal age at birth and longer leukocyte telomere length in the offspring which may indicate postponed replicative senescence (Prescott et al. 2012), but supportive studies are rather limited.

5.4 Conclusion

In this chapter, we discussed factors responsible for the higher cancer risk in the more developed world, and for concurrent increases in cancer risk and longevity in association with economic progress and a Western lifestyle. We suggested that in populations of developed countries, the proportion of individuals more susceptible to cancer may be higher than in less developed regions of the world, and that this may contribute to the typically higher overall cancer risk in such countries.

We provided evidence from human and animal studies suggesting that several factors associated with advanced economic development and a Western lifestyle could favor an increase in the proportion of individuals who are more susceptible to cancer in the respective populations. Such factors include (but are not limited to) dramatic improvements in medical care and living conditions, which may lead to a "relaxation" of environmental selection and improved survival of frail individuals in affluent societies. They may also lead to an insufficiently/inadequately trained immune system early in life, which may make an individual more vulnerable to cancer later in life. Other factors (e.g., some new foods, medicines, and other chemicals) may also increase a person's susceptibility to cancer. Not being directly or individually carcinogenic, these factors may affect the metabolism of established carcinogens, or act synergistically, and through this influence cancer risks. *Quantities* and variety of the chemicals rather than their individual qualities may also play an important role. People in developed countries are exposed to many new chemicals. We hypothesized that the large number of unfamiliar compounds may reduce the efficiency of the xenobiotic processing system, and through this increase the body's vulnerability to cancer, even in the absence of particular oncogenic effects of individual compounds. This potential mechanism deserves further investigation.

Some of the factors associated with economic prosperity and a Western lifestyle may influence both aging and vulnerability to cancer, sometimes oppositely. Current evidence supports a possibility of trade-offs between cancer and aging-related phenotypes (Ukraintseva et al. 2016), which could be influenced by delayed reproduction and exposures to growth factors (Levine et al. 2014; Li et al. 2014; Sun et al. 2015). The latter may be particularly beneficial at very old age. This is because the higher levels of growth factors may attenuate some phenotypes of physical senescence, such as decline in regenerative and healing ability, sarcopenia, frailty, elderly fractures and heart failure due to muscles athrophy. They may also increase the body's vulnerability to cancer, e.g., through growth promoting and anti-apoptotic effects (Pollak 2008; Ukraintseva et al. 2016). The increase in vulnerability to cancer due to growth factors can be compatible with extreme longevity because cancer is a major contributor to mortality mainly before age 85, while senescence-related causes (such as physical frailty) become major contributors to mortality at oldest old ages (85+). In this situation, the impact of growth factors on vulnerability to death could be more deleterious in middle-to-old life (~before 85) and more beneficial at older ages (85+).

The complex relationships between aging, cancer, and longevity are challenging. This complexity warns against simplified approaches to extending longevity without taking into account the possible trade-offs between phenotypes of physical aging and various health disorders, as well as the differential impacts of such trade-offs on mortality risks at different ages (e.g., Ukraintseva and Yashin 2003a; Yashin et al. 2009; Ukraintseva et al. 2010, 2016).

Acknowledgement Research reported in this chapter was partly supported by the National Institute on Aging of the National Institutes of Health under Award Numbers R01AG046860 and P01AG043352. The content is solely the responsibility of the authors and does not necessarily represent the official views of the National Institutes of Health.

References

Ahn, J., Sinha, R., Pei, Z., Dominianni, C., Wu, J., Shi, J., Goedert, J. J., Hayes, R. B., & Yang, L. (2013). Human gut microbiome and risk for colorectal cancer. *Journal of the National Cancer Institute, 105*(24), 1907–1911. doi:10.1093/jnci/djt300. PubMed PMID: 24316595, PubMed Central PMCID: PMC3866154, Epub 2013 Dec 6.

Akushevich, I., Kravchenko, J., Ukraintseva, S., Arbeev, K., & Yashin, A. I. (2012). Age patterns of incidence of geriatric disease in the U.S. elderly population: Medicare-based analysis. *Journal of American Geriatrics Society, 60*(2), 323–327. doi:10.1111/j.1532-5415.2011.03786.x. PubMed PMID: 22283485, PubMed Central PMCID: PMC3288526, Epub 2012 Jan 27.

Ananthakrishnan, A. N., Du, M., Berndt, S. I., Brenner, H., Caan, B. J., Casey, G., Chang-Claude, J., Duggan, D., Fuchs, C. S., Gallinger, S., Giovannucci, E. L., Harrison, T. A., Hayes, R. B., Hoffmeister, M., Hopper, J. L., Hou, L., Hsu, L., Jenkins, M. A., Kraft, P., Ma, J., Nan, H., Newcomb, P. A., Ogino, S., Potter, J. D., Seminara, D., Slattery, M. L., Thornquist, M., White, E., Wu, K., Peters, U., & Chan, A. T. (2015). Red meat intake, NAT2, and risk of colorectal cancer: A pooled analysis of 11 studies. *Cancer Epidemiology, Biomarkers and Prevention, 24*(1), 198–205. doi:10.1158/1055-9965. EPI-14-0897. Epub 2014 Oct 23. PubMed PMID: 25342387; PubMed Central PMCID: PMC4294960.

Anisimov, V. N., & Gvardina, O. E. (1995). N-nitrosomethylurea-induced carcinogenesis in the progeny of male rats of different ages. *Mutation Research, 316*(3), 139–145.

Arnaldez, F. I., & Helman, L. J. (2012). Targeting the insulin growth factor receptor 1. *Hematology/Oncology Clinics of North America, 26*(3), 527–542. doi:10.1016/j.hoc.2012.01.004. PubMed PMID: 22520978, PubMed Central PMCID: PMC3334849, vii-viii. Epub 2012 Feb 28. Review.

Astrakianakis, G., Seixas, N. S., Ray, R., Camp, J. E., Gao, D. L., Feng, Z., Li, W., Wernli, K. J., Fitzgibbons, E. D., Thomas, D. B., & Checkoway, H. (2007). Lung cancer risk among female textile workers exposed to endotoxin. *Journal of the National Cancer Institute, 99*(5), 357–364.

Baghurst, P., Robson, S., Antoniou, G., Scheil, W., & Bryce, R. (2014). The association between increasing maternal age at first birth and decreased rates of spontaneous vaginal birth in South Australia from 1991 to 2009. *The Australian & New Zealand Journal of Obstetrics & Gynaecology, 54*(3), 237–243. doi:10.1111/ajo.12182. Epub 2014 Feb 8.

Bartke, A., Chandrashekar, V., Dominici, F., Turyn, D., Kinney, B., Steger, R., & Kopchick, J. J. (2003). Insulin-like growth factor 1 (IGF-1) and aging: Controversies and new insights. *Biogerontology, 4*(1), 1–8. Review.

Batty, G. D., Shipley, M. J., Gunnell, D., Huxley, R., Kivimaki, M., Woodward, M., Lee, C. M., & Smith, G. D. (2009). Height, wealth, and health: An overview with new data from three longitudinal studies. *Economics and Human Biology, 7*(2), 137–152. doi:10.1016/j.ehb.2009.06.004. Epub 2009 Jun 28. Review.

Bruchim, I., Attias, Z., & Werner, H. (2009). Targeting the IGF1 axis in cancer proliferation. *Expert Opinion on Therapeutic Targets, 13*(10), 1179–1192. doi:10.1517/14728220903201702. Review.

Bujnowski, D., Xun, P., Daviglus, M. L., Van Horn, L., He, K., & Stamler, J. (2011). Longitudinal association between animal and vegetable protein intake and obesity among men in the United States: The Chicago Western Electric Study. *Journal of the American Dietetic Association, 111*(8), 1150–1155.e1. doi:10.1016/j.jada.2011.05.002. PubMed PMID: 21802560; PubMed Central PMCID: PMC3158996.

Canudas-Romo, V. (2010). Three measures of longevity: Time trends and record values. *Demography, 47*(2), 299–312. PubMed PMID: 20608098, PubMed Central PMCID: PMC3000019.

Carr, P. R., Walter, V., Brenner, H., & Hoffmeister, M. (2015). Meat subtypes and their association with colorectal cancer: Systematic review and meta-analysis. *International Journal of Cancer*. doi:10.1002/ijc.29423 [Epub ahead of print].

Cedergreen, N. (2014). Quantifying synergy: A systematic review of mixture toxicity studies within environmental toxicology. *PLoS ONE, 9*(5), e96580. doi:10.1371/journal.pone.0096580.eCollection2014. PubMed PMID: 24794244, PubMed Central PMCID: PMC4008607.

CI5 (1966–2013). *Cancer incidence in five continents* (Vol. I–X) (1966–2013). *IARC Scientific Publications*. Lyon: IARC (International Agency for Research on Cancer). http://ci5.iarc.fr/

CI5plus/Default.aspx, http://ci5.iarc.fr/CI5I-X/Pages/references.aspx, Last assessed on 28 Sept 2014 (The reference time period for Volumes I–X: 1965–2007).

Collaborative Group on Hormonal Factors in Breast Cancer. (2012). Menarche, menopause, and breast cancer risk: Individual participant meta-analysis, including 118 964 women with breast cancer from 117 epidemiological studies. *Lancet Oncology, 13*(11), 1141–1151. doi:10.1016/S1470-2045(12)70425-4. PubMed PMID: 23084519, PubMed Central PMCID: PMC3488186, Epub 2012 Oct 17.

Conti, E., Musumeci, M. B., De Giusti, M., Dito, E., Mastromarino, V., Autore, C., & Volpe, M. (2011). IGF-1 and atherothrombosis: Relevance to pathophysiology and therapy. *Clinical Science, 120*, 377–402.

Coussens, L. M., & Werb, Z. (2002). Inflammation and cancer. *Nature, 420*(6917), 860–867. Review. PubMed PMID: 12490959; PubMed Central PMCID: PMC2803035.

Coussens, L. M., Zitvogel, L., & Palucka, A. K. (2013). Neutralizing tumor-promoting chronic inflammation: A magic bullet? *Science, 339*(6117), 286–291. doi:10.1126/science.1232227. Review. Erratum in: Science. 2013 Mar 29;339(6127):1522. PubMed PMID: 23329041; PubMed Central PMCID: PMC3591506.

Davis, E., Jacoby, P., de Klerk, N. H., Cole, C., & Milne, E. (2011). Western Australian children with acute lymphoblastic leukemia are taller at diagnosis than unaffected children of the same age and sex. *Pediatric Blood & Cancer, 56*(5), 767–770. doi:10.1002/pbc.22832. Epub 2011 Jan 18.

De Coster, S., & van Larebeke, N. (2012). Endocrine-disrupting chemicals: Associated disorders and mechanisms of action. *Journal of Environment Public Health, 2012*, 713696. PubMed PMID: 22991565, PubMed Central PMCID: PMC3443608, Epub 2012 Sep 6. Review.

De Kleijn, M. J., van der Schouw, Y. T., Verbeek, A. L., Peeters, P. H., Banga, J. D., & van der Graaf, Y. (2002). Endogenous estrogen exposure and cardiovascular mortality risk in postmenopausal women. *American Journal of Epidemiology, 155*(4), 339–345.

Diggs, D. L., Myers, J. N., Banks, L. D., Niaz, M. S., Hood, D. B., Roberts, L. J., 2nd, & Ramesh, A. (2013). Influence of dietary fat type on benzo(a)pyrene [B(a)P] biotransformation in a B(a)P-induced mouse model of colon cancer. *Journal of Nutrition and Biochemistry, 24*(12), 2051–2063. doi:10.1016/j.jnutbio.2013.07.006. PubMed PMID: 24231098, PubMed Central PMCID: PMC3904801.

Dong, X., Chang, G., Ji, X. F., Tao, D. B., & Wang, Y. X. (2014). The relationship between serum insulin-like growth factor I levels and ischemic stroke risk. *PLoS ONE, 9*(4), e94845. doi:10.1371/journal.pone.0094845.eCollection2014. PubMed PMID: 24728374, PubMed Central PMCID: PMC3984250.

Dratva, J., Gómez Real, F., Schindler, C., Ackermann-Liebrich, U., Gerbase, M. W., Probst-Hensch, N. M., Svanes, C., Omenaas, E. R., Neukirch, F., Wjst, M., Morabia, A., Jarvis, D., Leynaert, B., & Zemp, E. (2009). Is age at menopause increasing across Europe? Results on age at menopause and determinants from two population-based studies. *Menopause, 16*(2), 385–394. doi:10.1097/gme.0b013e31818aefef.

Duron, E., Funalot, B., Brunel, N., Coste, J., Quinquis, L., Viollet, C., Belmin, J., Jouanny, P., Pasquier, F., Treluyer, J. M., Epelbaum, J., le Bouc, Y., & Hanon, O. (2012). Insulin-like growth factor-I and insulin-like growth factor binding protein-3 in Alzheimer's disease. *Journal for Clinical Endocrinology and Metabolism, 97*(12), 4673–4681. doi:10.1210/jc.2012-2063. Epub 2012 Sep 26.

Edwards, B. K., Noone, A.-M., Mariotto, A. B., Simard, E. P., Boscoe, F. P., Henley, S. J., Jemal, A., Cho, H., Anderson, R. N., Kohler, B. A., Eheman, C. R., & Ward, E. M. (2014). Annual Report to the Nation on the status of cancer, 1975–2010, featuring prevalence of comorbidity and impact on survival among persons with lung, colorectal, breast, or prostate cancer. *Cancer, 120*, 1290–1314. doi:10.1002/cncr.28509.

Ewertz, M., Duffy, S. W., Adami, H. O., Kvåle, G., Lund, E., Meirik, O., Mellemgaard, A., Soini, I., & Tulinius, H. (1990). Age at first birth, parity and risk of breast cancer: A meta-analysis of 8 studies from the Nordic countries. *International Journal of Cancer, 46*(4), 597–603.

Falk, P. G., Hooper, L. V., Midtvedt, T., & Gordon, J. I. (1998). Creating and maintaining the gastrointestinal ecosystem: What we know and need to know from gnotobiology. *Microbiology and Molecular Biology Reviews, 62*(4), 1157–1170. PubMed PMID: 9841668, PubMed Central PMCID: PMC98942, Review.

Ferlay, J., Soerjomataram, I., Ervik, M., Dikshit, R., Eser, S., Mathers, C., Rebelo, M., Parkin, D. M., Forman, D., & Bray, F. (2013). *GLOBOCAN 2012 v1.0, cancer incidence and mortality worldwide: IARC CancerBase No. 11 [Internet]*. Lyon: IARC (International Agency for Research on Cancer). Available from: http://globocan.iarc.fr. Accessed on 12/25/2014.

Finch, C. E., Beltrán-Sánchez, H., & Crimmins, E. M. (2014). Uneven futures of human lifespans: Reckonings from gompertz mortality rates, climate change, and air pollution. *Gerontology, 60*(2), 183–188. doi:10.1159/000357672. PubMed PMID: 24401556, PubMed Central PMCID: PMC4023560, Epub 2013 Dec 24.

Franceschi, S., La Vecchia, C., Booth, M., Tzonou, A., Negri, E., Parazzini, F., Trichopoulos, D., & Beral, V. (1991). Pooled analysis of 3 European case-control studies of ovarian cancer: II. Age at menarche and at menopause. *International Journal of Cancer, 49*(1), 57–60.

Francescone, R., Hou, V., & Grivennikov, S. I. (2014). Microbiome, inflammation, and cancer. *Cancer Journal, 20*(3), 181–189. doi:10.1097/PPO.0000000000000048. PubMed PMID: 24855005, PubMed Central PMCID: PMC4112188.

Gao, H., Yang, B. J., Li, N., Feng, L. M., Shi, X. Y., Zhao, W. H., & Liu, S. J. (2015). Bisphenol a and hormone-associated cancers: Current progress and perspectives. *Medicine (Baltimore), 94*(1), e211.

Garrido-Latorre, F., Lazcano-Ponce, E. C., López-Carrillo, L., & Hernández-Avila, M. (1996). Age of natural menopause among women in Mexico City. *International Journal of Gynaecology and Obstetrics, 53*(2), 159–166.

Giovannucci, E., Rimm, E. B., Liu, Y., & Willett, W. C. (2004). Height, predictors of C-peptide and cancer risk in men. *International Journal of Epidemiology, 33*(1), 217–225.

Hajdarbegovic, E., Verkouteren, J., & Balak, D. (2012). Non-melanoma skin cancer: The hygiene hypothesis. *Medical Hypotheses, 79*(6), 872–874. doi:10.1016/j.mehy.2012.09.012. Epub 2012 Oct 13.

Hanahan, D., & Weinberg, R. A. (2000). The hallmarks of cancer. *Cell, 100*(1), 57–70. Review.

Hanahan, D., & Weinberg, R. A. (2011). Hallmarks of cancer: The next generation. *Cell, 144*(5), 646–674. Review.

Hardy, R., & Kuh, D. (2005). Social and environmental conditions across the life course and age at menopause in a British birth cohort study. *BJOG, 112*(3), 346–354.

Health U.S. (2013). National Center for Health Statistics. Health, United States, 2013: With special feature on prescription drugs. Hyattsville, MD. 2014.

Helle, S., Lummaa, V., & Jokela, J. (2005). Are reproductive and somatic senescence coupled in humans? Late, but not early, reproduction correlated with longevity in historical Sami women. *Proceedings of the Biological Sciences, 272*(1558), 29–37. PubMed PMID: 15875567, PubMed Central PMCID: PMC1634941.

Hemminki, K., Kyyrönen, P., & Vaittinen, P. (1999). Parental age as a risk factor of childhood leukemia and brain cancer in offspring. *Epidemiology, 10*(3), 271–275.

Hodgson, M. E., Newman, B., & Millikan, R. C. (2004). Birthweight, parental age, birth order and breast cancer risk in African-American and white women: A population-based case-control study. *Breast Cancer Research, 6*(6), R656–R667. PubMed PMID: 15535848, PubMed Central PMCID: PMC1064078, Epub 2004 Sep 22.

Howlader, N., Noone, A. M., Krapcho, M., Garshell, J., Miller, D., Altekruse, S. F., Kosary, C. L., Yu, M., Ruhl, J., Tatalovich, Z., Mariotto, A., Lewis, D. R., Chen, H. S., Feuer, E. J., & Cronin, K. A. (Eds.). (2014). *SEER cancer statistics review, 1975–2011,* Bethesda: National Cancer Institute. http://seer.cancer.gov/csr/1975_2011/, based on November 2013 SEER data submission, posted to the SEER web site, April 2014. Accessed 15 Sept 2014.

IARC Monographs on the Evaluation of Carcinogenic Risks to Humans. Volumes 1–108, and Supplements 1–7, published during 1972–2014. IARC Working Group on the Evaluation of

Carcinogenic Risks to Humans. Lyon: IARC Press. The link to the monographs on the IARC/WHO web site: http://monographs.iarc.fr/ENG/Monographs/PDFs/index.php. Accessed 5 Jan 2015.

Ikeno, Y., Hubbard, G. B., Lee, S., Cortez, L. A., Lew, C. M., Webb, C. R., Berryman, D. E., List, E. O., Kopchick, J. J., & Bartke, A. (2009). Reduced incidence and delayed occurrence of fatal neoplastic diseases in growth hormone receptor/binding protein knockout mice. *Journals of Gerontology. Series A, Biological Sciences and Medical Sciences, 64*(5), 522–529. doi:10.1093/gerona/glp017. PubMed PMID: 19228785, PubMed Central PMCID: PMC2667132, Epub 2009 Feb 19.

Jansen, S. C., Temme, E. H., & Schouten, E. G. (2002). Lifetime estrogen exposure versus age at menopause as mortality predictor. *Maturitas, 43*(2), 105–112.

Jemal, A., Thun, M. J., Ries, L. A., Howe, H. L., Weir, H. K., Center, M. M., Ward, E., Wu, X. C., Eheman, C., Anderson, R., Ajani, U. A., Kohler, B., & Edwards, B. K. (2008). Annual report to the nation on the status of cancer, 1975–2005, featuring trends in lung cancer, tobacco use, and tobacco control. *Journal of the National Cancer Institute, 100*(23), 1672–1694. doi:10.1093/jnci/djn389. PubMed PMID: 19033571, PubMed Central PMCID: PMC2639291, Epub 2008 Nov 25.

Jemal, A., Bray, F., Center, M. M., Ferlay, J., Ward, E., & Forman, D. (2011). Global cancer statistics. *CA: A Cancer Journal for Clinicians, 61*(2), 69–90. doi:10.3322/caac.20107. Epub 2011 Feb 4. Erratum in: CA Cancer J Clin. 2011 Mar-Apr;61(2):134.

Jemal, A., Simard, E. P., Dorell, C., Noone, A. M., Markowitz, L. E., Kohler, B., Eheman, C., Saraiya, M., Bandi, P., Saslow, D., Cronin, K. A., Watson, M., Schiffman, M., Henley, S. J., Schymura, M. J., Anderson, R. N., Yankey, D., & Edwards, B. K. (2013). Annual Report to the Nation on the Status of Cancer, 1975–2009, featuring the burden and trends in human papillomavirus(HPV)-associated cancers and HPV vaccination coverage levels. *Journal of the National Cancer Institute, 105*(3), 175–201. doi:10.1093/jnci/djs491. Epub 2013 Jan 7. PubMed PMID: 23297039; PubMed Central PMCID: PMC3565628.

Johnsen, S. P., Hundborg, H. H., Sørensen, H. T., Orskov, H., Tjønneland, A., Overvad, K., & Jørgensen, J. O. (2005). Insulin-like growth factor (IGF) I, -II, and IGF binding protein-3 and risk of ischemic stroke. *Journal of Clinical Endocrinology and Metabolism, 90*(11), 5937–5941. Epub 2005 Aug 30.

Josephs, D. H., Spicer, J. F., Corrigan, C. J., Gould, H. J., & Karagiannis, S. N. (2013). Epidemiological associations of allergy. IgE and cancer. *Clinical and Experimental Allergy, 43*(10), 1110–1123. doi:10.1111/cea.12178. Review.

Kabat, G. C., Anderson, M. L., Heo, M., Hosgood, H. D., 3rd, Kamensky, V., Bea, J. W., Hou, L., Lane, D. S., Wactawski-Wende, J., Manson, J. E., & Rohan, T. E. (2013). Adult stature and risk of cancer at different anatomic sites in a cohort of postmenopausal women. *Cancer Epidemiology, Biomarkers and Prevention, 22*(8), 1353–1363. doi:10.1158/1055-9965.EPI-13-0305. Epub 2013 Jul 25.

Kagawa, Y. (1978). Impact of Westernization on the nutrition of Japanese: Changes in physique, cancer, longevity and centenarians. *Preventive Medicine, 7*(2), 205–217.

Kaklamani, V. G., Linos, A., Kaklamani, E., Markaki, I., Koumantaki, Y., & Mantzoros, C. S. (1999). Dietary fat and carbohydrates are independently associated with circulating insulin-like growth factor 1 and insulin-like growth factor-binding protein 3 concentrations in healthy adults. *Journal of Clinical Oncology, 17*(10), 3291–3298.

Key, T. J., Appleby, P. N., Reeves, G. K., Roddam, A. W., & Endogenous Hormones Breast Cancer Collaborative Group. (2010). Insulin-like growth factor 1 (IGF1), IGF binding protein 3 (IGFBP3), and breast cancer risk: Pooled individual data analysis of 17 prospective studies. *Lancet Oncology, 11*(6), 530–542. doi:10.1016/S1470-2045(10)70095-4. PubMed PMID: 20472501, PubMed Central PMCID: PMC3113287, Epub 2010 May 14.

Krämer, U., Heinrich, J., Wjst, M., & Wichmann, H. E. (1999). Age of entry to day nursery and allergy in later childhood. *Lancet, 353*(9151), 450–454.

Kriplani, A., & Banerjee, K. (2005). An overview of age of onset of menopause in northern India. *Maturitas, 52*(3-4), 199–204.

Lange, J. H., Rylander, R., Fedeli, U., & Mastrangelo, G. (2003). Extension of the "hygiene hypothesis" to the association of occupational endotoxin exposure with lower lung cancer risk. *Journal of Allergy and Clinical Immunology, 112*(1), 219–220.

Larsson, S. C., Wolk, K., Brismar, K., & Wolk, A. (2005). Association of diet with serum insulin-like growth factor I in middle-aged and elderly men. *American Journal of Clinical Nutrition, 81*(5), 1163–1167.

Levine, M. E., Suarez, J. A., Brandhorst, S., Balasubramanian, P., Cheng, C. W., Madia, F., Fontana, L., Mirisola, M. G., Guevara-Aguirre, J., Wan, J., Passarino, G., Kennedy, B. K., Wei, M., Cohen, P., Crimmins, E. M., & Longo, V. D. (2014). Low protein intake is associated with a major reduction in IGF-1, cancer, and overall mortality in the 65 and younger but not older population. *Cell Metabolism, 19*(3), 407–417. doi:10.1016/j.cmet.2014.02.006. PubMed PMID: 24606898; PubMed Central PMCID: PMC3988204.

Li, Z., Gu, M., & Cen, Y. (2014). Age at first birth and melanoma risk: A meta-analysis. *International Journal of Clinical and Experimental Medicine, 7*(12), 5201–5209. eCollection 2014. PubMed PMID: 25664022; PubMed Central PMCID: PMC4307469.

Locatelli, V., & Bianchi, V. E. (2014). Effect of GH/IGF-1 on bone metabolism and osteoporsosis. *International Journal of Endocrinology, 2014*, 235060. doi:10.1155/2014/235060. PubMed PMID: 25147565, PubMed Central PMCID: PMC4132406, Epub 2014 Jul 23. Review.

MacMahon, B., & Worcester, J. (1966). Age at menopause. United States – 1960–1962. *Vital Health Statistics Series, 11*(19), 1–20.

Mastrangelo, G., Grange, J. M., Fadda, E., Fedeli, U., Buja, A., & Lange, J. H. (2005). Lung cancer risk: Effect of dairy farming and the consequence of removing that occupational exposure. *American Journal of Epidemiology, 161*(11), 1037–1046.

McArdle, P. F., Pollin, T. I., O'Connell, J. R., Sorkin, J. D., Agarwala, R., Schäffer, A. A., Streeten, E. A., King, T. M., Shuldiner, A. R., & Mitchell, B. D. (2006). Does having children extend life span? A genealogical study of parity and longevity in the Amish. *Journals of Gerontology. Series A, Biological Sciences and Medical Sciences, 61*(2), 190–195.

McKinlay, S., Jefferys, M., & Thompson, B. (1972). An investigation of the age at menopause. *Journal of Biosocial Science, 4*(2), 161–173.

McLay, R. N., Maki, P. M., & Lyketsos, C. G. (2003). Nulliparity and late menopause are associated with decreased cognitive decline. *Journal of Neuropsychiatry and Clinical Neurosciences, 15*(2), 161–167.

Mellemkjær, L., Christensen, J., Frederiksen, K., Baker, J. L., Olsen, A., Sørensen, T. I., & Tjønneland, A. (2012). Leg length, sitting height and postmenopausal breast cancer risk. *British Journal of Cancer, 107*(1), 165–168. doi:10.1038/bjc.2012.244. PubMed PMID: 22677900, PubMed Central PMCID: PMC3389429, Epub 2012 Jun 7.

Modi, S. R., Collins, J. J., & Relman, D. A. (2014). Antibiotics and the gut microbiota. *Journal of Clinical Investigation, 124*(10), 4212–4218. doi:10.1172/JCI72333. Epub 2014 Oct 1.

Mondul, A. M., Rodriguez, C., Jacobs, E. J., & Calle, E. E. (2005). Age at natural menopause and cause-specific mortality. *American Journal of Epidemiology, 162*(11), 1089–1097. Epub 2005 Oct 12.

Moore, S. C., Rajaraman, P., Dubrow, R., Darefsky, A. S., Koebnick, C., Hollenbeck, A., Schatzkin, A., & Leitzmann, M. F. (2009). Height, body mass index, and physical activity in relation to glioma risk. *Cancer Research, 69*(21), 8349–8355. doi:10.1158/0008-5472.CAN-09-1669. PubMed PMID: 19808953, PubMed Central PMCID: PMC2783605, Epub 2009 Oct 6.

Morabia, A., & Costanza, M. C. (1998). International variability in ages at menarche, first livebirth, and menopause. World Health Organization Collaborative Study of Neoplasia and Steroid Contraceptives. *American Journal of Epidemiology, 148*(12), 1195–1205. Erratum in: Am J Epidemiol 1999 Sep 1;150(5):546.

Mosby, T. T., Cosgrove, M., Sarkardei, S., Platt, K. L., & Kaina, B. (2012). Nutrition in adult and childhood cancer: Role of carcinogens and anti-carcinogens. *Anticancer Research, 32*(10), 4171–4192. Review.

Musaro, A. (2012). To the heart of the problem. mIGF-1: Local effort for global impact. *Aging, 4*, 377–378.

Oikonomopoulou, K., Brinc, D., Kyriacou, K., & Diamandis, E. P. (2013). Infection and cancer: Revaluation of the hygiene hypothesis. *Clinical Cancer Research, 19*(11), 2834–2841. doi:10.1158/1078-0432.CCR-12-3661. Epub 2013 Mar 27. Review.

Ossewaarde, M. E., Bots, M. L., Verbeek, A. L., Peeters, P. H., van der Graaf, Y., Grobbee, D. E., & van der Schouw, Y. T. (2005). Age at menopause, cause-specific mortality and total life expectancy. *Epidemiology, 16*(4), 556–562.

Parsonnet, J. (Ed.). (1999). *Microbes and malignancy: Infection as a cause of human cancers*. New York: Oxford University Press. 465 pp. ISBN: 0-19-510401-3.

Pearce, N., & Douwes, J. (2006). The global epidemiology of asthma in children. *The International Journal of Tuberculosis and Lung Disease, 10*(2), 125–132. Review.

Pollak, M. (2008). Insulin and insulin-like growth factor signaling in neoplasia. *Nature Reviews Cancer, 8*(12), 915–928. doi:10.1038/nrc2536. Review. Erratum in: Nat Rev Cancer. 2009 Mar;9(3):224.

Prescott, J., Du, M., Wong, J. Y., Han, J., & De Vivo, I. (2012). Paternal age at birth is associated with offspring leukocyte telomere length in the nurses' health study. *Human Reproduction, 27* (12), 3622–3631. doi:10.1093/humrep/des314. PubMed PMID: 22940768, PubMed Central PMCID: PMC3501241, Epub 2012 Aug 30.

Rahman, I., Åkesson, A., & Wolk, A. (2015). Relationship between age at natural menopause and risk of heart failure. *Menopause, 22*(1), 12–16. doi:10.1097/GME.0000000000000261.

Ravdin, P. M., Cronin, K. A., Howlader, N., Berg, C. D., Chlebowski, R. T., Feuer, E. J., Edwards, B. K., & Berry, D. A. (2007). The decrease in breast-cancer incidence in 2003 in the United States. *New England Journal of Medicine, 356*(16), 1670–1674.

Renehan, A. G., Zwahlen, M., Minder, C., O'Dwyer, S. T., Shalet, S. M., & Egger, M. (2004). Insulin-like growth factor (IGF)-I, IGF binding protein-3, and cancer risk: Systematic review and meta-regression analysis. *Lancet, 363*, 1346–1353.

Ries, L. A., Wingo, P. A., Miller, D. S., Howe, H. L., Weir, H. K., Rosenberg, H. M., Vernon, S. W., Cronin, K., & Edwards, B. K. (2000). The annual report to the nation on the status of cancer, 1973–1997, with a special section on colorectal cancer. *Cancer, 88*(10), 2398–2424.

Rose, M. R., & Charlesworth, B. (1981). Genetics of life history in Drosophila melanogaster. II. Exploratory selection experiments. *Genetics, 97*(1), 187–196. PubMed PMID: 6790341, PubMed Central PMCID: PMC1214383.

Roubenoff, R., Parise, H., Payette, H. A., Abad, L. W., D'Agostino, R., Jacques, P. F., Wilson, P. W., Dinarello, C. A., & Harris, T. B. (2003). Cytokines, insulin-like growth factor 1, sarcopenia, and mortality in very old community-dwelling men and women: The Framingham Heart Study. *American Journal of Medicine, 115*(6), 429–435.

Ruiz-Torres, A., & Soares de Melo Kirzner, M. (2002). Ageing and longevity are related to growth hormone/insulin-like growth factor-1 secretion. *Gerontology, 48*(6), 401–407.

Savage, T., Derraik, J. G., Miles, H. L., Mouat, F., Hofman, P. L., & Cutfield, W. S. (2013). Increasing maternal age is associated with taller stature and reduced abdominal fat in their children. *PLoS ONE, 8*(3), e58869. doi:10.1371/journal.pone.0058869. PubMed PMID: 23527040, PubMed Central PMCID: PMC3604016, Epub 2013 Mar 20.

Sheflin, A. M., Whitney, A. K., & Weir, T. L. (2014). Cancer-promoting effects of microbial dysbiosis. *Current Oncology Reports, 16*(10), 406. doi:10.1007/s11912-014-0406-0. PubMed PMID: 25123079, PubMed Central PMCID: PMC4180221.

Sing, D., & Sing, C. F. (2010). Impact of direct soil exposures from airborne dust and geophagy on human health. *International Journal of Environmental Research and Public Health, 7*(3), 1205–1223. doi:10.3390/ijerph7031205. PubMed PMID: 20617027, PubMed Central PMCID: PMC2872320, Epub 2010 Mar 19. Review.

Singh, G. K., & van Dyck, P. C. (2010). *Infant mortality in the United States, 1935–2007: Over seven decades of progress and disparities*. A 75th Anniversary Publication. Health Resources and Services Administration, Maternal and Child Health Bureau. Rockville: U.S. Department of Health and Human Services. http://www.mchb.hrsa.gov/

Sloan, D. A., Fleiszer, D. M., Richards, G. K., Murray, D., & Brown, R. A. (1983). Increased incidence of experimental colon cancer associated with long-term metronidazole therapy. *American Journal of Surgery, 145*(1), 66–70.

Smith, M. A., Simon, R., Strickler, H. D., McQuillan, G., Ries, L. A., & Linet, M. S. (1998). Evidence that childhood acute lymphoblastic leukemia is associated with an infectious agent linked to hygiene conditions. *Cancer Causes Control, 9*(3), 285–298.

Snowdon, D. A., Kane, R. L., Beeson, W. L., Burke, G. L., Sprafka, J. M., Potter, J., Iso, H., Jacobs, D. R., Jr., & Phillips, R. L. (1989). Is early natural menopause a biologic marker of health and aging? *American Journal of Public Health, 79*(6), 709–714. PubMed PMID: 2729468, PubMed Central PMCID: PMC1349628.

Sonntag, W. E., Deak, F., Ashpole, N., Toth, P., Csiszar, A., Freeman, W., & Ungvari, Z. (2013). Insulin-like growth factor-1 in CNS and cerebrovascular aging. *Frontiers of Aging Neurosciences, 5*(July 2), 27. doi:0.3389/fnagi.2013.00027. eCollection 2013. PubMed PMID: 23847531; PubMed Central PMCID: PMC3698444.

Sun, F., Sebastiani, P., Schupf, N., Bae, H., Andersen, S. L., McIntosh, A., Abel, H., Elo, I. T., & Perls, T. T. (2015). Extended maternal age at birth of last child and women's longevity in the Long Life Family Study. *Menopause, 22*(1), 26–31. doi:10.1097/GME.0000000000000276. PubMed PMID: 24977462, PubMed Central PMCID:PMC4270889.

Te Morenga, L., Mallard, S., & Mann, J. (2012). Dietary sugars and body weight: Systematic review and meta-analyses of randomised controlled trials and cohort studies. *BMJ, 346*, e7492. doi:10.1136/bmj.e7492. Review.

The 2013 Berlaymont Declaration on Endocrine Disrupters http://www.ipcp.ethz.ch/IPCP_Berlaymont.html. Accessed 5 Jan 2015.

Thornton, M. J. (2013). Estrogens and aging skin. *Dermatoendocrinol, 5*(2), 264–270. doi:10.4161/derm.23872. PubMed PMID: 24194966, PubMed Central PMCID: PMC3772914, Review.

Tierney, M. C., Ryan, J., Ancelin, M. L., Moineddin, R., Rankin, S., Yao, C., & MacLusky, N. J. (2013). Lifelong estrogen exposure and memory in older postmenopausal women. *Journal of Alzheimer's Disease, 34*(3), 601–608. doi:10.3233/JAD-122062.

Torella, D., Rota, M., Nurzynska, D., Musso, E., Monsen, A., Shiraishi, I., Zias, E., Walsh, K., Rosenzweig, A., Sussman, M. A., Urbanek, K., Nadal-Ginard, B., Kajstura, J., Anversa, P., & Leri, A. (2004). Cardiac stem cell and myocyte aging, heart failure, and insulin-like growth factor-1 overexpression. *Circulation Research, 94*(4), 514–524. Epub 2004 Jan 15.

Travier, N., Gridley, G., De Roos, A. J., Plato, N., Moradi, T., & Boffetta, P. (2002). Cancer incidence of dry cleaning, laundry and ironing workers in Sweden. *Scandinavian Journal of Work, Environment and Health, 28*(5), 341–348.

Trejo, J. L., Piriz, J., Llorens-Martin, M. V., Fernandez, A. M., Bolós, M., LeRoith, D., Nuñez, A., & Torres-Aleman, I. (2007). Central actions of liver-derived insulin-like growth factor I underlying its pro-cognitive effects. *Molecular Psychiatry, 12*(12), 1118–1128. Epub 2007 Sep 11.

Ukraintseva, S. V., & Yashin, A. I. (2003a). Individual aging and cancer risk: How are they related? *Demo Research, 9–8*, 163–196. doi:10.4054/DemRes.2003.9.8.

Ukraintseva, S. V., & Yashin, A. I. (2003b). Opposite phenotypes of cancer and aging arise from alternative regulation of common signaling pathways. *Annals of the New York Academy of Sciences, 1010*, 489–492. Review.

Ukraintseva, S. V., & Yashin, A. I. (2005). Treating cancer with embryonic stem cells: Rationale comes from aging studies. *Frontiers in Bioscience, 10*, 588–595.

Ukraintseva, S. V., Arbeev, K. G., & Yashin, A. I. (2008). Epidemiology of hormone-associated cancers as a reflection of age. *Advances in Experimental Medicine and Biology, 630*, 57–71. Review.

Ukraintseva, S. V., Arbeev, K. G., Akushevich, I., Kulminski, A., Arbeeva, L., Culminskaya, I., Akushevich, L., & Yashin, A. I. (2010). Trade-offs between cancer and other diseases: Do they exist and influence longevity? *Rejuvenation Research, 13*(4), 387–396. PubMed PMID: 20426618, PubMed Central PMCID: PMC2959185.

Ukraintseva, S., Yashin, A., Arbeev, K., Kulminski, A., Akushevich, I., Wu, D., Joshi, G., Land, K. C., & Stallard, E. (2016). Puzzling role of genetic risk factors in human longevity: "Risk alleles" as pro-longevity variants. *Biogerontology, 17*(1), 109–127. doi:10.1007/s10522-015-9600-1. PubMed PMID: 26306600, PubMed Central PMCID: PMC4724477, Epub 2015 Aug 26.

van der Veeken, J., Oliveira, S., Schiffelers, R. M., Storm, G., van Bergen En Henegouwen, P. M., & Roovers, R. C. (2009). Crosstalk between epidermal growth factor receptor-and insulin-like growth factor-1 receptor signaling: Implications for cancer therapy. *Current Cancer Drug Targets, 9*(6), 748–760. Review.

Vasan, R. S., Sullivan, L. M., D'Agostino, R. B., Roubenoff, R., Harris, T., Sawyer, D. B., Levy, D., & Wilson, P. W. F. (2003). Serum insulin-like growth factor I and risk for heart failure in elderly individuals without a previous myocardial infarction: The Framingham Heart Study. *Annals of Internal Medicine, 139*, 642–648.

Vaupel, J. W., Carey, J. R., Christensen, K., Johnson, T. E., Yashin, A. I., Holm, N. V., Iachine, I. A., Kannisto, V., Khazaeli, A. A., Liedo, P., Longo, V. D., Zeng, Y., Manton, K. G., & Curtsinger, J. W. (1998). Biodemographic trajectories of longevity. *Science, 280*(5365), 855–860. Review.

Veldhuis, J. D., Anderson, S. M., Patrie, J. T., & Bowers, C. Y. (2004). Estradiol supplementation in postmenopausal women doubles rebound-like release of growth hormone (GH) triggered by sequential infusion and withdrawal of somatostatin: Evidence that estrogen facilitates endogenous GH-releasing hormone drive. *Journal of Clinical Endocrinology and Metabolism, 89*(1), 121–127.

Vinciguerra, M., Musaro, A., & Rosenthal, N. (2010). Regulation of muscle atrophy in aging and disease. In N. Tavernarakis (Ed.), *Protein metabolism and homeostasis in aging* (pp. 211–233).

Watson, A. J., & Collins, P. D. (2011). Colon cancer: A civilization disorder. *Digestive Diseases, 29*(2), 222–228. doi:10.1159/000323926. Epub 2011 Jul 5. Review.

Willett, W. (1989). The search for the causes of breast and colon cancer. *Nature, 338*(6214), 389–394. Review.

Wohlfahrt, J., & Melbye, M. (2001). Age at any birth is associated with breast cancer risk. *Epidemiology, 12*(1), 68–73.

Yang, J., Anzo, M., & Cohen, P. (2005). Control of aging and longevity by IGF-I signaling. *Experimental Gerontology, 40*(11), 867–872. Epub 2005 Sep 8. Review.

Yashin, A. I., Ukraintseva, S. V., De Benedictis, G., Anisimov, V. N., Butov, A. A., Arbeev, K., Jdanov, D. A., Boiko, S. I., Begun, A. S., Bonafe, M., & Franceschi, C. (2001). Have the oldest old adults ever been frail in the past? A hypothesis that explains modern trends in survival. *Journals of Gerontology. Series A, Biological Sciences and Medical Sciences, 56*(10), B432–B442. Review.

Yashin, A. I., Ukraintseva, S. V., Akushevich, I. V., Arbeev, K. G., Kulminski, A., & Akushevich, L. (2009). Trade-off between cancer and aging: what role do other diseases play? Evidence from experimental and human population studies. *Mechanisms of Ageing and Development, 130*(1-2), 98–104. PubMed PMID: 18452970, PubMed Central PMCID: PMC2708086.

Yeap, B. B., Paul Chubb, S. A., Lopez, D., Ho, K. K., Hankey, G. J., & Flicker, L. (2013). Associations of insulin-like growth factor-I and its binding proteins and testosterone with frailty in older men. *Clinical Endocrinology (Oxford), 78*(5), 752–759. doi:10.1111/cen.12052.

Yoshimoto, N., Nishiyama, T., Toyama, T., Takahashi, S., Shiraki, N., Sugiura, H., Endo, Y., Iwasa, M., Fujii, Y., & Yamashita, H. (2011). Genetic and environmental predictors, endogenous hormones and growth factors, and risk of estrogen receptor-positive breast cancer in Japanese women. *Cancer Science, 102*(11), 2065–2072. doi:10.1111/j.1349-7006.2011.02047.x. Epub 2011 Aug 24.

Zhang, Y., Kreger, B. E., Dorgan, J. F., Cupples, L. A., Myers, R. H., Splansky, G. L., Schatzkin, A., & Ellison, R. C. (1999). Parental age at child's birth and son's risk of prostate cancer. The Framingham Study. *American Journal of Epidemiology, 150*(11), 1208–1212.

Zota, A. R., Aschengrau, A., Rudel, R. A., & Brody, J. G. (2010). Self-reported chemicals exposure, beliefs about disease causation, and risk of breast cancer in the Cape Cod Breast Cancer and Environment Study: A case-control study. *Environmental Health, 9*, 40. doi:10.1186/1476-069X-9-40. PubMed PMID: 20646273, PubMed Central PMCID: PMC2918587.

Chapter 6
Medical Cost Trajectories and Onset of Age-Associated Diseases

Igor Akushevich, Julia Kravchenko, Konstantin G. Arbeev, Svetlana V. Ukraintseva, Kenneth C. Land, and Anatoliy I. Yashin

6.1 Introduction

The proportion of older adults in the U.S. population is growing. This raises important questions about the increasing prevalence of aging-related diseases, multimorbidity issues, and disability among the elderly population. Aging-related declines in health, in turn, raise questions about the medical costs associated with treatment and rehabilitation and how these can be minimized, thus making evaluation of national trends in the burden of disease and associated health expenditures a major public health concern and an important issue for policymakers and governmental institutions. To forecast such trends, it is important to understand the key factors driving the progression of aging-related diseases and how such progression could result in changes of associated medical costs of governmental health insurance programs such as Medicare and Medicaid. In 2009, 46.3 million people were covered by Medicare: 38.7 million of them were aged 65 years and older, and 7.6 million were disabled (HI and SMI 2010). By 2031, when the baby-boomer generation will be completely enrolled, Medicare is expected to reach 77 million individuals (HI and SMI 2009). Because the Medicare program covers 95 % of the nation's aged population (Klees et al. 2009), the prediction of future Medicare costs based on these data can be an important source of health care planning.

Detailed and comprehensive analyses have been performed to evaluate aggregate spending on Medicare Part A and B programs for the U.S. elderly population in their final years of life (Lubitz 2005; Lubitz and Riley 1993; Miller 2001). The relationships between Medicare costs and disability and morbidity were investigated by Goldman and colleagues (Goldman and RAND Corporation 2004): they developed the Future Elderly Model (FEM) that predicts medical costs and health status for the elderly. However, they did not investigate the characteristics of individual histories of changing health status and the relationships of such changes

A.I. Yashin et al., *Biodemography of Aging*, The Springer Series on Demographic Methods and Population Analysis 40, DOI 10.1007/978-94-017-7587-8_6

to the dynamics of Medicare expenditures as individuals become older. To open new possibilities for forecasting population health and medical costs, studies of the effects of disease onset on individual medical cost trajectories and of the behavior of individual health trajectories in the presence of comorbid and concurrent disorders are required. The results of such studies can help estimate the extent to which cumulative individual medical costs can determine future changes in the elderly patient's health status. New results in this area will open new possibilities for population health and medical cost forecasting, allowing for the development of an empirical base for assessing the impact of new biotechnologies on increasing the years of minimally disabled life (Pardes et al. 1999).

Three essential components (which could be also referred as sub-models) need to be developed to construct a modern model of forecasting of population health and associated medical costs: (i) a model of medical cost projections conditional on each health state in the model, (ii) health state projections, and (iii) a description of the distribution of initial health states of a cohort to be projected (Goldman et al. 2006; Goldman and RAND Corporation 2004; Goldman et al. 2005, 2009). In making medical cost projections, two major effects should be taken into account: the dynamics of the medical costs during the time periods comprising the date of onset of chronic diseases and the increase of medical costs during the last years of life. In this chapter, we investigate and model the first of these two effects. Note that the latter component has been intensively investigated in prior literature (Lubitz 2005; Lubitz and Riley 1993; Miller 2001). In part, we follow our paper (Akushevich et al. 2011b) and use a much more statistically powerful dataset for estimates, perform several new analyses, and extend the discussion of possible forecasting approaches based on our previously developed modeling approach. Patterns of Medicare expenditures for the entire U.S. elderly population and for disability- and comorbidity-specific subpopulations can be estimated from analyses of medical cost trajectories for the time period of health change (e.g., at the date of an onset of a chronic disease). In addition, the approach developed in this chapter generalizes the approach known as "life tables with covariates" (Akushevich et al. 2005; Manton et al. 1992), resulting in a new family of forecasting models with covariates such as comorbidity indexes or medical costs.

In sum, this chapter develops a model of the relationships between individual cost trajectories following the onset of aging-related chronic diseases. The model has demographically interpretable parameters and thus can serve as a building block in constructing a precise and comprehensive forecasting model of medical costs (including Medicare spending) at the population level. The underlying methodological idea is to aggregate the health state information into a single (or several) covariate(s) that can be determinative in predicting the risk of a health event (e.g., disease incidence) and whose dynamics could be represented by the model assumptions. An advantage of such an approach is its substantial reduction of the degrees of freedom compared with existing forecasting models (e.g., the FEM model, Goldman and RAND Corporation 2004).

6.2 Data and Methods

6.2.1 Data

Two datasets from which statistical estimates for older adults at the national level can be made are the Surveillance, Epidemiology, and End Results (SEER) Registry data linked to the Medicare Files of Service Use (SEER-Medicare), and the National Long Term Care Survey (NLTCS-Medicare) which is also linked to the MFSU. These extensive data sources facilitate identification of disease incidence and long-term remission/recovery events through the development and validation of specific computational algorithms.

The SEER-Medicare data is the primary dataset analyzed in this chapter. The expanded SEER registry covers approximately 26 % of the U.S. population. In total, the Medicare records for 2,154,598 individuals are available in the SEER-M including individuals (1) with diagnosed cancers of breast (n = 353,285), colon (n = 222,659), lung (n = 342,961), prostate (n = 448,410), and skin melanoma (n = 101,123); and (2) from a random 5 % sample of Medicare beneficiaries residing in the SEER areas who had none of the above mentioned cancers. For the majority of persons, we have continuous records of use of Medicare services since 1991 (or from the time the person passed age 65 after 1990) to the patient's death. A small fraction of individuals (e.g., new patients diagnosed with cancer in 2003–2005) had Medicare records from 1998. Medicare records are available for each institutional (MedPAR, outpatient, hospice, or home health agency HHA) and non-institutional (carrier-physician-supplier and durable medical equipment providers) claim type.

The second dataset used for analysis—the NLTCS-Medicare data—contains two of the six NLTCS waves: namely, the cohorts of 1994 and 1999. These specific waves were chosen because high quality Medicare follow-up data are available only since 1991 and the complete 5-year follow-up after the NLTCS interview for years later than 1991 is available only for these two waves. The NLTCS contains data on hundreds of variables including age, sex, and (instrumental) activities of daily living (ADL/IADL) allowing for disability measurements. The same data collection agency, the U.S. Census Bureau, was employed for conducting the NLTCS interviews over all of the waves. Hence, the training methods and materials, survey administration and management procedures, field operations, computer processing, and editing procedures were consistent across the surveys. Also, the high response rate (95 %) across all NLTCS waves minimizes the bias in trend estimates. As was found in Akushevich et al. (2011b), the results of interest (i.e., parameters describing medical cost trajectories) are similar for cohorts formed in 1994 and 1999 (this will be discussed further in the Discussion section). The NLTCS uses a sample of individuals drawn from the national Medicare enrollment files. The 1982–2004 NLTCS files include information on 49,258 different individuals, and 34,077 of them were in the 1994–2004 waves. The national population estimates were produced using screener weights released with the NLTCS (for a recent discussion, see Akushevich et al. (2013a)).

6.2.2 Date of Disease Onset Definitions

Unlike mortality, the onset time of chronic disease is difficult to define with high precision due to the large variety of disease-specific criteria for onset/incident case identification (e.g., changing criteria of diagnosis of incidence of acute coronary heart disease or stroke (NIH/NHLBI 2006)) used in clinical practice and epidemiological and population-based analyses. Therefore, there is always some arbitrariness in defining the date of chronic disease onset, and a unified definition of date of onset is necessary for population studies with a long-term follow-up. In this study, the date of disease onset was identified based on information from the Medicare Claims files. The scheme used in this chapter was based on an overview of several approaches to the definition of the disease onset (Nattinger et al. 2004, 2006; Sloan et al. 2003). This unified scheme was applied to all of the diseases considered herein and is useful for comparative analyses of the effects of different diseases on medical costs and appropriate for prediction purposes.

To reconstruct the ages at onsets of all diseases from the Medicare service use data, we began by reconstructing individual medical histories of each of the diseases to be studied using the Medicare files: all records were combined with their respective ICD-9 codes. Twenty diseases representing the major groups of chronic diseases in the elderly were considered: (i) circulatory (myocardial infarction, angina pectoris, heart failure, and stroke), (ii) cancer (breast, prostate, pancreatic, kidney, lung, and colon cancers, and skin melanoma), (iii) neurodegenerative (Parkinson's and Alzheimer's diseases), (iv) endocrine and metabolic (diabetes mellitus), (v) pulmonary (emphysema, and asthma), and (vi) several others (hip fracture, chronic renal diseases, ulcer, and arthritis). The diagnostic ICD-9 codes for these diseases are presented in Table 3.1 of Chap. 3 of this monograph. We excluded from analysis all individuals who had a history of one of the 20 diseases before the date of interview in 1994 or in 1999. The detailed individual records in the Medicare files are available since 1991; therefore, we had a sufficient period of time prior to 1994 or 1999 to reject the prevalence cases. The date of a Medicare record (referred to as "*this record*" below in (i) and (ii)) is identified with the date of onset of an applicable diseases when both conditions listed below are met:

(i) *This record* is the earliest record with one of the ICD codes as a primary diagnosis in one of four Medicare sources (inpatient care, outpatient care, physician services, and skilled nursing facilities);
(ii) In addition to *this record*, there is another record with the same ICD code as the primary diagnosis from one of the four Medicare sources listed in (i), which appeared with a date different from the date of *this record* and no later than 0.3 years after *this record*.

The first condition allows for identifying the first occurrence of a disease code, and the second condition is required for confirmation of the disease presence. This algorithm was used in several prior studies such as a study of recovery after stroke (Yashin et al. 2010), a study of age patterns of age-related (Akushevich et al. 2012) and circulatory (Akushevich et al. 2013a) diseases, a study of medical cost trajectories before and after disease onset (Akushevich et al. 2011b), and a study of the role of behavior factors in cancer risk (Akushevich et al. 2011a). The algorithm was implemented using SAS (SAS Institute, Inc., Cary, NC).

This definition of the age at disease onset completes our operational definition of disease incidence. Since the date of onset of a chronic disease is a quantity not defined as precise as mortality, some assumptions are required to identify the date of onset from individual records collected in administrative data. The specifications used in this chapter (e.g., choice of the four Medicare sources in item (i) and time period of 0.3 year in item (ii)) are in accordance with the general practice of reconstruction of the date at onset from Medicare data (Nattinger et al. 2004, 2006).

6.2.3 Medical Cost Trajectories

For each disease, individuals whose date of disease onset occurred during the follow-up period (i.e., during the 5-year period after the date of interview for NLTCS-Medicare, and until the end of 2005 from the date of enrollment for SEER-Medicare) were selected. Information about two types of cost-related variables is available in Medicare data: the total costs of medical procedures that have been performed and the total Medicare payments for these procedures. The first variable (i.e., total costs) better describes the costs of medical services, and the second variable (i.e., total Medicare payments) is free from biases that could be potentially caused by multiple counting of costs for the same procedure when several bills were submitted but only one bill was paid by Medicare. In this study, we focus on the total Medicare payment. Trajectories of the costs of medical procedures were analyzed by Akushevich et al. (2011b).

For each of the 20 diseases, means and standard errors of the distributions of medical cost spending per month per capita were estimated within 20 months before and after disease onset. The empirical estimates demonstrated that 20 months is a sufficient period of time for disease "stabilization" after disease onset by reaching a plateau in the mean of the medical cost trajectories. These month patterns (or medical cost trajectories) were subject to analysis, mutual comparison, and modeling. All costs were presented in terms of the dollar value from year 2000, adjusted for inflation using the Medical Care Consumer Price Index provided by the Bureau of Labor Statistics (BLS 2009).

6.3 Results

Empirical estimates of the cost trajectories are presented in Fig. 6.1. The shapes of the majority of medical cost trajectories in the time range of 20 months before and after the date of onset of a focal disease have the same structure. They can be

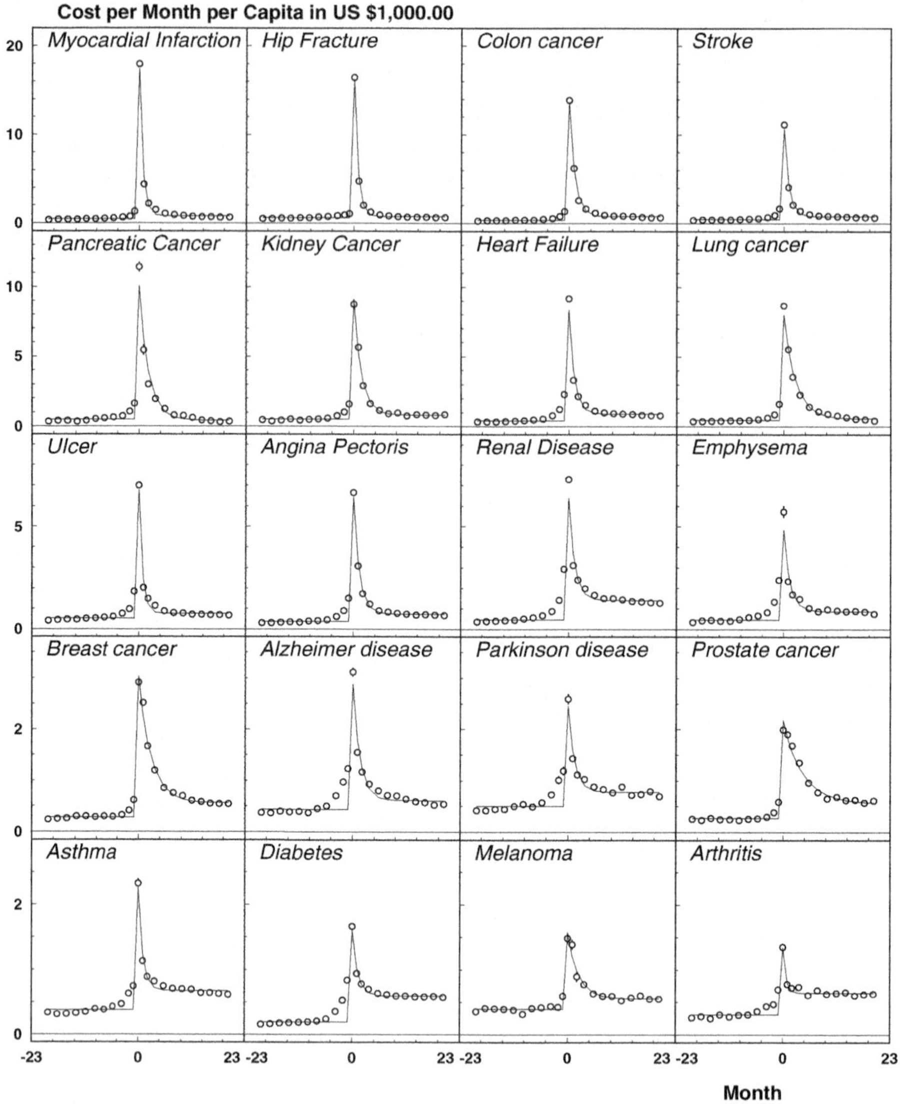

Fig. 6.1 Empirical estimates *(dots)* and model predictions *(solid lines)* of costs per month per capita obtained using SEER-Medicare data. The diseases are ordered according to the cost of onset. Note that the scale of the vertical axes is not the same for different rows of plots

described in terms of the four components sketched in Fig. 6.2. The first one is the *cost of the pre-diagnosis period/pre-focal-disease treatment cost*: this variable likely reflects an *initial comorbidity* (De Groot et al. 2003). The second component is the *cost of the disease onset/cost peak associated with onset of the focal-disease and its treatment*. The third variable characterizes the rate of the reduction of

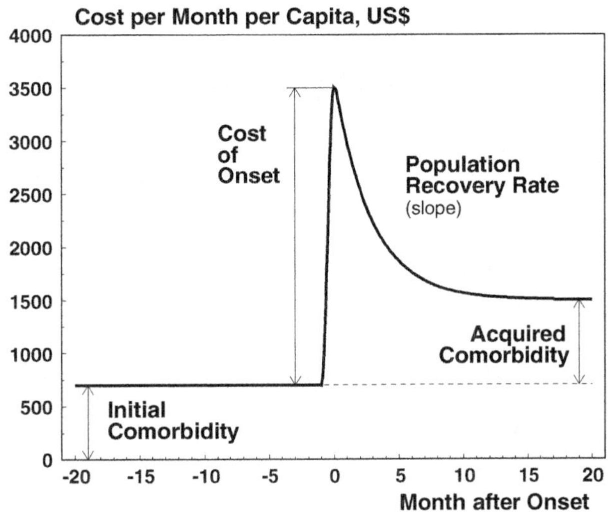

Fig. 6.2 Schematic representation of the pattern of costs per month per capita and notation for the parameters estimated in the four plots below (Fig. 6.3) using the dynamic model of changes in medical costs accompanying the onset of chronic disease

medical expenses associated with a focal disease during the period after the diagnosis was made, that is, the *reduction of medical costs during the period after diagnosis*; this variable could be interpreted as a *population recovery rate*. And, finally, the fourth variable is the difference between the post- and pre-diagnosis cost levels that characterizes an *acquired comorbidity cost* due to a focal disease, that is, the *cost of continuing/follow-up care*.

Based on these empirical patterns, a model for the monthly patterns of the medical cost trajectories with four parameters was constructed as follows. Before the month of disease onset, all trajectories demonstrated a plateau; therefore, this region can be described by a single parameter c associated with the comorbidity of the studied population group. In the month of onset, the trajectories had a sharp peak associated with the cost of onset, which was modeled by a single parameter P. During the months after onset, medical costs decreased and the decline was relatively exponential; therefore, this decline was modeled by an exponential function with a slope r characterizing population recovery in terms of medical costs. The level to which the trajectories converge by leveling-off could also be associated with comorbidity; this level differs from the initial one, c, by a quantity δ that reflects the contribution of the focal disease to an elevated comorbidity level. Thus, the analytical expression for medical cost per month per capita $C(m)$ could be presented as

$$C(m) = c + (\delta + (P - \delta)\exp(-rm))I(m \geq 0) \qquad (6.1)$$

where m is the time (in months) since onset of the focal disease (i.e., time before the onset m is negative), and is the indicator function ($I = 1$ for $m \geq 0$ and $I = 0$ otherwise). The four model parameters correspond exactly to the components presented in Fig. 6.2. Three of them—i.e., the pre- and post-diagnosis costs associated with initial and acquired comorbidity (c and δ), and the cost of the focal disease onset P are all denominated in dollars, while the metric of the slope of the population recovery rate r is months^{-1}.

The model was applied to the data for each of the 20 diseases identified above and estimated by non-linear least squares for the U.S. elderly population of patients who had an onset of a considered disease during the period of observation. The resulting curves are shown in Fig. 6.1. Estimates of the model parameters with the standard errors for all diseases using NLTCS-Medicare and SEER-Medicare are shown in Fig. 6.3. Comparisons of the model estimates allowed us to reveal the properties of the model components described below.

The first component, the pre-disease treatment cost level associated with initial comorbidity, c, describes the plateau in the cost trajectories that appeared before the disease onset. In the majority of trajectories, this is truly a plateau without a significant time trend. Since only individuals with disease onset for 1 of the 20 diseases were selected for constructing cost trajectories, the magnitude of the plateau (i.e., the value of the cost per month per capita) reflects the mean comorbidity index measured in terms of medical costs associated with the diseases. In other words, the magnitude of the estimates of the initial comorbidity depends on how strongly the risk of a focal disease is determined by comorbidity. The stronger this association, the higher the mean comorbidity index is. In Akushevich et al. (2011b), this hypothesis was tested directly by using a separate analysis of each subpopulation with the Charlson comorbidity index (Charlson et al. 1987; Quan et al. 2005) which was estimated for a specific month using Medicare information for the previous 12 months. A positive correlation between the Charlson index and the initial comorbidity was found for all diseases. The strongest associations were detected for stroke, ulcer, lung cancer, and diabetes. Overall, the estimates of the initial comorbidities for trajectories generated by different diseases are similar and, on average, represent mean comorbidity level measured by medical costs.

The second component, P, measures the cost peak at the date of disease onset (i.e., for month zero in Fig. 6.1): its height reflects the disease-specific cost at onset. The diseases shown in Fig. 6.1 are ordered by the decline of this component. High variability of P for a specific disease results from different medical procedures performed at the time of onset (to make a diagnosis and to treat a disease). Therefore, the variability of this component likely reflects variability of disease severity. This hypothesis is tested in the next section. Note that although for some diseases (e.g., asthma, Alzheimer's disease, diabetes, and renal disease) an increment in cost for several months before disease onset is visible (likely due to expenses for pre-diagnosis procedures), in the model this effect is neglected. In further refinements of the model, the cost of onset can be modeled using a normal distribution with a finite variance rather than the single parameter P.

The third component, r, characterizes the rate of reduction of medical expenses associated with a disease during the period after diagnosis (the population recovery rate). This quantity is defined as positive, i.e., the larger the estimate for this

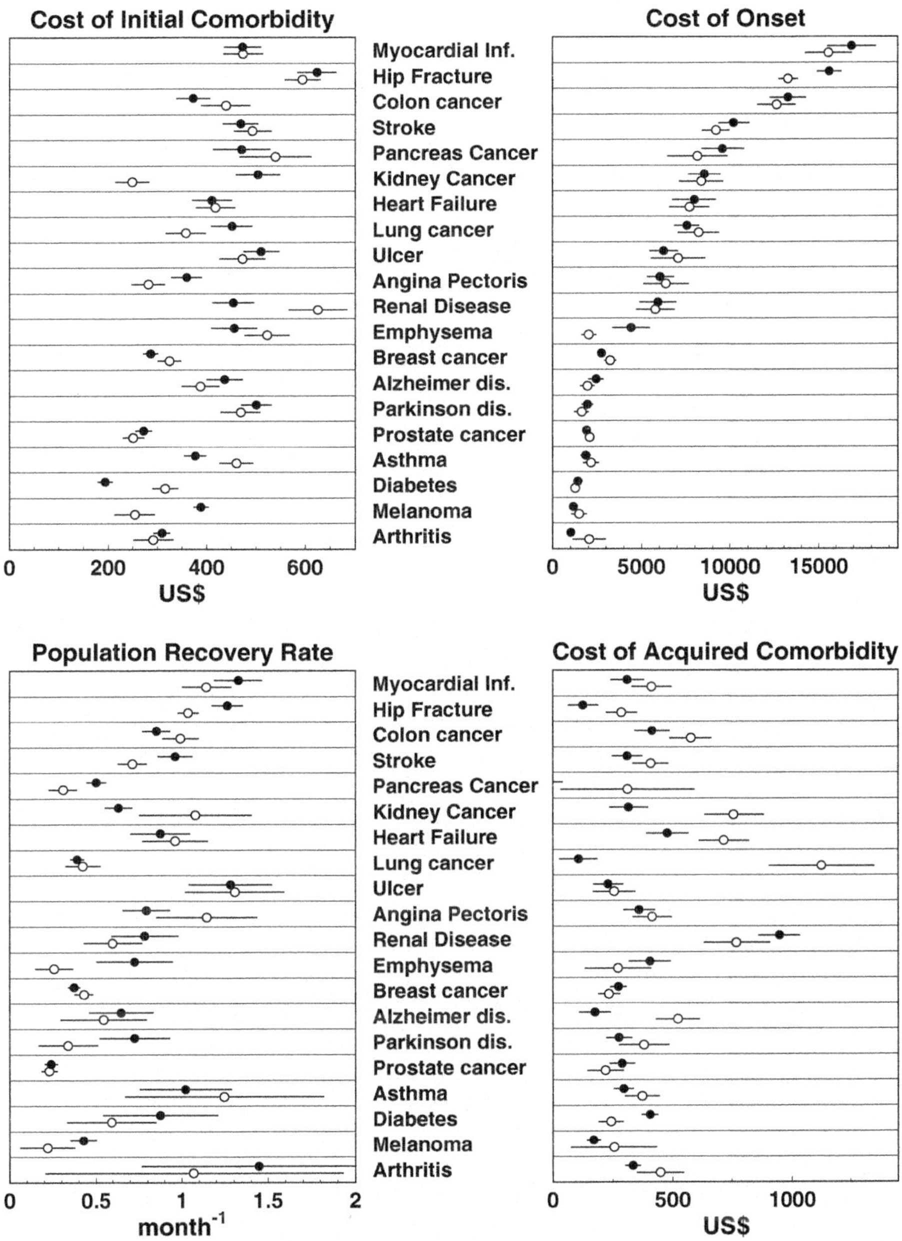

Fig. 6.3 The model parameters (as sketched in Fig. 6.2) were estimated using SEER-Medicare (*closed dots*) and NLTCS-Medicare (*open dots*) within the 20-month period before and after onset of each of the 12 chronic conditions (four lower plots): (i) cost of initial comorbidity in U.S. dollars, i.e., the mean cost per month per capita before onset, (ii) cost of onset in U.S. dollars, i.e., the mean expenditures in the month of onset, (iii) population recovery rate in 1/month, i.e., the speed of approach to a new steady-state in medical expenditures, and (iv) cost of acquired comorbidity in U.S. dollars, i.e., the excess in expenditures in a new steady-state compared to those before disease onset. Horizontal bars denote the standard errors of the nonlinear least squares parameter estimates. The diseases are ordered according to the cost of onset

component the higher the population recovery, or, in other words, the faster the decline in medical expenses associated with the disease. Statistically significant estimates of this component were obtained for all of the 20 diseases (see Fig. 6.3). From a clinical point of view, there are certain diseases (e.g., diabetes, Alzheimer's disease) for which complete clinical recovery cannot be observed at the individual level. For these diseases, a reduction of the medical costs (i.e., positive moderate effect of r) could be explained by the costs of medical procedures around the time of diagnosis and the partial contribution of acute events at the disease onset that required specific treatment. A high variability in estimates of the component r was detected for asthma, arthritis, diabetes, Parkinson's disease, and Alzheimer's disease: these diseases are primarily chronic and can be defined as "permanent conditions with nonreversible pathologic alterations" that generally cannot be cured but rather can have periods of short- or/and long-term remission (CDC 2004).

The fourth component, δ, represents the cost of continuing/followup care, which we termed the acquired comorbidity cost resulting from the onset of a focal disease (measured by the difference between post- and pre-diagnosis cost levels). As can be seen in Fig. 6.3, this component is disease-specific. For some diseases (e.g., renal disease), the estimate of δ can exceed the pre-disease cost level by a factor 2. For the majority of the other diseases, the costs of the initial and acquired comorbidities are comparable. For some diseases (e.g., melanoma), the estimate of δ is small. As expected and shown in Akushevich et al. (2011b), for all diseases the acquired comorbidity was larger for those who died during the first 2.5 years after disease onset. These associations were strongly significant.

Figure 6.3 shows that estimates obtained from the two datasets used in the present analyses are consistent. For several diseases (e.g., kidney cancer, renal disease, diabetes, and melanoma), the differences between the estimates of the pre-disease cost levels from the two datasets are statistically significant. Estimates of the costs of disease onset are similar for both datasets, except for hip fracture and emphysema, for which the difference is not dramatic. Also, there are no strong differences detected for the population recovery rate. The differences for hip fracture, stroke, and pancreatic cancer were at the level of $p \approx 0.05$. Similarly, the estimates of acquired comorbidity costs also are compatible for the two datasets, except for kidney and lung cancers, Alzheimer's disease, and diabetes. The differences obtained for certain parameters and certain diseases can be caused by the different structure of populations represented by the two datasets: NLTCS-Medicare represents the entire U.S. elderly population, while the SEER-Medicare represents the population of the SEER areas only. The age and sex distribution of the total SEER population is similar to the non-SEER areas, though the SEER areas have fewer whites, more urban residents, and fewer areas with low socioeconomic status compared to the non-SEER areas (Warren et al. 2002). Also, a difference in circulatory disease incidence has been found for the NLTCS-Medicare and SEER-Medicare datasets (Akushevich et al. 2012).

6.3.1 Medical Cost as Measure of Disease Severity

Medical costs correlate with comorbidity patterns; they have been used to construct comorbidity indices: e.g., the Shwartz comorbidity index was constructed using a regression model to predict costs (de Groot et al. 2003; Shwartz et al. 1996). The best way to measure comorbidity or multimorbidity from Medicare data is based on analyses of individual Medicare trajectories constructed using the ICD-9 diagnosis codes, e.g., using an approach described in Sect. 6.2.2. However, such measures do not provide information about the severity of diagnosed diseases. Therefore, we investigated whether medical costs could serve as a surrogate of severity of comorbid disease(s) which cannot be captured when comorbidity measures are derived traditionally—i.e., only from ICD-9 diagnostic codes. We used the Cox proportional hazard model for survival of patients with onsets of a given disease. In this model, the predictor was the disease-specific cost at onset. If the costs reflect the disease severity, then we expect a positive association between disease-specific costs at onset and death. The results in Table 6.1 show estimates of hazard ratios for total and disease-specific costs (i.e., the part of the total costs resulting from the records containing the diagnostic code of the given disease) per $1,000. All hazard

Table 6.1 Mortality hazard ratios (HRs) for total and disease specific costs (per $1000) estimated using Cox proportional hazard ratios for cohorts of patients with a disease onset. All estimates are statistically significant except four HRs for disease-specific costs for Alzheimer's disease and female breast cancer (marked by italics)

Disease	Total cost			Disease-specific costs		
Follow-up:	1 year	3 year	5 year	1 year	3 year	5 year
Myocardial infarction	1.010	1.009	1.009	1.009	1.008	1.007
Hip fracture	1.012	1.011	1.010	1.012	1.012	1.011
Colon cancer	1.010	1.009	1.008	1.012	1.010	1.009
Stroke	1.029	1.027	1.027	1.041	1.045	1.046
Pancreatic cancer	1.036	1.038	1.038	1.074	1.065	1.056
Kidney cancer	1.093	1.088	1.084	1.157	1.174	1.155
Heart failure	1.011	1.009	1.008	1.013	1.011	1.010
Lung cancer	1.016	1.015	1.015	1.014	1.013	1.013
Ulcer	1.017	1.017	1.018	1.029	1.026	1.024
Angina pectoris	1.019	1.020	1.020	1.081	1.060	1.054
Chronic renal	1.036	1.033	1.032	1.053	1.048	1.042
Emphysema	1.046	1.033	1.030	1.042	1.032	1.030
Breast cancer	1.003	1.002	1.001	1.003	1.001	*1.000*
Alzheimer's disease	1.014	1.013	1.012	*1.010*	*1.006*	*1.005*
Parkinson's disease	1.023	1.022	1.022	1.022	1.021	1.021
Prostate cancer	1.010	1.010	1.010	1.008	1.008	1.008
Asthma	1.009	1.009	1.009	1.011	1.011	1.010
Diabetes	1.006	1.005	1.006	1.002	1.002	1.003
Melanoma	1.047	1.039	1.030	1.054	1.066	1.052
Arthritis	1.007	1.006	1.007	1.008	1.009	1.009

ratios are greater than one, thus supporting the hypothesis that medical costs in the month when a diagnosis was made reflect the severity of the diagnosed diseases.

6.3.2 Forecasting Models

The approach for modeling cost trajectories presented above allows for developing comprehensive forecasting models based on microsimulation of individual trajectories. Also, in certain cases, the model developed for cost trajectories results in simple forecasting models that provide projections in an analytical form. Indeed, in many specific cases, an averaging over individual trajectories can be performed analytically by reducing the results to aggregated characteristics observed at the population level. As an example, consider a cohort of individuals under a risk of a certain disease. Let the disease survival function $S(x)$ be known from other studies. This survival function (or corresponding hazard rate $h(x) = -[\log S(x)]_x^{\prime}$ or density function $f(x) = h(x)S(x)$ or probability distribution $F(x) = 1 - S(x)$) can be estimated from Medicare data as well (Akushevich et al. 2013b) (reviewed in a Chap. 3 of this monograph). Assume also that during the follow-up the individuals are not subject to another health event, including death. The medical costs for the cohort of individuals at age x can be predicted by summing (or integrating over) individual cost trajectories given by Eq. 6.1:

$$C_{tot}(x) = cS(x) + (c+\delta)F(x) + (P-\delta)\int_0^x \exp(-r(x-u))f(u)du. \tag{6.2}$$

The first term on the right hand side of Eq. (6.2) reflects the contribution of healthy individuals, i.e., those that have not yet developed this disease. The mean of their costs is characterized by initial comorbidity c, and their fraction equals $S(x)$. The last two terms characterize the contribution of unhealthy people. These terms involve integration of individual trajectories $C(u)$ over different times of onsets denoted by u. The second term in Eq. (6.2) describes the acquired comorbidity and the third term reflects the costs of treatment after onset. Below we consider three specific models for which the integration can be performed analytically.

The first model has a constant hazard rate (μ_0). Many chronic diseases have an age pattern (e.g., melanoma, (Akushevich et al. 2012)) which can be considered approximately constant. The model is a representation of the contribution of this disease to total costs. In this case, the cost is

$$\begin{aligned} C_{tot}(x) = \quad & c + \delta(1-\exp(-\mu_0 x)) + (P-\delta)\mu_0 \frac{\exp(-\mu_0 x) - \exp(-rx)}{r-\mu_0} \approx \\ & c + \delta(1-\exp(-\mu_0 x)) + (P-\delta)\mu_0 \frac{1-\exp(-rx)}{r} \end{aligned} \tag{6.3}$$

The last approximation is obtained assuming that the hazard rates μ_0 are much lower than the disease burden rates (or, in other words, the population recovery rates), r. This assumption is justified by their numerical values: i.e., $\mu_0 \approx 0.000001$ -0.01 year^{-1} for the diseases we are studying, while $r \approx 7$–15 year^{-1}.

A second model represents the situation in which individuals in a cohort were exposed to a specific risk factor (e.g., antigen or ionizing radiation) and part of the cohort develops a disease after a latent period. Examples include: (i) contact with infected individuals and forthcoming onset of infection disease, e.g., hepatitis B or C; and (ii) acute exposure to ionizing radiation and a subsequent increased rate of leukemia. The density function can be chosen as $f(x) = p_0 f_G(x; \bar{x}, \sigma)$ where p_0 is the probability of developing the disease and $f_G(x, ; \bar{x}, \sigma)$ is the Gaussian distribution with mean $\bar{x}$ and variance σ^2. The Gaussian distribution was chosen to characterize the typical situation where the lag period represented by $\bar{x}$ is known and the variance of the lag in population is not large. The cost when the peak incidence is passed is estimated as:

$$C_{tot}(x) = c + p_0\delta + p_0(P - \delta)\exp\left(-r(x - \bar{x}) + \frac{1}{2}r^2\sigma^2\right) \tag{6.4}$$

The third model deals with the situation where the survival function for a disease is known from empirical analysis, e.g. is represented by the Kaplan-Meier estimator. In this case, it can be characterized by the set of its decrements $\Delta S(x_i)$ that occurred at times x_i. After integration by parts, Eq. (6.2) can be rewritten in terms of the survival function as:

$$\begin{aligned} C_{tot}(x) = c + \delta(1 - S(x)) - (P - \delta)(S(x) - \exp(-rx) \\ - r\int_0^x \exp(-r(x - u))S(u)du). \end{aligned} \tag{6.5}$$

The integral can be calculated analytically and presented in terms of the decrements:

$$C_{tot}(x) = c + \delta(1 - S(x)) - (P - \delta)\sum\nolimits_i \Delta S(x_i)\exp(-r(x - x_i)). \tag{6.6}$$

The base model (6.1) as well as the models for which analytical solutions (6.2, 6.3, 6.4, 6.5 and 6.6) exist might be further generalized to describe the entire chain of health events describing the evolution of individual health from a healthy state at younger ages to multimorbidity and death at older ages. A useful property of these models is that they have inputs and outputs represented by the same single quantity: comorbidity measured by medical costs of continuing care. This property allows researchers to use the base model as a building block in the construction of more detailed models. This property also allows different chronic diseases to be incorporated into the same approach without increasing the dimensionality of the model.

At a certain stage of model development, analytic solutions may no longer be possible. Instead, a microsimulation approach might be used. Several further generalizations might also be required to generate a more comprehensive microsimulation model. One important generalization is a model of mortality risks based on assumptions other than those used in model (6.1). For example, the assumptions could allow changing costs as the main predictive variable. Given the model estimated, the simulation of individual trajectories is naturally generalized by considering two competing risks (i.e., the risk of disease onset and the risk of death) which can be dependent or, more specifically, conditionally independent given the value of a covariate (i.e., the medical cost level). Other directions for model generalization could include (i) adjustment to the effect of a second health event that occurred before complete recovery from the previous one, (ii) adjustment to possible recurrence of the disease diagnosed earlier, (iii) implementation of period and cohort effects, (iv) implementation of generalized models of the risks of the health events with dependence on covariates incorporated in addition to age dependence, (v) incorporation of the effects of increasing medical expenditures before death, (vi) modeling and implementation of the distribution of the covariate, including the distribution conditional on a specific value in the previous time period. This approach will produce population projections of health and associated medical costs under the assumption that current conditions (i.e., those which can be captured by available data) will continue during the projection period. Specific scenarios regarding the future healthcare environment developed by panels of experts (Goldman and Rand Corporation 2004) can also be incorporated into the simulation model. In all these developments, models (6.1) and (6.2) serve as baselines that must be reproduced numerically or analytically with simplified versions of more comprehensive forecasting models.

6.4 Discussion

In this chapter, a model of the relationships between individual cost trajectories around the onset of aging-related diseases was developed and applied empirically. In total, 20 diseases were analyzed. The main methodological idea was to develop a mathematical model to predict medical care costs for these diseases for the time period around the date of the disease onset (identified from data on the medical care costs associated with treatment of the disease) and create a methodological background for development of forecasting models of dynamic changes of the health state and associated medical costs. The empirical results obtained are important for the U.S. elderly population, because the diseases included in the analyses have high prevalence and are associated with high medical costs. An innovative approach was developed for selecting individuals with disease onset and used for identification of the age at onset. We found that the time patterns of medical cost trajectories were similar for all diseases considered and can be described in terms of four components having the meanings of (i) the pre-diagnosis cost associated with initial comorbidity represented by medical expenditures, (ii) the cost peak associated with the onset of each disease, (iii) the decline/reduction in medical expenditures after the disease onset, and (iv) the difference between post- and pre-diagnosis cost levels associated with an acquired comorbidity. The description of the trajectories was formalized by a model which explicitly involves four parameters reflecting these four components.

The results of these analyses extend those presented earlier in Akushevich et al. (2011b), which were on analysis of the effect of costs of medical procedures (not of Medicare payments as in this chapter), and used stratified subgroups of patients by certain indices such as Charlson's comorbidity index (calculated using Medicare data), a disability index (measured in screener interviews, see (Manton and Gu 2001)), survival status in a 2.5 year follow-up period, and age at diagnosis. The Charlson comorbidity index was calculated according to specifications described in Charlson et al. (1987) and Quan et al. (2005), as a weighted sum of chronic conditions that appeared in individual medical records during the year prior to the date of interview. These parameters were evaluated for the entire U.S. population as well as for stratified subgroups of patients. The most important conclusions from the analyses of the stratified populations were the following: (1) pre-disease cost levels can be associated with initial comorbidity/disease treatment; (2) there are no strong dependences of the disease-specific costs at onset among the studied strata, except survival status (i.e., for 2.5-year survivors the cost was significantly lower for a majority of diseases); (3) the associations of population recovery with comorbidity/continuing care and disability showed no essential dependences on these indices; and (4) for all diseases the acquired comorbidity was significantly higher for those who died during the first 2.5 years after disease onset.

The patterns of medical expenditures evaluated in this chapter could help clarify which of the model components is responsible for the effects of medical costs on health and mortality and which of them is more (or less) sensitive to subpopulation specifications. All medical cost trajectories were considered for the U.S. elderly population, as well as for subgroups stratified by disability, comorbidity, age, and survival (for 2.5 years after the onset). The model of medical cost trajectories was applied to all empirically verified patterns, and parameters of the model were statistically estimated and compared. These analyses revealed the basic properties of the medical cost trajectories. The most important were the following. The differences in estimates of pre-disease cost level for different diseases were moderate but not identical (Fig. 6.3); since the medical cost trajectories were considered to be conditional on disease-specific incidence, the detected differences likely reflected variations in disease risk depending on comorbidity. In contrast, the cost of disease onset was essentially disease-specific, and the diversity was likely due to disease-specific diagnostic procedures and initial therapies at the disease onset. The diseases considered in our study included (i) those with possible clinical recovery (e.g., ACHD, stroke, and ulcer) and (ii) those with unlikely clinical recovery (e.g., diabetes and Alzheimer's disease). Estimates of population recovery (i.e., the rate of reduction of post-diagnosis cost level) reflected potential opportunities for recovery from aging-related diseases. Positive estimates were detected for all diseases; however, the statistical significance of the estimates for diseases with unlikely recovery (e.g., for diabetes mellitus or Alzheimer's disease) was lower (or non-significant), especially in subpopulations stratified by disability or comorbidity (Fig. 6.3). The acquired comorbidity/continuing care cost (i.e., the difference between pre- and post-diagnosis cost levels) was disease-specific and strongly dependent on the survival status of the patients after disease onset (Fig. 6.3, and Table 6.5 of Akushevich et al. (2011b)). Parameter estimates (Fig. 6.3) confirmed that the model parameters are chosen so that the effects of multiple diseases on their estimates are minimal. The first parameter measures comorbidity before disease onset/pre-disease treatment costs and represents the effects of multiple comorbidities. The costs of disease onset and acquired comorbidity/followup care are defined as the cost level above the mean level of comorbidity. The rate of population recovery, i.e., the rate of reduction of medical expenses after a diagnosis, is reflected by reduction of costs for the focal disease, while changes in costs due to other diseases (i.e., comorbid conditions) are less essential (at least in the first approximation).

Typically, medical costs associated with specific chronic diseases are analyzed and projected for a certain period of time after disease onset or health-related event (e.g., hospitalization) (Yabroff et al. 2009). Often, analyses are performed for specific population groups such as a subpopulation of disabled or comorbid individuals (Goldman and Rand Corporation 2004; Noyes et al. 2008; Yabroff et al. 2008). Recently, the Episode Treatment Group (ETG) approach has been adopted by Medicare for estimation of disease episode-based medical costs (Forthman et al. 2000): detailed information is collected for each disease episode for about 600 clinically homogeneous groups adjusted for patient's severity, age, complications, comorbidities, and major surgeries. Despite being a very useful tool for direct comparison of treatment patterns among providers within the ETG, this approach was not intended to be the basis for population-level analysis. Compared to this approach, the method developed in this chapter has fewer details on each disease episode, but allows for inclusion in the analysis of all patient-related information on comorbidities (i.e., information not related to only one specific disease) and disabilities. That makes the whole model more flexible and non-dependent from the pre-selected diseases in the ETG episode-related conditions. In our approach, only data-driven information was incorporated into the model, and "human factor"-related issues, such as episode-specific information on disease-specific procedures, disabilities, and comorbidities were avoided.

Models of medical cost projections usually are based on regression models estimated with the majority of independent predictors describing demographic status of the individual, patient's health state, and level of functional limitations, as well as their interactions (Goldman and Rand Corporation 2004; Pope et al. 2004). If the health states needs to be described by a number of simultaneously manifested diseases, then detailed stratification over the categorized variables or use of multivariate regression models allows for a better description of the health states. However, it can result in an abundance of model parameters to be estimated. One way to overcome these difficulties is to use an approach in which the model components are demographically-based aggregated characteristics that mimic the effects of specific states.

The model developed in this chapter is an example of such an approach: the use of a comorbidity index rather than of a set of correlated categorical regressor variables to represent the health state allows for an essential reduction in the degrees of freedom of the problem. The medical costs of both the first months and the last months of the trajectories are associated with comorbidity. Since the complete individual trajectory of health changes can be simplified in terms of subsequent incidence events, this model of medical costs before and after an incident event can serve as a building block for constructing the complete individual trajectory. Many uncertainties typical for existing models are overcome with such an approach. Thus, this model of the dynamics of medical costs before and after the onset of a chronic disease can serve as a key component of a model for projecting medical expenditures.

The results obtained are new and important, both substantively and methodologically. Substantively, the trajectories of medical costs evaluated at the disease onset in the U.S. elderly population provide new information of potential interest for planning public health expenditures. Our analysis showed that these trajectories could be well described by a model with four well-defined and interpretable components which were estimated for each of the studied diseases. Interestingly, all of the aging-related diseases studied had very similar structures of cost trajectories. The model was validated for several population groups and demonstrated a good ability to describe cost trajectories for different levels of disability and comorbidity (Akushevich et al. 2011b). This model could be extended to a level of even higher practical importance, such as forecasting health/incidence, mortality, and associated medical costs in the U.S. elderly population using this limited set of parameters (and with a great potential for improvements when more detailed data become available).

Methodologically, the model leads to a general comprehensive microsimulation forecasting model of medical expenditures that can be formulated as follows. The population dynamics are represented by random trajectories in a covariate space. The end of each trajectory is associated with individual death. To simulate an individual trajectory means to evaluate covariates for all time points between the beginning of a follow-up period (e.g., at age 65 years old) and the age of death. At each time point, an individual is at risk of a disease onset and death. The model can be Markov and non-Markov. In the former case, the risks and dynamics are defined by the current health status represented by covariates and age. The model developed in this chapter (or its generalizations) can be used to simulate the dynamics of the covariate (i.e., the comorbidity index represented by medical costs aggregated during a certain time period) before and after disease onset. An auxiliary model of the risks of disease onset and mortality associated with the covariate and age (e.g., the model described in Chaps. 15 and 16) can then be used to simulate these events. An important property of the model (6.1) is that it has an input and output represented by the same single quantity: comorbidity measured by medical cost, and this property allows the researchers to use the base model (6.1) as a building block in simulating a life history as a sequence of such blocks associated with disease onsets. This property also allows for including different age-related diseases within the same approach without increasing the dimensionality of the model. Note that the risks of the diseases as well as the associations of these risks with potential covariates such as comorbidity, disability indices, and age groups can be roughly estimated using the numerical estimates presented in Table 6.1 (a detailed investigation of the model for health state projections estimated with Medicare data will be presented in a separate publication).

If the general comprehensive microsimulation model is specified as a Markov model, the past history of an individual does not contribute to the probabilities of future events or, in other words, current covariates and age are taken to represent sufficient information for the description of health states and future event probabilities. By reducing the dimensionality of the model, we are able to better estimate the covariate-specific effects; however, the model becomes less precise. Therefore, a model with a specific set of covariates always represents an approximation to reality. This is a limitation of all Markov models. Specifically, model (6.1) needs to be improved when a second disease onset occurs almost immediately after the first one. Partly, this can be accommodated by using the comprehensive microsimulation model: if the simulation is performed on a month-by-month basis, the onset of the second disease can be simulated for any time after onset, including the time period when recovery is not completed. Greater values of the covariate then will be associated with higher probabilities of such an event. The approach's precision can be estimated by developing individual trajectories for a pair of disease onsets using an approach close to that described in this chapter. Another limitation of the modeling approach is that the model (6.1) is not capable of describing all types of diseases equally well: for example, several months before their onset, asthma, Alzheimer's and Parkinson's diseases, and melanoma are not described very well by the model. This effect could be explained by diagnostic tests/procedures performed before the actual clinical diagnoses, and, therefore, this effect was not deemed critical for the modeling approach.

Acknowledgements The research reported in this chapter was supported by the National Institute on Aging grants R01AG027019, R01AG030612, R01AG030198, R01AG032319, and R01AG046860.

References

Akushevich, I., Kulminski, A., & Manton, K. (2005). Life tables with covariates: Dynamic model for nonlinear analysis of longitudinal data. *Mathematical Population Studies, 12*(2), 51–80.

Akushevich, I., Kravchenko, J., Akushevich, L., Ukraintseva, S., Arbeev, K., & Yashin, A. (2011a). Cancer risk and behavioral factors, comorbidities, and functional status in the U.S. elderly population. *ISRN Oncology* 2011:Article ID 415790.

Akushevich, I., Kravchenko, J., Akushevich, L., Ukraintseva, S., Arbeev, K., & Yashin, A. I. (2011b). Medical cost trajectories and onsets of cancer and noncancer diseases in U.S. elderly population. *Computational and Mathematical Methods in Medicine* 2011:Article ID 857892.

Akushevich, I., Kravchenko, J., Ukraintseva, S., Arbeev, K., & Yashin, A. I. (2012). Age patterns of incidence of geriatric disease in the U.S. elderly population: Medicare-based analysis. *Journal of the American Geriatrics Society, 60*(2), 323–327.

Akushevich, I., Kravchenko, J., Ukraintseva, S., Arbeev, K., & Yashin, A. I. (2013a). Circulatory diseases in the U.S. Elderly in the linked national long-term care survey-medicare database: Population-based analysis of incidence, comorbidity, and disability. *Research on Aging, 35*(4), 437–458.

Akushevich, I., Kravchenko, J., Ukraintseva, S., Arbeev, K., & Yashin, A. I. (2013b). Recovery and survival from aging-associated diseases. *Experimental Gerontology, 48*(8), 824–830.

BLS. (2009). Consumer price index. Bureau of Labor Statistics.

CDC. (2004). *The burden of chronic diseases and their risk factors*. Atlanta: US Department of Health and Human Services, Centers for Disease Control and Prevention.

Charlson, M. E., Pompei, P., Ales, K. L., & Mackenzie, C. R. (1987). A new method of classifying prognostic co-morbidity in longitudinal-studies – development and validation. *Journal of Chronic Diseases, 40*(5), 373–383.

de Groot, V., Beckerman, H., Lankhorst, G. J., & Bouter, L. M. (2003). How to measure comorbidity: A critical review of available methods. *Journal of Clinical Epidemiology, 56*(3), 221–229.

Forthman, M. T., Dove, H. G., & Wooster, L. D. (2000). Episode Treatment Groups (ETGs): A patient classification system for measuring outcomes performance by episode of illness. *Topics in Health Information Management, 21*(2), 51–61.

Goldman, D. P., & Rand Corporation. (2004). *Health status and medical treatment of the future elderly: Final report*. Santa Monica: RAND.

Goldman, D. P., Shang, B., Bhattacharya, J., Garber, A. M., Hurd, M., Joyce, G. F., Lakdawalla, D. N., Panis, C., & Shekelle, P. G. (2005). Consequences of health trends and medical innovation for the future elderly. *Health Aff*: hlthaff.w5.r5.

Goldman, D. P., Cutler, D. M., Shang, B, & Joyce, G. F. (2006). The value of elderly disease prevention. *Forum for Health Economics & Policy, 9*, article 2 (Biomedical Research and the Economy). Available online at http://www.bepress.com/fhep/biomedical_research/1

Goldman, D. P., Zheng, Y. H., Girosi, F., Michaud, P. C., Olshansky, S. J., Cutler, D., & Rowe, J. W. (2009). The benefits of risk factor prevention in americans aged 51 years and older. *American Journal of Public Health, 99*(11), 2096–2101.

HI and SMI. (2009). 2009 Annual report of the Boards of Trustees of the Federal Hospital Insurance and Federal Supplementary Medical Insurance Trust Funds. Washington, DC.

HI and SMI. (2010). 2010 annual report of the Boards of Trustees of the Federal Hospital Insurance and Federal Supplementary Medical Insurance Trust Funds. Washington, DC.

Klees, B. S., Wolfe, C. J., & Curtis, C. A. (2009). Brief summaries of medicare and medicaid. In *Health care financing review/2009 statistical supplement*.

Lubitz, J. (2005). Health, technology, and medical care spending. *Health Affairs, 24*(6), W5r81–W85r85.

Lubitz, J. D., & Riley, G. F. (1993). Trends in medicare payments in the last year of life. *New England Journal of Medicine, 328*(15), 1092–1096.

Manton, K. G., & Gu, X. L. (2001). Changes in the prevalence of chronic disability in the United States black and nonblack population above age 65 from 1982 to 1999. *Proceedings of the National Academy of Sciences of the United States of America, 98*(11), 6354–6359.

Manton, K. G., Stallard, E., & Singer, B. (1992). Projecting the future size and health status of the U.S. elderly population. *International Journal of Forecasting, 8*(3), 433–458.

Miller, T. (2001). Increasing longevity and medicare expenditures. *Demography, 38*(2), 215–226.

Nattinger, A. B., Laud, P. W., Bajorunaite, R., Sparapani, R. A., & Freeman, J. L. (2004). An algorithm for the use of medicare claims data to identify women with incident breast cancer. *Health Services Research, 39*(6), 1733–1749.

Nattinger, A. B., Laud, P. W., Bajorunaite, R., Sparapani, R. A., & Freeman, J. L. (2006). Clarification note to an algorithm for the use of medicare claims data to identify women with incident breast cancer (vol 39, pg 6, 2004). *Health Services Research, 41*(1), 302–302.

NIH/NHLBI. (2006). *Incidence and prevalence: 2006 chart book on cardiovascular and lung diseases*. Bethesda: National Institutes of Health, National Heart, Lung, and Blood Institute.

Noyes, K., Liu, H. S., & Temkin-Greener, H. (2008). Medicare capitation model, functional status, and multiple comorbidities: Model accuracy. *American Journal of Managed Care, 14*(10), 679–690.

Pardes, H., Manton, K. G., Lander, E. S., Tolley, H. D., Ullian, A. D., & Palmer, H. (1999). Effects of medical research on health care and economy. *Science, 283*(5398), 36–37.

Pope, G. C., Kautter, J., Ellis, R. P., Ash, A. S., Ayanian, J. Z., Iezzoni, L. I., Ingber, M. J., Levy, J. M., & Robst, J. (2004). Risk adjustment of medicare capitation payments using the CMS-HCC model. *Health Care Financing Review, 25*(4), 119–141.

Quan, H. D., Sundararajan, V., Halfon, P., Fong, A., Burnand, B., Luthi, J. C., Saunders, L. D., Beck, C. A., Feasby, T. E., & Ghali, W. A. (2005). Coding algorithms for defining comorbidities in ICD-9-CM and ICD-10 administrative data. *Medical Care, 43*(11), 1130–1139.

Shwartz, M., Iezzoni, L. I., Moskowitz, M. A., Ash, A. S., & Sawitz, E. (1996). The importance of comorbidities in explaining differences in patient costs. *Medical Care, 34*(8), 767–782.

Sloan, F. A., Brown, D. S., Carlisle, E. S., Ostermann, J., & Lee, P. P. (2003). Estimates of incidence rates with longitudinal claims data. *Archives of Ophthalmology, 121*(10), 1462–1468.

Warren, J. L., Klabunde, C. N., Schrag, D., Bach, P. B., & Riley, G. F. (2002). Overview of the SEER-medicare data: Content, research applications, and generalizability to the United States elderly population. *Medical Care, 40*(8), IV-3-IV-18.

Yabroff, K. R., Lamont, E. B., Mariotto, A., Warren, J. L., Topor, M., Meekins, A., & Brown, M. L. (2008). Cost of care for elderly cancer patients in the United States. *Journal of the National Cancer Institute, 100*(9), 630–641.

Yabroff, K. R., Warren, J. L., Banthin, J., Schrag, D., Mariotto, A., Lawrence, W., Meekins, A., Topor, M., & Brown, M. L. (2009). Comparison of approaches for estimating prevalence costs of care for cancer patients what is the impact of data source? *Medical Care, 47*(7), S64–S69.

Yashin, A., Akushevich, I., Ukraintseva, S., Akushevich, L., Arbeev, K., & Kulminski, A. (2010). Trends in survival and recovery from stroke evidence from the national long-term care survey/ medicare data. *Stroke, 41*(3), 563–565.

Chapter 7
Indices of Cumulative Deficits

Alexander M. Kulminski, Kenneth C. Land, and Anatoliy I. Yashin

7.1 Introduction

Despite broad interest in the mechanisms responsible for human aging and numerous efforts to identify factors contributing to morbidity, biological senescence, and longevity, these processes still remain elusive. This makes the systemic description of aging-related changes embedded in data from different studies a difficult task. Indeed, observational studies typically measure not only major changes in health and well-being captured by well-defined risk factors (e.g., physiological measurements), but also various aging-related changes spread throughout hundreds of distinct variables. The connection between such variables as well as between each of these variables and health or survival outcomes is unclear and often cannot be evaluated statistically with acceptable accuracy. This is due to the fact that the number of these variables is typically large, while the effect of each on health and survival is small, so most estimates of effect parameters in corresponding statistical models are statistically non-significant. This chapter describes a line of analysis that is based on the premise that, by taking such "mild-effect" variables into account, the description of aging-related deterioration in health and well-being in humans can be substantially improved without costly investments in collecting new data. To realize this potential, new statistical methods are required.

One promising approach was suggested by Rockwood and Mitnitski (Mitnitski et al. 2004; Rockwood et al. 2004). These authors developed a cumulative index (called a frailty index) arguing that health and well-being disorders (e.g., signs, symptoms, impairments, abnormal lab tests, diseases, etc.) accumulated by individuals during their life course can be considered as indicators of physiological frailty. The rationale behind this concept is that degradation and decline of neuroendocrine, immune, and other functions of an organism can result in a wide spectrum of adverse health and well-being disorders (Ferrucci et al. 2003; Goggins et al. 2005; Vanitallie 2003). On the cellular level, frailty can be associated with a process of gradual accumulation of damage in cellular tissues (Kirkwood 2002).

A.I. Yashin et al., *Biodemography of Aging*, The Springer Series on Demographic Methods and Population Analysis 40, DOI 10.1007/978-94-017-7587-8_7

This approach can be generalized to characterize the overall process of health deterioration with age that can be extended to younger ages. The level of health deterioration can then be described by a composite index, called an *index of cumulative deficits or deficits index* (*DI*). The DI is based on the Rockwood and Mitnitski (Mitnitski et al. 2004; Rockwood et al. 2004) premise that mild-effect traits, which individually have small impacts on risks of adverse health, can collectively substantially impact morbidity and survival chances. Accordingly, DIs can become reliable predictors of health and survival as well as the level of aging-related health deterioration (Goggins et al. 2005; Kulminski et al. 2007a, b, c; Mitnitski et al. 2002; Yashin et al. 2007a, b).

7.2 Conceptualization of the Deficits Index

The DI is conceptualized as the proportion of failed (e.g., definitive deficits) or abnormal (e.g., doubtful deficits) health traits an individual has experienced by age x—that is, as a summary measure of the average level of deterioration at age x. Thus, an empirical estimate of this proportion in a given individual, i.e., the $DI(x)$, can be calculated by selecting a sub-set of M units out of a full set of N such units. Specifically, by summing the number of failed or abnormal units from the selected set by age x, $m(x)$, an empirical estimate of the DI can be calculated as $DI(x) = m(x)/M$. For example, if an individual has been administered 30 questions and responded positively (there is a deficit) to five and negatively (no deficit) to 21, then her/his DI is 5/26. Thus, based on a large and diverse array of deficits, the DI is a quasi-continuous quantity ranging theoretically between 0 (no deficits or perfect health) and 1 (pure ill-health) or, equivalently, between 0 % and 100 %. An important property of the DI is that it is weakly sensitive to the specific array of deficits for which responses are obtained as long as a wide spectrum is considered (Kulminski et al. 2011; Rockwood et al. 2006; Searle et al. 2008).

7.3 Cross-Sectional Age Patterns of the Deficits Index as Characteristics of Aging-Related Processes

Deficits Indices have being intensively studied for their utility in characterizing aging-related processes in various populations and settings. One of the important characteristics of DIs is that they are robustly associated with age; despite the specifics of a given population or setting, clear trends of increase of DIs with age are observed in different settings (e.g., Gu et al. 2009; Kulminski et al. 2011; Mitnitski et al. 2005; Yu et al. 2012).

Figure 7.1 displays examples of cross-sectional patterns of the DI in three qualitatively different populations. The DI in the National Long Term Care Survey

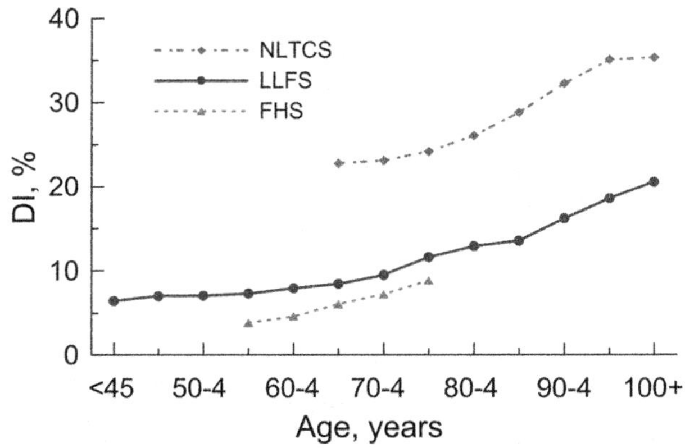

Fig. 7.1 Age patterns of *DI*s in population samples characterized by differential success in aging. The *NLTCS* sample includes 24,213 person-observations at five examinations conducted in 1982, 1984, 1989, 1994, and 1999. The *FHS* sample represents 3833 individuals examined at the ninth examination in 1964. The *LLFS* sample includes 4954 subjects comprising long-living parents and their children, as well as their spouses, examined in 2010

(NLTCS) sample is comprised of 32 deficits characterizing disabilities (e.g., difficulty with eating, dressing, walking around), diseases (e.g., arthritis, Parkinson's disease, glaucoma, diabetes), and problems with vision, hearing, and teeth (Kulminski et al. 2007a). The NLTCS sample represents a selected population of ages 65+ individuals who are mostly of less than average health, characterized by the presence of chronic disability. The DI in the Framingham Heart Study (FHS) was constructed using 37 deficits covering milder deficits than those in the NLTCS (e.g., abnormal laboratory tests and mild health traits) because of the younger ages of individuals in this sample (Kulminski et al. 2008a). The DI in the FHS characterizes an unselected population of individuals developing health problems according to a "normal" life course. The DI in the Long Life Family Study (LLFS) includes 85 deficits covering milder and heavier deficits including disabilities, morbidities, mental health, depression, abnormal laboratory tests, etc. (Kulminski et al. 2011). The LLFS sample represents a population of individuals selected for long life, their children, and spouses. Despite substantial differences in the population settings, all DIs show increases with age, implying that these patterns capture systemic changes during the life course of an aging body (Kulminski et al. 2006).

7.4 Deficits Indices and Age as Indicators of Aging-Related Processes

To better understand the potential of Deficits Indices for characterizing aging-related processes, we investigate to what extent DIs are distinct from age and whether they are a better indicator of aging-related processes than age. We conduct systematic comparative analyses of DI- and age-specific patterns for a number of statistics including frequency distributions, correlations, time to death patterns, mortality rates, and relative risks of death. Because one of the key characteristics of aging is mortality, we consider to what extent a DI can discriminate the population at risk of death, compared to age. The results are presented for a DI constructed in the NLTCS sample using 32 deficits characterizing disabilities, diseases, and problems with vision, hearing, and teeth (Kulminski et al. 2007a).

7.4.1 Frequency Distributions

Summary statistics (samples sizes, means, standard deviations, kurtosis, and skewness) of the marginal distributions of the DI (Panel A) and age (Panel B) along with bar charts are shown in Fig. 7.2 for two groups of NLTCS individuals: (1) those who survived 1 year and (2) those who died within 1 year after interview.

Figure 7.2 shows that the NLTCS sample is relatively old and unhealthy, as evidenced by the relatively large mean values of age and the DI in this sample (Fig. 7.2, insets). Decedents are older and have larger mean DI levels compared to survivors. Similarly, the DI and age frequency patterns of the survivors have larger skewness (shape) parameters than decedents. That is, compared to decedents, survivors are less likely to have large DI values and to be at the oldest ages. This implies that the frequency patterns for survivors are closer in shape to a gamma distribution, while those for deceased individuals are bell-shaped being closer to a normal distribution (Fig. 7.2). The kurtosis parameters of the frequency

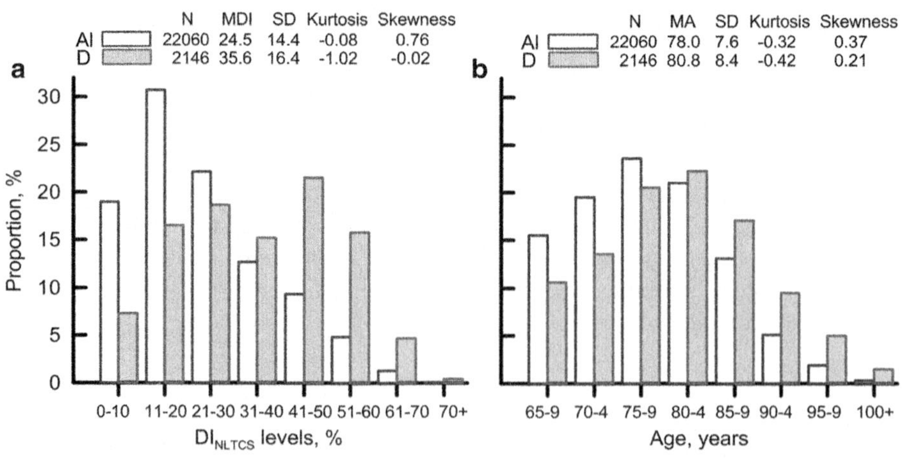

Fig. 7.2 The DI (Panel **a**) and age (Panel **b**) frequency patterns for NLTCS participants who were examined in 1982, 1984, 1989, 1994, and 1999 and either survived 1 year (*Al*) or died within 1 year (*D*) after the date of the interview. Insets show characteristics of the population marginal distributions, including the number of person-observations (*N*), mean DI (*MDI*), mean age (*MA*), standard deviation (*SD*), Kurtosis, and Skewness

distributions show that the DI- and age-specific patterns for decedents are flatter than those for survivors.

7.4.2 *Correlation of the DI and Age*

The correlation between the DI and age is small. Specifically, the Pearson correlation coefficient between the DI and age in the entire NLTCS sample is $r_{DI\text{-}Age}=0.193$ $(p<0.01)$. It is even smaller for decedents, $r_{DI\text{-}Age}=0.127$ $(p<0.01)$ and of the same magnitude for survivors, $r_{DI\text{-}Age}=0.183$ $(p<0.01)$. These weak correlations between the DI and age imply that a small DI is not necessarily found only in younger individuals and that a large DI is not necessarily found only among the oldest members of the sample. This important result shows that deficits can be accumulated in a person independently of age.

7.4.3 *DI-Specific Age Patterns for Decedents and Survivors*

If individuals can accumulate deficits at any age, this measure can be used as a health index, implying that accumulation of deficits is associated with higher chances of death. Then decedents should have larger DI values than survivors at any age. Figure 7.3 shows that this is the case; the entire pattern for decedents is shifted up toward larger DI values, i.e., individuals with smaller numbers of deficits live longer regardless of age. By contrast, having a large DI can be predictive of death regardless of age as well. Figure 7.3 also shows that the DI age pattern for decedents is flatter than that for survivors. This is likely caused by a saturation

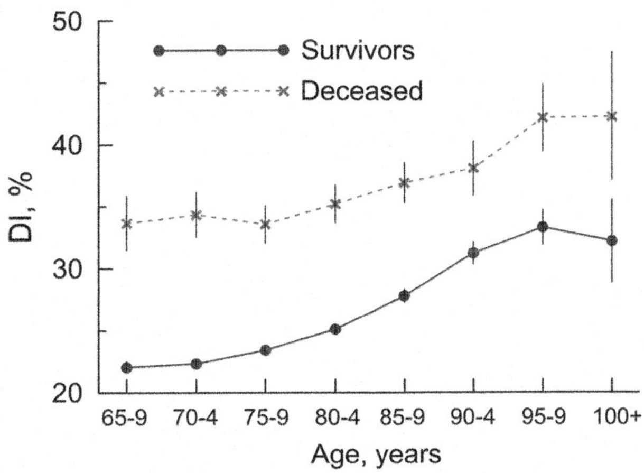

Fig. 7.3 Five-year age patterns of the DI for NLTCS participants who either survived or died within 1 year after the respective interview. Bars showing 95 % confidence intervals ensure that differences in DI levels for survivors and deceased individuals are highly significant in each age group

effect, given the limited abilities of an organism to cope with multiple health problems, that is compatible with the existence of an upper limit in deficit accumulation (Rockwood and Mitnitski 2006). This is further supported by the finding that the average number of 15 major causes of death in multiple cause of death mortality data for decedents aged 65 years or older is relatively constant (~1.9–2.1) across 5-year age groups up to age 99 (Stallard 2002, Table 7).

7.4.4 The DI and Age Patterns of Time to Death

To empirically evaluate the relationship of the DI and age with survival, we considered the distribution of time to death (TTD) among NLTCS participants of different ages or levels of accumulated deficits. Figure 7.4 shows the mean TTD (the number of years individuals stay alive after interview) for individuals in different age groups and DI levels. Obviously, younger individuals have longer life spans. A striking result is that the DI-specific distribution of the TTD resembles that for the age despite the small correlation of the DI with age. This finding provides further evidence that individuals have longer life spans when they have smaller DI values, virtually independent of age.

Figure 7.5a displays the mean values of the DI and age (vertical axis) for individuals who survive the indicated number of years after interview (horizontal axis). These patterns suggest that the DI can better characterize the chances of death, especially during a short follow-up period, than age. Indeed, Fig. 7.5b shows that correlation of DI with TTD ($r_{DI\text{-}TTD}$) is much stronger than that of age with TTD ($r_{Age\text{-}TTD}$) within shorter follow-up periods. For longer follow-up periods, this difference diminishes and the correlation coefficients $r_{Age\text{-}TTD}$ and $r_{DI\text{-}TTD}$ become similar.

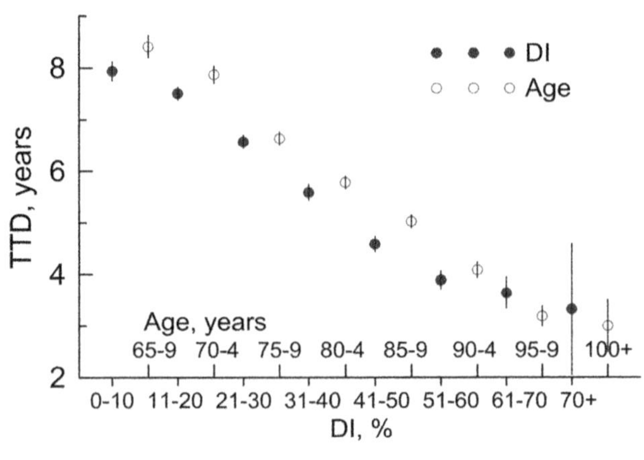

Fig. 7.4 Time to death (*TTD*) distributions by DI levels and age groups for NLTCS participants. *Bars* show 95 % confidence intervals

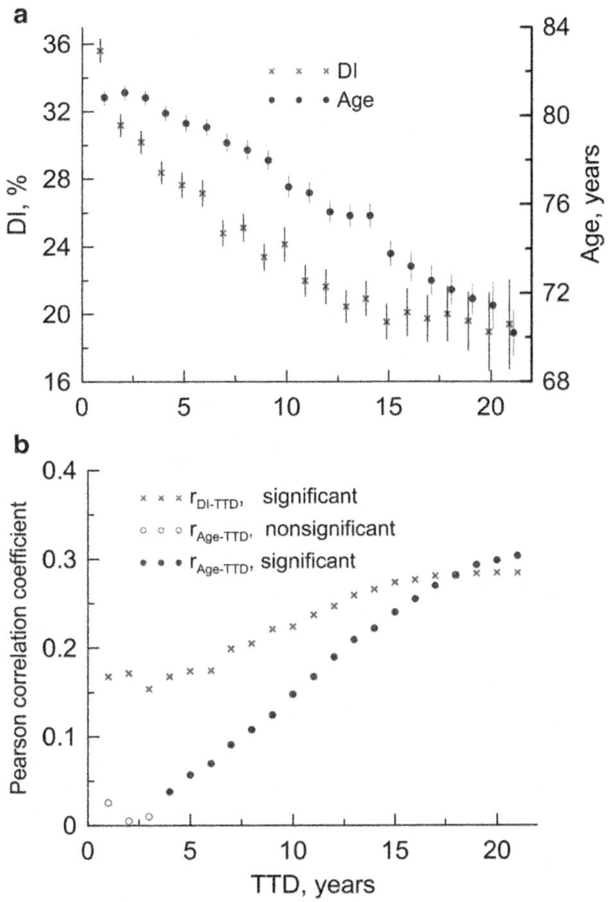

Fig. 7.5 Panel **a** Time to death (*TTD*) patterns of the DI (*left axis*) and age (*right axis*) for NLTCS participants. *Bars* show 95 % confidence intervals. Panel **b** Pearson correlation coefficients among DI and TTD ($r_{DI\text{-}TTD}$) and age and TTD ($r_{Age\text{-}TTD}$) for individuals who survive at least the number of years after respective interview shown in the x axis

7.4.5 The DI and Age Specific Mortality Rates

To ascertain whether the foregoing conjecture is manifested in mortality patterns, we calculated 1-year age- and DI-specific mortality rates. Figure 7.6a explicitly shows that the DI is associated with mortality. Individuals having small DIs also have smaller chances of death within 1 year after interview. These chances increase in an accelerated pattern as the DI increases. At large DIs (about 50 % and larger), the increase in the mortality rates decelerates likely reflecting a limit in the deficit accumulation (Rockwood and Mitnitski 2006).

The accelerated increase in mortality rates with increasing DI (Fig. 7.6a) is exponential up to the 51–60 % level and then flattens, which, as indicated, likely reflects a limit in deficit accumulation. By comparison, the accelerated increase in the age-specific mortality rates (Fig. 7.6b) is exponential up to the oldest age interval (100+), which is consistent with the Strehler-Mildvan model of mortality and age (Strehler and Mildvan 1960). Note that because the NLTCS sample has an

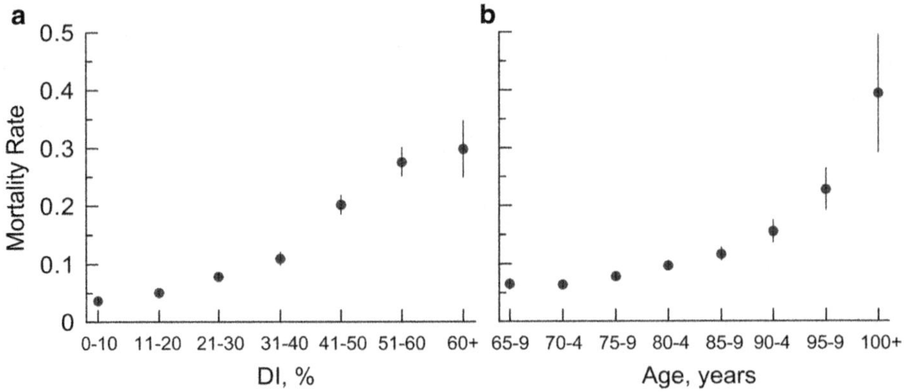

Fig. 7.6 The DI-specific (Panel **a**) and age-specific (Panel **b**) patterns of the mortality rate for NLTCS participants. *Bars* show 95 % confidence intervals

over-representation of disabled individuals, the age-specific mortality rates are larger than those for the general U.S. elderly population aged 65–85 years (Akushevich et al. 2005).

7.4.6 Relative Risks of Death

To better assess the potential of the DI for characterizing the chances of death compared to age, we evaluated the relative risk (RR) of death for the DI and age. The analyses were conducted using univariate (i.e., either the DI or age were included as predictors) and multivariate (both the DI and age were included as predictors) Cox regression models. Table 7.1 shows the results of the analyses for the occurrence of death within 1–4 years of follow-up. Univariate analyses using either the DI or age as a predictor variable show that the RRs resemble those from multivariate analyses with both the DI and age included (Table 7.1). In other words, the DI and age are largely independently associated with mortality risks.

Table 7.1 shows that the RR of death associated with a 1 % increment in the DI (RR_{DI}) is larger than that associated with the 1-year increment in age (RR_{Age}) for a 1-year follow up in the multivariate model. For the longer follow-up times, the RR_{Age} increases whereas the RR_{DI} declines. These opposite dynamics imply a diminishing contribution of specific health factors to the risk of death compared to the non-specific ones which are likely associated with the inherent process of biological senescence.

Because the range of variation in the DI and age are different, we also evaluated the cumulative risk of death due to accumulated deficits (CR_{DI}) and age (CR_{Age}). Specifically, while the DI theoretically ranges between 0 % and 100 %, we defined the range of the DI as zero to the empirical maximum for the DIs in our data and elsewhere (Rockwood and Mitnitski 2006), i.e., 70 %. The range of age in the

Table 7.1 Logarithm of the relative risks (beta) of death within 1–4 years of follow up after examination of NLTCS participants

	Univariate				Multivariate						
Follow up, years	$Beta_{Age}$	SE	$Beta_{DI}$	SE	$Beta_{Age}$	SE	$Beta_{DI}$	SE	CR_{Age}	CR_{DI}	CR_{DI}/CR_{Age}
1	0.048	2.7×10^{-3}	0.042	1.3×10^{-3}	0.033	2.7×10^{-3}	0.039	1.3×10^{-3}	3.7	15.3	4.1
2	0.053	1.9×10^{-3}	0.036	9.1×10^{-4}	0.041	1.9×10^{-3}	0.033	9.3×10^{-4}	5.2	10.1	2.0
3	0.056	1.6×10^{-3}	0.035	7.5×10^{-4}	0.044	1.6×10^{-3}	0.031	7.7×10^{-4}	5.8	8.8	1.5
4	0.056	1.4×10^{-3}	0.033	6.7×10^{-4}	0.046	1.4×10^{-3}	0.029	6.8×10^{-4}	6.3	7.6	1.2

The Cox regression models were adjusted for sex. The $Beta_{Age}$ column contains estimates of the logarithm of the relative risk of death for a 1-year increment in age. The $Beta_{DI}$ contains estimates of the logarithm of the relative risk of death for a 1 % increment in the DI level. CR = cumulative risk of death as defined in the text. $P < 0.001$ for all estimates

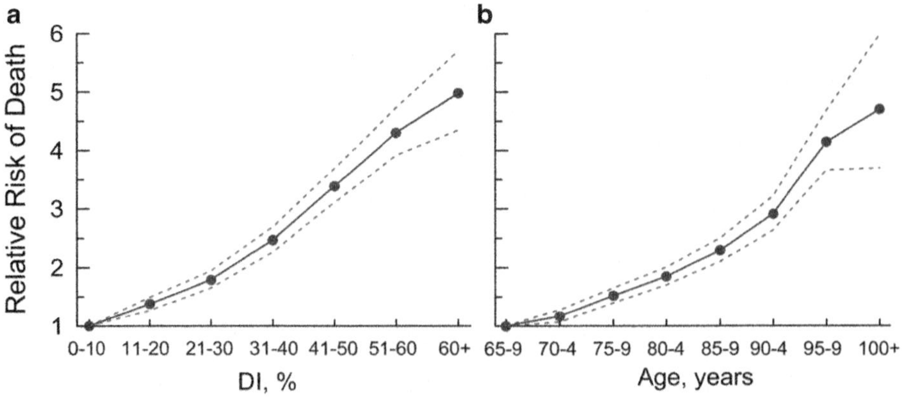

Fig. 7.7 Relative risks of death within 4 years of follow up for NLTCS participants stratified by DI levels (Panel **a**) and age (Panel **b**). *Dashed lines* show 95 % confidence intervals

NLTCS was from 65 years to about 105 years, a span of 40 years. The cumulative risk was calculated for a 70 % DI increment as, e.g., $CR_{DI} = \exp(70 \times 0.039) = 15.3$ for the 1-year follow-up. Similarly, we have for the 40-year age span, e.g., $CR_{Age} = \exp(40 \times 0.033) = 3.7$ for the 1 year follow-up. Table 7.1 shows that the DI predicts the cumulative chances of death better than age, especially within a short-time horizon when the ratio CR_{DI}/CR_{Age} is particularly large. These observations support our conclusion from the analyses in Fig. 7.5, that the DI can better characterize the chances of death than can age.

Given a possible nonlinear effect of the DI on the risk of death (see Fig. 7.3), we evaluated the RRs of stratified DI and age groups for death events occurring within the 4-year follow-up. The risks of death were contrasted to the lowest (0–10 %) DI level and the youngest (65–69) age group (Fig. 7.7). The RR_{DI} increases in a nonlinear fashion similar to that of the RR_{Age}. At smaller DI levels and younger ages, there is an accelerated growth of the risks which decelerates at large DIs and old ages.

The results of these comparative analyses of the roles of the DI and age as indicators of the aging-related processes provide evidence in support of the following three major conclusions. First, the DI can be a useful summary of the aging phenotype in models of mortality, aging, and survival. Second, the DI can characterize aging-related processes independently of age. Third, the DI is a better indicator of these processes than age.

7.5 Longitudinal Analyses: The DI as an Indicator and Predictor of Long Life

Analyses using cross-sectional data can be biased by differential exposures of different birth cohorts to changing environments. More precise information on the role of the DI as a characteristic of the aging-related processes and chances of living long lives is available in longitudinal data collected on the same individuals as they age. The connections of the DI with aging-related processes in longitudinal settings were analyzed using the sample of the U.S. elderly individuals who participated in four waves of the NLTCS conducted in 1984, 1989, 1994, and 1999. The problem addressed in this section is the ability of the DI to differentiate long- and short-life phenotypes in this longitudinal context.

7.5.1 Construction of Long- and Short-Life Phenotypes

Three cohorts of NLTCS respondents with distinct survival profiles were selected. The first cohort comprised individuals who lived short lives and died when they were between 65 and 74 years of age; the *short-lived cohort*, denoted SL. The SL cohort was between 65 and 74 years of age at baseline in the 1984 survey with mean age and standard deviation (MA $\pm$ SD) equal to 68.4 ± 2.4. The second cohort consists of respondents who lived long lives but died when they were 85+ years of age; the *long-lived and deceased cohort*, denoted LLD. The LLD cohort was between 66 and 110 years of age ($\text{MA} \pm \text{SD} = 82.5 \pm 6.5$) at baseline. The third cohort comprises individuals who lived long lives and were alive at the end of the observation in August, 2003; the *long-lived and alive cohort*, denoted LLA. The LLA cohort was between 65 and 92 years of age ($\text{MA} \pm \text{SD} = 70.6 \pm 4.4$) at baseline. The choice of the age cutoffs distinguishing these cohorts is flexible – an analysis of different age cutoffs adjusted to available sample sizes showed that the results are consistent across these methodological decisions.

7.5.2 Longitudinal Changes of the Mean DI in the SL, LLD, and LLA Cohorts

Figure 7.8 shows the longitudinal changes in mean DI in the SL, LLD, and LLA cohorts across the subsequent examinations. These analyses included individuals who were alive in a given NLTCS wave and had no missing information on DI. It can be seen that the DIs for individuals from the SL cohort are larger than those for individuals from the LLD cohort, and the DIs for the LLD cohort are substantially larger than the DIs for individuals from the LLA cohort.

Because individuals with higher DI levels are more likely to die (Fig. 7.6a), the longitudinal patterns of the DI for the SL and LLD cohorts in Fig. 7.8 represent a superposition of two major processes. One is selective survival of robust individuals (i.e., those with low DI levels) who have a greater chance of surviving to older ages. This selection process lowers the mean DI of the surviving members of the cohorts over time. The second process is an accumulation of aging-related health deficits in members of the cohorts. This process increases the mean DI of the cohort

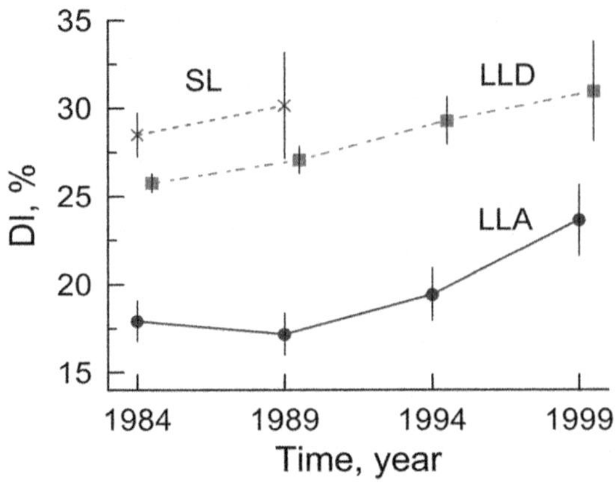

Fig. 7.8 Longitudinal changes of the mean DI among NLTCS respondents in the *SL*, *LLD*, and *LLA* cohorts since the 1984-wave baseline. *Bars* show 95 % confidence intervals

members over time. To delineate compositional (e.g., due to death) and aging-related changes in the DI over time, we followed the same individuals longitudinally. For these analyses, we selected individuals from the SL and LLD cohorts who had no missing observations and were alive in the 1984 and 1989 waves (for the SL) and in the 1984, 1989, and 1994 waves (for the LLD). By construction, all members of the LLA cohort were living in each wave. Figure 7.9a shows that to live long lives individuals should have low DIs and keep this low level over time.

Given the admixture of various birth cohorts in the SL, LLD, and LLA samples and the potential sensitivity of different cohorts to changing social, economic, and health care environments, we examined the dynamics of the mean DI in more homogeneous 10-year birth cohorts. Figure 7.9b shows that the DI patterns for the 1910–1919 birth cohort resemble those in the entire samples shown in Fig. 7.9a, implying no substantial non-additive modulation by non-overlapping birth cohorts. Further, the DI patterns in Fig. 7.9b ensure that long life is indeed characterized by low DI levels over time in relatively homogeneous birth cohorts. Figure 7.9b also shows that older individuals from early birth cohorts from the LLD and LLA samples tend to accumulate deficits with more acceleration than younger individuals from the later birth cohorts.

The results of the longitudinal analyses of the DI variation over time presented in this section confirm our inferences from the cross-sectional analyses that the DI is a sensitive summary of aging-related processes in older individuals. These results show that the DI can robustly differentiate individuals with respect to their chances of living long lives.

7.5.3 *The DI as an Indicator of Frailty*

Because the DI can characterize aging-related processes and lifespan in humans, it can be used as a reliable indicator of frailty. Indeed, frailty is traditionally

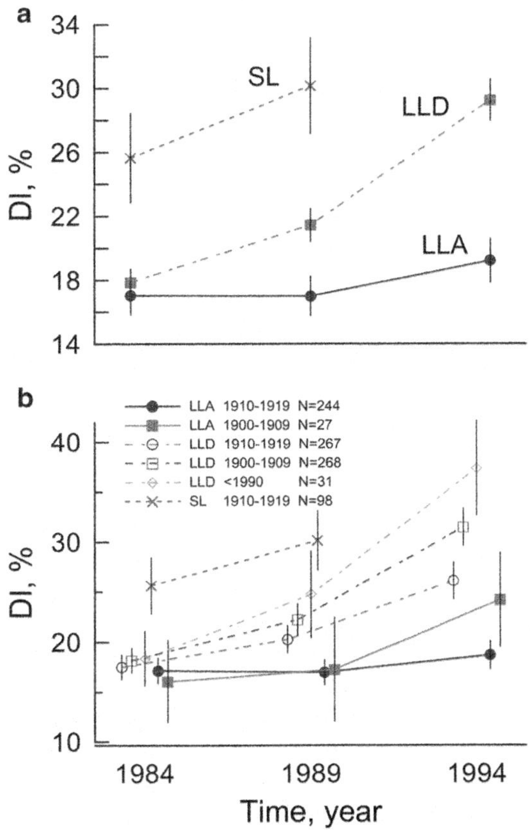

Fig. 7.9 Aging-related dynamics of the mean DI among NLTCS respondents in the SL, LLD, and LLA cohorts who were alive and examined in the 1984 and 1989 waves (for the SL cohort) and in the 1984, 1989, and 1994 waves (for the LLD and LLA cohorts). Panel **a** Dynamics for all birth cohorts combined. Panel **b** Dynamics for 10-year birth cohorts when applicable (shown in the *inset* along with the sample sizes)

considered as a physiological state that results from the general decline of an organism's reserves and deregulation of multiple physiologic systems which is likely associated with biological aging. Frail individuals are believed to have increased non-specific vulnerability and are more susceptible to various adverse health outcomes including death, disability, and hospitalization (Ferrucci et al. 2003; Fried et al. 2001; Mitnitski et al. 2005; Newman et al. 2001; Puts et al. 2005; Woo et al. 2006). The problem is how to define frailty (Bergman et al. 2007; Bortz 2002; Fisher 2005; Lally and Crome 2007; Levers et al. 2006; Rockwood and Mitnitski 2011; Rothman et al. 2008).

One widely-used operational definition of frailty proposed by Fried et al. (2001) distinguishes *phenotypic frailty as a clinical syndrome*, i.e., a set of signs and symptoms that tend to occur together thus characterizing a specific medical condition. This phenotypic frailty definition rests on selected indicators of physical frailty, i.e., unintentional weight loss, exhaustion, weakened grip strength, slow walking, and low physical activity. It is believed that physical frailty is due to physiological aging (the basic cause) and disease (serving as a risk factor) and results in an inability to cope with everyday stresses of life and, thus, in an increased vulnerability to adverse health outcomes (Bergman et al. 2007; Fisher 2005; Fried et al. 2001).

Another widespread approach, that articulated by Rockwood and Mitnitski (Mitnitski et al. 2004; Rockwood et al. 2004), is based on the characterization of *frailty as a non-specific multifactorial state that is better characterized by the quantity rather than the quality of deficits accumulated by individuals during their life courses*, i.e., *by the DI*. According to the DI approach, frailty reflects the impact of physiological aging and results in increased vulnerability to adverse health outcomes, including death. Deficits, however, are considered as non-specific and equally weighted markers of frailty rather than as risk factors. To better understand the rationale behind this (at first glance, oversimplified) approach, let us assume that the impact of physiological aging on the risk of death outweighs the impact of a particular disease at advanced ages. Indeed, while the total and some cause-specific (particularly acute) mortality risks continue to increase among the oldest-old, the *relative* risks of deaths due to particular causes (e.g., fractures, heart disease, cancer, etc.) seem to decline (see, e.g., Center et al. 1999; Forsen et al. 1999; Horiuchi and Wilmoth 1997; Richmond et al. 2003). That is, aging-associated increases in mortality may happen regardless of the specific health disorders yet be accompanied by increasing numbers of deficits (of diverse nature) in individuals.

What is the potential of each of these two approaches for characterizing non-specific vulnerability to death? To address this question, we focused on comparative analyses of the DI approach and the originally defined phenotypic frailty (Fried et al. 2001) in the same dataset: the main cohort of the Cardiovascular Health Study (CHS) (Kulminski et al. 2008b).

7.5.4 The Phenotypic Frailty Index (PFI) and the DI

The PFI was defined using five components: weight loss, exhaustion, low activity, slowness, and grip strength following Fried et al. (2001) as described in Kulminski et al. (2008b). According to this criterion, Fried et al. (2001) defined three frailty phenotypes: robust (no positive components for frailty), pre-frail (1–2 positive components), and frail (3+ positive components) arguing that groups "with three components positive for frailty had significantly worse survival than those with two components, or the "no frailty" groups."

The DI was defined based on counts of 48 deficits including pulmonary diseases; nervous/emotional disorder; high blood pressure; hearing problems; vision problems; heart disease; diabetes; arthritis; cancer; difficulty walking; feeling about life; life satisfaction; people to talk to when lonely; walking for exercise; household chores; mowing lawn; raking lawn; gardening; exercise cycle; dancing; calisthenic exercises; pulmonary embolus; sleep on 2+ pillows to help breathe; awakened by trouble breathing; swelling of feet/ankles; pain in leg; pneumonia; asthma; cough; shortness of breath; palpitations; dizziness; fatigue; weakness; nausea; indigestion;

diarrhea; groggy; trouble falling asleep; walking 0.5 mile, ten steps; difficulties lifting, reaching out, gripping; bleeding; problems staying; hypotension, and major ECG abnormality (Kulminski et al. 2008b).

7.5.5 The PFI and DI as Predictors of Death

Because the DI is a quasi-continuous index ranging theoretically from 0 % to 100 %, whereas the PFI is categorical one, we need an appropriate strategy for comparing their effects on survival. One approach is to categorize the DI into three categories (Kulminski et al. 2008b) which have to be similar to those for the PFI. To do so, the following strategy was adopted. A preliminary categorization was performed arbitrarily. Then it was refined in order to have *the same* estimates of the relative risks of death in the Cox regression model when it includes both the 3-level PFI (PFI3) and 3-level DI (DI3). This procedure yields the estimates for the DI3 and PFI3 shown in Table 7.2 (column "RR") and results in the DI categorization as shown in Table 7.2 (column "Cut-offs") with the frequency of subjects shown in column "N".

The sample then was stratified by categories of the PFI3 and DI3 into nine sub-groups as shown in Table 7.3. Table 7.3 and Fig. 7.10, respectively, show the relative risks of death and survival functions for each subgroup evaluated using the Cox proportional hazard regression model within a 5 year follow-up period after the baseline examination. Individuals who were recognized as frail by both the phenotypic frailty and deficit definitions (Table 7.3; N = 274) have the lowest survival prospects and die faster than those in the other sub-groups. For sub-groups 6 and 8, the RRs and survival are nearly the same. Sub-group 8 is recognized as frail by the PFI and as pre-frail by the DI (Table 7.3; N = 69). Individuals in this sub-group, however, have the same risk of dying as in group 6 (pre-frail by the PFI; N = 879). Thus, the PFI underestimates risks of mortality for 879 individuals, while the DI underestimates the risk for 69 persons. The PFI also recognizes 18 persons

Table 7.2 Relative risks (RR) of death according to the three-level deficit index (DI3) and three-level phenotypic frailty index (PFI3) in the sample of CHS participants

Index	Condition	Cut-offs	N	RR	95 % CI
PFI3	Robust	0 of 5	2008	Reference	
	Prefrail	1–2 of 5	2352	1.55	1.26–1.90
	Frail	3+ of 5	361	2.46	1.85–3.26
DI3	Robust	DI $\leq$22 %	1764	Reference	
	Prefrail	20 % $<$ DI $\leq$31 %	1528	1.51	1.19–1.91
	Frail	DI $>$31 %	1429	2.50	1.97–3.15

CI denotes confidence interval

The models were adjusted for sex and age. To balance the follow-up period and the number of deceased persons, a 5-year follow-up period for survival was used in these analyses

Table 7.3 The number of all individuals (N_{tot}) in each sub-group of CHS participants defined according to the same-risk three-level DI3 and PFI3 (see Table 7.2) and those who died (N_{died}) within a 5-year follow-up period along with relative risks (RR) of death for each category

		DI3-robust			ID	DI3-prefrail			ID	DI3-frail		
PHI3	ID	$N_{tot}(N_{died})$	RR	95 % CI		$N_{tot}(N_{died})$	RR	95 % CI		$N_{tot}(N_{died})$	RR	95 % CI
Robust	1	1098 (54)	Reference		2	634 (50)	1.78	1.21–2.61	3	276 (39)	3.57	2.36–5.40
Pre-frail	4	648 (68)	2.08	1.46–2.98	5	825 (114)	2.65	1.91–3.68	6	879 (164)	4.23	3.09–5.80
Frail	7	18 (1)	1.23	0.17–8.87	8	69 (18)	4.45	2.59–7.63	9	274 (89)	7.10	4.99–10.1

ID is the same as in Fig. 7.10. CI denotes confidence interval

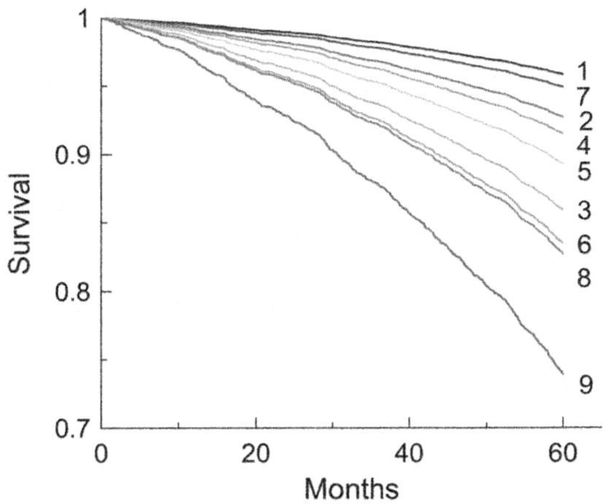

Fig. 7.10 Survival curves smoothed by Cox regression for each of the nine selected sub-groups defined on the basis of categorization of the phenotypic frailty index (PFI) and the deficit index (DI) into three categories (see Table 7.3) as robust, pre-frail, and frail: (1) $PFI3_{robust}$ and $DI3_{robust}$; (2) $PFI3_{robust}$ and $DI3_{pre\text{-}frail}$; (3) $PFI3_{robust}$ and $DI3_{frail}$; (4) $PFI3_{pre\text{-}frail}$ and $DI3_{robust}$; (5) $PFI3_{pre\text{-}frail}$ and $DI3_{pre\text{-}frail}$; (6) $PFI3_{pre\text{-}frail}$ and $DI3_{frail}$; (7) $PFI3_{frail}$ and $DI3_{robust}$; (8) $PFI3_{frail}$ and $DI3_{pre\text{-}frail}$; and (9) $PFI3_{frail}$ and $DI3_{frail}$

(sub-group 7) as frail, while only one of them died within at least 5 years and this group is recognized as robust by the DI.

The results of the comparative analyses of the roles of the DI and PFI as a characteristic of frailty and their ability to discriminate chances of death suggest the following:

1. The PFI has clear advantages for clinical operationalization, since only five substantive characteristics for each person are considered. This is also a weak point of this measure, because this operationalization considerably restricts the flexibility of the approach. Specifically, the foregoing analyses show that, for the proposed scale of robust, pre-frail, and frail phenotypes (Fried et al. 2001), the PFI underestimates the chances of death for 879 persons (who are defined by the PFI3 as pre-frail; ID = 6), while the DI does so for 69 persons (who are defined by the DI3 as pre-frail; ID = 8) under the same categorization. Clearly, the DI can be categorized more finely to more precisely evaluate chances of death.
2. The DI identifies 274 individuals (ID = 9) as frail out of 361 individuals (ID = 7 + 8 + 9) recognized as frail by the phenotypic frailty definition. This observation, along with possible connection of the PFI with a frailty syndrome (Fried et al. 2001), indicates that the DI is also frailty-related.
3. The foregoing results suggest that an integration of both approaches is highly promising for increasing the precision of mortality risk discrimination, especially among the most vulnerable part of the elderly. This means that from the health and well-being history of an individual, the survival chances of elderly individuals can be evaluated more precisely by using both a measure of health/well-being and a more specific measure. This conclusion seems to be intuitively

clear, especially for clinicians, but it has not been formally stated and demonstrated. The DI provides a reasonable alternative for operationalizing this intuitive understanding using an appropriate measure of health/well-being and linking it to the aging-associated processes in an organism (Kulminski et al. 2006, 2007a; Yashin et al. 2007b). A possible disadvantage of the DI associated with problems with its clinical translation is mitigated by wide-spread informational technologies used in clinical practice. From these, the whole-life health and well-being history of an individual can be readily made available to clinicians. Linked with standardized procedures for construction of cumulative indices (Searle et al. 2008), this approach may become a powerful alternative to the phenotypic frailty approach.

7.5.6 *Mid-to-Late Life DIs and Physiological Indices as Characteristics of Long-Term Survival*

Historically, studies of the properties of the DI were largely limited to elderly individuals. Contemporary priorities of aging research include the identification of the most important factors contributing to a long and healthy life throughout the entire life course. Consequently, a focus on a wide spectrum of potential determinants of long and healthy life, as well as on early-life conditions is of importance (Hadley and Rossi 2005). Does the DI retain its predictive power for younger (e.g., middle-aged) individuals? How efficiently can the DI predict survival? Can the DI compete with traditional (e.g., physiological) risk factors? Answers to these questions can contribute greatly to understanding aging-associated processes because the underlying tool, the DI, has the potential of bringing into the analysis additional health dimensions typically ignored due to their small, inconsistent or non-significant effects on survival.

To address these questions, we focus on the DI and a set of five physiological indices (termed endophenotypes) that are among traditional cardiovascular risk factors consistently assessed and documented in the Framingham Heart Study (FHS) (Gagnon et al. 1994; Wilson et al. 1998), namely, systolic blood pressure (SBP, mm Hg), total cholesterol (TC, mg/100 ml), blood glucose (BG, mg/100 ml), body mass index (BMI, kg/m^2), and hematocrit (Htc, %). The analyses are based on data from two representative examinations in the FHS performed in 1964 (9th) and 1974 (14th) using the same 39 deficits (Kulminski et al. 2008b) with comparable diagnostic procedures across time. We use the Cox regression model with a backward likelihood ratio elimination technique to examine the contributions of the DI and the endophenotypes, measured in mid-to-late-life, to long-term survival of the FHS participants. Long-term survival was defined by the currently maximal time horizon available for participants of the 14th examination, i.e., 34 years (the last known vital status assessment was in 2008).

Analyses were initially performed for each examination separately to ensure that the estimates were not affected by possible secular trends. Since the results were comparable across these examinations, we pooled the data to increase the statistical power. We selected 10-year birth cohorts of the same chronological age in the 9th and 14th examinations which did not overlap between the examinations (because they were 10-years apart). Excluding individuals with missing values for either covariate, the sample comprised five birth cohorts with participants aged: (1) younger than 50 years ($N_{9th} = 448$, $N_{14th} = 0$), (2) 50–59 years ($N_{9th} = 1301$, $N_{14th} = 449$), (3) 60–69 years ($N_{9th} = 995$, $N_{14th} = 1234$), (4) 70–79 years ($N_{9th} = 463$, $N_{14th} = 708$), and (5) 80 years and older ($N_{9th} = 0$, $N_{14th} = 215$) at the 9th and 14th examinations. The models were adjusted for sex, age, and smoking status.

7.5.7 *The DI, Endophenotypes, and Long-Term Survival in the FHS*

The analyses of the risks of survival within the entire 34-year follow up period show that the DI and each endophenotype can characterize the risk of death individually, i.e., in a univariate model with one predictor variable (i.e., the DI or endophenotype) included (Table 7.4). However, these effects are sensitive to age. For example, the DI and all five endophenotypes confer significant risks of death only for the 50–59 age group. Multivariate analyses show that the most significant predictors of death are the DI, SBP, and BG. They confer risks of death virtually independently, i.e., additively. The SBP outperforms the DI in terms of significance of the estimates in younger age group (<50 years), whereas the DI is the only highly significant predictor in the oldest group (80+ years). To ensure that deaths that occurred within a shorter follow-up period do not alter inferences for long-term survival, we estimated the models with subjects who died within first 10 years of follow-up excluded. The estimates resemble those for the multivariate models (Table 7.4).

The analyses of the role of the DI and five traditional cardiovascular risk factors (i.e., SBP, BG, BMI, TC, and Htc) measured in mid-to-late life as predictors of long-term survival show that DI, SBP, and BG are the most significant. Further, these three factors show additive effects implying that they characterize non-overlapping health dimensions contributing to mortality through different pathways. This finding is in agreement with additive effects of major physiological risk factors evidenced in prior studies (e.g., Grundy et al. 1999; Wilson et al. 1998). The highly significant effect of the DI on long-term survival of the oldest-old study participants (i.e., 80+ years) suggests that the DI can be a sensitive and particularly informative predictor of longevity for the oldest-old.

Table 7.4 Logarithm of the relative risks (beta) of death for the DI and five physiological endophenotypes in a sample of FHS participants

Age				DI		SBP		BG		BMI		TC		Htc	
Years	N_{total}	N_{died}	Model	Beta	SE	Beta	SE	Beta	SE	Beta	SE	Beta	SE	Beta	SE
<50	448	201	Uni	0.71^{***}	0.19	$0.19^{\dagger}$	0.04			0.28^{**}	0.09	0.04^{**}	0.02		
	448	201	Multi	0.60^{**}	0.20	$0.16^{\dagger}$	0.04					$0.03^{\#}$	0.02		
	426	179	10-years	0.47^{*}	0.22	$0.18^{\dagger}$	0.04								
50–59	1750	1278	Uni	$0.58^{\dagger}$	0.06	$0.14^{\dagger}$	0.01	$0.06^{\dagger}$	0.01	$0.15^{\dagger}$	0.03	0.02^{*}	0.01	0.13^{**}	0.04
	1750	1278	Multi	$0.49^{\dagger}$	0.06	$0.12^{\dagger}$	0.01	$0.05^{\dagger}$	0.01					0.09^{*}	0.04
	1573	1103	10-years	$0.40^{\dagger}$	0.07	$0.11^{\dagger}$	0.02	0.04^{***}	0.01	$0.07^{\#}$	0.04			0.14^{**}	0.05
60–69	2229	2093	Uni	$0.50^{\dagger}$	0.04	$0.09^{\dagger}$	0.01	$0.05^{\dagger}$	0.01	0.07^{*}	0.03				
	2229	2093	Multi	$0.46^{\dagger}$	0.04	$0.08^{\dagger}$	0.01	$0.05^{\dagger}$	0.01						
	1727	1591	10-years	$0.37^{\dagger}$	0.05	$0.08^{\dagger}$	0.01	$0.05^{\dagger}$	0.01						
70–79	1171	1168	Uni	$0.48^{\dagger}$	0.05	$0.09^{\dagger}$	0.02	$0.06^{\dagger}$	0.01						
	1171	1168	Multi	$0.47^{\dagger}$	0.05	$0.08^{\dagger}$	0.02	$0.05^{\dagger}$	0.01	-0.09^{*}	0.04			$-0.08^{\#}$	0.04
	662	660	10-years	$0.35^{\dagger}$	0.07	0.06^{**}	0.02	0.05^{**}	0.02			$-0.02^{\#}$	0.01		
80+[a]	215	215	Uni	$0.50^{\dagger}$	0.11	0.08^{*}	0.04	0.06^{*}	0.03						
	215	215	Multi	$0.49^{\dagger}$	0.11	$0.06^{\#}$	0.04	$0.05^{\#}$	0.03						

Ten-years: the model for the 34-year follow up with deaths occurring within the first 10 years of follow up excluded
SE denotes standard error. Blank cells denote the estimates with $p > 0.1$
The risks were estimated for a 10 % change in the DI, 10 mmHg change in SBP, 10 mg/100 ml change in BG and TC, 5 kg/m^2 change in BMI, and 5 % change in Htc for better visibility of the coefficients
Uni univariate model with one predictor variable (i.e., the DI or endophenotype) included
Multi multivariate model with all predictor variables included
$^{\#}0.05 < p \leq 0.1$, $^{*}0.01 < p \leq 0.05$, $^{**}0.001 < p \leq 0.01$, $^{***}10^{-4} < p \leq 0.001$, $^{\dagger}p \leq 10^{-4}$
[a]The model with deaths occurring within the first 10 years of follow-up excluded for the age group 80+ years was not fitted because too few individuals survived to ages 90+ years

7.6 Conclusion

The results of the analyses presented in this chapter help in conceptualizing a new instrument for measuring aging-related processes in humans, the DI, which is promising for applications in population and clinical settings. Specifically, comparative analyses of statistical properties of various age- and DI-specific outcomes, including frequency distributions, time to death, and mortality, strongly support the role of the DI as a tool to characterize aging-related processes in the elderly independently of age. Accordingly, the DI is not merely a substituent for age, but an alternative summary of the aging-related processes. The ability of the DI to robustly characterize aging-related processes and disentangle phenotypes of short and long lives is confirmed in the analyses using longitudinal data on the same individuals. The DI appears to be a better measure of phenotypic frailty in elderly individuals than an alternative tool developed by Fried and colleagues (2001). An important finding is that the DI can robustly predict longevity and long-term survival of mid-aged individuals which is of inherent concern in public health applications. Our analyses show that the DI in this context outperforms many conventional characteristics such as BMI, lipids, and hematocrit.

Acknowledgements This chapter was partly supported by the National Institute on Aging of the National Institutes of Health under Award Numbers R01AG030198, R01AG032319, R01AG030612, R01AG046860, P01AG043352, and U01 AG023746. The content is solely the responsibility of the authors and does not necessarily represent the official views of the National Institutes of Health. The Framingham Heart Study (FHS) is conducted and supported by the National Heart, Lung, and Blood Institute (NHLBI) in collaboration with Boston University (Contract No. N01-HC-25195). This chapter was not prepared in collaboration with investigators of the FHS and does not necessarily reflect the opinions or views of the FHS, Boston University, or NHLBI. Funding for SHARe Affymetrix genotyping was provided by NHLBI Contract N02-HL-64278. SHARe Illumina genotyping was provided under an agreement between Illumina and Boston University. This chapter was prepared using a limited access dataset obtained from the NHLBI and the Framingham SHARe data obtained through dbGaP.

References

Akushevich, I., Manton, K., Kulminski, A. (2005). *Human mortality and chronic disease incidence at extreme ages: New data and analysis*. Presented at 2005 Population Association of America annual meeting, 31 March–2 April, Philadelphia, USA.

Bergman, H., Ferrucci, L., Guralnik, J., Hogan, D. B., Hummel, S., Karunananthan, S., & Wolfson, C. (2007). Frailty: An emerging research and clinical paradigm – Issues and controversies. *The Journals of Gerontology. Series A, Biological Sciences and Medical Sciences, 62*(7), 731–737.

Bortz, W. M., 2nd. (2002). A conceptual framework of frailty: A review. *The Journals of Gerontology. Series A, Biological Sciences and Medical Sciences, 57*(5), M283–M288.

Center, J. R., Nguyen, T. V., Schneider, D., Sambrook, P. N., & Eisman, J. A. (1999). Mortality after all major types of osteoporotic fracture in men and women: An observational study. *Lancet, 353*(9156), 878–882.

Ferrucci, L., Guralnik, J. M., Cavazzini, C., Bandinelli, S., Lauretani, F., Bartali, B., Repetto, L., & Longo, D. L. (2003). The frailty syndrome: A critical issue in geriatric oncology. *Critical Reviews in Oncology/Hematology, 46*(2), 127–137.

Fisher, A. L. (2005). Just what defines frailty? *Journal of the American Geriatrics Society, 53*(12), 2229–2230.

Forsen, L., Sogaard, A. J., Meyer, H. E., Edna, T., & Kopjar, B. (1999). Survival after hip fracture: Short- and long-term excess mortality according to age and gender. *Osteoporosis International, 10*(1), 73–78.

Fried, L. P., Tangen, C. M., Walston, J., Newman, A. B., Hirsch, C., Gottdiener, J., Seeman, T., Tracy, R., Kop, W. J., Burke, G., & McBurnie, M. A. (2001). Frailty in older adults: Evidence for a phenotype. *The Journals of Gerontology. Series A, Biological Sciences and Medical Sciences, 56*(3), M146–M156.

Gagnon, D. R., Zhang, T. J., Brand, F. N., & Kannel, W. B. (1994). Hematocrit and the risk of cardiovascular disease – The Framingham study: A 34-year follow-up. *American Heart Journal, 127*(3), 674–682.

Goggins, W. B., Woo, J., Sham, A., & Ho, S. C. (2005). Frailty index as a measure of biological age in a Chinese population. *The Journals of Gerontology. Series A, Biological Sciences and Medical Sciences, 60*(8), 1046–1051.

Grundy, S. M., Pasternak, R., Greenland, P., Smith, S., Jr., & Fuster, V. (1999). Assessment of cardiovascular risk by use of multiple-risk-factor assessment equations: A statement for healthcare professionals from the American Heart Association and the American College of Cardiology. *Circulation, 100*(13), 1481–1492.

Gu, D., Dupre, M. E., Sautter, J., Zhu, H., Liu, Y., & Yi, Z. (2009). Frailty and mortality among Chinese at advanced ages. *The Journals of Gerontology. Series B, Psychological Sciences and Social Sciences, 64*(2), 279–289.

Hadley, E. C., & Rossi, W. K. (2005). Exceptional survival in human populations: National Institute on aging perspectives and programs. *Mechanisms of Ageing and Development, 126*(2), 231–234.

Horiuchi, S., & Wilmoth, J. R. (1997). Age patterns of the life table aging rate for major causes of death in Japan, 1951–1990. *The Journals of Gerontology. Series A, Biological Sciences and Medical Sciences, 52*(1), B67–B77.

Kirkwood, T. B. (2002). Molecular gerontology. *Journal of Inherited Metabolic Disease, 25*(3), 189–196.

Kulminski, A., Yashin, A., Ukraintseva, S., Akushevich, I., Arbeev, K., Land, K., & Manton, K. (2006). Accumulation of health disorders as a systemic measure of aging: Findings from the NLTCS data. *Mechanisms of Ageing and Development, 127*(11), 840–848.

Kulminski, A., Ukraintseva, S. V., Akushevich, I., Arbeev, K. G., Land, K., & Yashin, A. I. (2007a). Accelerated accumulation of health deficits as a characteristic of aging. *Experimental Gerontology, 42*(10), 963–970.

Kulminski, A., Yashin, A., Arbeev, K., Akushevich, I., Ukraintseva, S., Land, K., & Manton, K. (2007b). Cumulative index of health disorders as an indicator of aging-associated processes in the elderly: Results from analyses of the National Long Term Care Survey. *Mechanisms of Ageing and Development, 128*(3), 250–258.

Kulminski, A. M., Ukraintseva, S. V., Akushevich, I. V., Arbeev, K. G., & Yashin, A. I. (2007c). Cumulative index of health deficiencies as a characteristic of long life. *Journal of the American Geriatrics Society, 55*(6), 935–940.

Kulminski, A. M., Arbeev, K. G., Ukraintseva, S. V., Culminskaya, I. V., Land, K., & Yashin, A. I. (2008a). Changes in health status among participants of the Framingham Heart Study from the 1960s to the 1990s: Application of an index of cumulative deficits. *Annals of Epidemiology, 18* (9), 696–701.

Kulminski, A. M., Ukraintseva, S. V., Kulminskaya, I. V., Arbeev, K. G., Land, K., & Yashin, A. I. (2008b). Cumulative deficits better characterize susceptibility to death in elderly people than

phenotypic frailty: Lessons from the Cardiovascular Health Study. *Journal of the American Geriatrics Society, 56*(5), 898–903.
Kulminski, A. M., Arbeev, K. G., Christensen, K., Mayeux, R., Newman, A. B., Province, M. A., Hadley, E. C., Rossi, W., Perls, T. T., Elo, I. T., & Yashin, A. I. (2011). Do gender, disability, and morbidity affect aging rate in the LLFS? Application of indices of cumulative deficits. *Mechanisms of Ageing and Development, 132*(4), 195–201.
Lally, F., & Crome, P. (2007). Understanding frailty. *Postgraduate Medical Journal, 83*(975), 16–20.
Levers, M. J., Estabrooks, C. A., & Ross Kerr, J. C. (2006). Factors contributing to frailty: Literature review. *Journal of Advanced Nursing, 56*(3), 282–291.
Mitnitski, A., Graham, J., Mogilner, A., & Rockwood, K. (2002). Frailty, fitness and late-life mortality in relation to chronological and biological age. *BMC Geriatrics, 2*, 1.
Mitnitski, A., Song, X., & Rockwood, K. (2004). The estimation of relative fitness and frailty in community-dwelling older adults using self-report data. *The Journals of Gerontology. Series A, Biological Sciences and Medical Sciences, 59*(6), M627–M632.
Mitnitski, A., Song, X., Skoog, I., Broe, G. A., Cox, J. L., Grunfeld, E., & Rockwood, K. (2005). Relative fitness and frailty of elderly men and women in developed countries and their relationship with mortality. *Journal of the American Geriatrics Society, 53*(12), 2184–2189.
Newman, A. B., Gottdiener, J. S., McBurnie, M. A., Hirsch, C. H., Kop, W. J., Tracy, R., Walston, J. D., & Fried, L. P. (2001). Associations of subclinical cardiovascular disease with frailty. *The Journals of Gerontology. Series A, Biological Sciences and Medical Sciences, 56*(3), M158–M166.
Puts, M. T., Lips, P., & Deeg, D. J. (2005). Sex differences in the risk of frailty for mortality independent of disability and chronic diseases. *Journal of the American Geriatrics Society, 53*(1), 40–47.
Richmond, J., Aharonoff, G. B., Zuckerman, J. D., & Koval, K. J. (2003). Mortality risk after hip fracture. *Journal of Orthopaedic Trauma, 17*(1), 53–56.
Rockwood, K., & Mitnitski, A. (2006). Limits to deficit accumulation in elderly people. *Mechanisms of Ageing and Development, 127*(5), 494–496.
Rockwood, K., & Mitnitski, A. (2011). Frailty defined by deficit accumulation and geriatric medicine defined by frailty. *Clinics in Geriatric Medicine, 27*(1), 17–26.
Rockwood, K., Mogilner, A., & Mitnitski, A. (2004). Changes with age in the distribution of a frailty index. *Mechanisms of Ageing and Development, 125*(7), 517–519.
Rockwood, K., Mitnitski, A., Song, X., Steen, B., & Skoog, I. (2006). Long-term risks of death and institutionalization of elderly people in relation to deficit accumulation at age 70. *Journal of the American Geriatrics Society, 54*(6), 975–979.
Rothman, M. D., Leo-Summers, L., & Gill, T. M. (2008). Prognostic significance of potential frailty criteria. *Journal of the American Geriatrics Society, 56*(12), 2211–2116.
Searle, S. D., Mitnitski, A., Gahbauer, E. A., Gill, T. M., & Rockwood, K. (2008). A standard procedure for creating a frailty index. *BMC Geriatrics, 8*, 24.
Stallard, E. (2002). Underlying and multiple cause mortality at advanced ages: United States 1980-1998. *North American Actuarial Journal, 6*(3), 64–87.
Strehler, B. L., & Mildvan, A. S. (1960). General theory of mortality and aging. *Science, 132*, 14–21.
Vanitallie, T. B. (2003). Frailty in the elderly: Contributions of sarcopenia and visceral protein depletion. *Metabolism, 52*(10 Suppl 2), 22–26.
Wilson, P. W., D'Agostino, R. B., Levy, D., Belanger, A. M., Silbershatz, H., & Kannel, W. B. (1998). Prediction of coronary heart disease using risk factor categories. *Circulation, 97*(18), 1837–1847.
Woo, J., Goggins, W., Sham, A., & Ho, S. C. (2006). Public health significance of the frailty index. *Disability and Rehabilitation, 28*(8), 515–521.
Yashin, A. I., Arbeev, K. G., Kulminski, A., Akushevich, I., Akushevich, L., & Ukraintseva, S. V. (2007a). Cumulative index of elderly disorders and its dynamic contribution to mortality and longevity. *Rejuvenation Research, 10*(1), 75–86.

Yashin, A. I., Arbeev, K. G., Kulminski, A., Akushevich, I., Akushevich, L., & Ukraintseva, S. V. (2007b). Health decline, aging and mortality: How are they related? *Biogerontology, 8*(3), 291–302.

Yu, P., Song, X., Shi, J., Mitnitski, A., Tang, Z., Fang, X., & Rockwood, K. (2012). Frailty and survival of older Chinese adults in urban and rural areas: Results from the Beijing Longitudinal Study of Aging. *Archives of Gerontology and Geriatrics, 54*(1), 3–8.

Chapter 8
Dynamic Characteristics of Aging-Related Changes as Predictors of Longevity and Healthy Lifespan

Anatoliy I. Yashin, Konstantin G. Arbeev, Svetlana V. Ukraintseva, Liubov S. Arbeeva, Igor Akushevich, Julia Kravchenko, Alexander M. Kulminski, Irina Culminskaya, Deqing Wu, and Kenneth C. Land

8.1 Introduction

Individual age trajectories of physiological indices are the product of a complicated interplay among genetic and non-genetic (environmental, behavioral, stochastic) factors that influence the human body during the course of aging. Accordingly, they may differ substantially among individuals in a cohort. Despite this fact, the average age trajectories for the same index follow remarkable regularities. As an illustration, Fig. 8.1a shows the age trajectories of the mean values for selected physiological indices using the data on the original cohort of the Framingham Heart Study (FHS).

It can be seen from Figs. 8.1a and 8.1b that some indices tend to change monotonically with age: the level of blood glucose (BG) increases almost monotonically; pulse pressure (PP) increases from age 40 until age 85, then levels off and shows a tendency to decline only at later ages. The age trajectories of other indices are non-monotonic: they tend to increase first and then decline. Body mass index (BMI) increases up to about age 70 and then declines, diastolic blood pressure (DBP) increases until age 55–60 and then declines, systolic blood pressure (SBP) increases until age 75 and then declines, serum cholesterol (SCH) increases until age 50 in males and age 70 in females and then declines, ventricular rate (VR) increases until age 55 in males and age 45 in females and then declines. With small variations, these general patterns are similar in males and females.

The shapes of the age-trajectories of the physiological variables also appear to be similar for different genotypes. For example, Fig. 8.1b shows the effect of the APOE e4 allele on average age trajectories of eight physiological indices. While the mean values of BMI and cholesterol differ for e4 and non-e4 carriers (at ages above 55), the pattern of the age-related changes per se remains stable. Figures 8.1a and 8.1b also shows that gender influences the age trajectories of physiological variables more substantially than genetic differences in the APOE allele. This may

A.I. Yashin et al., *Biodemography of Aging*, The Springer Series on Demographic Methods and Population Analysis 40, DOI 10.1007/978-94-017-7587-8_8

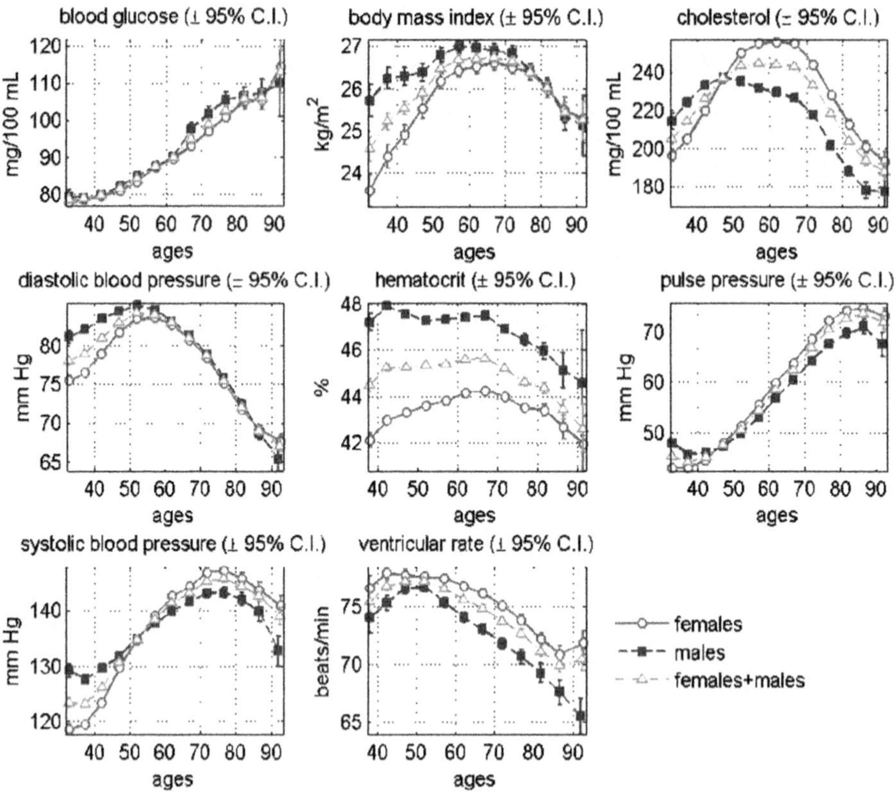

Fig. 8.1a Mean values (± s.e.) of physiological indices among participants of the original cohort of the Framingham Heart Study (FHS) (pooled data of available measurements from exams 1–28)

indicate that factors regulating changes in gene and protein expression during ontogeny may play a more significant role in shaping physiological trajectories than pure genetic variability.

The effects of these physiological indices on mortality risk were studied in Yashin et al. (2006), who found that the effects are gender and age specific. They also found that the dynamic properties of the individual age trajectories of physiological indices may differ dramatically from one individual to the next. The fact that their age dependence affects the shape of the mortality risk function also provides important insights into the mechanisms by which the aging process affects the decline in stress resistance in individuals (Ukraintseva and Yashin 2003; Yashin et al. 2007, 2008, 2009, 2010b).

Researchers continue to study the determinants of the aging rate and the possible contribution of this rate to life span and healthy life span (Arbeev et al. 2005; Austad 2005; Colman et al. 2009; De Martinis et al. 2005; Doubal and Klemera 1990; Nakamura and Miyao 2007, 2008; Nussey et al. 2007; Ruiz-Torres et al. 1994; Ukraintseva and Yashin 2001; Vasto et al. 2010). Since the rate of

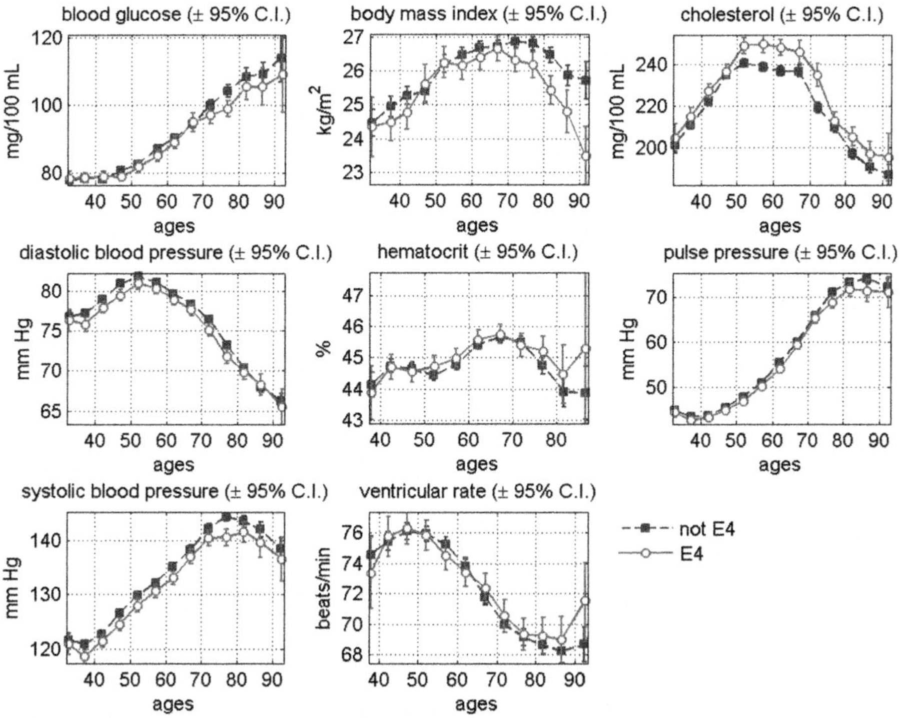

Fig. 8.1b Mean values (± s.e.) of physiological indices (males and females combined) in carriers versus non-carriers of APOE e4 allele, in the FHS original cohort

aging literally means the rate of change with age, it is reasonable to assume that individual differences in the aging rate are manifested in the variability of the dynamic properties of individual age trajectories of the physiological indices. And, if the individual aging rate affects life span and healthy life span, then one can expect that the dynamic characteristics of such trajectories will affect morbidity and mortality risks. A number of studies also highlight the importance of using dynamic properties of individual age-associated changes in physiological indices as characteristics of the aging process that predict morbidity and mortality risks, in addition to the use of the age-specific baseline measurements (Andres et al. 1993; Kerstjens et al. 1997; Pekkanen et al. 1994; Rantanen et al. 2003; Ryan et al. 1999; Sircar et al. 2007; Zureik et al. 1997).

In this chapter, we investigate the effects of parameters describing the dynamic relationships of the age trajectories of eight physiological indices to subsequent morbidity and mortality risks among participants of the FHS Original Cohort.

8.2 Data and Methods

The FHS data and phenotypes of interest are described in Chap. 2. We investigate the dynamic properties of individual age trajectories for the eight physiological indices described above with the objective of identifying "dynamic" risk factors capable of affecting mortality risk and the onset of an unhealthy life. BG was excluded from the list of indices for analyses of the onset of unhealthy life because in the FHS data the onset of diabetes is specifically defined from the values of BG. In the analyses performed in this chapter, we use longitudinal data from the Original Cohort of the Framingham Heart Study (FHS).

8.2.1 Definitions and Evaluation of Dynamic Risk Factors

We first evaluate the effects of the rate of change in physiological indices at ages 40–60 on mortality risk and risk of onset of unhealthy life at ages 60+. For this purpose, we approximated the individual trajectories of those physiological indices that show nearly linear dynamics (for females and/or males) at ages 40–60 (BG, BMI, H, SBP, and PP) by a linear function of the form $y(x) = a_{40-60} + b_{40-60}(x - 40)$, where x is age and y is the value of the physiological index at age x. Individuals having less than five observations of an index at ages 40–60 were excluded from the analyses. Consequently, we have estimates of three risk factors for each individual and each index: an initial value of the index at age 40 (i.e., a_{40-60}, referred to as "**Intercept$_{\mathbf{40-60}}$**" in Tables 8.1 and 8.2 and the text below), the rate of change in the physiological index at ages 40–60 (b_{40-60}, "**Slope$_{\mathbf{40-60}}$**"), and the mean of the absolute values of the residuals, i.e., deviations of observed values of an index from those approximated by a linear function at ages 40–60 ("**Variability$_{\mathbf{40-60}}$**"). The joint effects of these risk factors on mortality and the incidence of unhealthy life were estimated (separately for each physiological index) by a Cox proportional hazards model with delayed entry (the left truncation time was defined as the maximum of the age at the first FHS exam and 60). Individuals with ages at death (onset of unhealthy life)/censoring below age 60 were excluded from the analyses. Note that although the use of linear functions for describing individual aging-related changes is a rough approximation to monotonic changes, it captures an important dynamic risk factor – the average rate of change of each individual index during the age interval between 40 and 60 years.

Second, we evaluated the effects of dynamic characteristics of physiological indices with non-monotonic age trajectories on mortality risk and risk of onset of "unhealthy life." For this purpose, we approximated the age trajectories of such indices (BMI, DBP, SBP, VR, H, and SCH) by two linear functions. The first approximates the increase in the trajectory at the initial interval $[x_L, x_{\max}] : y(x) = a_L + b_L(x - x_L)$, where x is age and y is the value of a physiological index at age x. The second approximates the subsequent decline in the

Table 8.1 Effects of "dynamic" risk factors calculated from individual trajectories of physiological indices at ages 40–60 on ***mortality risk*** at ages 60+ in the Framingham Heart Study (Original Cohort) estimated by Cox proportional hazards regression models

Physiological index	Risk factor (RF)	Mean RF (St. Dev.)	Cox regression model Parameter (S.E.)	Hazard ratio (95 % C.I.)
BG ($N=2135$, $N_e=1946$, $N_c=189$)	**Intercept$_{40-60}$**	77.750 (19.789)	0.005^{*} (0.002)	**1.081 (1.010, 1.157)**
	Slope$_{40-60}$	0.552 (1.953)	$0.068^{\#}$ (0.026)	**1.103 (1.025, 1.188)**
	Variability$_{40-60}$	8.529 (6.826)	$0.017^{\S}$ (0.005)	**1.083 (1.035, 1.133)**
	Sex	0.453 (0.498)	$0.500^{\dagger}$ (0.046)	1.648 (1.505, 1.804)
BMI ($N=742$, $N_e=600$, $N_c=142$)	Intercept$_{40-60}$	25.382 (4.149)	0.014 (0.012)	1.071 (0.952, 1.206)
	Slope$_{40-60}$	0.071 (0.173)	−0.161 (0.267)	0.973 (0.889, 1.064)
	Variability$_{40-60}$	0.639 (0.466)	$0.279^{\#}$ (0.092)	**1.125 (1.042, 1.215)**
	Sex	0.439 (0.497)	$0.453^{\dagger}$ (0.088)	1.573 (1.324, 1.870)
SBP ($N=3025$, $N_e=2800$, $N_c=225$)	**Intercept$_{40-60}$**	127.201 (22.004)	$0.013^{\dagger}$ (0.001)	**1.331 (1.259, 1.408)**
	Slope$_{40-60}$	0.553 (1.242)	$0.147^{\dagger}$ (0.021)	**1.209 (1.146, 1.275)**
	Variability$_{40-60}$	6.358 (2.884)	$0.042^{\dagger}$ (0.007)	**1.141 (1.091, 1.194)**
	Sex	0.433 (0.496)	$0.513^{\dagger}$ (0.039)	1.671 (1.549, 1.802)
PP ($N=3025$, $N_e=2800$, $N_c=225$)	**Intercept$_{40-60}$**	44.377 (12.927)	$0.022^{\dagger}$ (0.002)	**1.304 (1.236, 1.375)**
	Slope$_{40-60}$	0.512 (0.851)	$0.274^{\dagger}$ (0.032)	**1.284 (1.213, 1.359)**
	Variability$_{40-60}$	4.820 (2.038)	$0.038^{\S}$ (0.011)	**1.094 (1.041, 1.151)**
	Sex	0.433 (0.496)	$0.535^{\dagger}$ (0.039)	1.707 (1.582, 1.842)
H ($N=2027$, $N_e=1813$, $N_c=214$)	**Intercept$_{40-60}$**	45.208 (4.634)	$0.073^{\dagger}$ (0.010)	**1.511 (1.353, 1.688)**
	Slope$_{40-60}$	0.004 (0.285)	$0.876^{\dagger}$ (0.141)	**1.300 (1.197, 1.413)**
	Variability$_{40-60}$	1.447 (0.629)	$0.102^{\#}$ (0.038)	**1.084 (1.022, 1.150)**
	Sex	0.455 (0.498)	$0.229^{\S}$ (0.062)	1.257 (1.113, 1.419)

Notes: $^{*}0.01 \le p < 0.05$, $^{\#}0.001 \le p < 0.01$, $^{\S}0.0001 \le p < 0.001$, $^{\dagger}p < 0.0001$, for other estimates: $p \ge 0.05$; **Sex**: 1 male, 0 female; the other **Risk Factors** are continuous and calculated as described in the "Data and Methods" section (Sect. 8.2); N denotes the total number of individuals; N_e is the total number of events (deaths); N_c is the total number of censored individuals; **Hazard Ratios** for continuous risk factors are for an increase from the first quartile to the third quartile of respective empirical distributions

trajectory at the interval $[x_{\max}, x_R]$ after reaching the maximum value $y_{\max} = a_L + b_L(x_{\max} - x_L)$ at age $x_{\max} : y(x) = a_R + b_R(x - x_{\max})$. The intervals $[x_L, x_R]$ for the fit were defined empirically for each index and sex as follows: [35, 55] for VR (females); [40, 60] for VR (males) and SCH (males); [45, 65] for BMI (males) and DBP (females and males); [50, 70] for SCH (females); [55, 75] for BMI (females); [65, 85] for SBP (females and males), and [55, 75] for H. Note that the following restrictions on parameters were used in the estimation procedures: $b_L > 0$, $b_R > 0$, and $a_R = a_L + b_L(x_{\max} - x_L)$, to ensure the appropriate shape of the fit. Individuals having less than six observations of the index at ages $[x_L - 5, x_R + 5]$ and those having estimates of b_L, b_R at the boundary of allowable values (i.e., nearly zero) were excluded from the analyses. As a result, we have estimates of six risk factors for each individual and each index: an initial value of an index at age x_L (i.e., a_L, referred to as "**Intercept$_{2L}$**" in Tables 8.3, and 8.4 and the text below), the rate of

Table 8.2 Effects of "dynamic" risk factors calculated from individual trajectories of physiological indices at ages 40–60 on risk of onset of unhealthy life at ages 60+ in the Framingham Heart Study (Original Cohort) estimated by Cox proportional hazards regression models

Physiological index	Risk factor (RF)	Mean RF (St. Dev.)	Cox regression model Parameter (S.E.)	Hazard ratio (95 % C.I.)
BMI ($N=545$, $N_e=464$, $N_c=81$)	**Intercept$_{40-60}$**	25.159 (3.987)	0.030* (0.015)	**1.147 (1.003, 1.311)**
	Slope$_{40-60}$	0.079 (0.165)	0.426 (0.318)	1.071 (0.969, 1.185)
	Variability$_{40-60}$	0.599 (0.438)	−0.030 (0.122)	0.989 (0.901, 1.085)
	Sex	0.421 (0.494)	0.548† (0.103)	1.729 (1.413, 2.116)
SBP ($N=2318$, $N_e=1929$, $N_c=389$)	**Intercept$_{40-60}$**	125.695 (20.931)	0.009† (0.002)	**1.221 (1.142, 1.305)**
	Slope$_{40-60}$	0.552 (1.189)	0.137† (0.026)	**1.184 (1.112, 1.260)**
	Variability$_{40-60}$	6.127 (2.727)	0.039† (0.009)	**1.120 (1.061, 1.182)**
	Sex	0.409 (0.492)	0.567† (0.047)	1.762 (1.607, 1.933)
PP ($N=2318$, $N_e=1929$, $N_c=389$)	Intercept$_{40-60}$	43.871 (12.501)	0.014† (0.003)	1.181 (1.108, 1.260)
	Slope$_{40-60}$	0.480 (0.818)	0.251† (0.039)	1.250 (1.169, 1.337)
	Variability$_{40-60}$	4.668 (1.953)	0.049§ (0.013)	1.120 (1.055, 1.189)
	Sex	0.409 (0.492)	0.591† (0.047)	1.805 (1.646, 1.979)
H ($N=1519$, $N_e=1278$, $N_c=241$)	Intercept$_{40-60}$	44.898 (4.633)	0.063† (0.011)	1.417 (1.254, 1.602)
	Slope$_{40-60}$	0.012 (0.286)	0.894† (0.163)	1.308 (1.188, 1.439)
	Variability$_{40-60}$	1.441 (0.637)	0.100* (0.045)	1.083 (1.009, 1.162)
	Sex	0.425 (0.495)	0.371† (0.070)	1.449 (1.262, 1.663)

Notes: *$0.01 \le p < 0.05$, #$0.001 \le p < 0.01$, §$0.0001 \le p < 0.001$, †$p < 0.0001$, for other estimates: $p \ge 0.05$; **Sex**: 1 male, 0 female; the other **Risk Factors** are continuous and calculated as described in the "Data and Method" section (Sect. 8.2); N denotes the total number of individuals; N_e is the total number of events (onset of unhealthy life); N_c is the total number of censored individuals; **Hazard Ratios** for continuous risk factors are for an increase from the first quartile to the third quartile of respective empirical distributions

increase in the physiological index at ages $[x_L, x_{\max}]$ (b_L, "**Left Slope**"), the maximal value of the index approximated by two linear functions describing the increase and decline in the respective individual indices ($y_{\max}$, "**Max Index**"), age at reaching the maximal value of the index ($x_{\max}$, "**Age Max**"), the rate of decline in the index at ages $[x_{\max}, x_R]$ (b_R, "**Right Slope**," see also Fig. 8.2 for an illustration), and the mean of the absolute values of residuals, i.e., the deviations of the observed values of an index from those approximated by two linear functions at ages $[x_L, x_R]$ ("**Variability$_{2L}$**"). The joint effects of these risk factors on mortality and incidence of unhealthy life were estimated (separately for each physiological index) by the Cox proportional hazards model with delayed entry (the left truncation time was defined as the maximum of the age at the first FHS exam and x_R). Individuals with ages at death (onset of unhealthy life) or censoring below x_R were excluded from the analyses. If x_R was different for females and males for some index, then the maximum of the two values was used in the (sex-adjusted) model applied to that index. Note that all calculations were performed for individual age trajectories of the indices. As a result, each individual was characterized by a vector of dynamic parameters.

Table 8.3 Effects of "dynamic" risk factors calculated from individual trajectories of physiological indices with non-monotonic patterns on all-cause ***mortality risk*** in the Framingham Heart Study (Original Cohort) estimated by Cox proportional hazards regression models

			Cox regression model	
Physiological index	Risk factor (RF)	Mean RF (St. Dev.)	Parameter (S.E.)	Hazard ratio (95 % C.I.)
BMI ($N=1428$, $N_e=1231$, $N_c=197$)	Age max	65.722 (7.457)	−0.006 (0.005)	0.911 (0.790, 1.050)
	Max index	27.724 (4.439)	−0.035 (0.019)	0.835 (0.690, 1.011)
	Intercept$_{2L}$	25.953 (4.068)	0.041^{*} (0.020)	1.224 (1.012, 1.481)
	Left slope	0.229 (0.413)	−0.023 (0.100)	0.994 (0.946, 1.045)
	Right slope	−0.327 (2.076)	$-0.097^{\dagger}$ (0.011)	0.977 (0.972, 0.982)
	Variability$_{2L}$	0.733 (0.436)	$0.360^{\dagger}$ (0.072)	1.177 (1.105, 1.254)
	Sex	0.384 (0.486)	$0.396^{\dagger}$ (0.061)	1.486 (1.319, 1.674)
DBP ($N=2982$, $N_e=2774$, $N_c=208$)	Age max	55.045 (7.104)	−0.002 (0.003)	0.974 (0.893, 1.063)
	Max index	86.958 (10.530)	$0.016^{\dagger}$ (0.004)	1.229 (1.118, 1.352)
	Intercept$_{2L}$	81.001 (11.704)	0.002 (0.004)	1.020 (0.936, 1.111)
	Left slope	0.855 (1.787)	−0.009 (0.019)	0.991 (0.958, 1.026)
	Right slope	−1.114 (2.732)	$-0.031^{\dagger}$ (0.005)	0.963 (0.951, 0.975)
	Variability$_{2L}$	3.994 (1.388)	$0.100^{\dagger}$ (0.015)	1.181 (1.125, 1.239)
	Sex	0.424 (0.494)	$0.459^{\dagger}$ (0.039)	1.582 (1.466, 1.707)
SBP ($N=1316$, $N_e=1124$, $N_c=192$)	Age max	77.113 (6.781)	−0.003 (0.006)	0.964 (0.831, 1.117)
	Max index	152.515 (17.810)	−0.002 (0.004)	0.958 (0.814, 1.127)
	Intercept$_{2L}$	137.922 (16.383)	0.007 (0.004)	1.173 (0.996, 1.382)
	Left slope	1.481 (1.469)	0.007 (0.030)	1.011 (0.925, 1.104)
	Right slope	−3.007 (7.775)	$-0.024^{\dagger}$ (0.003)	0.930 (0.915, 0.946)
	Variability$_{2L}$	8.323 (2.979)	0.022 (0.011)	1.086 (0.996, 1.184)
	Sex	0.305 (0.461)	$0.367^{\dagger}$ (0.066)	1.444 (1.269, 1.643)
VR ($N=1479$, $N_e=1280$, $N_c=199$)	Age max	47.705 (7.207)	−0.001 (0.005)	0.994 (0.882, 1.119)
	Max index	81.312 (10.886)	$0.019^{\dagger}$ (0.004)	1.299 (1.184, 1.426)
	Intercept$_{2L}$	68.735 (20.786)	$-0.006^{\#}$ (0.002)	0.901 (0.832, 0.975)
	Left slope	1.635 (3.135)	−0.025 (0.016)	0.957 (0.905, 1.013)
	Right slope	−1.064 (2.126)	0.006 (0.017)	1.007 (0.964, 1.052)
	Variability$_{2L}$	5.027 (2.001)	0.035^{*} (0.015)	1.091 (1.014, 1.173)
	Sex	0.546 (0.498)	$0.569^{\dagger}$ (0.064)	1.766 (1.558, 2.001)
SCH ($N=2182$, $N_e=2082$, $N_c=100$)	Age max	55.736 (8.264)	0.003 (0.004)	1.037 (0.941, 1.143)
	Max index	261.990 (42.515)	0.001 (0.001)	1.057 (0.955, 1.170)
	Intercept$_{2L}$	231.096 (51.812)	−0.000 (0.001)	0.977 (0.891, 1.071)
	Left slope	4.683 (6.591)	0.001 (0.006)	1.006 (0.949, 1.067)
	Right slope	−4.609 (10.450)	−0.003 (0.002)	0.984 (0.966, 1.003)
	Variability$_{2L}$	13.711 (6.309)	$0.011^{\#}$ (0.004)	1.081 (1.025, 1.140)
	Sex	0.390 (0.488)	$0.466^{\dagger}$ (0.065)	1.593 (1.402, 1.810)
H ($N=2193$, $N_e=2004$, $N_c=189$)	Age max	65.879 (7.058)	$-0.013^{\S}$ (0.004)	0.837 (0.762, 0.919)
	Max index	46.740 (3.235)	0.016 (0.011)	1.069 (0.975, 1.172)
	Intercept$_{2L}$	44.200 (4.013)	0.016 (0.009)	1.073 (0.991, 1.162)
	Left slope	0.369 (0.697)	0.032 (0.039)	1.012 (0.983, 1.042)

(continued)

Table 8.3 (continued)

Physiological index	Risk factor (RF)	Mean RF (St. Dev.)	Cox regression model	
			Parameter (S.E.)	Hazard ratio (95 % C.I.)
	Right slope	−0.520 (2.041)	$-0.028^{\S}$ (0.008)	0.989 (0.983, 0.995)
	Variability$_{2L}$	1.370 (0.541)	$0.126^{\#}$ (0.043)	1.084 (1.027, 1.145)
	Sex	0.413 (0.492)	$0.325^{\dagger}$ (0.055)	1.385 (1.242, 1.543)

Notes: $^{*}0.01 \leq p < 0.05$, $^{\#}0.001 \leq p < 0.01$, $^{\S}0.0001 \leq p < 0.001$, $^{\dagger}p < 0.0001$, for other estimates: $p \geq 0.05$; **Sex**: 1 male, 0 female; the other **Risk Factors** are continuous and calculated as described in the "Data and Method" section (Sect. 8.2); N denotes the total number of individuals; N_e is the total number of events (deaths); N_c is the total number of censored individuals; **Hazard Ratios** for continuous risk factors are for an increase from the first quartile to the third quartile of respective empirical distributions

We also evaluated empirical (Kaplan-Meier) estimates of survival functions (and probabilities of staying free of the diseases defining the onset of unhealthy life) for individuals with different values of the dynamic risk factors based on the indices with non-monotonic trajectories (separately for females and males). For each physiological index and each dynamic risk factor ("**Age Max**," "**Max Index**," "**Intercept$_{2L}$**," "**Left Slope**," "**Right Slope**," and "**Variability$_{2L}$**"), we calculated the values of the risk factor for all eligible individuals from the sample using the procedure described above. Then we evaluated the medians of the resulting empirical distributions of risk factors, separately for females and males. These median values were used to define the sex-specific strata for estimation of survival curves. We assigned individuals of each sex to one of two strata depending on whether the value of the risk factor for this individual was below (this stratum is denoted as "lower half" in Figs. 8.3, 8.4, 8.5, and 8.6) or above (denoted as "upper half" in Figs. 8.3, 8.4, 8.5, and 8.6) the (sex-specific) median value. In case of an odd number of individuals, the individual with the value of the risk factor equal to the median was assigned to the upper stratum. Then we calculated the Kaplan-Meier estimates of the survival curves (conditional on the sex- and index-specific ages x_R) for individuals in these two strata. Note that individuals with ages at death (onset of unhealthy life) or censoring below x_R were excluded from the analyses, as described above.

The graphs resulting from these calculations are shown in Figs. 8.3, 8.4, 8.5, and 8.6. For example, the median value of the right slopes calculated for BMI in females equals −0.485. Hence, individuals from the stratum denoted as "lower half" in the upper left graph of Fig. 8.3 are females with values of the right slope of BMI smaller than −0.485. Individuals belonging to the stratum named "upper half" in the upper left graph of Fig. 8.3 are females with values of the right slope of BMI larger than −0.485. The conditional survival curves for the two strata presented in this figure deal with individuals who survived to age 75 years or older, which is the value of x_R for BMI in females, as described above.

Table 8.4 Effects of "dynamic" risk factors calculated from individual trajectories of physiological indices with non-monotonic patterns on risk of onset of unhealthy life in the Framingham Heart Study (Original Cohort) estimated by Cox proportional hazards regression models

Physiological index	Risk factor (RF)	Mean RF (St. Dev.)	Cox regression model Parameter (S.E.)	Hazard ratio (95 % C.I.)
BMI ($N=642$, $N_e=461$, $N_c=181$)	Age max	65.530 (7.377)	0.006 (0.008)	1.095 (0.873, 1.373)
	Max index	27.131 (4.081)	0.061* (0.031)	1.383 (1.001, 1.911)
	Intercept_{2L}	25.289 (3.797)	−0.042 (0.032)	0.824 (0.621, 1.094)
	Left slope	0.251 (0.506)	−0.185 (0.150)	0.953 (0.883, 1.029)
	Right slope	−0.193 (0.556)	0.137 (0.111)	1.025 (0.986, 1.067)
	Variability_{2L}	0.707 (0.441)	0.091 (0.126)	1.040 (0.936, 1.154)
	Sex	0.286 (0.452)	0.478† (0.106)	1.614 (1.310, 1.988)
DBP ($N=1984$, $N_e=1588$, $N_c=396$)	Age max	55.312 (7.084)	0.004 (0.005)	1.054 (0.936, 1.187)
	Max index	85.795 (10.039)	0.006 (0.005)	1.074 (0.954, 1.209)
	Intercept_{2L}	79.737 (11.383)	0.009 (0.005)	1.109 (0.991, 1.241)
	Left slope	0.817 (1.441)	0.050 (0.030)	1.048 (0.992, 1.106)
	Right slope	−0.964 (1.882)	−0.011 (0.013)	0.988 (0.959, 1.017)
	Variability_{2L}	3.892 (1.321)	0.029 (0.020)	1.047 (0.983, 1.115)
	Sex	0.388 (0.487)	0.504† (0.052)	1.655 (1.495, 1.833)
SBP ($N=378$, $N_e=188$, $N_c=190$)	Age max	77.962 (6.693)	−0.032* (0.016)	0.696 (0.489, 0.991)
	Max index	152.156 (17.512)	0.023# (0.009)	1.757 (1.156, 2.670)
	Intercept_{2L}	135.743 (16.885)	−0.007 (0.008)	0.861 (0.591, 1.254)
	Left slope	1.529 (1.446)	−0.030 (0.068)	0.959 (0.796, 1.156)
	Right slope	−3.065 (9.409)	0.001 (0.008)	1.002 (0.956, 1.050)
	Variability_{2L}	8.028 (2.899)	−0.013 (0.030)	0.956 (0.782, 1.167)
	Sex	0.214 (0.410)	0.350 (0.179)	1.000 (1.000, 1.000)
VR ($N=1087$, $N_e=937$, $N_c=150$)	Age max	47.356 (7.273)	0.011 (0.006)	1.151 (0.984, 1.345)
	Max index	80.437 (10.362)	0.005 (0.005)	1.067 (0.948, 1.200)
	Intercept_{2L}	67.857 (18.848)	0.004 (0.003)	1.071 (0.956, 1.201)
	Left slope	1.642 (2.434)	0.029 (0.024)	1.056 (0.967, 1.154)
	Right slope	−0.897 (1.381)	0.002 (0.030)	1.002 (0.936, 1.073)
	Variability_{2L}	4.997 (2.054)	0.006 (0.018)	1.016 (0.933, 1.106)
	Sex	0.521 (0.500)	0.479† (0.075)	1.615 (1.395, 1.869)
CH ($N=1241$, $N_e=961$, $N_c=280$)	Age max	56.754 (8.208)	−0.004 (0.006)	0.949 (0.809, 1.113)
	Max index	262.260 (42.990)	0.001 (0.001)	1.029 (0.889, 1.190)
	Intercept_{2L}	229.538 (49.338)	0.000 (0.001)	1.017 (0.890, 1.163)
	Left slope	4.882 (6.869)	−0.002 (0.007)	0.991 (0.914, 1.075)
	Right slope	−4.602 (10.158)	−0.002 (0.003)	0.990 (0.960, 1.021)
	Variability_{2L}	13.490 (6.290)	0.007 (0.006)	1.051 (0.968, 1.141)
	Sex	0.314 (0.464)	0.549† (0.099)	1.732 (1.428, 2.101)
H ($N=1054$, $N_e=749$, $N_c=305$)	Age max	65.735 (7.040)	−0.000 (0.006)	0.995 (0.853, 1.160)
	Max index	46.176 (3.164)	0.024 (0.016)	1.108 (0.967, 1.269)
	Intercept_{2L}	43.672 (3.760)	−0.000 (0.014)	0.999 (0.895, 1.116)
	Left slope	0.396 (0.919)	0.014 (0.044)	1.005 (0.973, 1.039)

(continued)

Table 8.4 (continued)

Physiological index	Risk factor (RF)	Mean RF (St. Dev.)	Cox regression model: Parameter (S.E.)	Hazard ratio (95 % C.I.)
	Right slope	−0.485 (1.590)	−0.021 (0.025)	0.993 (0.976, 1.010)
	Variability$_{2L}$	1.363 (0.544)	0.109 (0.071)	1.072 (0.981, 1.171)
	Sex	0.338 (0.473)	0.360† (0.088)	1.434 (1.207, 1.704)

Notes: $^{*}0.01 \leq p < 0.05$, $^{\#}0.001 \leq p < 0.01$, $^{\S}0.0001 \leq p < 0.001$, $^{\dagger}p < 0.0001$, for other estimates: $p \geq 0.05$; **Sex**: 1 male, 0 female; the other **Risk Factors** are continuous and calculated as described in the "Data and Methods" section (Sect. 8.2); N denotes the total number of individuals; N_e is the total number of events (onset of unhealthy life); N_c is the total number of censored individuals; **Hazard Ratios** for continuous risk factors are for an increase from the first quartile to the third quartile of respective empirical distributions

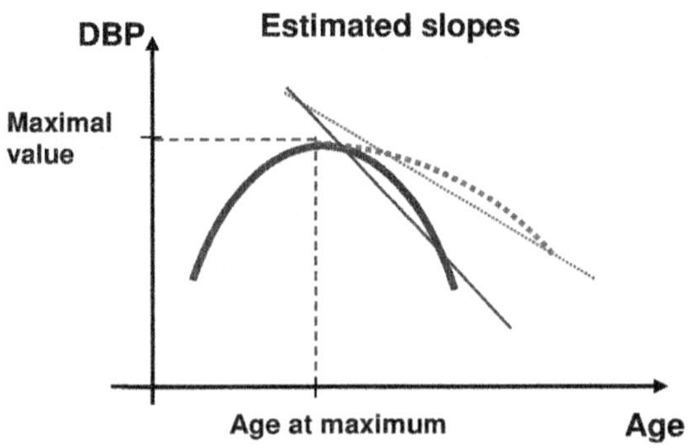

Fig. 8.2 Dynamic characteristics of a hypothetical non-monotonically changing physiological index (denoted here "*DBP*") considered as potential risk factors: (1) Maximum value; (2) Age at which the maximum has been reached; (3) Average rate of decline after reaching the maximum. The figure illustrates evaluation of average rates of decline in two individuals having the same pattern of increase until reaching the maximum, and different patterns of decline after reaching the maximum: (a) the *thick solid line* for a rapidly declining index and the thin *solid line* for its approximation by a straight line; (b) the *thick dotted line* for a slowly declining index and the thin *dotted line* for its linear approximation. The slopes of respective *straight lines* are considered as risk factors for mortality and onset of "unhealthy life"

8.2.2 Statistical Analyses

Statistical analyses and graphic output were performed with SAS/STAT (© SAS Institute Inc.) and MATLAB (© MathWorks Inc.) software packages. P-values for the regression parameters in the tables were calculated using the Wald chi-square statistic with respect to a chi-square distribution with one degree of freedom using SAS/STAT PROC PHREG. The log-rank test was used to test the null hypotheses about the equality of the empirical survival curves in the strata. The p-values are shown in Figs. 8.3, 8.4, 8.5, and 8.6 (SAS/STAT PROC LIFETEST was used for these purposes).

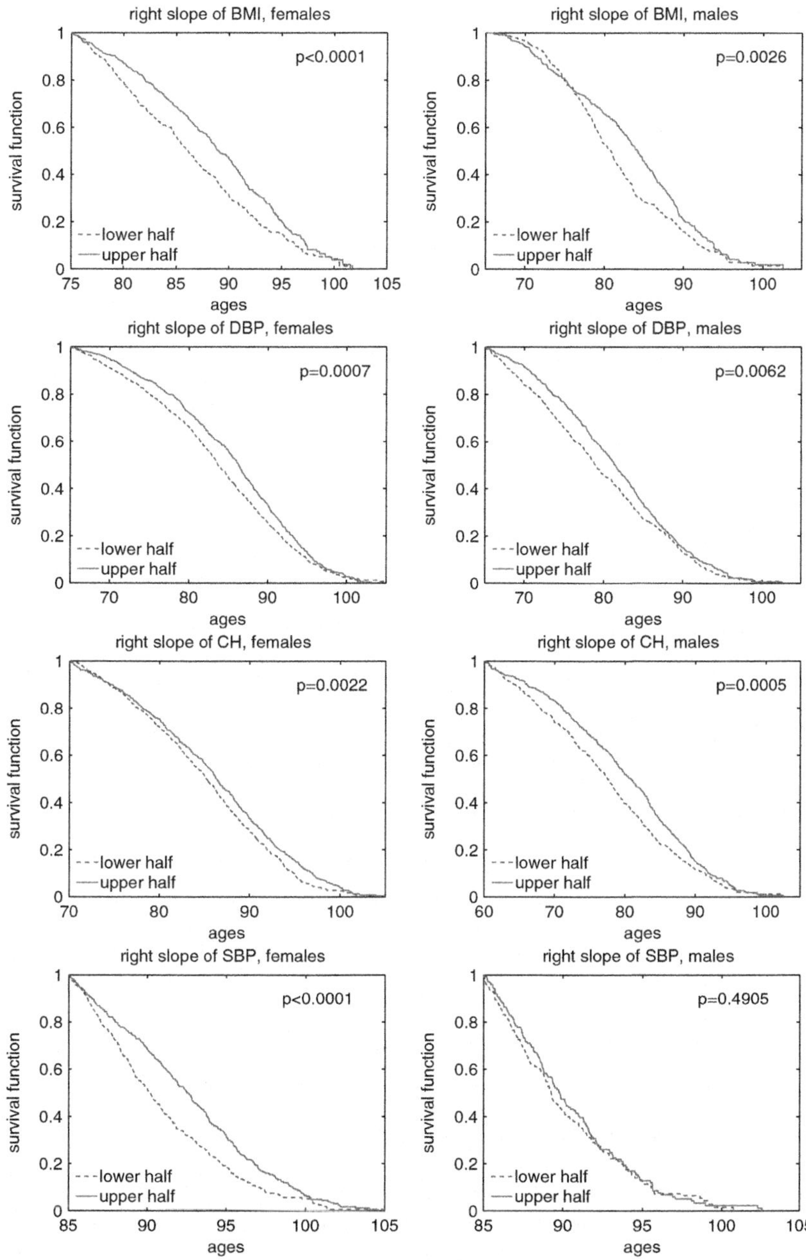

Fig. 8.3 Kaplan-Meier estimates of survival functions for females (*left* columns) and males (*right* columns) having the average rate of decline of different physiological indices after reaching the maximum ("right slope," see "**Data and Methods**" section (Sect. 8.2)) from the lower and upper halves of the empirical distributions of this risk factor for the respective indices; *p* denotes *p*-value for the null hypotheses about the equality of the survival curves in the strata evaluated by the log-rank test

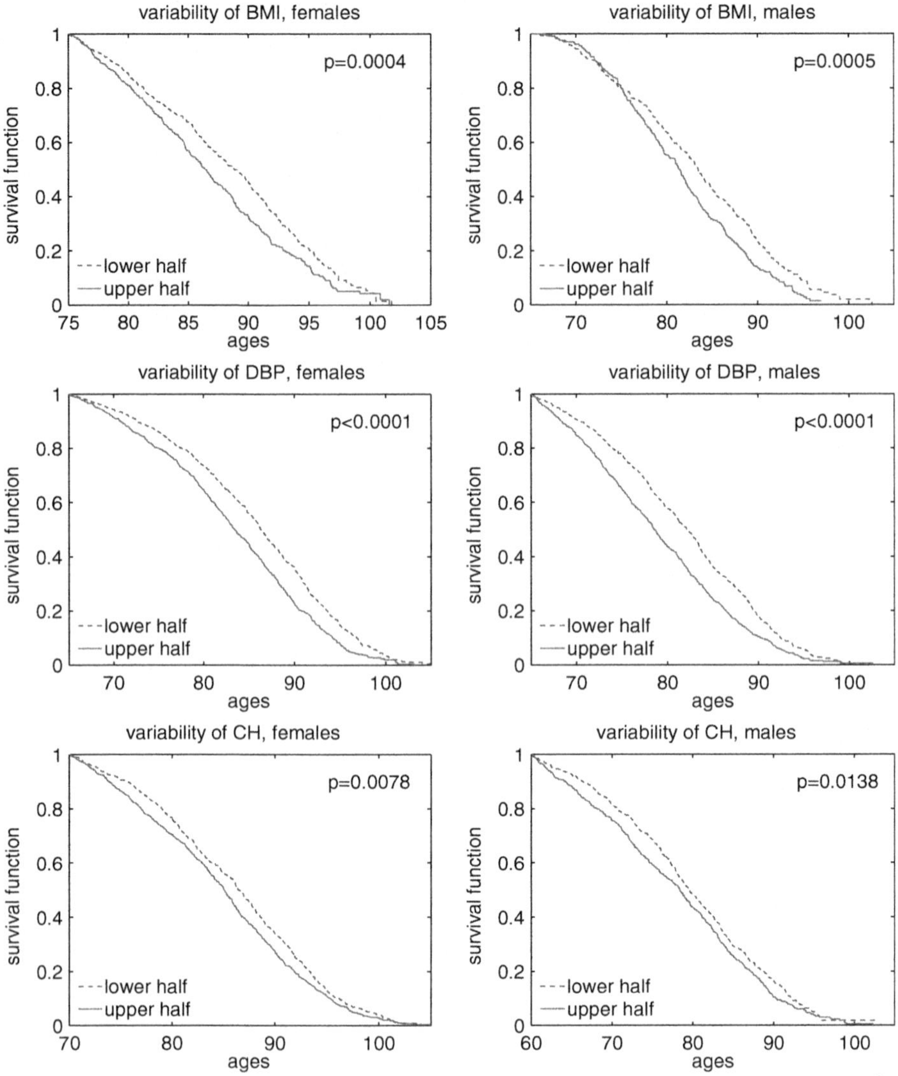

Fig. 8.4 Kaplan-Meier estimates of survival functions for females (*left* columns) and males (*right* columns) having "variability" of different physiological indices (the mean of absolute values of residuals, i.e., deviations of observed values of an index from those approximated by two linear functions at respective age intervals, see "**Data and Methods**" section (Sect. 8.2)) from the lower and upper halves of the empirical distributions of this risk factor for the respective indices; p denotes p-value for the null hypotheses about the equality of the survival curves in the strata evaluated by the log-rank test

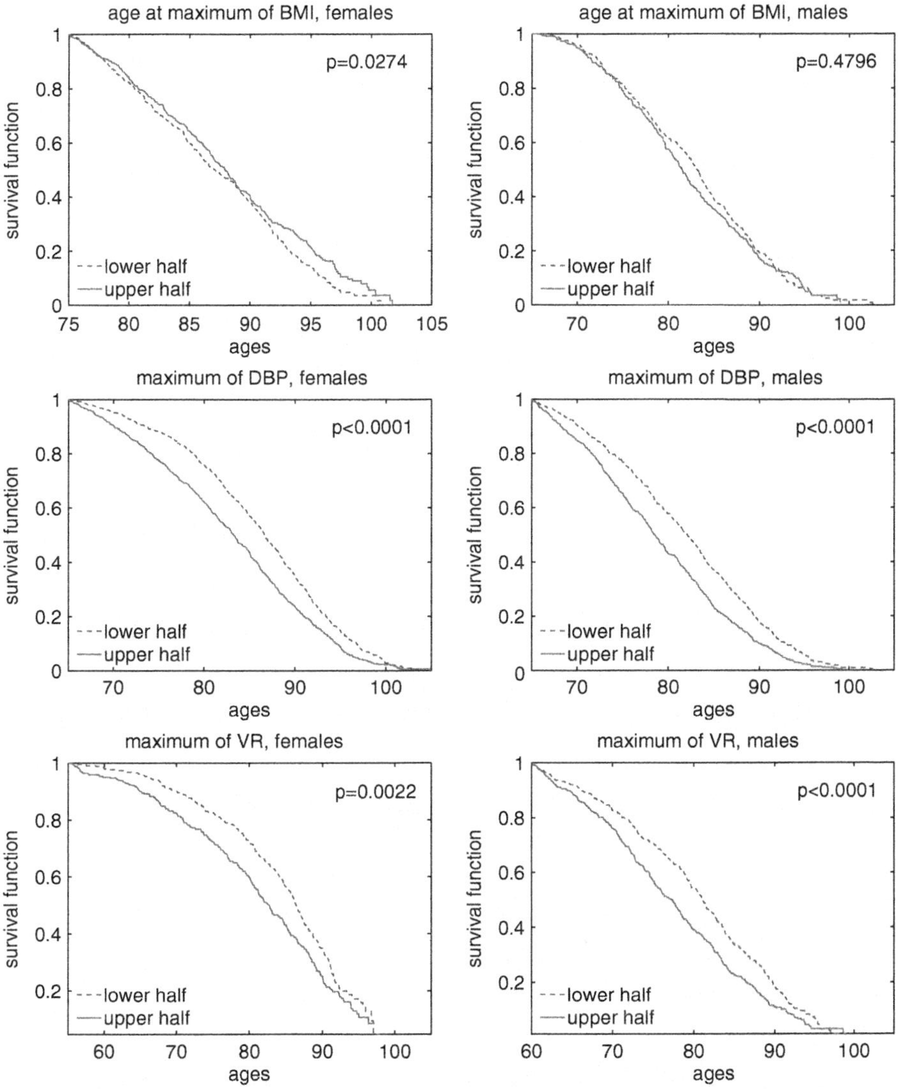

Fig. 8.5 Kaplan-Meier estimates of survival functions for females (*left* columns) and males (*right* columns) having ages at reaching the maximum and the estimated maximal value (see "**Data and Methods**" section (Sect. 8.2)) of different physiological indices from the lower and upper halves of the empirical distributions of these risk factors for the respective indices; *p* denotes *p*-value for the null hypotheses about the equality of the survival curves in the strata evaluated by the log-rank test

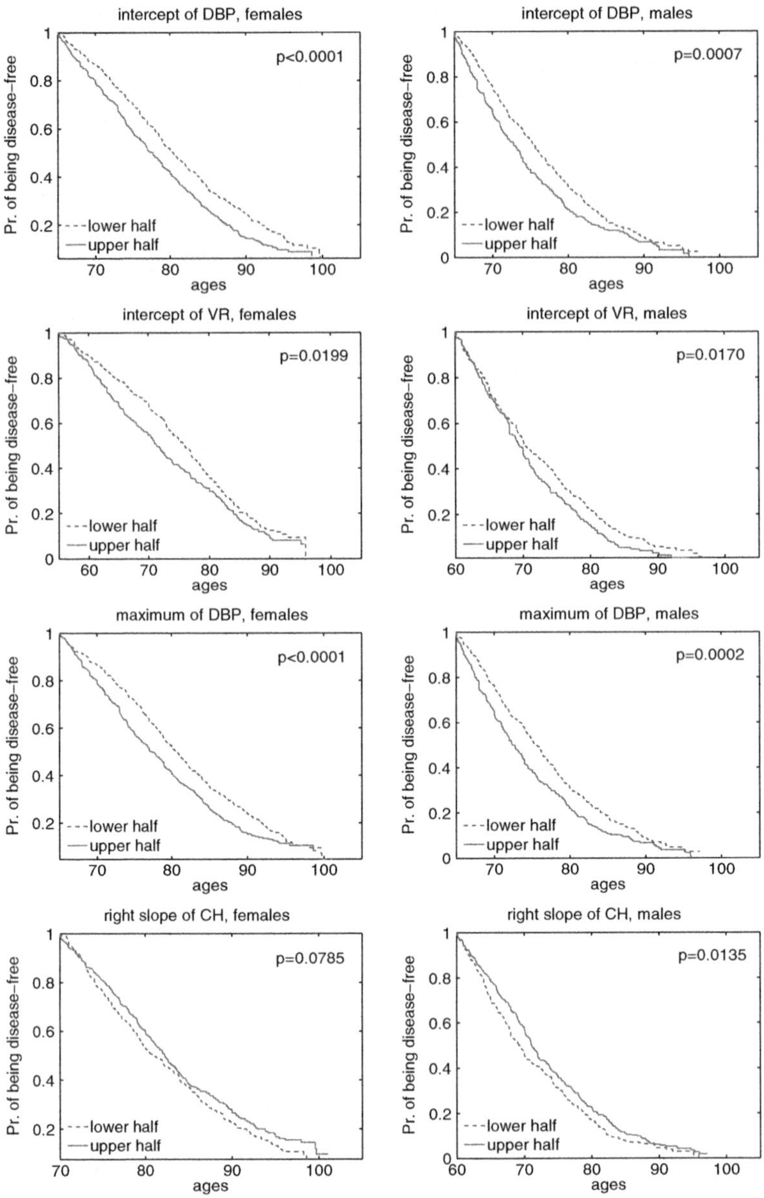

Fig. 8.6 Kaplan-Meier estimates of probabilities of staying free of the diseases defining the onset of unhealthy life for females (*left* columns) and males (*right* columns) having initial values of diastolic blood pressure (DBP) at age 65, initial values of ventricular rate (VR) at age 55 (females) and 60 (males), the estimated maximal values of DBP, and the average rates of decline in total cholesterol (CH) after reaching the maximum ("intercepts," "maximum," and "right slope," respectively, see "**Data and Methods**" section (Sect. 8.2)) from the lower and upper halves of the empirical distributions of these risk factors; *p* denotes *p*-value for the null hypotheses about the equality of the survival curves in the strata evaluated by the log-rank test

8.3 Results

8.3.1 *Effects of Individual Dynamics of Physiological Indices at Ages 40–60 on Mortality Risk and Risk of Onset of "Unhealthy Life" at Ages 60+*

As described in the "**Data and Methods**" section (Sect. 8.2), we evaluated the effects of individual dynamics of physiological indices at ages 40–60 on mortality risk and risk of onset of unhealthy life at ages 60+ for those indices that have a nearly linear pattern of change during the age interval 40–60 for both females and males. Table 8.1 shows the estimates of the joint effects of these risk factors on mortality as evaluated by Cox proportional hazards models. The variability around the average linear trajectory ("$\mathbf{Variability_{40-60}}$") is a significant risk factor for mortality for all indices and the average rate of change between ages 40 and 60 ("$\mathbf{Slope_{40-60}}$") is a significant risk factor for BG, SBP, and PP. The significance is highest ($p < 0.0001$) for the slopes of SBP and PP. The initial value of the index at age 40 ("$\mathbf{Intercept_{40-60}}$") is also a highly significant ($p < 0.0001$) risk factor for mortality for SBP and PP (i.e., higher values of the index at age 40 correspond to higher risk of death compared to smaller values of this index) but not significant for BMI.

The effect of these dynamic characteristics on the incidence of unhealthy life is similar (see Table 8.2). However, the variability is not significant for BMI. The effect of "**Sex**" on both mortality and risk of onset of unhealthy life is highly significant and that the risks for males are higher than those for females.

8.3.2 *Effects of Dynamic Characteristics of Physiological Indices with Non-monotonic Age Trajectories on Mortality Risk and Risk of Onset of "Unhealthy Life"*

For indices with non-monotonic age trajectories, we evaluated the maximum value of each index, the age at which this maximum is reached, the intercept, and the left and right slopes of the linear functions approximating the increase and decline of the index, as described in the "**Data and Methods**" section (Sect. 8.2). Tables 8.3, and 8.4 show the estimates of the joint effects of these risk factors on mortality and incidence of unhealthy life as evaluated by the Cox proportional hazards model. The effect of the rate of decline in the index after reaching the maximum ("**Right Slope**") on mortality risk is highly significant ($p < 0.0001$) for mortality for BMI, DBP, and SBP, but not significant for SCH and VR. In this case, a faster decline in an index after reaching the maximum corresponds to a significant increase in mortality risk (note that the values of "**Right Slope**" are negative by definition, see "**Data and Methods**" section (Sect. 8.2)). By comparison, the rate of increase in the index before reaching the maximum ("**Left Slope**") is not a significant risk factor for mortality for any variable, and the initial value from which the increase commenced is not significant for DBP, SBP, and CH, but highly significant for VR. The estimated maximal value of an index is also a significant risk factor for mortality for DBP and VR (both have $p < 0.0001$). This means that larger (sex-adjusted) maximal values of these indices (reached at "**Age Max**") correspond to significant increases in mortality risk. The variability in all indices except SBP significantly affects mortality risks (larger variability corresponds to higher mortality risks).

The effects of these dynamic characteristics on the risk of onset of unhealthy life are less pronounced than their effects on mortality risks. Table 8.4 shows that the rate of decline after reaching the maximum ("**Right Slope**") and the variability are non-significant for all variables. The maximal value reached is significant only for SBP ($p < 0.0101$) and BMI ($p < 0.05$). The effect of "**Sex**" on both mortality and risk of onset of unhealthy life is significant for all variables except SBP and the risk for males is higher than for females.

8.3.3 *Effects of Dichotomized Dynamic Characteristics of Physiological Indices with Non-monotonic Age Trajectories*

We also evaluated Kaplan-Meier estimates of survival functions for individuals with different values of the dynamic risk factors based on the indices with non-monotonic trajectories by dividing the entire sex-specific samples into strata consisting of individuals with values of each index in the lower and upper halves of the empirical distribution of the index (see "**Data and Methods**" (Sect. 8.2)).

Figure 8.3 shows the estimates of survival functions for females and males having the average rate of decline of different physiological indices after reaching the maximum ("**Right Slope**") from the lower and upper halves of the empirical distributions of each index. It can be seen from this figure that the lower absolute values of the slope (i.e., the lower rates of the post-maximum decline) in individuals from the upper half of the distribution are associated with better survival for all indices except SBP for males (non-significant results for VR for both sexes are not shown). The highest level of statistical significance ($p < 0.0001$) is observed for BMI and SBP in females.

Figure 8.4 presents the corresponding estimates for the "variability" of the different physiological indices (the mean of the absolute values of the residuals, (i.e., the deviations of observed values of the index from those approximated by the two linear functions at the two age intervals; see the "**Data and Methods**" section (Sect. 8.2)). The higher variability of the trajectories of BMI, DBP, and SCH for individuals from the upper half of the distribution are associated with worse survival for both females and males (non-significant results for SBP and VR are not shown). The highest level of statistical significance ($p < 0.0001$) is observed for DBP for both females and males.

Later ages at reaching the maximal value of BMI in females from the upper half of the distribution are associated with better survival (Fig. 8.5); however this was not observed for males. The higher estimated maximal values of DBP and VR in individuals from the upper half of the distribution correspond to worse survival for both females and males. No other indices produced any significant results and none are shown in Fig. 8.5.

Similar calculations for the probabilities of staying free of diseases defining the onset of unhealthy life revealed a more mosaic picture. The most consistent results were observed for DBP (see Fig. 8.6).

Higher initial values of DBP at age 45 and VR at ages 35 (females) and 40 (males) from the upper halves of the distributions are associated with worse chances of staying free of unhealthy life for both sexes. Higher estimated values of DBP reached in individuals from the upper halves of the distributions are associated with worse chances of staying free of unhealthy life for both sexes. Lower rates of post-maximum declines of SCH in males, but not females, from the upper half of the distribution correspond to better chances of staying free of unhealthy life (Fig. 8.6). In addition, lower rates of the post-maximum declines in individuals from the upper half of the distribution of DBP in females ($p = 0.0477$), smaller estimated maximal values of SCH ($p = 0.0054$) and VR ($p = 0.0010$) for males, and a smaller initial value of BMI at age 55 ($p = 0.0107$) for females and a smaller initial value of SCH at age 40 ($p = 0.0100$) for males are associated with better chances of staying free of unhealthy life. Higher "variability" of the trajectories of BMI ($p = 0.0187$) and SBP ($p = 0.0075$) for females, and DBP ($p = 0.0256$) for males in individuals from the upper half of the distribution result in worse chances of staying free of unhealthy life.

8.3.4 Sensitivity Analyses

Questions about the effects of the robustness of the estimates are important given that at most 11 observations for the monotone indices or 15 observations for the non-monotone indices were used (note that for non-monotone indices data from the 30-year intervals $[x_L - 5, x_R + 5]$, where $x_R - x_L = 20$, were used for calculating the dynamic risk factors). These observations were used to estimate two parameters (those of the linear regression) for monotone indices and four parameters (age at reaching the maximal value of the index, intercept and two slopes) for non-monotone indices. To reduce the effect of a poor fit due to a small number of longitudinal observations, we removed those individuals having fewer than five (fewer than six for non-monotone indices) observations from the analyses. The empirical findings could change if different numbers were used for the minimal allowable numbers of observations. However, our sensitivity analyses (Yashin et al. 2010a) showed that the choice of the minimal allowable number of observations does not affect the conclusions.

8.4 Discussion

An increase in the mortality rate with age is traditionally associated with the process of aging. This influence is mediated by aging-associated changes in thousands of biological and physiological variables, some of which have been measured in aging studies. The fact that the age trajectories of some of these variables differ among individuals with short and long life spans and healthy life spans indicates that dynamic properties of the indices affect life history traits. Our analyses of the FHS data clearly demonstrate that the values of physiological indices at age 40 are significant contributors both to life span and healthy life span (as shown by the estimates of $\mathbf{Intercept_{40-60}}$ in Tables 8.1 and 8.2), suggesting that normalizing these variables around age 40 is important for preventing age-associated morbidity and mortality later in life. Two dynamic parameters, $\mathbf{Slope_{40-60}}$ and $\mathbf{Variability_{40-60}}$, also have significant effects on mortality risks (the former being a more important predictor in case of healthy life span).

These results suggest that keeping physiological indices stable over the years of life could be as important as their normalizing around age 40. If so, then *a more advanced dynamic phenotype combining information on both the initial value of an index and the rate and variability of its changes afterwards could potentially be a better predictor of longevity and healthspan.* In Yashin et al. (2009), we provided indirect support for using a combination of dynamic and static aging phenotypes for predicting longevity. In that study, FHS participants who survived to the oldest-old age (>= 90) had not only better (lower) average levels of random blood glucose (BG) at the intercept (circa age 45), but their levels also changed at a slower rate over the life course. Those who nearly reached the oldest-old age (lifespan (LS) = 80–89 years) but did not survive to extreme old age (>= 90) had similarly good (low) BG levels; however, they changed these levels faster compared to the longest-living individuals. If the average BG levels are compared between the LS > =90 and LS = 80–89 groups at circa age 45, almost no differences are evident; thus, information on BG levels at the intercept alone did not predict longevity. However, combining information on both the initial value of BG circa age 45 and the rate of change afterwards could form a better predictor.

The results of that earlier study and the current study, taken together, also indicate that, in the quest of identifying longevity genes, it may be important to look for candidate genes with *pleiotropic* effects on more than one dynamic characteristic of the age-trajectory of a physiological variable, such as genes that may influence both the initial value of a trait (intercept) and the rates of its changes over age (slopes).

Table 8.3 shows that dynamic properties of indices that change non-monotonically with age contribute significantly to mortality risks and further demonstrates the importance of maintaining the stability of physiological states in aging humans: a slower rate of decline in an index after reaching the age at maximum has a more beneficial effect on all-cause mortality.

The finding that the effects of the dynamic characteristics of non-monotonic trajectories of physiological indices on the risk of onset of unhealthy life (Table 8.4) are less significant than on all-cause mortality risks (Table 8.3) indicates that these dynamic characteristics may reflect aging-related processes in the body that result in increasing non-specific vulnerability to death with age rather than in increasing vulnerability to a particular disease.

Findings from prior studies are consistent with the findings reported in this chapter with respect to the importance of taking into account longitudinal changes in physiological indices when evaluating/predicting morbidity and mortality risks. There are, however, few prior studies of the impact and comparative contributions of the dynamic parameters (left and/or right slopes, variability, intercept) on mortality risks.

As mentioned above, in our earlier publications we demonstrated that individuals who have different rates of aging-related changes in BG levels also differ in longevity (Yashin et al. 2009, 2010b). In Arbeev et al. (2011) and Yashin et al. (2012), we evaluated how hidden processes accompanying human aging (such as declines in resistance to stresses and adaptive capacity, age dependent physiological "norms") can be evaluated from longitudinal data. We showed how these components of the aging process can lead to an increase in the risk of death and the risk of onset of unhealthy life with age. Our findings (Arbeev et al. 2012; Yashin et al. 2013) strongly indicate the presence of a genetic component in aging-related mechanisms. Such differences may contribute to the patterns of allele- and sex-specific mortality and incidence rates.

Van Vliet and colleagues (2010) described the dynamics of traditional metabolic risk factors in association with mortality in old age in the Leiden 85-plus Study, a prospective population-based study of 599 participants initially aged 85 years. Participants were annually assessed during a 5-year follow-up period and observed for mortality for 10 years. The authors found that larger declines in BMI, total cholesterol, and diastolic blood pressure, and weaker increases in HDL cholesterol levels, between ages 85 and 90 years, were all associated with increased mortality.

The effects of aging-associated changes in serum cholesterol on coronary and all-cause mortality were evaluated in the Finnish Cohorts of the Seven Countries Study (Pekkanen et al. 1994). Men with the greatest declines in cholesterol levels had increased cardiovascular and all-cause mortality compared with men with the least change in the levels. In the Paris Prospective Study (Zureik et al. 1997), it was shown that not only a low baseline total cholesterol level, but also its decline over time, was associated with a higher cancer mortality.

A study of two independent French male cohorts (Benetos et al. 1999) found that longitudinal changes in systolic and diastolic BP may be more accurate determinants of cardiovascular risks than baseline BP measures. In both cohorts, the group with a long-term increase in systolic and a decrease in diastolic BP (i.e., with an increase in pulse pressure) had the highest relative risk of mortality from CVD compared to the group with no changes in either systolic or diastolic BP, independently of absolute values of BP or other risk factors. Since this study included only males, it is important to note that changes in pulse pressure may have different effects on mortality risks in males and females (Cooper et al. 1994; Hall 1990).

The prognostic importance of baseline values of the heart rate (HR) as well as its variability during 24-h HR monitoring in patients with heart disease and in the general population is generally recognized (Bigger et al. 1992; Huikuri et al. 1998; Kleiger et al. 1987). However, the prognostic role of the long-term and age-related dynamics of HR is not sufficiently investigated and the existing studies are limited. Research on the effects of HR at baseline, final HR, and HR change during follow-up, on the survival of patients attending the Glasgow Blood Pressure Clinic revealed that the highest risk of all-cause mortality was in patients who had increased their HR by $\geq$5 bpm at the end of the follow-up, as compared with those who had a consistently high (high-high) or low (low-low) HR. Paul and colleagues (Paul et al. 2010) found that the change in HR during the follow-up is a better predictor of mortality risk in hypertensive patients than the baseline or final HR.

Body mass index (BMI) is, probably, the most intensively studied index in connection with health and survival. Over recent decades, many studies have addressed the effect of BMI dynamics on morbidity and mortality, especially the effect of losing body weight in overweight/obese people on risk factors for CVD and diabetes. It was shown that overweight adults who lost weight over 9 years had more favorable (lower) total and LDL cholesterol levels compared to normal-weight controls, but less favorable BG levels (Truesdale, Stevens and Cai 2005). In other studies, weight loss was associated with *excess* mortality (when compared with weight stability), even when controlled for confounding due to diseases known to cause both weight loss and increased mortality (Ostergaard et al. 2010; Sorensen 2003). Weight stability was associated with a lower mortality risk as compared with weight change (gain or loss) in Lee and Paffenbarger (1992) and Somes et al. (2002). Nilsson et al. (2002) showed that, for men with decreasing BMIs during 16 years of follow-up, the non-cancer mortality risk was higher than for BMI-stable men. The authors hypothesized that involuntary weight loss in otherwise healthy people could be a sign of premature aging, which in turn causes a non-specific increase in mortality risk. In other studies, baseline weight and weight change had independent effects on total mortality, with both associations exhibiting U-shaped relationships (Kulminski et al. 2008; Mikkelsen et al. 1999).

The eight physiological indices used in this chapter do not exhaust the list of all possible physiological risk factors for mortality and morbidity. Therefore, the dynamic characteristics calculated from these particular indices cannot explain the entire variability of human life span and healthy life span. The association of other indices and risk factors with mortality/morbidity risk can be explored if measurements of such indices are available in a longitudinal study for a substantially long time period. See, for example, Willcox et al. (2006) where midlife risk factors were investigated for a cohort of Japanese American men with 40 years of follow-up.

8.5 Conclusion

Our results indicate that the dynamic characteristics of age-related changes in physiological variables are important predictors of morbidity and mortality risks in aging individuals.

We showed that the initial value (*intercept*), the rate of changes (*slope*), and the *variability* of a physiological index, in the age interval 40–60 years, significantly influenced both mortality risk and onset of unhealthy life at ages 60+ in our analyses of the Framingham Heart Study data. That is, these dynamic characteristics may serve as good predictors of late life morbidity and mortality risks. The results also suggest that physiological changes taking place in the organism in middle life may affect longevity through promoting or preventing diseases of old age.

For non-monotonically changing indices, we found that having a later age at the peak value of the index (*age max*), a lower peak value (*max index*), a slower rate of decline in the index at older ages (*right slope*), and less *variability* in the index over time, can be beneficial for longevity. Also, the dynamic characteristics of the physiological indices were, overall, associated with *mortality* risk more significantly than with onset of unhealthy life. This was especially true for the rate of old age decline in the indices (*right slope*), and their *variability*. The results of this study also indicate that dynamic risk factors, such as slopes, might be even better predictors of longevity and healthspan if they would be considered together with the indices describing the age-specific physiological state (such as intercept) in the framework of a single index.

Previously published epidemiological findings are generally consistent with our results, which indicate the need for further detailed studies of the dynamic parameters of aging-related changes in the human body with further application of these principles to prevention strategies.

Senescence is a key player in physiological changes observed in aging humans. The dynamic properties of these changes are likely to contain important information about individual aging processes. This information, however, can be masked by other factors, such as the effects of compensatory adaptation and remodeling that develop in response to the primary aging process. Studying mechanisms of such adaptation and their connection to morbidity and mortality risks is important for a better understanding of factors shaping the age trajectories of physiological indices as well as incidence and mortality rates.

Acknowledgements The research reported in this chapter was supported by the National Institute on Aging grants R01AG027019, R01AG030612, R01AG030198, 1R01AG046860, and P01AG043352. The content is solely the responsibility of the authors and does not necessarily represent the official views of the National Institute on Aging or the National Institutes of Health. The Framingham Heart Study (FHS) is conducted and supported by the National Heart, Lung and Blood Institute (NHLBI) in collaboration with the FHS Investigators. This chapter was prepared using a limited access dataset obtained from the NHLBI and does not necessarily reflect the opinions or views of the FHS or the NHLBI.

References

Andres, R., Muller, D. C., & Sorkin, J. D. (1993). Long-term effects of change in body weight on all-cause mortality. A review. *Annals of Internal Medicine, 119*(7), 737–743.

Arbeev, K. G., Ukraintseva, S. V., Arbeeva, L. S., & Yashin, A. I. (2005). Mathematical models for human cancer incidence rates. *Demographic Research, 12*(10), 237–271.

Arbeev, K. G., Ukraintseva, S. V., Akushevich, I., Kulminski, A. M., Arbeeva, L. S., Akushevich, L., Culminskaya, I. V., & Yashin, A. I. (2011). Age trajectories of physiological indices in relation to healthy life course. *Mechanisms of Ageing and Development, 132*(3), 93–102.

Arbeev, K. G., Ukraintseva, S. V., Kulminski, A. M., Akushevich, I., Arbeeva, L. S., Culminskaya, I. V., Wu, D., & Yashin, A. I. (2012). Effect of the APOE polymorphism and age trajectories of physiological variables on mortality: Application of genetic stochastic process model of aging. *Scientifica, 2012*, 568628.

Austad, S. N. (2005). Diverse aging rates in metazoans: Targets for functional genomics. *Mechanisms of Ageing and Development, 126*(1), 43–49.

Benetos, A., Rudnichi, A., Thomas, F., Safar, M., & Guize, L. (1999). Influence of heart rate on mortality in a French population – Role of age, gender, and blood pressure. *Hypertension, 33*(1), 44–52.

Bigger, J. T., Jr., Fleiss, J. L., Steinman, R. C., Rolnitzky, L. M., Kleiger, R. E., & Rottman, J. N. (1992). Frequency domain measures of heart period variability and mortality after myocardial infarction. *Circulation, 85*(1), 164–171.

Colman, R. J., Anderson, R. M., Johnson, S. C., Kastman, E. K., Kosmatka, K. J., Beasley, T. M., Allison, D. B., Cruzen, C., Simmons, H. A., Kemnitz, J. W., & Weindruch, R. (2009). Caloric restriction delays disease onset and mortality in rhesus monkeys. *Science, 325*(5937), 201–204.

Cooper, L. T., Cooke, J. P., & Dzau, V. J. (1994). The vasculopathy of aging. *Journal of Gerontology, 49*(5), B191–B196.

De Martinis, M., Franceschi, C., Monti, D., & Ginaldi, L. (2005). Inflamm-ageing and lifelong antigenic load as major determinants of ageing rate and longevity. *FEBS Letters, 579*(10), 2035–2039.

Doubal, S., & Klemera, P. (1990). Influence of aging rate change on mortality curves. *Mechanisms of Ageing and Development, 54*(1), 75–85.

Hall, P. M. (1990). Hypertension in women. *Cardiology, 77*(Suppl. 2), 25–30.

Huikuri, H. V., Makikallio, T. H., Airaksinen, K. E. J., Seppanen, T., Puukka, P., Raiha, I. J., & Sourander, L. B. (1998). Power-law relationship of heart rate variability as a predictor of mortality in the elderly. *Circulation, 97*(20), 2031–2036.

Kerstjens, H. A., Rijcken, B., Schouten, J. P., & Postma, D. S. (1997). Decline of FEV1 by age and smoking status: Facts, figures, and fallacies. *Thorax, 52*(9), 820–827.

Kleiger, R. E., Miller, J. P., Bigger, J. T., Jr., & Moss, A. J. (1987). Decreased heart rate variability and its association with increased mortality after acute myocardial infarction. *American Journal of Cardiology, 59*(4), 256–262.

Kulminski, A. M., Arbeev, K. G., Kulminskaya, I. V., Ukraintseva, S. V., Land, K., Akushevich, I., & Yashin, A. I. (2008). Body mass index and nine-year mortality in disabled and nondisabled older U.S. Individuals. *Journal of the American Geriatrics Society, 56*(1), 105–110.

Lee, I. M., & Paffenbarger, R. S., Jr. (1992). Change in body weight and longevity. *JAMA: Journal of the American Medical Association, 268*(15), 2045–2049.

Mikkelsen, K. L., Heitmann, B. L., Keiding, N., & Sorensen, T. I. (1999). Independent effects of stable and changing body weight on total mortality. *Epidemiology, 10*(6), 671–678.

Nakamura, E., & Miyao, K. (2007). A method for identifying biomarkers of aging and constructing an index of biological age in humans. *Journals of Gerontology Series A: Biological Sciences and Medical Sciences, 62*(10), 1096–1105.

Nakamura, E., & Miyao, K. (2008). Sex differences in human biological aging. *Journals of Gerontology Series A: Biological Sciences and Medical Sciences, 63*(9), 936–944.

Nilsson, P. M., Nilsson, J. A., Hedblad, B., Berglund, G., & Lindgarde, F. (2002). The enigma of increased non-cancer mortality after weight loss in healthy men who are overweight or obese. *Journal of Internal Medicine, 252*(1), 70–78.

Nussey, D. H., Kruuk, L. E. B., Morris, A., & Clutton-Brock, T. H. (2007). Environmental conditions in early life influence ageing rates in a wild population of red deer. *Current Biology, 17*(23), R1000–R1001.

Ostergaard, J. N., Gronbaek, M., Schnohr, P., Sorensen, T. I. A., & Heitmann, B. L. (2010). Combined effects of weight loss and physical activity on all-cause mortality of overweight men and women. *International Journal of Obesity, 34*(4), 760–769.

Paul, L., Hastie, C. E., Li, W. L. S., Harrow, C., Muir, S., Connell, J. M. C., Dominiczak, A. F., McInnes, G. T., & Padmanabhan, S. (2010). Resting heart rate pattern during follow-up and mortality in hypertensive patients. *Hypertension, 55*(2), 567–574.

Pekkanen, J., Nissinen, A., Vartiainen, E., Salonen, J. T., Punsar, S., & Karvonen, M. J. (1994). Changes in serum cholesterol level and mortality: A 30-year follow-up. The Finnish cohorts of the seven countries study. *American Journal of Epidemiology, 139*(2), 155–165.

Rantanen, T., Volpato, S., Ferrucci, L., Heikkinen, E., Fried, L. P., & Guralnik, J. M. (2003). Handgrip strength and cause-specific and total mortality in older disabled women: Exploring the mechanism. *Journal of the American Geriatrics Society, 51*(5), 636–641.

Ruiz-Torres, A., Gimeno, A., & Munoz, F. J. (1994). Estimation of the aging rates of populations through the prevalence of age-related diseases. *Archives of Gerontology and Geriatrics, 19*(3), 235–242.

Ryan, G., Knuiman, M. W., Divitini, M. L., James, A., Musk, A. W., & Bartholomew, H. C. (1999). Decline in lung function and mortality: The Busselton Health Study. *Journal of Epidemiology and Community Health, 53*(4), 230–234.

Sircar, K., Hnizdo, E., Petsonk, E., & Attfield, M. (2007). Decline in lung function and mortality: Implications for medical monitoring. *Occupational and Environmental Medicine, 64*(7), 461–466.

Somes, G. W., Kritchevsky, S. B., Shorr, R. I., Pahor, M., & Applegate, W. B. (2002). Body mass index, weight change, and death in older adults: The systolic hypertension in the elderly program. *American Journal of Epidemiology, 156*(2), 132–138.

Sorensen, T. I. (2003). Weight loss causes increased mortality: Pros. *Obesity Reviews, 4*(1), 3–7.

Truesdale, K. P., Stevens, J., & Cai, J. (2005). The effect of weight history on glucose and lipids: The Atherosclerosis Risk in Communities Study. *American Journal of Epidemiology, 161*(12), 1133–1143.

Ukraintseva, S. V., & Yashin, A. I. (2001). How individual age-associated changes may influence human morbidity and mortality patterns. *Mechanisms of Ageing and Development, 122*(13), 1447–1460.

Ukraintseva, S. V., & Yashin, A. I. (2003). Individual aging and cancer risk: How are they related? *Demographic Research, 9*(8), 163–196.

van Vliet, P., Oleksik, A. M., van Heemst, D., de Craen, A. J. M., & Westendorp, R. G. J. (2010). Dynamics of traditional metabolic risk factors associate with specific causes of death in old age. *Journals of Gerontology Series A: Biological Sciences and Medical Sciences, 65*(5), 488–494.

Vasto, S., Scapagnini, G., Bulati, M., Candore, G., Castiglia, L., Colonna-Romano, G., Lio, D., Nuzzo, D., Pellicano, M., Rizzo, C., Ferrara, N., & Caruso, C. (2010). Biomarkers of aging. *Frontiers in Bioscience (Scholar Edition), S2*(2), 392–402.

Willcox, B. J., He, Q. M., Chen, R., Yano, K., Masaki, K. H., Grove, J. S., Donlon, T. A., Willcox, D. C., & Curb, J. D. (2006). Midlife risk factors and healthy survival in men. *JAMA: Journal of the American Medical Association, 296*(19), 2343–2350.

Yashin, A. I., Akushevich, I. V., Arbeev, K. G., Akushevich, L., Ukraintseva, S. V., & Kulminski, A. (2006). Insights on aging and exceptional longevity from longitudinal data: Novel findings from the Framingham Heart Study. *Age, 28*(4), 363–374.

Yashin, A. I., Arbeev, K. G., Akushevich, I., Kulminski, A., Akushevich, L., & Ukraintseva, S. V. (2007). Stochastic model for analysis of longitudinal data on aging and mortality. *Mathematical Biosciences, 208*(2), 538–551.

Yashin, A. I., Arbeev, K. G., Akushevich, I., Kulminski, A., Akushevich, L., & Ukraintseva, S. V. (2008). Model of hidden heterogeneity in longitudinal data. *Theoretical Population Biology, 73*(1), 1–10.

Yashin, A. I., Ukraintseva, S. V., Arbeev, K. G., Akushevich, I., Arbeeva, L. S., & Kulminski, A. M. (2009). Maintaining physiological state for exceptional survival: What is the normal level of blood glucose and does it change with age? *Mechanisms of Ageing and Development, 130*(9), 611–618.

Yashin, A. I., Arbeev, K. G., Akushevich, I., Arbeeva, L., Kravchenko, J., Il'yasova, D., Kulminski, A., Akushevich, L., Culminskaya, I., Wu, D., Ukraintseva, S. V. (2010a). Dynamic determinants of longevity and exceptional health. *Current Gerontology and Geriatrics Research*, Article ID 381637, *2010*, 1–13.

Yashin, A. I., Arbeev, K. G., Akushevich, I., Ukraintseva, S. V., Kulminski, A., Arbeeva, L. S., & Culminskaya, I. (2010b). Exceptional survivors have lower age trajectories of blood glucose: Lessons from longitudinal data. *Biogerontology, 11*(3), 257–265.

Yashin, A. I., Arbeev, K. G., Ukraintseva, S. V., Akushevich, I., & Kulminski, A. (2012). Patterns of aging related changes on the way to 100: An approach to studying aging, mortality, and longevity from longitudinal data. *North American Actuarial Journal, 16*(4), 403–433.

Yashin, A. I., Arbeev, K. G., Wu, D., Arbeeva, L. S., Kulminski, A., Akushevich, I., Culminskaya, I., Stallard, E., Ukraintseva, S. (2013) How lifespan associated genes modulate aging changes: lessons from analysis of longitudinal data. *Frontiers in Genetics, 4*(3), 1–20.

Zureik, M., Courbon, D., & Ducimetiere, P. (1997). Decline in serum total cholesterol and the risk of death from cancer. *Epidemiology, 8*(2), 137–143.

Chapter 9
The Complex Role of Genes in Diseases and Traits in Late Life: An Example of the Apolipoprotein E Polymorphism

Alexander M. Kulminski, Anatoliy I. Yashin, Irina Culminskaya, Kenneth C. Land, and Svetlana V. Ukraintseva

9.1 Genes and Diseases in Late Life

Decades of studies of candidate genes show that they are not linked to aging-related traits in a straightforward manner (Finch and Tanzi 1997; Martin 2007). Recent genome-wide association studies (GWAS) have reached fundamentally the same conclusion by showing that the traits in late life likely are controlled by a relatively large number of common genetic variants (e.g., Teslovich et al. 2010). Further, GWAS often show that the detected associations are of tiny effect (Stranger et al. 2011).

The primary reason for complex actions of genes on traits in late life is the lack of evolutionary-programmed direct mechanisms linking genes, which are inherited from parents at conception, to aging-related traits, which occur in the post-reproductive period (Di Rienzo and Hudson 2005; Vijg and Suh 2005). Therefore, the complexity of gene actions on traits in late life appears to be inherent. For example, recent studies have demonstrated that genes could show antagonistic pleiotropy (postulated by Williams 1957) whereby the same gene could be advantageous for fitness traits in early life but became detrimental for diseases at old ages (Alexander et al. 2007; Kulminski et al. 2010; Martin 2007; Schnebel and Grossfield 1988; Summers and Crespi 2010; Williams and Day 2003). Another example is that the effect of the same allele on the same trait in late life can be different at different ages (Bergman et al. 2007; De Benedictis et al. 1998; Ilveskoski et al. 1999; Martin 2007; Ukraintseva 2005; Yashin et al. 2001).

The inherent complexity of gene actions on traits in late life can well explain why many genetic signals appear to be weak. Indeed, the weak effect of genes on traits in late life can be not only because they confer small risks having small penetrance but because they confer large risks but in a complex fashion (Kulminski et al. 2010). Accordingly, aging-related processes can be the key to a better understanding of the nature of weak genetic effects, and, consequently, the genetic origin of traits in late life.

A.I. Yashin et al., *Biodemography of Aging*, The Springer Series on Demographic Methods and Population Analysis 40, DOI 10.1007/978-94-017-7587-8_9

In this chapter, we consider several examples of complex modes of gene actions, including genetic tradeoffs, antagonistic genetic effects on the same traits at different ages, and variable genetic effects on lifespan. The analyses focus on the *APOE* common polymorphism.

The Study Population We focus on the participants of the original FHS cohort and the FHS offspring (FHSO) cohort (Cupples et al. 2009; Dawber 1980; Gail and Johnson 1989; Govindaraju et al. 2008; Splansky et al. 2007). In short, the FHS includes N = 5209 respondents aged 28–62 years at baseline who have been biennially followed for about 60 years. The FHSO respondents (N = 5124) aged 5–70 years at baseline were the biological descendants (N = 3514), their spouses (N = 1576), and adopted offspring (N = 34) of the FHS participants, who have been followed for about 36 years. The FHS/FHSO participants have been monitored for the onset of CVD, cancer, and death through regular examinations at the FHS clinic, surveillance of hospital admissions, and death registries (Govindaraju et al. 2008; Splansky et al. 2007), currently through 2008. Biospecimens were mostly collected in the late 1980s and through 1990s from surviving participants (Lahoz et al. 2001; Myers et al. 1996). These genotyped FHS and FHSO participants represent demographically unbiased samples of aged populations (Kulminski et al. 2013). The procedure used for the *APOE* genotyping is described by Lahoz et al. (2001). The data available for this study include information on the *APOE* e2/3/4 polymorphism for the 1258 FHS and 3924 FHSO participants.

Genotypes Following Kulminski et al. (2011), the analyses focus on the effect of the *APOE* e4 (risk; e2/4, e3/4, and e4/4) allele contrasted to the non-e4 allele (e2/2, e2/3, and e3/3) genotypes.

Phenotypes To better understand the age-related complexity of the effects of the APOE gene, we focus on ages at onset of major diseases in humans, i.e., cardiovascular disease (CVD) and cancer as well as on age at death. We consider all CVDs, including diseases of heart and stroke. For cancer, we consider all sites except skin.

Methods Associations of the *APOE* polymorphism with survival and risks of CVD and cancer in genotyped survivors are characterized by the Kaplan-Meier estimator and the Cox proportional hazards regression model. Age at event (i.e., death or onset of the disease) or age at censoring in 2008 is a time variable in these analyses. Because the genotyped FHS and FHSO participants represent demographically unbiased samples of aged populations (Kulminski et al. 2013), the analyses focus on the baseline FHS/FHSO participants to maximize the sample size. The Cox regression model was adjusted for age at baseline and sex, when applicable, if not explicitly stated. We use the robust sandwich estimator of variances in the Cox model to account for potential clustering (e.g., familial) (Lee et al. 1992).

9.2 The Antagonistic Role of the APOE Gene and Two Types of Sexually Dimorphic Tradeoffs: The Case of CVD and Cancer

9.2.1 The FHSO: Tradeoffs in the Effects of the APOE Polymorphism on the Ages at Onset of CVD and Cancer

Survival/Time-to-Event Analyses The results of survival/time-to-event analyses of the probability of remaining free of either CVD or cancer for carriers and non-carriers of the e4 allele in the FHSO cohort are shown in Fig. 9.1. These results show that the same e4 allele can confer an increased risk of CVD but also be protective of cancer, particularly of cancer with onset at older ages.

Sex stratification shows that the CVD-conferring effect mostly occurs among women, whereas the cancer-protective effect mostly occurs among men (Fig. 9.2).

Relative Risks of Diseases To quantify the descriptive observations in Figs. 9.1 and 9.2, we conducted Cox regression analyses. These analyses show that e4 allele carriers have significantly larger relative risks (RRs) of CVD, i.e., RR = 1.22, $p = 0.01$ (Table 9.1). The RR is significant in women (RR = 1.35, $p = 0.011$). The e4 male carriers are also at increased risk of CVD, although the RR does not attain statistical significance.

Contrary to CVD, the e4 allele carriers have smaller risks of cancer. This effect occurs mostly among men, although the RR does not attain statistical significance (Table 9.1).

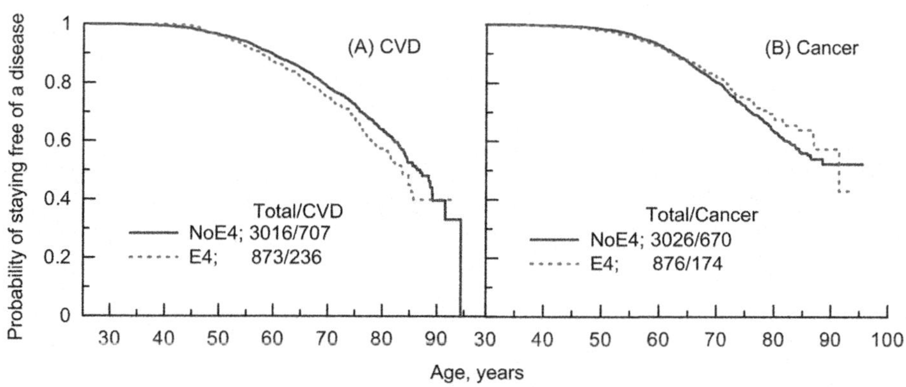

Fig. 9.1 Descriptive age patterns of the probability of remaining free of **a** CVD and **b** cancer for the FHSO carriers (E4) and non-carriers (NoE4) of the *APOE* e4 allele. The numbers in the *insets* show the total number of genotyped subject with non-missing information and the number of CVD or cancer cases among them

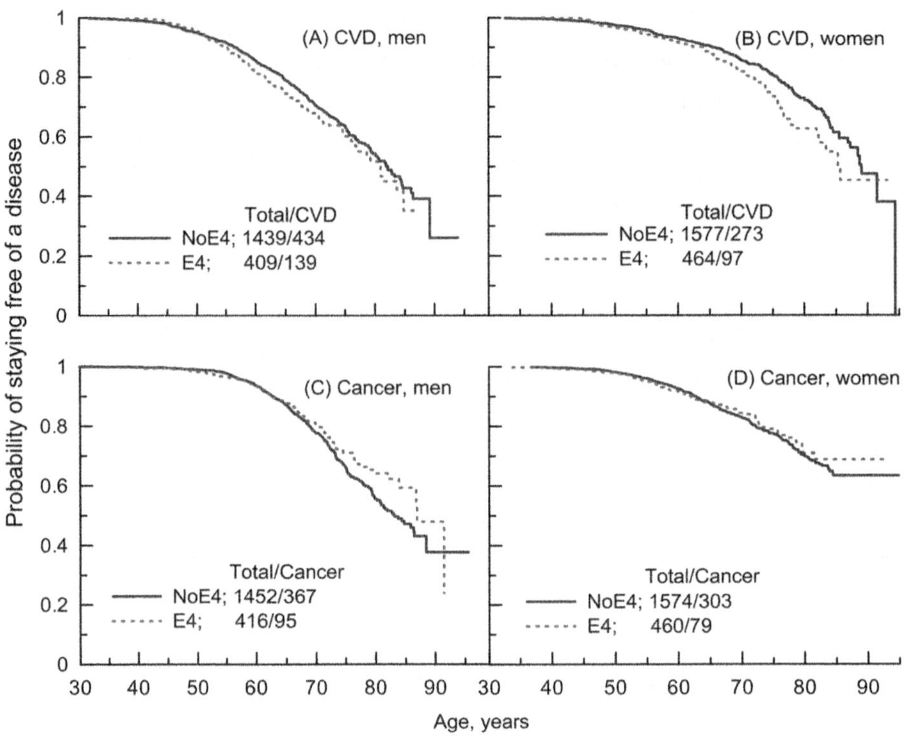

Fig. 9.2 Descriptive age patterns of the probability of remaining free of (**a**, **b**) CVD and (**c**, **d**) cancer for the FHSO (**a**, **c**) men and (**b**, **d**) women carrying (E4) and not carrying (NoE4) the *APOE* e4 allele. The numbers in the *insets* show the total number of genotyped subject with non-missing information and the number of CVD or cancer cases among them

Table 9.1 Relative risks of CVD and cancer in the genotyped participants of the FHSO cohort

	CVD			Cancer		
Sample	RR	*p*	95 % CI	RR	*p*	95 % CI
Men and women	1.22	0.010	1.05–1.41	0.88	0.129	0.74–1.04
Men	1.14	0.170	0.94–1.39	0.83	0.114	0.66–1.05
Women	1.35	0.011	1.07–1.71	0.91	0.447	0.71–1.17

The number of subjects is shown in Figs. 9.1 and 9.2
RR denotes relative risk, *CI* denotes Confidence Interval

Risks of Diseases: The Role of Aging-Related Processes Survival/time-to-event analyses (Fig. 9.2) suggest that the effect of the e4 allele is disproportionally shifted to onsets at older ages, i.e., the effect of the e4 allele is sensitive to aging-related processes. As a result, the Cox regression model with proportional hazards likely provides underpowered estimates in Table 9.1. The most effective way to address the observed disproportionality in the framework of the Cox proportional hazards

Table 9.2 Relative risks of CVD and cancer in more homogeneous "younger" (≤60 years) and "older" (>60 years) groups of the genotyped participants of the FHSO cohort

Sample	Group	CVD				Cancer			
		N_{tot}/N_{CVD}	RR	p	95 % CI	N_{tot}/N_{canc}	RR	p	95 % CI
Men and women	≤60 year	964/404	1.04	0.747	0.83–1.30	802/259	1.09	0.522	0.83–1.44
Men	≤60 year	518/263	1.03	0.840	0.78–1.35	379/111	1.12	0.603	0.73–1.71
Women	≤60 year	446/141	1.07	0.713	0.74–1.57	423/148	1.10	0.605	0.77–1.56
Men and women	>60 year	2925/539	1.21	0.062	0.99–1.48	3100/585	0.78	0.019	0.64–0.96
Men	>60 year	1330/310	1.07	0.633	0.82–1.39	1489/351	0.76	0.045	0.59–0.99
Women	>60 year	1595/229	1.44	0.016	1.07–1.94	1611/234	0.79	0.170	0.57–1.11

RR denotes relative risk, *CI* denotes Confidence interval, N_{tot} and N_{CVD} or N_{canc} denote the total number of genotyped individuals and the number of CVD or cancer cases among them, respectively

regression model is to stratify the sample according to ages at onset of diseases (note that stratification by age at baseline does not solve this problem).

Accordingly, we defined two more homogeneous groups in the FHSO as:

- "younger", those developing CVD or cancer in early life or being censored at younger ages (representatively, 60 years and younger at the end of follow up in 2008);
- "older", those developing CVD or cancer in late life or being censored at older ages (60 years and older at the end of follow up in 2008).

Analyses of the "younger group" show no effect of the e4 allele either on onset of CVD or on onset of cancer. On the other hand, analyses of the "older group" reveal a significant effect of the e4 allele on risk of CVD in older women. They also show significant protective effects of the same allele in decreasing the risks of cancer in men and women combined and in men only at older ages (Table 9.2).

9.2.2 The FHS: The Antagonistic Role of the APOE Polymorphism in CVD and Its Tradeoffs with Cancer

Survival/Time-to-Event Analyses The results of survival/time-to-event analyses of the probabilities of remaining free of either CVD or cancer for carriers and non-carriers of the e4 allele in the FHS original cohort are shown in Fig. 9.3. The pattern for ages at onset of cancer resembles that in the FHSO (Fig. 9.1a); it tends to be protective at older ages. The pattern for onset of CVD is, however, more complex – showing antagonistic effects on risks of CVD at younger and older ages.

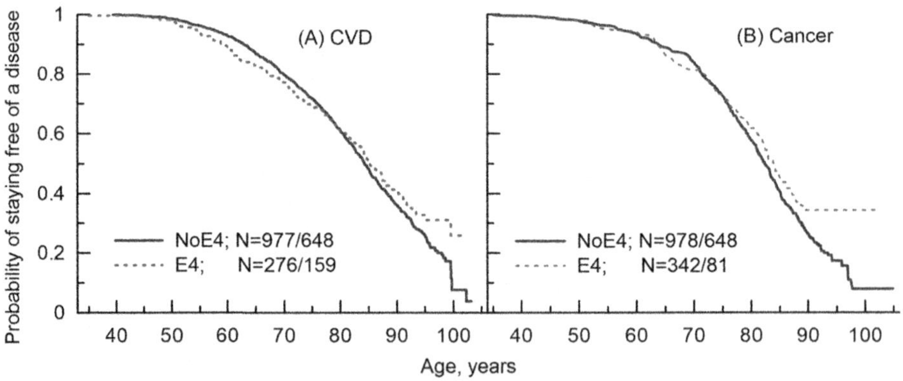

Fig. 9.3 Descriptive age patterns of the probability of remaining free of (**a**) CVD and (**b**) cancer for the FHS carriers (E4) and non-carriers (NoE4) of the *APOE* e4 allele. The numbers in the *insets* show the total number of genotyped subject with non-missing information and the number of CVD or cancer cases among them

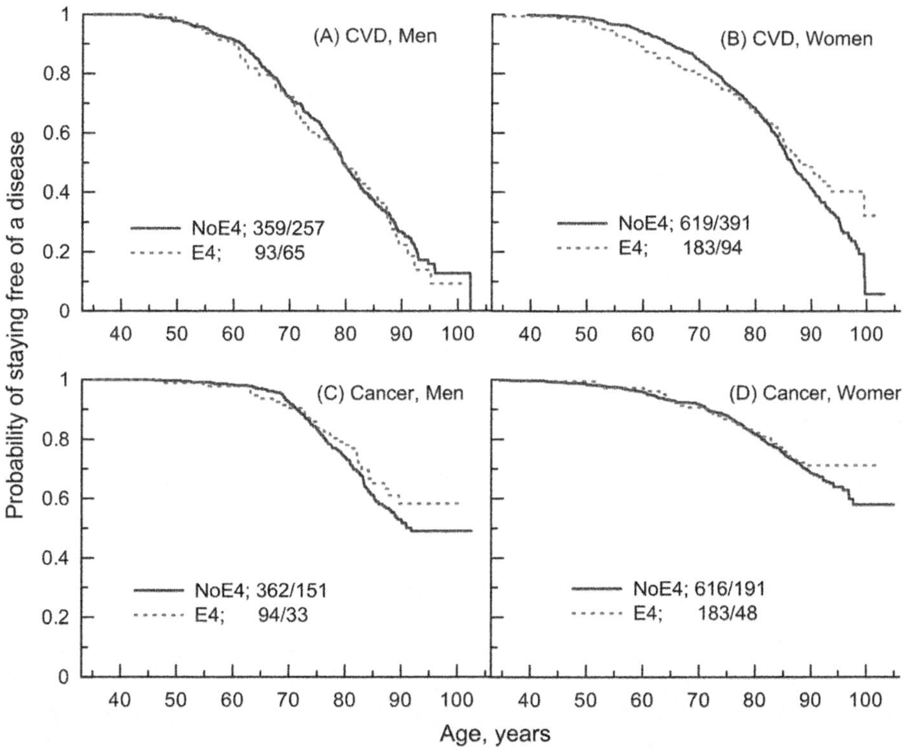

Fig. 9.4 Relative risks of CVD and cancer in more homogeneous "younger" (≤60 years) and "older" (>60 years) groups of the genotyped participants of the FHS cohort

Sex-specific analyses show that the protective effect for cancer is more prevalent among men, whereas the complex antagonistic CVD pattern is more prevalent among women (Fig. 9.4).

Relative Risks of Diseases The descriptive analyses of Figs. 9.3 and 9.4 were quantified using the Cox regression model. Traditional analysis disregarding aging-related complexity results, at best, in small and non-significant associations of the e4 allele with onsets of CVD and cancer in the FHS sample of men and women combined and in each sex (Table 9.3). Descriptive patterns clearly show, however, that the effects are small, not because of small penetrance of the e4 allele, but because this allele exhibits aging-related complexity.

Risks of Diseases: The Role of Aging-Related Processes To address the problem of differential roles of the same allele over the life course, we again stratified the sample by ages at onset of diseases. Specifically, we defined more homogeneous groups in the FHS original cohort as:

Table 9.3 Relative risks of CVD and cancer in the genotyped participants of the FHS original cohort

	CVD			Cancer		
Sample	RR	*p*	95 % CI	RR	*p*	95 % CI
Men and women	0.92	0.387	0.77–1.10	0.85	0.203	0.67–1.09
Men	1.08	0.547	0.83–1.41	0.83	0.333	0.57–1.21
Women	0.84	0.143	0.66–1.06	0.87	0.382	0.63–1.20

The number of subjects is shown in Figs. 9.3 and 9.4
RR denotes relative risk, *CI* denotes Confidence Interval

- "younger", those developing CVD or cancer in early life or being censored at younger ages (representatively, 65 years and younger at the end of follow up in 2008);
- "older", those developing CVD or cancer in late life or being censored at older ages (representatively, 65 years and older at the end of follow up in 2008).

Analyses of risks of CVD in the "younger group" reveal a significant effect of the e4 allele on risks of CVD in younger women (Table 9.4). The adverse effect of the e4 allele in men in this group is less pronounced than in women and it does not attain significance. Given the same direction of the effect in men and women, the adverse effect of this allele in the "younger group" of men and women combined is also significant. Non-significant protective effects of the e4 allele are seen for cancer in the younger group of men and women combined and for women only.

Analyses of risks of CVD in the "older group" (Table 9.4) reveal a highly significant protective effect of the e4 allele in older women and, as a consequence, in men and women combined. No effect is seen in men in this group. The protective effect for the e4 allele for cancer attains marginal significance in men and women combined in the older group. Unlike the younger group, the major contribution in the older group is due to the protective effect in men.

9.2.3 The FHS and the FHSO: Aging-Related Heterogeneity in a Changing Environment

Relative Risks of Diseases An attempt to improve power by pooling data from the FHS original and the FHSO cohorts and disregarding aging-related heterogeneity may not help because this procedure also increases heterogeneity of the sample in parallel. Specifically, Table 9.5 shows no significant effect of the e4 allele on the onset of CVD. The protective effect of this allele for cancer attains only marginal significance in the sample of men and women combined and in the sample of men only.

Table 9.4 Relative risks of CVD and cancer in more homogeneous "younger" (≤65 years) and "older" (>65 years) groups of the genotyped participants of the FHS original cohort

		CVD				Cancer			
Sample	Group	N_{tot}/N_{CVD}	RR	p	95 % CI	N_{tot}/N_{canc}	RR	p	95 % CI
Men and women	≤65 years	176/176	1.49	0.018	1.07–2.06	72/72	0.65	0.109	0.38–1.10
Men	≤65 years	83/83	1.42	0.131	0.90–2.24	16/16	0.97	0.953	0.35–2.69
Women	≤65 years	93/93	1.64	0.030	1.05–2.57	56/56	0.59	0.101	0.32–1.11
Men and women	>65 years	1077/631	0.80	0.035	0.65–0.99	1183/351	0.79	0.092	0.60–1.04
Men	>65 years	369/239	1.03	0.851	0.76–1.39	440/168	0.76	0.173	0.50–1.13
Women	>65 years	708/392	0.69	7.8×10^{-3}	0.52–0.91	743/183	0.83	0.311	0.57–1.20

RR denotes relative risk, *CI* denotes Confidence interval, N_{tot} and N_{CVD} or N_{canc} denote the total number of genotyped individuals and the number of CVD or cancer cases among them, respectively

Table 9.5 Relative risks of CVD and cancer in the genotyped participants of the FHS original and the FHSO cohorts

	CVD				Cancer			
Sample	N_{tot}/N_{CVD}	RR	p	95 % CI	N_{tot}/N_{canc}	RR	p	95 % CI
Men and women	5142/1750	1.08	0.217	0.96–1.21	5157/1267	0.87	0.051	0.76–1.00
Men	2300/895	1.12	0.170	0.95–1.31	2324/646	0.83	0.066	0.68–1.01
Women	2842/855	1.03	0.706	0.87–1.22	2833/621	0.89	0.266	0.73–1.09

The models are adjusted for inter-cohort difference
RR denotes relative risk, *CI* denotes Confidence interval, N_{tot} and N_{CVD} or N_{canc} denote the total number of genotyped individuals and the number of CVD or cancer cases among them, respectively

Risks of Diseases: The Role of Aging-Related Processes in a Changing Environment Clearly (see Figs. 9.2 and 9.4 and Tables 9.2 and 9.4), the lack of an effect for CVD and the weak association signal for cancer are the results of complex aging-related heterogeneity of the sample. Given the same direction of the effect for the e4 allele on onset of cancer at older ages, we can safely pool the FHS and FHSO data to increase power. By selecting more homogeneous groups with onset of cancer at older ages (older than 65 years), the role of the e4 allele in the etiology of cancer becomes much more compelling (Table 9.6). It is also clear that the differential effect of the e4 allele on onset of CVD across generations (e.g., protective effect in the FHS and detrimental effect in the FHSO at older ages, see Tables 9.2 and 9.4) makes attempts to improve power of the estimates by pooling the FHS and FHSO samples in this case useless (Table 9.6).

Conclusions on the Antagonistic Role of the APOE Gene in Diseases Careful analyses addressing the role of aging-related heterogeneity help to better characterize the puzzling complexity of gene actions on risks of CVD and cancer and their sensitivity to gender, ages, and environment associated with differences in human generations.

The analyses reported in this chapter suggest that the e4 allele can be protective against cancer with a more pronounced role in men. This protective effect is more characteristic of cancers at older ages and it holds in both the parental and offspring generations of the FHS participants.

Unlike cancer, the effect of the e4 allele on risks of CVD is more pronounced in women. The analyses suggest that the role of this allele in the etiology of CVD can be sensitive to age and generation. In the parental generation of the FHS participants, we observe the antagonistic action of the e4 allele on onset of CVD in women across the ages: the e4 allele can confer risks of CVD in younger women but protect against CVD in older women. In the offspring generation, the e4 allele can confer risks of CVD primarily in older women.

These results provide two important insights on the role of genes in traits in late life. First, they explicitly show that the same allele can change its role on risks of

Table 9.6 Relative risks of CVD and cancer in more homogeneous "younger" (≤65 years) and "older" (>65 years) groups of the genotyped participants of the FHS original and the FHSO cohorts

		CVD				Cancer			
Sample	Group	N_{tot}/N_{CVD}	RR	p	95 % CI	N_{tot}/N_{canc}	RR	p	95 % CI
Men and women	≤65 years	1843/724	1.04	0.650	0.88–1.24	1590/496	1.10	0.399	0.88–1.37
Men	≤65 years	931/445	1.15	0.192	0.93–1.41	735/218	1.29	0.114	0.94–1.78
Women	≤65 years	912/279	0.92	0.579	0.68–1.24	855/278	0.98	0.913	0.73–1.32
Men and women	>65 years	3299/1026	0.94	0.405	0.80–1.09	3567/771	0.78	7.3×10^{-3}	0.65–0.94
Men	>65 years	1369/450	1.00	0.994	0.80–1.25	1589/428	0.75	0.020	0.59–0.96
Women	>65 years	1930/576	0.89	0.282	0.72–1.10	1978/343	0.82	0.155	0.62–1.08

The models are adjusted for inter-cohort difference

RR denotes relative risk, *CI* denotes Confidence interval, N_{tot} and N_{CVD} or N_{canc} denote the total number of genotyped individuals and the number of CVD or cancer cases among them, respectively

CVD in an antagonistic fashion from detrimental in women with onsets at younger ages to protective in women with onsets at older ages.

Second, the analyses suggest two modes of sexually-dimorphic genetic tradeoffs. One mode is observed in the FHSO generation, wherein the e4 allele confers risk of CVD primarily in women and this allele can protect against cancer primarily in men *of the same age*. The other mode is highlighted in the FHS generation. The genetic tradeoff is seen in *different age groups*: a protective role of the e4 allele against cancer is observed in older men (as well as in men and women combined) from the FHS and FHSO cohorts, whereas the e4 allele shows a detrimental role in CVD in younger FHS women.

Both of these insights suggest the key role of aging-related processes and a changing environment in genetic susceptibility to traits in late life.

9.3 Tradeoffs in the Effects of *APOE* on Risks of CVD and Cancer Influence Human Lifespan

9.3.1 The FHS and FHSO: Survival

Time-to-Event Analyses The results of descriptive survival analyses of the FHS and FHSO participants for carriers and non-carriers of the e4 allele are shown in Fig. 9.5. These results suggest that e4 allele carriers have worse survival compared to non-e4 carriers in each cohort. The detrimental role of the e4 allele is more pronounced in older FHSO participants. However, the longest-living individuals aged about 95 years and older in the FHS cohort show no e4-specific survival differences.

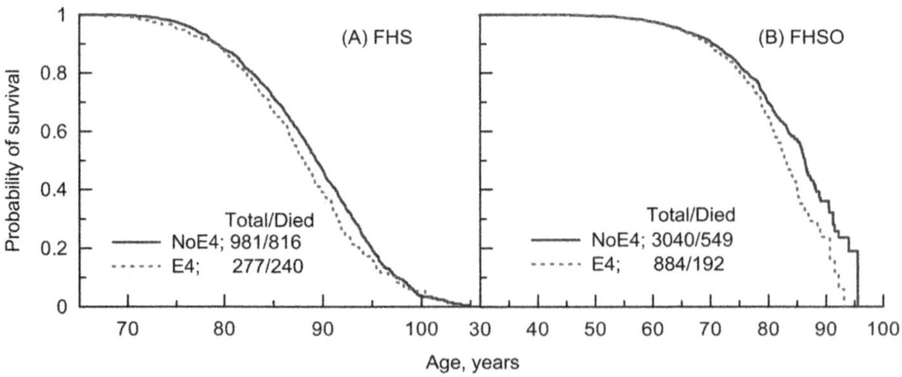

Fig. 9.5 Descriptive survival age patterns for genotyped participants of (**a**) FHS and (**b**) FHSO cohorts who carry (E4) and do not carry (NoE4) *APOE* e4 allele. The numbers in the *insets* show the total number of genotyped subject and the number of deaths among them

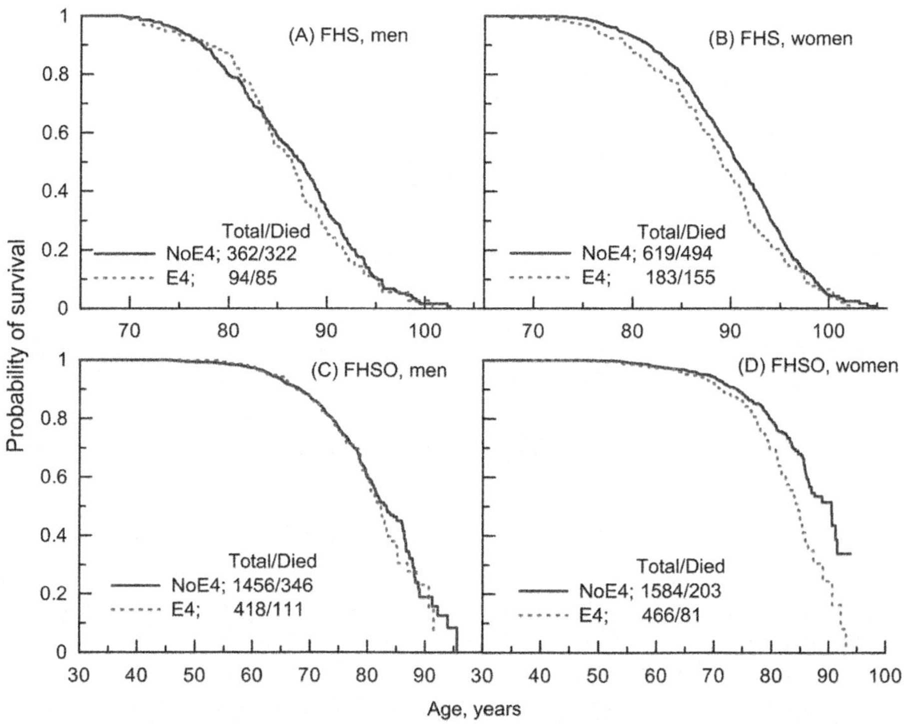

Fig. 9.6 Descriptive survival age patterns for (**a**, **c**) men and (**b**, **d**) women genotyped in (**a**, **b**) FHS and (**c**, **d**) FHSO cohorts who carry (E4) and do not carry (NoE4) the *APOE* e4 allele. The numbers in the *insets* show the total number of genotyped subject and the number of deaths among them

Sex stratification shows sexual dimorphism in the effect of the e4 allele on survival (Fig. 9.6) with the e4 female carriers, particularly, being more exposed to worse survival. Of note is that the role of the e4 allele diminishes in both the longest living men and women, i.e., it is sex insensitive.

Relative Risks of Death Cox regression analyses of risks of death of either FHS or FHSO participants shows that female e4-allele carriers are at higher risk of death (i.e., they have significantly shorter lifespan) compared to the non-e4-allele female carriers (Table 9.7). No significant effects are seen for men in these cohorts. The statistically significant RRs in the sample of men and women combined are largely attributed to women.

Because the FHS and FHSO participants were exposed to risks of death differentially at different ages (Figs. 9.5 and 9.6), we evaluated the relative risks in more homogeneous samples. Based on the foregoing empirical evidence (Fig. 9.5), these samples were defined in:

Table 9.7 Relative risks of death in the genotyped participants of the FHS and FHSO cohorts

	FHS			FHSO		
Sample	RR	*p*	95 % CI	RR	*p*	95 % CI
Men and women	1.22	1.1×10^{-2}	1.05–1.42	1.26	5.8×10^{-3}	1.07–1.48
Men	1.16	0.239	0.91–1.48	1.08	0.492	0.87–1.33
Women	1.25	2.7×10^{-2}	1.03–1.52	1.59	2.4×10^{-4}	1.24–2.05

The numbers of subjects are shown in Figs. 9.5 and 9.6
RR denotes relative risk, *CI* denotes Confidence interval

- the FHS as being younger than 95 years at death or the end of follow up in 2008;
- the FHSO as being 75 years and older at death or the end of follow up in 2008.

Table 9.8 shows substantial improvement in the estimated risks of death and their significance in these more homogeneous samples as compared to the entire samples shown in Table 9.7.

Because the e4 allele is associated with risks of CVD and cancer in complex ways and because these diseases are major causes of death in the U.S., the associations with survival in Table 9.8 can be modulated by those diseases. Accordingly, we next evaluated the risks of death for all genotyped participants of the FHS or FHSO cohorts in the regression models adjusted for CVD and cancer (Table 9.9). This analysis shows that neither CVD nor cancer explain the observed associations of the e4 allele with death (see Table 9.7), i.e., that they do not mediate the genetic effect on lifespan. On the contrary, they even improve the estimates in women and in men and women combined (Table 9.9). This implies that CVD and cancer modulate the effect of the e4 allele on survival rather than mediate it. Both CVD and cancer play seemingly minor modulating roles in men's survival.

Taking into account both the modulating role of CVD and cancer in the effect of the e4 allele on risks of death and aging-related heterogeneity (see the discussion of Table 9.8), we find that more realistic excesses of the risk of death in the female e4 carriers (Table 9.10) are even larger than in the case when either the aging-related heterogeneity (Table 9.8) or the modulating role of CVD and cancer (Table 9.9), or both of them (Table 9.7) are disregarded. This refinement of the analyses also reveals increasing risks of deaths in men and women combined and in men only.

Given the same direction of the effects for the e4 allele in each cohort, we pooled samples of genotyped participants of the FHS and FHSO cohorts. Table 9.11 shows that traditional analyses based on pooling samples from the different studies and cohorts and disregarding aging-related specifics can improve the statistical significance of the estimates in this case (compare Tables 9.7 and 9.11, FHS+FHSO). It is also clear that taking into account aging-related heterogeneity (Table 9.11, FHS+FHSO, <95 years) or the modulating role of CVD and cancer (Table 9.12, FHS+ FHSO) helps in unraveling stronger genetic effects. Refinement of the

Table 9.8 Relative risks of death in more homogenous groups of the genotyped participants of the FHS and FHSO cohorts

Sample	FHS, <95 years				FHSO, 75+ years			
	N_{tot}/N_{died}	RR	p	95 % CI	N_{tot}/N_{died}	RR	p	95 % CI
Men and women	1103/944	1.29	1.5×10^{-3}	1.10–1.51	1235/320	1.53	5.7×10^{-4}	1.20–1.96
Men	427/385	1.18	0.195	0.92–1.53	567/185	1.17	0.356	0.84–1.62
Women	676/559	1.37	1.7×10^{-3}	1.12–1.66	668/135	2.19	1.4×10^{-5}	1.54–3.11

RR denotes relative risk, *CI* denotes Confidence interval, N_{tot} and N_{died} denote the total number of genotyped individuals and the number of deaths among them, respectively

Table 9.9 Health-adjusted relative risks of death in the genotyped participants of the FHS and FHSO cohorts

	FHS			FHSO		
Sample	RR	*p*	95 % CI	RR	*p*	95 % CI
Men and women	1.27	3.2×10^{-3}	1.08–1.48	1.34	7.3×10^{-4}	1.13–1.58
Men	1.18	0.201	0.92–1.51	1.16	0.172	0.94–1.44
Women	1.35	2.9×10^{-3}	1.11–1.64	1.63	2.1×10^{-4}	1.26–2.11

Model is adjusted for the prevalence of CVD and cancer
The numbers of subjects are shown in Figs. 9.5 and 9.6
RR denotes relative risk, *CI* denotes Confidence interval

Table 9.10 Health-adjusted relative risks of death in more homogenous groups of the genotyped participants of the FHS and FHSO cohorts

	FHS, <95 years			FHSO, 75+ years		
Sample	RR	*p*	95 % CI	RR	*p*	95 % CI
Men and women	1.34	2.5×10^{-4}	1.15–1.57	1.60	2.1×10^{-4}	1.25–2.05
Men	1.21	0.133	0.94–1.56	1.22	0.247	0.87–1.69
Women	1.47	7.6×10^{-5}	1.21–1.78	2.28	4.0×10^{-6}	1.61–3.24

Model is adjusted for the prevalence of CVD and cancer
The number of subjects is shown in Table 9.8
RR denotes relative risk, *CI* denotes Confidence interval

analyses by taking into account both of these factors (i.e., aging-related heterogeneity and the modulating role of CVD and cancer) reveals impressively large and significant detrimental effects of the e4 allele on death in women only and in men and women combined, as well as marginally significant effects in men (Table 9.12, FHS+FHSO, <95 years).

Conclusions on the Role of the APOE Gene in Lifespan The results of these analyses provide two important insights into the role of genes in lifespan. First, they provide evidence on the key role of aging-related processes in genetic susceptibility to lifespan. For example, taking into account the specifics of aging-related processes gains 18 % in estimates of the RRs and five orders of magnitude in significance in the same sample of women (i.e., $RR = 1.61, p = 1.2 \times 10^{-9}$ (Table 9.12, FHS+FHSO, <95 years) vs. $RR = 1.36, p = 1.4 \times 10^{-4}$ (Table 9.11, FHS+FHSO)) without additional investments in increasing sample sizes and new genotyping. The second is that a detailed study of the role of aging-related processes in estimates of the effects of genes on lifespan (and healthspan) helps in detecting more homogeneous subsamples at excessive risks, such as those of death in women with shorter lifespan (i.e., less than 95 years) in the case of lifespan.

Table 9.11 Relative risks of death in the pooled sample of the genotyped participants of the FHS and FHSO cohorts

Sample	FHS+FHSO				FHS+FHSO, <95 years			
	N_{tot}/N_{died}	RR	*p*	95 % CI	N_{tot}/N_{died}	RR	*p*	95 % CI
Men and women	5182/1797	1.25	1.5×10^{-4}	1.11–1.39	5026/1684	1.29	2.1×10^{-5}	1.15–1.44
Men	2330/864	1.12	0.178	0.95–1.31	2300/841	1.12	0.169	0.95–1.32
Women	2852/933	1.36	1.3×10^{-4}	1.16–1.60	2726/843	1.47	1.9×10^{-6}	1.25–1.72

The models are adjusted for inter-cohort difference

RR denotes relative risk, *CI* denotes Confidence interval, N_{tot} and N_{died} denote the total number of genotyped individuals and the number of deaths among them, respectively

Table 9.12 Health-adjusted relative risks of death in the pooled sample of the genotyped participants of the FHS and FHSO cohorts

	FHS+FHSO			FHS+FHSO, <95 years		
Sample	RR	*p*	95 % CI	RR	*p*	95 % CI
Men and women	1.33	4.2×10^{-6}	1.18–1.49	1.37	2.3×10^{-7}	1.22–1.55
Men	1.16	0.103	0.97–1.37	1.18	6.6×10^{-2}	0.99–1.40
Women	1.50	1.0×10^{-6}	1.27–1.76	1.61	1.2×10^{-9}	1.38–1.88

Model is adjusted for inter-cohort difference and the prevalence of CVD and cancer
The number of subjects is shown in Table 9.11
RR denotes relative risk, *CI* denotes Confidence interval

9.4 Conclusion

The results of the analyses presented in this chapter are indicative of the complex role of genes in healthspan and lifespan. Accordingly, adequate methods are necessary to disentangle this role and gain further insights into genetic origin of such complex traits. An important immediate consequence of such analyses is that their results are crucial for the efficient translation of genetic discoveries into strategies aiming to improve population health.

Acknowledgements This chapter was partly supported by the National Institute on Aging of the National Institutes of Health under Award Numbers R01AG030198, R01AG032319, R01AG030612, R01AG046860, and P01AG043352. The content is solely the responsibility of the authors and does not necessarily represent the official views of the National Institutes of Health. The Framingham Heart Study (FHS) is conducted and supported by the National Heart, Lung, and Blood Institute (NHLBI) in collaboration with Boston University (Contract No. N01-HC-25195). This chapter was not prepared in collaboration with investigators of the FHS and does not necessarily reflect the opinions or views of the FHS, Boston University, or NHLBI. Funding for SHARe Affymetrix genotyping was provided by NHLBI Contract N02-HL-64278. SHARe Illumina genotyping was provided under an agreement between Illumina and Boston University. This work was prepared using a limited access dataset obtained from the NHLBI and the Framingham SHARe data obtained through dbGaP.

References

Alexander, D. M., Williams, L. M., Gatt, J. M., Dobson-Stone, C., Kuan, S. A., Todd, E. G., Schofield, P. R., Cooper, N. J., & Gordon, E. (2007). The contribution of apolipoprotein e alleles on cognitive performance and dynamic neural activity over six decades. *Biological Psychology, 75*, 229–238.

Bergman, A., Atzmon, G., Ye, K., MacCarthy, T., & Barzilai, N. (2007). Buffering mechanisms in aging: A systems approach toward uncovering the genetic component of aging. *PLoS Computational Biology, 3*, e170.

Cupples, L. A., Heard-Costa, N., Lee, M., & Atwood, L. D. (2009). Genetics analysis workshop 16 problem 2: The Framingham heart study data. *BMC Proceedings, 3*(Suppl 7), S3.

Dawber, T. R. (1980). *The Framingham study: The epidemiology of atherosclerotic disease.* Cambridge, MA: Harvard University Press.

De Benedictis, G., Carotenuto, L., Carrieri, G., De Luca, M., Falcone, E., Rose, G., Yashin, A. I., Bonafe, M., & Franceschi, C. (1998). Age-related changes of the 3′apob-vntr genotype pool in ageing cohorts. *Annals of Human Genetics, 62*, 115–122.

Di Rienzo, A., & Hudson, R. R. (2005). An evolutionary framework for common diseases: The ancestral-susceptibility model. *Trends in Genetics, 21*, 596–601.

Finch, C. E., & Tanzi, R. E. (1997). Genetics of aging. *Science, 278*, 407–411.

Gail, M. H., & Johnson, N. L. (1989). *Proceedings of the American statistical association: Sesquicentennial invited papers session*. Alexandria: American Statistical Association.

Govindaraju, D. R., Cupples, L. A., Kannel, W. B., O'Donnell, C. J., Atwood, L. D., D'Agostino, R. B., Sr., Fox, C. S., Larson, M., Levy, D., Murabito, J., Vasan, R. S., Splansky, G. L., Wolf, P. A., & Benjamin, E. J. (2008). Genetics of the Framingham heart study population. *Advances in Genetics, 62*, 33–65.

Ilveskoski, E., Perola, M., Lehtimaki, T., Laippala, P., Savolainen, V., Pajarinen, J., Penttila, A., Lalu, K. H., Mannikko, A., Liesto, K. K., Koivula, T., & Karhunen, P. J. (1999). Age-dependent association of apolipoprotein e genotype with coronary and aortic atherosclerosis in middle-aged men: An autopsy study. *Circulation, 100*, 608–613.

Kulminski, A. M., Culminskaya, I., Ukraintseva, S. V., Arbeev, K. G., Land, K. C., & Yashin, A. I. (2010). Beta2-adrenergic receptor gene polymorphisms as systemic determinants of healthy aging in an evolutionary context. *Mechanisms of Ageing and Development, 131*, 338–345.

Kulminski, A. M., Culminskaya, I., Ukraintseva, S. V., Arbeev, K. G., Arbeeva, L., Wu, D., & Yashin, A. I. (2011). Trade-off in the effects of the apolipoprotein E polymorphism on the ages at onset of CVD and cancer influences human lifespan. *Aging Cell, 10*(3), 533–541. doi:10.1111/j.1474-9726.2011.00689.x.

Kulminski, A. M., Culminskaya, I., Arbeev, K. G., Ukraintseva, S. V., Arbeeva, L., & Yashin, A. I. (2013). Trade-off in the effect of the APOE gene on the ages at onset of cardiocascular disease and cancer across ages, gender, and human generations. *Rejuvenation Research, 16*(1), 28–34. doi:10.1089/rej.2012.1362.

Lahoz, C., Schaefer, E. J., Cupples, L. A., Wilson, P. W., Levy, D., Osgood, D., Parpos, S., Pedro-Botet, J., Daly, J. A., & Ordovas, J. M. (2001). Apolipoprotein e genotype and cardiovascular disease in the Framingham heart study. *Atherosclerosis, 154*, 529–537.

Lee, E. W., Wei, L. J., Amato, D. A., & Leurgans, S. (1992). Cox-type regression-analysis for large numbers of small-groups of correlated failure time observations. In *Survival analysis: State of the art* (Vol. 211, pp. 237–247). Dordrecht: Springer.

Martin, G. M. (2007). Modalities of gene action predicted by the classical evolutionary biological theory of aging. *The Annals of the New York Academy of Sciences, 1100*, 14–20.

Myers, R. H., Schaefer, E. J., Wilson, P. W., D'Agostino, R., Ordovas, J. M., Espino, A., Au, R., White, R. F., Knoefel, J. E., Cobb, J. L., McNulty, K. A., Beiser, A., & Wolf, P. A. (1996). Apolipoprotein e epsilon4 association with dementia in a population-based study: The Framingham study. *Neurology, 46*, 673–677.

Schnebel, E. M., & Grossfield, J. (1988). Antagonistic pleiotropy – An interspecific drosophila-comparison. *Evolution, 42*, 306–311.

Splansky, G. L., Corey, D., Yang, Q., Atwood, L. D., Cupples, L. A., Benjamin, E. J., D'Agostino, R. B., Sr., Fox, C. S., Larson, M. G., Murabito, J. M., O'Donnell, C. J., Vasan, R. S., Wolf, P. A., & Levy, D. (2007). The third generation cohort of the national heart, lung, and blood institute's Framingham heart study: Design, recruitment, and initial examination. *American Journal of Epidemiology, 165*, 1328–1335.

Stranger, B. E., Stahl, E. A., & Raj, T. (2011). Progress and promise of genome-wide association studies for human complex trait genetics. *Genetics, 187*, 367–383.

Summers, K., & Crespi, B. J. (2010). Xmrks the spot: Life history tradeoffs, sexual selection and the evolutionary ecology of oncogenesis. *Molecular Ecology, 19*, 3022–3024.

Teslovich, T. M., Musunuru, K., Smith, A. V., Edmondson, A. C., Stylianou, I. M., Koseki, M., Pirruccello, J. P., Ripatti, S., Chasman, D. I., Willer, C. J., Johansen, C. T., Fouchier, S. W., Isaacs, A., Peloso, G. M., Barbalic, M., Ricketts, S. L., Bis, J. C., Aulchenko, Y. S.,

Thorleifsson, G., Feitosa, M. F., Chambers, J., Orho-Melander, M., Melander, O., Johnson, T., Li, X., Guo, X., Li, M., Shin Cho, Y., Jin Go, M., Jin Kim, Y., Lee, J. Y., Park, T., Kim, K., Sim, X., Twee-Hee Ong, R., Croteau-Chonka, D. C., Lange, L. A., Smith, J. D., Song, K., Hua Zhao, J., Yuan, X., Luan, J., Lamina, C., Ziegler, A., Zhang, W., Zee, R. Y., Wright, A. F., Witteman, J. C., Wilson, J. F., Willemsen, G., Wichmann, H. E., Whitfield, J. B., Waterworth, D. M., Wareham, N. J., Waeber, G., Vollenweider, P., Voight, B. F., Vitart, V., Uitterlinden, A. G., Uda, M., Tuomilehto, J., Thompson, J. R., Tanaka, T., Surakka, I., Stringham, H. M., Spector, T. D., Soranzo, N., Smit, J. H., Sinisalo, J., Silander, K., Sijbrands, E. J., Scuteri, A., Scott, J., Schlessinger, D., Sanna, S., Salomaa, V., Saharinen, J., Sabatti, C., Ruokonen, A., Rudan, I., Rose, L. M., Roberts, R., Rieder, M., Psaty, B. M., Pramstaller, P. P., Pichler, I., Perola, M., Penninx, B. W., Pedersen, N. L., Pattaro, C., Parker, A. N., Pare, G., Oostra, B. A., O'Donnell, C. J., Nieminen, M. S., Nickerson, D. A., Montgomery, G. W., Meitinger, T., McPherson, R., McCarthy, M. I., McArdle, W., Masson, D., Martin, N. G., Marroni, F., Mangino, M., Magnusson, P. K., Lucas, G., Luben, R., Loos, R. J., Lokki, M. L., Lettre, G., Langenberg, C., Launer, L. J., Lakatta, E. G., Laaksonen, R., Kyvik, K. O., Kronenberg, F., Konig, I. R., Khaw, K. T., Kaprio, J., Kaplan, L. M., Johansson, A., Jarvelin, M. R., Janssens, A. C., Ingelsson, E., Igl, W., Kees Hovingh, G., Hottenga, J. J., Hofman, A., Hicks, A. A., Hengstenberg, C., Heid, I. M., Hayward, C., Havulinna, A. S., Hastie, N. D., Harris, T. B., Haritunians, T., Hall, A. S., Gyllensten, U., Guiducci, C., Groop, L. C., Gonzalez, E., Gieger, C., Freimer, N. B., Ferrucci, L., Erdmann, J., Elliott, P., Ejebe, K. G., Doring, A., Dominiczak, A. F., Demissie, S., Deloukas, P., de Geus, E. J., de Faire, U., Crawford, G., Collins, F. S., Chen, Y. D., Caulfield, M. J., Campbell, H., Burtt, N. P., Bonnycastle, L. L., Boomsma, D. I., Boekholdt, S. M., Bergman, R. N., Barroso, I., Bandinelli, S., Ballantyne, C. M., Assimes, T. L., Quertermous, T., Altshuler, D., Seielstad, M., Wong, T. Y., Tai, E. S., Feranil, A. B., Kuzawa, C. W., Adair, L. S., Taylor, H. A., Jr., Borecki, I. B., Gabriel, S. B., Wilson, J. G., Holm, H., Thorsteinsdottir, U., Gudnason, V., Krauss, R. M., Mohlke, K. L., Ordovas, J. M., Munroe, P. B., Kooner, J. S., Tall, A. R., Hegele, R. A., Kastelein, J. J., Schadt, E. E., Rotter, J. I., Boerwinkle, E., Strachan, D. P., Mooser, V., Stefansson, K., Reilly, M. P., Samani, N. J., Schunkert, H., Cupples, L. A., Sandhu, M. S., Ridker, P. M., Rader, D. J., van Duijn, C. M., Peltonen, L., Abecasis, G. R., Boehnke, M., & Kathiresan, S. (2010). Biological, clinical and population relevance of 95 loci for blood lipids. *Nature, 466*, 707–713.

Ukraintseva S. V. (2005). "Bad" in the young – "good" in the old: Is this consistent with the antagonistic pleiotropy concept? In Martin G (Ed.), *How is the evolutionary biological theory of aging holding up against mounting attacks?* American Aging Association, 2012. http://www.americanaging.org/news/AGE%20George%20Martin%20Discussion%20April%202005.pdf

Vijg, J., & Suh, Y. (2005). Genetics of longevity and aging. *Annual Review of Medicine, 56*, 193–212.

Williams, G. C. (1957). Pleiotropy, natural-selection, and the evolution of senescence. *Evolution, 11*, 398–411.

Williams, P. D., & Day, T. (2003). Antagonistic pleiotropy, mortality source interactions, and the evolutionary theory of senescence. *Evolution, 57*, 1478–1488.

Yashin, A. I., Ukraintseva, S. V., De Benedictis, G., Anisimov, V. N., Butov, A. A., Arbeev, K., Jdanov, D. A., Boiko, S. I., Begun, A. S., Bonafe, M., & Franceschi, C. (2001). Have the oldest old adults ever been frail in the past? A hypothesis that explains modern trends in survival. *The Journals of Gerontology. Series A, Biological Sciences and Medical Sciences, 56*, B432–B442.

Chapter 10
Conclusions Regarding Empirical Patterns of Aging, Health, and Longevity

Alexander M. Kulminski, Anatoliy I. Yashin, Irina Culminskaya, Kenneth C. Land, and Svetlana V. Ukraintseva

Age is a major risk factor for phenotypes characterizing human health, well-being, and survival in late life. The risks of these phenotypes expressed in forms of pathological dysregulation of physiological functions, incidence or prevalence of diseases, case fatality, or mortality also change with age. This change integrates all challenges occurring in a human organism during the life course by a given age. Accordingly, the age patterns of various age-related phenotypes are a valuable source of information about health-related processes in human organisms.

Chapter 2 focused on phenotypes that are considered critical markers of physiological processes in an organism—including blood glucose, body mass index, blood pressure, lipids, hematocrit, and ventricular rate. Changes in these biomarkers (also called *endophenotypes*) are routinely monitored in clinical practice in order to detect premature dysregulation of the respective processes in an organism with age. It is believed that such dysregulation is a manifestation of pathological changes causing diseases later in life (e.g., cardiovascular diseases). Longitudinal studies collecting information on endophenotypes during long periods of human life are a unique source of such information available to the research community. In many cases, these studies have important advantages over clinical observations because they often collect information on initially healthy individuals.

Having a clear understanding of the research goals and the data supporting the analyses, an array of fundamental questions regarding health-related processes with age can be addressed. Specifically, Chap. 2 investigated whether age patterns of endophenotypes exhibit some regularity patterns, whether these patterns are gender-specific, whether they differ between the specific population of the Framingham Heart Study and the general U.S. population, whether genetic and non-genetic factors associated with lifespan can modulate these patterns, whether there are hidden (unobserved) components of biological mechanisms regulating the dynamics of age-related changes, and how they are linked to observed endophenotypes.

The results of the empirical analyses in Chap. 2 showed that gender-specific age patterns of endophenotypes can be modulated by behavioral, socio-economic, and

A.I. Yashin et al., *Biodemography of Aging*, The Springer Series on Demographic Methods and Population Analysis 40, DOI 10.1007/978-94-017-7587-8_10

genetic factors. Analyses of groups of short-lived and long-living individuals showed apparent differences in age patterns of endophenotypes between these groups. These patterns are also different between groups of individuals having short and long healthspans. These patterns are affected by events characterizing individuals' health status (such as the presence or absence of certain age-related diseases) and, eventually, they can pinpoint groups of individuals at excess risk of death.

However, these analyses make it clear that a better understanding of the biological mechanisms for coping with deleterious influences of aging and harmful external factors requires more rigorous analyses and the development and application of sophisticated mathematical and computer models. These models are useful for delineating the effects of fundamental process of biological aging from those associated with compositional changes due to the process of mortality selection. The biological aging-related processes, in turn, can be decomposed into components dealing with the senescence process per se, changes due to the ontogenetic program, changes in response to persistent exogenous stresses, and the effects of compensatory mechanisms that try to maintain an organism's functioning despite all these forces.

The analyses of age patterns of endophenotypes in Chap. 2 were further extended by the analyses of age patterns of major human diseases and mortality in Chap. 3. The importance of these analyses is twofold. First, they are important for gaining insights into key factors driving the onset and progression of age-related chronic diseases. Second, they inform policymakers and governmental institutions how to better address the health demands of the elderly and to reduce the associated economic burdens on society. Identification of disease age patterns with sufficient precision requires large population-based databases that are costly to collect. However, there is an important and readily-available resource in the form of administrative health data which is routinely generated through the administration of health care programs. This resource is the Medicare Standard Analytic Files for service use for the entire Medicare-enrollee population of adults aged 65 years and older. Analysis of age patterns of health and survival risks using administrative data requires appropriate analytic strategies adapted to such data.

Chapter 3 focused on a series of epidemiologic and biodemographic measures that can be studied using Medicare data. The topics discussed included age patterns of morbidity and mortality, recovery or long-term remission (when appropriate), comorbidity and multimorbidity, risk factors of disease incidence and mortality, and projection modeling of health and mortality. The analyses were performed using Medicare data linked to the Surveillance, Epidemiology, and End Results (SEER) Registry (SEER-M), and the National Long Term Care Survey (NLTCS-M) data.

The results of the analyses showed that the administrative data can be used to adequately evaluate the national age patterns of incidence of a large array of diseases in late life including cardiovascular diseases, cancers at different sites, neurodegenerative diseases, and asthma, among the others. It was also concluded that remission/recovery rates for age-related diseases and their time trends were detectable using Medicare data; the patients for whom there were large periods without reported ICD-9 diagnoses were the healthier subcohort. The date of onset of a disease can be identified using information collected in the Medicare data with specific assumptions used to define the associated calculation algorithm.

Chapter 3 also discussed methodological and substantive issues of the analyses of associations between hundreds of variables measured in the NLTCS-M and risks of all-cause mortality and morbidity extracted from the Medicare data. The chapter presented a new multimorbidity index—AMMI—for the U.S. older adult population which reflected recent innovations in prevention and treatment and which was estimated using information from the Medicare data.

The substantive results of the analyses in Chap. 3 consisted of estimated statistical associations between different risks and characteristics representing distinct features of human aging. Methodological results included evaluation and comparison of several popular approaches of dealing with high dimensional categorical measurements and investigation of their power in predicting associations with the considered risks. These analyses highlighted a number of factors predicting disease incidence including physical activity, smoking, comorbidity, demographics, health insurance, and medical care providers. The same approach can be used to detect the most influential preventable behavioral risk factors using data from other studies with more detailed description of the risk factors. These approaches can be used to develop individualized forecasts and prevention strategies.

Despite the common understanding that age is a risk factor of not just one but a large portion of human diseases in late life, each specific disease is typically considered as a standalone trait. Independence of diseases was a plausible hypothesis in the era of infectious diseases caused by different strains of microbes. Unlike those diseases, the exact etiology and precursors of diseases in late life are still elusive. It is clear, however, that the origin of these diseases differs from that of infectious diseases and that age-related diseases reflect a complicated interplay among ontogenetic changes, senescence processes, and damages from exposures to environmental hazards. Studies of the determinants of diseases in late life provide insights into a number of risk factors, apart from age, that are common for the development of many health pathologies. The presence of such common risk factors makes chronic diseases and hence risks of their occurrence interdependent. This means that the results of many calculations using the assumption of disease independence should be used with care.

Chapter 4 argued that disregarding potential dependence among diseases may seriously bias estimates of potential gains in life expectancy attributable to the control or elimination of a specific disease and that the results of the process of coping with a specific disease will depend on the disease elimination strategy, which may affect mortality risks from other diseases. Therefore, any strategy for the reduction of the burden of a disease or its complete eradication has to take the underlying mechanisms of disease dependence into account.

The chapter provided evidence of dependence among various diseases considering a large microlevel dataset on Multiple Causes of Death (MCD). MCD data were used to evaluate correlations among mortalities from cancer and other major health disorders, including heart disease, stroke, diabetes, Alzheimer's and Parkinson's diseases, and asthma. The types of dependence that were uncovered included a form of antagonism (trade-off) with significant negative correlations, e.g., as found between cancer and other selected diseases. Consideration was given to possible mechanisms, including pleiotropic effects of genetic factors, as well as appropriate mathematical methods of dealing with dependence among diseases in the analyses of data on aging, health, and longevity. The study of mechanisms of dependence opens new opportunities for improving population health by developing proper preventive measures and adequate treatment strategies which minimize the chances of harmful side effects.

Chapter 5 dealt with the facts that the cancer incidence rate for all sites combined and life expectancy have increased over time in many countries around the world. These increases are concurrent with the economic progress and spread of the Western lifestyle. What caused this global increase in cancer risk, beyond known carcinogenic exposures? Could life in affluent societies make people more susceptible to cancer? And could an increase in cancer risk and longevity be favored by the same factors linked to economic prosperity? This chapter reviewed the global epidemiological evidence and results of human and animal studies to show that the higher overall cancer risk in the more developed world might be a result of a higher proportion of individuals in the populations more *susceptible* to cancer, rather than merely the result of a higher carcinogenic burden. This proportion could increase over time under the influence of several factors linked to the high economic development and Western lifestyle, including: improved medical and living conditions that allow for survival of people with less efficient immune systems; novel exposures that are not carcinogenic themselves but may increase one's vulnerability to established carcinogens; and others. Some of the factors associated with the Western lifestyle (e.g., food enriched with growth factors and delayed childbirth) may favor both longevity and vulnerability to cancer. This suggests that trade-offs between cancer and aging may contribute to concurrent increases in longevity and cancer risks in modern human populations.

A complex array of problems of increasing risks of various health traits with age along with possible dependencies among some of them in aging populations worldwide leads to increasing governmental concerns on how to achieve further compression of morbidity in the most efficient manner. This objective highlights an important economic component: i.e., medical costs associated with treatment and rehabilitation to improve well-being and reduce the burden on the economies. This problem requires evaluation of trends in disease burden and associated health expenditures. To forecast such trends one needs to understand the key factors driving the progression of age-related diseases and how such progression could result in changes in associated medical costs. In the U.S., this is a primary concern for the two main governmental health insurance programs, Medicare and Medicaid.

To open new possibilities for forecasting population health and medical costs, studies of the effects of disease onset on individual medical cost patterns and of the behavior of individual health patterns in the presence of comorbid and concurrent disorders are required. The results of such studies can help to estimate to what extent cumulative individual medical costs can determine future changes in elderly patients' health status.

Chapter 6 focused on developing a model capable of generating a quantitative description of the relationships between individual cost patterns accompanying the onset of age-related diseases. The model is designed to have demographically interpretable parameters and to serve as a building block in constructing a precise and comprehensive forecasting model of medical costs (including Medicare spending) at the population level. The underlying methodological idea is that one can aggregate health state information into a small number of covariates which can be pivotal in predicting the risk of a health event (e.g., disease incidence) and whose dynamics can be determined within the model constraints.

The new model was applied to 20 diseases. The results are important for the whole U.S. elderly population because the list of diseases includes those with high prevalence and with high medical costs associated with their treatment. It was found that the time patterns of the medical cost trajectories were similar for all selected diseases and could be described in terms of four components: (i) the pre-diagnosis costs associated with initial comorbidity represented by medical expenditures, (ii) the costs associated with the onset of each disease, (iii) the rate of reduction in medical expenditures after the disease onset, and (iv) the difference between post- and pre-diagnosis cost levels associated with an acquired comorbidity. The description of the trajectories was formalized by a model which explicitly involved four parameters reflecting these four components. The model was validated for several population groups and demonstrated the ability to describe cost trajectories for different levels of disability and comorbidity. This model could be further extended to forecast health/incidence, mortality, and associated medical costs in the U.S. elderly population using more limited sets of parameters derived from more broadly available data sources.

Physiological changes in the aging human body are manifested in changes in various biomarkers which are routinely collected in clinical settings. Some studies of human health, aging, and lifespan also collect that information, often, over relatively long time periods. The other major phenotypes accompanying aging are the various diseases. It is apparent that these two types of phenotypes (biomarkers and diseases) do not exhaust all possible changes in an aging human organism because human life is also accompanied by small and moderate changes occurring with age, e.g., signs, symptoms, minor impairments, etc. Many studies of human health, aging, and lifespan collect this information. However, it is rare that this information is used for published analyses. This is mainly because important information about age-related processes in an organism is spread through hundreds of mild-effect phenotypes. The connections among such phenotypes, as well as between each of them and diseases and/or survival, are unclear. To effectively use this information, either substantially larger samples or new methods are needed. A cost-efficient solution would be attained if one could improve the methods for working with existing information.

A promising approach would be to construct a measure which would aggregate such mild-effect phenotypes. This idea was suggested in the literature in 1990s and was promoted in studies presented Chap. 7. Following those ideas, an index of cumulative deficits (DI) was constructed using information on mild-effect traits available from several surveys and studies in the U.S., including the National Long Term Care Survey, the Framingham Heart Study, and the Long Life Family Study.

The importance of this index is twofold. First, it can characterize health conditions with elusive expressions such as the geriatric syndrome of frailty. Second, given its inherent nature of gathering mild-effect phenotypes, it can characterize the general process of age-related health decline which is believed to be associated with aging. The results of the analyses were presented in Chap. 7 where they provided strong evidence that the DI should be considered as a promising tool for applications in population and clinical settings. An important finding was that the DI could be considered as a tool to characterize age-related processes in the elderly, independently of age. This result characterized the DI not merely as a substituent of age but as an alternative to age in the summary of age-related processes. Another important result was that the DI could be a better tool for geriatricians working with old-aged patients to measure phenotypic frailty than an alternative tool developed by L. Fried and colleagues.

Endophenotypes characterizing the physiological state of an individual are a product of a complicated interplay between genetic and non-genetic (environmental, behavioral, or stochastic) factors. Because the effects of these factors may change during the life course, endophenotypes may differ substantially at different ages in different individuals. Despite these intra- and inter-individual differences, the average age trajectories for the same endophenotype follow remarkable regularities. Epidemiology suggests that the endophenotypes may influence the risks of morbidity and mortality. However, most studies in the field are focused on the connections of the risks of events with endophenotypes at a given age. The dynamic nature of endophenotypes is often ignored. If, however, static endophenotypes may influence risks of morbidity and mortality, it is also logical to expect that the dynamics of these endophenotypes may also influence those risks. This is reasonable because the dynamic properties of individual trajectories of physiological biomarkers over calendar age may reflect the rate of individual aging. And, if that aging rate affects lifespan and healthspan, then one can expect that the dynamic characteristics of such trajectories will affect morbidity and mortality risks.

Chapter 8 analyzed individual trajectories of aging changes in key physiological biomarkers measured in participants of the FHS and focused on establishing connections between characteristics of dynamic trajectories (across time) and human lifespan and healthspan. They considered a broad range of such dynamic characteristics including, e.g., the rate of change, the rate of increase or decrease, the mean of residuals, the maximal value, etc. They found that these dynamic variables may influence longevity and exceptional health more substantially than the variables describing static physiological states. The major conclusion of the analyses was that such dynamic variables can be important targets for prevention aiming to postpone onsets of complex diseases and increase lifespan. In order to see a clearer picture of the relationships between physiological biomarkers, healthspan, and lifespan, the dynamic variables should be more actively employed in aging and health research, and routinely used together with the variables describing static physiological states.

The aging of populations in developed countries requires effective strategies to extend healthspan. A promising solution could be to yield insights into the genetic predispositions for endophenotypes, diseases, well-being, and survival. It was thought that genome-wide association studies (GWAS) would be a major breakthrough in this endeavor. Various genetic association studies including GWAS assume that there should be a deterministic (unconditional) genetic component in such complex phenotypes. However, the idea of unconditional contributions of genes to these phenotypes faces serious difficulties which stem from the lack of direct evolutionary selection against or in favor of such phenotypes. In fact, evolutionary constraints imply that genes should be linked to age-related phenotypes in a complex manner through different mechanisms specific for given periods of life. Accordingly, the linkage between genes and these traits should be strongly modulated by age-related processes in a changing environment, i.e., by the individuals' life course. The inherent sensitivity of genetic mechanisms of complex health traits to the life course will be a key concern as long as genetic discoveries continue to be aimed at improving human health.

Given this rationale, Chap. 9 presented the results of detailed analyses of the effect of the *APOE* common polymorphism on age patterns of risks of major human diseases such as CVD and cancer, as well as on survival. The analyses were focused on two generations of humans participating in the FHS. The results provided examples of complex modes of *APOE* actions including genetic trade-offs, antagonistic genetic effects on the same traits at different ages, and changes in genetic effects on lifespan at different ages.

The results on the antagonistic roles of *APOE* in diseases provided two important insights. First, they explicitly showed that the same allele can change its role in the risks of CVD in an antagonistic fashion across age. Second, these genetic trade-offs can be strongly affected by sex. Both of these insights underscore the key role of the life course in genetic susceptibility to age-related phenotypes.

The results of the analyses of the role of *APOE* in lifespan provided additional evidence about the key role of age-related processes in genetic susceptibility to phenotypes in late life. They also showed that detailed ascertainment of the role of age-related processes in estimating the effects of genes on lifespan and healthspan could help in detecting more homogeneous subsamples at excess risks.

Part II
Statistical Modeling of Aging, Health, and Longevity

Chapter 11
Approaches to Statistical Analysis of Longitudinal Data on Aging, Health, and Longevity: Biodemographic Perspectives

Konstantin G. Arbeev, Igor Akushevich, Alexander M. Kulminski, Kenneth C. Land, and Anatoliy I. Yashin

11.1 Introduction

Longitudinal data play a pivotal role in discovering different aspects of knowledge related to aging, health, and longevity. There are many statistical methods for the analysis of longitudinal data, which is one of the most prolific areas of statistical science. Different types of research questions require different analytic approaches: for example, a researcher would apply different approaches depending on the type of longitudinal data at hand (categorical or continuous) and whether the primary research interest is in the longitudinal outcomes alone or in combined analyses of longitudinal data and survival (or, generally, time-to-event) outcomes. The goal of this chapter is not to give a comprehensive overview of various approaches that can be used in such analyses, because it is impossible to cover all of them in any substantial detail in a single chapter. Details on a broad range of state-of-the-art statistical methods of longitudinal data analysis can be found in recent books aimed at a general audience (e.g., Fitzmaurice et al. 2009, 2011) or specifically for researchers in aging, health, and social sciences (Newsom et al. 2012). Rather, the focus of this chapter is narrower: to provide a brief discussion of approaches to statistical analyses of longitudinal data on aging relevant to the major topic of this monograph, *Biodemography of Aging*, and relate this discussion to the subsequent chapters in Part II.

Biodemography is a multidisciplinary branch of science that unites under its umbrella various analytic approaches aimed at integrating biological knowledge and methods and traditional demographic analyses to shed more light on variability in mortality and health across populations and between individuals. *Biodemography of aging* is a special subfield of biodemography that focuses on understanding the impact of processes related to aging on health and longevity. Although it is a relatively young discipline, biodemography in general, and biodemography of aging in particular, have quickly evolved into the one of the most innovative and fastest growing areas of demography with substantial

A.I. Yashin et al., *Biodemography of Aging*, The Springer Series on Demographic Methods and Population Analysis 40, DOI 10.1007/978-94-017-7587-8_11

achievements to date and with great opportunities and new challenges for the future (Carey 2008; Carey and Vaupel 2005; Christensen 2008; Crimmins et al. 2010; Kaplan and Gurven 2008; Vasunilashorn and Crimmins 2008; Vaupel 2010; Wachter 2008). The rapid progress of this field is fueled in particular by the rapid increase in the number of large-scale studies that collect various biomarkers that can be incorporated into demographic analyses (Crimmins et al. 2008, 2010; Weinstein et al. 2007).

Aging, as this word connotes, is a process that develops with age. Therefore, data on various relevant biomarkers measured at different ages for the same individuals are necessary to develop knowledge about the mechanisms and dynamics of the process of aging. The potential and value of biodemographic approaches is now realized due to the availability of longitudinal biomarkers in existing studies and in those that will develop multiple waves with longitudinal biomarker data in the foreseeable future (Crimmins et al. 2010).

Mortality rates as a function of age are a cornerstone of many demographic analyses. The longitudinal age trajectories of biomarkers add a new dimension to the traditional demographic analyses: the mortality rate becomes a function of not only age but also of these biomarkers (with additional dependence on a set of socio-demographic variables). Such analyses should incorporate dynamic characteristics of trajectories of biomarkers to evaluate their impact on mortality or other outcomes of interest. Traditional analyses using baseline values of biomarkers (e.g., Cox proportional hazards or logistic regression models) do not take into account these dynamics. One approach to the evaluation of the impact of biomarkers on mortality rates is to use the Cox proportional hazards model with time-dependent covariates; this approach is used extensively in various applications and is available in all popular statistical packages. In such a model, the biomarker is considered a time-dependent covariate of the hazard rate and the corresponding regression parameter is estimated along with standard errors to make statistical inference on the direction and the significance of the effect of the biomarker on the outcome of interest (e.g., mortality). However, the choice of the analytic approach should not be governed exclusively by its simplicity or convenience of application. It is essential to consider whether the method gives meaningful and interpretable results relevant to the research agenda. In the particular case of biodemographic analyses, the Cox proportional hazards model with time-dependent covariates is not the best choice. This is due to features of the analyses and data being analyzed, as discussed below.

Longitudinal studies of aging present special methodological challenges due to inherent characteristics of the data that need to be addressed in order to avoid biased inference. The challenges are related to the fact that the populations under study (aging individuals) experience substantial dropout rates related to death or poor health and often have co-morbid conditions related to the disease of interest. The standard assumption made in longitudinal analyses (although usually not explicitly mentioned in publications) is that dropout (e.g., death) is not associated with the outcome of interest. While this can be safely assumed in many general longitudinal studies (where, e.g., the main causes of dropout might be the administrative end of the study or moving out of the study area, which are presumably not related to the studied outcomes), the very nature of the longitudinal outcomes (e.g., measurements of some physiological biomarkers) analyzed in a longitudinal study of aging assumes that they are (at least hypothetically) related to the process of aging. Because the process of aging leads to the development of diseases and, eventually, death, in longitudinal studies of aging an assumption of non-association of the reason for dropout and the outcome of interest is, at best, risky, and usually is wrong. As an illustration, we found that the average trajectories of different physiological indices of individuals dying at earlier ages markedly deviate from those of long-lived individuals, both in the entire Framingham original cohort (see, e.g., Chap. 2 and Yashin et al. (2012b)) and also among carriers of specific alleles (Arbeev et al. 2012). In such a situation, panel compositional changes due to attrition affect the averaging procedure and modify the averages in the total sample.

Furthermore, biomarkers are subject to measurement error and random biological variability. They are usually collected intermittently at examination times which may be sparse and typically biomarkers are not observed at event times. It is well known in the statistical literature that ignoring measurement errors and biological variation in such variables and using their observed "raw" values as time-dependent covariates in a Cox regression model may lead to biased estimates and incorrect inferences (Prentice 1982; Sweeting and Thompson 2011). For example, as Sweeting and Thompson (2011) showed, the Cox regression model with time-dependent covariates severely underestimates associations between the current underlying longitudinal value and the event hazard. When biomarkers are measured at sparse examinations or with a long time interval before an outcome event such bias can worsen. This is because the Cox model must have values of such variables at different time points, which is usually achieved in software implementations by assuming that the values of the time-dependent covariates are constant between observations (exams) and that the hazard at some future point is associated with the extrapolated value of the covariate at this time point.

That said, it is clear that standard methods of longitudinal data analyses such as mixed-effects models (Laird and Ware 1982) or generalized estimating equations (Liang and Zeger 1986) are not appropriate in analyses of longitudinal data on aging because they assume non-informative dropout. Standard methods of survival analysis such as the Cox proportional hazards model (Cox 1972) with time-dependent covariates should be avoided in analyses of biomarkers measured with errors because they can lead to biased estimates.

The need to use appropriate statistical methods to take into account challenges associated with analyses of longitudinal data on aging is recognized in the gerontological literature (Murphy et al. 2011). Which statistical methods are then appropriate for analyses of longitudinal data on aging depends on the actual research aims (Kurland et al. 2009). Although, as noted above, the field of biodemography encompasses a diverse research agenda, we will specifically focus in this chapter on analyses of mortality (or, generally, time-to-event) data and longitudinal measurements of biomarkers. That is, we will focus on approaches that include models for both the time-to-event and longitudinal outcomes (thus omitting methods that concentrate on the longitudinal outcome and treat the time-to-event data as a nuisance factor to be adjusted for, and approaches that do not include time-to-event information, e.g., onset of a disease, but, say, include instead binary indicators such as prevalence of a disease in logistic regression). Our choice of joint analyses of the time-to-event and longitudinal outcomes is consistent with keeping the narrative focused on specific applications to biodemography of aging. There is a considerable literature on advanced methods of analyses of longitudinal data summarized, for example, in a recent book by Fitzmaurice et al. (2009).

Statistical methods aimed at analyses of time-to-event data jointly with longitudinal measurements have become known in the mainstream biostatistical literature as "joint models for longitudinal and time-to-event data" ("survival" or "failure time" are often used interchangeably with "time-to-event") or simply "joint models." This is an active and fruitful area of biostatistics with an explosive growth in recent years. Reviews of some earlier approaches to joint modeling of longitudinal and time-to-event data can be found in Hogan and Laird (1997), Troxel (2002), Tsiatis and Davidian (2004), and Yu et al. (2004). Recent developments are summarized in reviews by Diggle et al. (2008), Ibrahim et al. (2010), Sousa (2011), Wu et al. (2012), McCrink et al. (2013), Proust-Lima et al. (2014), and Gould et al. (2015). We refer readers to the references cited above and the book by Rizopoulos (2012) for a detailed and comprehensive overview of the theory and applications of joint models. In our recent paper (Arbeev et al. 2014) we reviewed both joint models and stochastic process models (see Chap. 12) with a particular focus on applications to prediction of health and survival. In the next section, we briefly present the basics of joint models and their various extensions suggested in the recent biostatistical literature and discuss them in the context of biodemographic applications.

11.2 Statistical Approaches to Joint Analysis of Longitudinal and Time-to-Event Outcomes

11.2.1 Standard Joint Models and Their Extensions

The standard joint model consists of two parts, the first representing the dynamics of longitudinal data (which is referred to as the "longitudinal sub-model") and the second one modeling survival or, generally, time-to-event data (which is referred to as the "survival sub-model"). The standard paradigm in this class of models postulates the dynamics of the "true" (unobserved) longitudinal process in terms of a vector of subject-specific random effects. The observed longitudinal data are the values of this "true" process that are collected intermittently at some time points (possibly different for different individuals) and are subject to measurement error. The survival sub-model typically assumes that the risk of an event at some age t depends on the value of the "true" longitudinal process at that age in a Cox proportional hazards context. For example, the standard model for continuous longitudinal data can be formulated using a *linear mixed-effects* (*LME*) *model* with normally distributed errors and random effects (Faucett and Thomas 1996; Rizopoulos 2012; Tsiatis and Davidian 2004; Wulfsohn and Tsiatis 1997):

$$Y_i(t) = X_i^T(t)\beta + Z_i^T(t)b_i + \varepsilon_i(t), \tag{11.1}$$

where $Y_i(t)$ denotes the longitudinal outcome for individual i at age t, $X_i^T(t)$ and $Z_i^T(t)$ are design vectors of fixed effects β and random effects b_i ("T" denotes transposition; here and below we will use column vectors if not stated otherwise), and $\varepsilon_i(t)$ is the error term. The error terms $\varepsilon_i(t)$ are assumed independent and normally distributed, $\varepsilon_i(t)\tilde{}N(0, \sigma^2)$ Random effects b_i are assumed to be independent of the error terms and also normally distributed, $b_i\tilde{}N(0, B)$. The expression for the hazard rate for the time-to-event outcome represents dependence of the risk of the event on the current "true" value of the longitudinal outcome:

$$\mu_i\left(t\middle|\overline{Y}_i(t), w_i\right) = \mu_0(t)\exp\left\{w_i^T\gamma + \alpha\overline{Y}_i(t)\right\}, \tag{11.2}$$

where $\mu_0(t)$ is the baseline hazard, w_i is a vector of baseline covariates with associated vector of regression coefficients γ, $\overline{Y}_i(t)$ stands for the "true" (unobserved) value of the longitudinal outcome:

$$\overline{Y}_i(t) = X_i^T(t)\beta + Z_i^T(t)b_i, \tag{11.3}$$

and α is the respective regression coefficient.

The model (11.1, 11.2 and 11.3) represents the joint model in its simplest form. Numerous extensions of this basic model have appeared in the joint modeling literature in recent decades, providing great flexibility in applications to a wide range of practical problems. The extensions involve different characteristics of both the longitudinal and the time-to-event sub-models, such as specification of trajectories, distribution of random effects, type of longitudinal data (continuous or categorical), alternative expressions for hazard rates, etc., as described in the remainder of this section.

In the longitudinal sub-model, more flexible specification of the individual trajectories, such as splines (Brown 2009; Brown et al. 2005; Ding and Wang 2008; Rizopoulos and Ghosh 2011; Rizopoulos et al. 2009; Yao 2007) or stochastic processes (Chiang 2011; Henderson et al. 2000, 2002; Struthers and McLeish 2011; Wang and Taylor 2001; Xu and Zeger 2001b), have been suggested. We will discuss the latter approach in more detail in the next section on the use of stochastic processes to model biological variation and heterogeneity in individual longitudinal trajectories. Another type of model with more flexible specification of longitudinal trajectories that allows for testing relevant biological hypotheses is change-point joint models (Dantan et al. 2011; Faucett et al. 2002; Garre et al. 2008; Ghosh et al. 2011; Jacqmin-Gadda et al. 2006; Pauler and Finkelstein 2002; Tapsoba et al. 2011b; Yu and Ghosh 2010). *Change-point joint models* extend the random effects specification of the longitudinal data by adding an unknown change-point which represents the time when the longitudinal age trajectory experiences a change in its pattern (e.g., an increase in the slope, or a change from a linear to a non-linear pattern). Biologically, such change-points may correspond to some internal processes in an organism that ultimately manifest themselves in the changes in behavior of the longitudinal trajectories. The change-points typically are assumed to be random variables with distributions (means) depending on observed covariates, thus allowing the investigation of their impact on the change in behavior of the longitudinal trajectory. The change-points can also be thought of as representing an onset of a pre-disease or pre-diagnosis state in the joint multistate model context (Dantan et al. 2011).

Incorrect specification of the distribution of random effects in joint models can, in principle, lead to biased inference. Several authors have considered relaxing the assumption of normality of random effects or making no parametric assumptions about the random effects distribution (Brown and Ibrahim 2003b; Song et al. 2002a; Song and Wang 2008; Tapsoba et al. 2011a; Tsiatis and Davidian 2001). Hsieh et al. (2006) showed in simulation studies that maximum likelihood estimates (MLEs) of parameters in joint models are robust against the violation of the normality assumption of random effects if information from the longitudinal data is rich enough. Rizopoulos et al. (2008) proved that the effect of misspecification of the random effects distribution diminishes as the number of longitudinal measurements per individual increases. However, as Huang et al. (2009) noted, "a relevant question is whether or not the available longitudinal information in a particular data set is rich enough to yield an MLE insensitive to model misspecification." They developed a diagnostic tool to reveal misspecification in the random effects model and provided a graphical method and test statistics to quantitatively assess the robustness of parameter estimators.

Standard joint models utilize LME models for longitudinal data. *Nonlinear mixed effects* (*NLME*) *models* in which more sophisticated nonlinear processes generating the longitudinal data are assumed (Davidian and Giltinan 1995) can be useful, and *generalized linear mixed models* (*GLMM*) (Diggle et al. 2002) can accommodate both continuous and categorical longitudinal outcomes (such as Gaussian, binomial or Poisson variables). Joint models incorporating NLME and GLMM have appeared recently in the literature (Huang et al. 2011; Rizopoulos and Ghosh 2011; Wu et al. 2008, 2010; Yao 2008). Such nonlinear models may be useful in many applications, but they are more computationally demanding and the nonlinearity of the longitudinal models may require special approaches to reduce computation time.

The most widely used approach in the joint models literature applies the *Cox proportional hazards* (*regression*) *model* to represent the relationship between the longitudinal outcomes and failure times. The distinct feature of the Cox model is a completely unspecified baseline hazard. Although such a specification is possible in the joint modeling context, it can result in underestimated standard errors of parameters (Hsieh et al. 2006), thus necessitating an explicit characterization of the baseline hazard. Models with flexible specifications of the baseline hazard include the use of piecewise constant hazards or approximations by splines, e.g., as implemented in the R package JM (Rizopoulos 2010, 2012).

There can be many situations in real-world applications in which the proportionality of hazards assumption does not hold and can be hard to justify biologically. In these cases, the use of models that assume the proportionality of hazards to capture the relationship between failure times and longitudinal outcomes can be misleading. The accelerated failure time (AFT) model (Cox and Oakes 1984) is considered an appealing alternative to the Cox model in aging research (Swindell 2009), often providing more intuitive interpretations of the effects of a covariate (e.g., treatment) on the survival outcome. Joint models with the AFT survival sub-model were considered in the literature (Hanson et al. 2011; Huang et al. 2011; Rizopoulos et al. 2010; Tseng et al. 2005; Vonesh et al. 2006; Wu et al. 2010). Another possible alternative with more flexible models for effects of covariates on survival data is the time-varying coefficient proportional hazards model (Zucker and Karr 1990) in which the regression coefficients may vary over time. Song and Wang (2008) applied the time-varying coefficient proportional hazards model in the joint model context.

The standard joint model (11.1, 11.2 and 11.3) is formulated for a single longitudinal outcome. Often several longitudinal markers that can be related to the time-to-event outcome are available in the study. Such biomarkers can represent different manifestations of underlying biological processes that work in concert so their joint analysis in the framework of a multidimensional model that would take dependence between the markers into account can be advantageous. Joint models for multiple longitudinal markers have been considered in the literature (Brown et al. 2005; Chi and Ibrahim 2006, 2007; Ibrahim et al. 2004; Lin et al. 2002; Rizopoulos and Ghosh 2011; Song and Wang 2008; Song et al. 2002b; Xu and Zeger 2001a). However, this advantage can be offset by computational difficulties related to the need for numerical integration over the random effects. In the case of several longitudinal outcomes, the dimensionality of the random effects can become prohibitively large for practical implementations.

Model (11.1, 11.2 and 11.3) focuses on a single failure type such as death or onset of a disease. Extensions of joint models to work with multiple failure times (competing risks) have been discussed in the literature. This literature is described in more detail in Chap. 13 which also reviews modifications of joint models, including an additional random variable in the time-to-event sub-model (typically referred to as "frailty") and joint models that accommodate latent subpopulations (latent classes) called "joint latent class models."

A special case of joint models is the class of models with a cure fraction or *joint cure models* (Abu Bakar et al. 2009; Brown and Ibrahim 2003a; Chen et al. 2004; Chi and Ibrahim 2007, 2006; Law et al. 2002; Song et al. 2012; Taylor et al. 2005; Yu and Ghosh 2010; Yu et al. 2004, 2008). Such models are relevant, for example, in applications to cancer research where they represent the natural setting of events (treatment and subsequent cure or recurrence of cancer). When such a "cured" group (not susceptible to the risk of the event) is present in the data, there will be a plateau in the survival function when there is a substantial follow-up period, i.e., the survival function will never reach zero. That is, standard models with proper survival functions are not appropriate in such applications, because they assume that every individual would eventually experience the event. Joint cure models also represent an interesting example of the incorporation of biological theories and concepts into statistical models.

As discussed in Chap. 6, modeling and analysis of medical cost trajectories in relation to the onset of aging-related diseases and death are important topics in research on aging. These are especially relevant nowadays when the population is aging and medical costs are increasing rapidly. For example, joint analyses of trajectories of medical costs and survival can be important in cost-effectiveness studies. Joint models provide an approach to performing such joint analyses of individual trajectories of medical costs and time-to-event data. Liu et al. (2007) suggested a joint model for monthly medical costs and survival time that takes into account the possible correlation between the medical costs trajectory and survival time and possible differential patterns of medical costs close to death. Such a correlation is introduced through a common random effect in the sub-models for survival and medical costs. This random effect in the survival sub-model represents "frailty" (see Chap. 13). Specification of a sub-model for medical costs takes into account the possibility of a changing pattern of costs at some time period before death which makes it similar to the change-point models discussed above. As Liu et al. (2007) showed in their simulation studies, ignoring the dependence of death times on medical costs results in biased estimates of the longitudinal model for medical costs, whereas the joint model produces correct estimates. Another point to consider in such applications is the assumption of non-informative observation times, i.e., that observation times do not carry information on the longitudinal measures (medical costs). This is the usual assumption in the joint models literature and it is relevant in most applications, but it is at least questionable in analyses of medical costs data. For example, patients at a more severe disease stage visit hospitals more often (i.e., have more densely distributed observation points), accrue medical costs faster, and have worse survival chances than those having a milder form of disease or no disease at all. Therefore, joint model for analyses of such data should account for both informative observation times and a dependent terminal event simultaneously. Liu et al. (2008a) proposed such a joint model that includes three components: a frailty model for the intensity of recurrent events (hospital admissions), a random effects model for repeated observations (costs) collected at these recurrent visits, and a proportional hazards model for the failure time. The model includes correlated random effects in all three sub-models to introduce dependence between the respective processes. Liu (2009) extended the approach to apply it to a more realistic situation with monthly medical costs, which are characterized by the presence of a large proportion of zero values and right skewness of non-zero values.

The standard parameterization of the joint model (11.2) assumes that the risk of the event at age t depends on the current "true" value of the longitudinal biomarker at this age. While this is a reasonable assumption in general, it may be argued that additional dynamic characteristics of the longitudinal trajectory can also play a role in the risk of death or onset of a disease. For example, if two individuals at the same age have exactly the same level of some biomarker at this age, but the trajectory for the first individual increases faster with age than that of the second one, then the first individual can have worse survival chances for subsequent years. We showed in analyses of the Framingham data that, indeed, the dynamic characteristics of individual trajectories (e.g., slopes, variability) are related to mortality risk and risk of onset of major aging-related diseases (Yashin et al. 2010, 2012b); see Chap. 8. Therefore, extensions of the basic parameterization of joint models allowing for dependence of the risk of an event on such dynamic characteristics of the longitudinal trajectory can provide additional opportunities for comprehensive analyses of relationships between the risks and longitudinal trajectories. Several authors have considered such extended models. For example, Yu et al. (2008) considered a joint cure model with the current value of a biomarker and its current slope included as time-dependent covariates in the specification of the hazard. Ye et al. (2008) proposed a semiparametric joint model with the hazard rate depending on both the current value of the underlying subject-specific trajectory and its rate of change (slope). Brown (2009) extended the semiparametric approach of Brown et al. (2005) to include the slope and integral of the cubic B-spline of the longitudinal trajectory as time-varying covariates in the hazard model. Gao et al. (2011) relaxed the standard assumption of a common (homogeneous) variance-covariance structure of random effects used in joint models assuming a linear mixed-effects model with individual-specific variances of random effects. Their model includes the random intercept and slope, as well as the logarithm of the individual-specific variance of random effects as covariates in the survival sub-model. This allows for testing hypotheses of the effects of individual-level differences in variability of longitudinal biomarkers on the time-to-event outcomes. Rizopoulos and Ghosh (2011) developed a semiparametric multivariate joint model with a flexible parameterization that, among other elements, includes derivatives of the longitudinal profile functions, thus permitting the risk of an event to depend not only on the true value of the longitudinal outcome but also on the dynamic characteristics (e.g., the slope and the curvature) of the true longitudinal trajectory at that time. This specification along with other generalizations of joint models such as those involving cumulative effects (integrals) or lagged effects are implemented in the R package *JM* (Rizopoulos 2010) and are thoroughly discussed in the book by Rizopoulos (2012).

One particular advantage of joint models is that they provide a natural framework for performing individual predictions of longitudinal and time-to-event outcomes that takes into account the dependence of the risk of an event on the longitudinal observations. Applications of joint models to make dynamic individual predictions (of both longitudinal and time-to-event outcomes) and the development of predictive accuracy measures have recently been discussed in the literature (Commenges et al. 2012; Garre et al. 2008; Hanson et al. 2011; Hatfield and Carlin 2012; Proust-Lima et al. 2014; Proust-Lima and Taylor 2009; Rizopoulos 2011; Sweeting and Thompson 2011; Taylor et al. 2005; Yu et al. 2008); see also Chap. 7 in the book by Rizopoulos (2012).

As this section illustrates, there is great variability in the specifications of joint models that may be suitable in different applications. Since joint models are computationally intensive and are sometimes prone to convergence problems so that the estimation algorithm requires "attendant nursing," as Wang and Taylor (2001) termed the required close monitoring of the iterative processes, their practical use depends critically on the implementation of the algorithms in available software packages. Different authors have used different software packages to fit specific versions of joint models. This includes implementations in both commercial software such as SAS (see, e.g., Gueorguieva et al. 2012; Guo and Carlin 2004; Liu 2009; Liu et al. 2008a; Vonesh et al. 2006; Ye et al. 2008), Stata (Crowther et al. 2013), and Mplus (Muthén and Muthén 1998–2012)—see Wang et al. (2012)—as well as implementations in freely available software such as WinBUGS (Lunn et al. 2000)—see Gao et al. (2011), Guo and Carlin (2004), Hatfield et al. (2011), Huang et al. (2011), Rizopoulos and Ghosh (2011), and Sweeting and Thompson (2011),—aML (Lillard and Panis 2003)—see Liu et al. (2008b)—and several packages in R (*JM* (Rizopoulos 2010), *JMbayes*, *JMLSD*, *joineR*, and *lcmm*). Although not all published papers on joint models provide software code for their estimation algorithms, which hinders their use in practical applications, the recent development of a flexible R package *JM* covering a wide range of joint models and the availability of a book providing comprehensive practical guidance on the use of the R packages (*JM* and *lcmm*) to fit joint models (Rizopoulos 2012) should facilitate their widespread application in different research areas.

We should also note one more advantage of joint models. They provide more efficient estimates of the effect of a covariate (e.g., treatment) on time-to-event outcomes in the case in which there is an effect of the covariate on the longitudinal trajectory of a biomarker. This means that joint analyses of longitudinal and time-to-event data in joint models may require smaller sample sizes to achieve comparable power with analyses based on time-to-event data alone and ignoring the longitudinal process can lead to biased estimates of the effect of a covariate and a potential loss of power (Chen et al. 2011). There is also an additional possibility for increasing the power of joint analyses of longitudinal and survival data related to the application of recent biodemographic methods (for more details, see Chap. 14).

11.2.2 The Use of Stochastic Processes to Capture Biological Variation and Heterogeneity in Longitudinal Patterns in Joint Models

The previous section outlined a wide spectrum of models that provide great flexibility in joint analyses of longitudinal and time-to-event outcomes that may be relevant in applications to various research questions in different scientific disciplines including biodemography. When it comes to biodemographic applications, however, one additional important point to consider is how to integrate biological knowledge and methods and statistical models used in joint analyses of longitudinal and survival data. The basic joint models (11.1, 11.2 and 11.3) have "standard" specifications of their components traditionally used in modern survival and longitudinal data analysis (such as the Cox proportional hazards model for the survival sub-model and the linear mixed-effects model for longitudinal data). These are motivated mostly by convenience of estimation and availability of statistical software. Such specifications of the models have a rather limited utility for investigating biological mechanisms leading to the observed dynamics of longitudinal biomarkers and the outcomes of interest (e.g., survival, onset of a disease, etc.) because they are not based on any substantive knowledge accumulated in prior research. To be useful for such applications, the models need to be the "biologically-based" ones. That is, they should take into account the complex dynamics of underlying biological processes, thus providing the possibility of estimating parameters that can be meaningfully interpreted from a biological point of view.

One possibility for incorporating individuals' biological backgrounds into joint models is to provide a more flexible form of longitudinal trajectories of biomarkers that would be more plausible and interpretable than just a linear function of age in the respective applications. For example, in applications to prostate cancer data, the longitudinal dynamics of prostate-specific antigen (PSA), which is an important disease progression marker, can be represented by a nonlinear exponential decay—an exponential growth model where the parameters have natural interpretations, e.g., the initial decline in PSA after radiation (Law et al. 2002; Taylor et al. 2005; Yu et al. 2008, 2004). Similarly, alternative specifications of the hazard rate can be more biologically interpretable in specific applications than the standard exponential proportional hazards used in the basic joint models. For example, the stochastic model of tumor recurrence by Yakovlev et al. (1993) formulates the hazard rate in terms of the mean number of clonogens (i.e., the surviving neoplastic cells that are capable of propagating into a newly detectable tumor) surviving the treatment and the probability density function of the distribution of progression times (i.e., the times for clonogens to produce a detectable tumor). The stochastic model of spontaneous carcinogenesis (Yakovlev and Tsodikov 1996) also expresses the hazard rate in a biologically-motivated fashion as a function of the intensity of non-repaired lesion formation (which can lead in the long run to an observable tumor) and the cumulative distribution function of progression times. These models can be incorporated into the joint modeling framework to provide an appealing possibility for a biological interpretation of the impact of observed covariates on the respective characteristics. Implementations and adaptations of Yakovlev's models to the joint models context have been discussed in the literature (Abu Bakar et al. 2009; Brown and Ibrahim 2003a; Chen et al. 2004; Chi and Ibrahim 2006, 2007; Song et al. 2012).

Standard joint models of the form (11.1, 11.2 and 11.3) specify some simple (e.g., linear) age patterns of longitudinal trajectories of biomarkers. This is a convenient approximation justifiable from a computational point of view. However, it ignores the biological variability of individual trajectories over time, a simplification that may be biologically implausible in specific applications. For example, longitudinal data on CD4 counts are widely used in applications of joint models to AIDS studies. The basic joint model with a linear growth curve (Tsiatis et al. 1995) assumes that individual trajectories of "true" (unobserved) CD4 counts are straight lines. While this may be true on average, as CD4 counts tend to decline over the course of HIV infection, it is known that the CD4 count is a highly variable immune system marker and its individual age trajectory cannot be captured by a simple linear function. Note that such variability of longitudinal biomarkers *within* individuals is different from measurement errors (which are given by the i.i.d. random variables), because the values of biomarkers are typically correlated over time. It has been proposed in the literature to use stochastic processes to better capture biological variation and heterogeneity in longitudinal trajectories of biomarkers in individuals. In this case, Eq. 11.1 includes an additional term representing a stochastic process modeling the correlation between measurements:

$$Y_i(t) = X_i^T(t)\beta + Z_i^T(t)b_i + W_i(t) + \varepsilon_i(t), \tag{11.4}$$

where $W_i(t)$ denotes a mean zero stochastic process which is assumed to be independent of the error terms $\varepsilon_i(t)$ and random effects b_i. When such a stochastic process is included in the model, individual longitudinal trajectories of biomarkers are considered as realizations of a stochastic process. Specifications of the process $W_i(t)$ differ in applications. One choice for modeling the longitudinal trajectories includes an integrated Ornstein-Uhlenbeck process (LaValley and DeGruttola 1996; Wang and Taylor 2001) as suggested in applications to modeling longitudinal CD4 counts by Taylor et al. (1994). Another option for representing longitudinal trajectories is to use the *semiparametric stochastic mixed model* by Zhang et al. (1998) that includes a zero mean integrated Wiener stochastic process, as in Ye et al. (2008). Henderson et al. (2000) proposed modeling the joint distribution of longitudinal measurements and events via an unobserved (latent) stationary bivariate Gaussian process, so that the correlation between the two components of the process induces dependence between the longitudinal data and event times. A *zero-mean stationary Gaussian process* specification of the longitudinal model also was used in Xu and Zeger (2001b) in the framework of generalized linear models. Chiang (2011) generalized the correlation mechanism in the joint latent model of Henderson et al. (2000), considering it in a varying-coefficient model (Chiang et al. 2001; Hoover et al. 1998). Struthers and McLeish (2011) used a generalized Ornstein-Uhlenbeck process with observed covariates included in parameters of this process.

The *Ornstein-Uhlenbeck (OU) process* (Uhlenbeck and Ornstein 1930) is one type of stochastic processes that can be used to represent the longitudinal trajectories of biomarkers. It is widely used in physics and financial mathematics and it is also of a particular interest from the biodemographic perspective, because it has some appealing properties especially relevant for biological interpretation. The OU process is stationary, Gaussian, and Markovian (the only nontrivial process having these three conditions). The stochastic differential equation of the OU process has the form:

$$dY(t) = a(f_1 - Y(t))dt + \sigma dW(t), \tag{11.5}$$

where $W(t)$ is the Wiener process, and $a > 0$, f_1, and $\sigma > 0$ are parameters. The parameter f_1 corresponds to the long-term mean value of the OU process $Y(t)$, σ is the diffusion coefficient representing the degree of volatility around the mean, and the parameter a controls the rate at which the process reverts to the mean. The current value of the OU process defines the direction of the drift of the process. If the current value of the process is less than the mean value f_1 then the drift will be positive (i.e., towards the mean) and if the current value of $Y(t)$ is greater than f_1 then the drift will be negative (i.e., again towards the mean f_1). That is, in the long run, the OU process tends to drift towards its long-term mean f_1. This remarkable "mean-reverting" property of the OU process has a natural biological interpretation in terms of homeostatic regulation of an organism, and it makes this process the method of choice for modeling age trajectories of biomarkers. This process naturally models homeostatic regulation, a fundamental property of living organisms that tends to maintain stability by returning or restoring a biological subsystem to an "equilibrium state" in case of disturbances of various kinds. The use of such stochastic processes in the specification of statistical models provides a vivid example of how biological knowledge can be incorporated into statistical analysis. This is a key feature of biodemographic models that will be discussed in the next section and subsequent chapters of this monograph.

11.3 Bringing Biology to Statistics: Biodemographic Models for Analysis of Longitudinal Data on Aging, Health, and Longevity

An important challenge to consider in the context of biodemographic analyses is that such analyses aim at integrating biological knowledge and theories with demographic analyses. In particular, for the biodemography of aging, this means incorporating knowledge and theories about the processes of aging into analytic approaches. Substantial knowledge about mechanisms of aging-related changes has been accumulated in the prior research literature and different concepts of aging have been formulated. Individual measurements of biomarkers represent a "snapshot" of an individual's physiological state at a particular age. Longitudinal data on aging, health, and longevity that contain measurements of biomarkers observed in the same individual at different ages along with his/her health and survival status are especially valuable for biodemographic analyses. Such data contain information not only about one's physiological state at a given age, but also about its *dynamics*, which may be associated with the process of aging which leads to the development of aging-related diseases and, eventually, death. However, a conceptual analytic framework is necessary that incorporates available knowledge about mechanisms of aging-related changes that may be hidden in the individual longitudinal trajectories of biomarkers in order to analyze their indirect impact on the risks of diseases and death. The joint models reviewed in the previous sections are of limited use in this respect. Although some of them are based on sound biological theories relevant to specific applications, e.g., cancer studies, they typically lack specific parameters or components of the models that can be biologically interpreted in the context of aging. However, standard joint models can be useful at the initial stages of analysis when the presence of specific effects has to be identified (for example, the effect of the values of a biomarker on the risk of death).

One possibility for bringing biological knowledge into statistical models is the use of stochastic processes that model the complex dynamics of underlying mechanisms (such as the Ornstein-Uhlenbeck process described above). This provides a structure for estimating parameters that can be meaningfully interpreted from the biological point of view. Such "biologically-based" models are more appropriate for understanding the biological mechanisms leading to the observed longitudinal dynamics of biomarkers and the outcomes of interest (e.g., survival, onset of a disease, etc.) than standard models based on conventional assumptions (e.g., proportionality of hazards or linearity of individual age trajectories of biomarkers).

Biodemographic analyses of mechanisms and regularities of aging in relation to mortality (or other time-to-event outcomes of interest) can be performed using a special type of statistical model which is known as the *stochastic process model of aging*. The specific version of this model that incorporates substantive knowledge about different aging-related concepts has been developed recently by the research team contributing to this monograph (Yashin et al. 2007) and has been extended in various ways and applied in different contexts; see, e.g., our recent review paper (Yashin et al. 2012a). This model incorporates a generalized version of the Ornstein-Uhlenbeck process of Eq. 11.5 with parameters depending on age (Yashin et al. 2007) and observed covariates (Yashin et al. 2012a), including genetic markers (Arbeev et al. 2009; Yashin et al. 2013). For specification of the hazard rate, the model uses a biologically plausible quadratic function of the value of each biomarker as justified by numerous epidemiological observations for different biomarkers (hence its alternative name, the *quadratic hazard model*).

There is one historical aspect related to the stochastic process model which is rarely recognized in the mainstream biostatistical literature on joint models. This model has its roots in the *random walk model* of Woodbury and Manton (1977). This conceptually important model with an elegant mathematical development (Aalen et al. 2008) has several important advantages (Martinussen and Keiding 1997). But the model, as well as its applications and extensions, remained rather unnoticed in the literature on joint models. Nevertheless, these works by Woodbury, Manton, Stallard, Vaupel, and Yashin (see, e.g., Woodbury and Manton 1977; Woodbury et al. 1979; Yashin et al. 1985, 1986a, b) predated the "classical" developments of joint models (those reviewed in Sect. 11.2) by several years.

The stochastic process model will be further discussed in more detail in Chap. 12 and its generalizations will be the topics of Chaps. 13, 14, 15, and 16.

Acknowledgements This chapter was partly supported by the National Institute on Aging of the National Institutes of Health under Award Numbers R01AG030198, R01AG032319, R01AG030612, R01AG046860, P01AG043352, and P30AG034424. The content is solely the responsibility of the authors and does not necessarily represent the official views of the National Institutes of Health.

References

Aalen, O. O., Borgan, O., & Gjessing, H. K. (2008). *Survival and event history analysis: A process point of view*. New York: Springer.

Abu Bakar, M. R., Salah, K. A., Ibrahim, N. A., & Haron, K. (2009). Bayesian approach for joint longitudinal and time-to-event data with survival fraction. *Bulletin of the Malaysian Mathematical Sciences Society, 32*(1), 75–100.

Arbeev, K. G., Akushevich, I., Kulminski, A. M., Arbeeva, L. S., Akushevich, L., Ukraintseva, S. V., Culminskaya, I. V., & Yashin, A. I. (2009). Genetic model for longitudinal studies of aging, health, and longevity and its potential application to incomplete data. *Journal of Theoretical Biology, 258*(1), 103–111.

Arbeev, K.G., Ukraintseva, S.V., Kulminski, A.M., Akushevich, I., Arbeeva, L.S., Culminskaya, I. V., Wu, D., & Yashin, A.I. (2012). Effect of the APOE polymorphism and age trajectories of physiological variables on mortality: Application of genetic stochastic process model of aging. *Scientifica* 2012:Article ID 568628.

Arbeev, K.G., Akushevich, I., Kulminski, A.M., Ukraintseva, S., & Yashin, A.I. (2014). Joint analyses of longitudinal and time-to-event data in research on aging: Implications for predicting health and survival. *Frontiers in Public Health 2*:article 228.

Brown, E. R. (2009). Assessing the association between trends in a biomarker and risk of event with an application in pediatric HIV/AIDS. *Annals of Applied Statistics, 3*(3), 1163–1182.

Brown, E. R., & Ibrahim, J. G. (2003a). Bayesian approaches to joint cure-rate and longitudinal models with applications to cancer vaccine trials. *Biometrics, 59*(3), 686–693.

Brown, E. R., & Ibrahim, J. G. (2003b). A Bayesian semiparametric joint hierarchical model for longitudinal and survival data. *Biometrics, 59*(2), 221–228.

Brown, E. R., Ibrahim, J. G., & DeGruttola, V. (2005). A flexible B-spline model for multiple longitudinal biomarkers and survival. *Biometrics, 61*(1), 64–73.

Carey, J. R. (2008). Biodemography: Research prospects and directions. *Demographic Research, 19*, 1749–1757.

Carey, J. R., & Vaupel, J. W. (2005). Biodemography. In D. Poston & M. Micklin (Eds.), *Handbook of population* (pp. 625–658). New York: Kluwer Academic/Plenum Publishers.

Chen, M. H., Ibrahim, J. G., & Sinha, D. (2004). A new joint model for longitudinal and survival data with a cure fraction. *Journal of Multivariate Analysis, 91*(1), 18–34.

Chen, L. M., Ibrahim, J. G., & Chu, H. (2011). Sample size and power determination in joint modeling of longitudinal and survival data. *Statistics in Medicine, 30*(18), 2295–2309.

Chi, Y. Y., & Ibrahim, J. G. (2006). Joint models for multivariate longitudinal and multivariate survival data. *Biometrics, 62*(2), 432–445.

Chi, Y.-Y., & Ibrahim, J. G. (2007). Bayesian approaches to joint longitudinal and survival models accommodating both zero and nonzero cure fractions. *Statistica Sinica, 17*(2), 445–462.

Chiang, C.-T. (2011). A more flexible joint latent model for longitudinal and survival time data. *Metrika, 73*(2), 151–170.

Chiang, C. T., Rice, J. A., & Wu, C. O. (2001). Smoothing spline estimation for varying coefficient models with repeatedly measured dependent variables. *Journal of the American Statistical Association, 96*(454), 605–619.

Christensen, K. (2008). Human biodemography: Some challenges and possibilities for aging research. *Demographic Research, 19*, 1575–1586.

Commenges, D., Liquet, B., & Proust-Lima, C. (2012). Choice of prognostic estimators in joint models by estimating differences of expected conditional Kullback-Leibler risks. *Biometrics, 68*(2), 380–387.

Cox, D. R. (1972). Regression models and life-tables. *Journal of the Royal Statistical Society, Series B (Statistical Methodology), 34*(2), 187–220.

Cox, D. R., & Oakes, D. (1984). *Analysis of survival data*. London: Chapman and Hall.

Crimmins, E., Vasunilashorn, S., Kim, J. K., & Alley, D. (2008). Biomarkers related to aging in human populations. *Advances in Clinical Chemistry, 46*, 161–216.

Crimmins, E., Kim, J. K., & Vasunilashorn, S. (2010). Biodemography: New approaches to understanding trends and differences in population health and mortality. *Demography, 47*(Supplement), S41–S64.

Crowther, M. J., Abrams, K. R., & Lambert, P. C. (2013). Joint modeling of longitudinal and survival data. *Stata Journal, 13*(1), 165–184.

Dantan, E., Joly, P., Dartigues, J. F., & Jacqmin-Gadda, H. (2011). Joint model with latent state for longitudinal and multistate data. *Biostatistics, 12*(4), 723–736.

Davidian, M., & Giltinan, D. M. (1995). *Nonlinear models for repeated measurements data*. London: Chapman & Hall.

Diggle, P. J., Heagerty, P., Liang, K.-Y., & Zeger, S. L. (2002). *Analysis of longitudinal data*. Oxford: Oxford University Press.

Diggle, P. J., Sousa, I., & Chetwynd, A. G. (2008). Joint modelling of repeated measurements and time-to-event outcomes: The fourth Armitage lecture. *Statistics in Medicine, 27*(16), 2981–2998.

Ding, J., & Wang, J.-L. (2008). Modeling longitudinal data with nonparametric multiplicative random effects jointly with survival data. *Biometrics, 64*(2), 546–556.

Faucett, C. L., & Thomas, D. C. (1996). Simultaneously modelling censored survival data and repeatedly measured covariates: A Gibbs sampling approach. *Statistics in Medicine, 15*(15), 1663–1685.

Faucett, C. L., Schenker, N., & Taylor, J. M. G. (2002). Survival analysis using auxiliary variables via multiple imputation, with application to AIDS clinical trial data. *Biometrics, 58*(1), 37–47.

Fitzmaurice, G., Davidian, M., Verbeke, G., & Molenberghs, G. (2009). *Longitudinal data analysis*. Boca Raton: Chapman and Hall/CRC.

Fitzmaurice, G. M., Laird, N. M., & Ware, J. H. (2011). *Applied longitudinal analysis*. Hoboken: Wiley.

Gao, F., Miller, J. P., Xiong, C., Beiser, J. A., Gordon, M., & The Ocular Hypertension Treatment Study. (2011). A joint-modeling approach to assess the impact of biomarker variability on the risk of developing clinical outcome. *Statistical Methods and Applications, 20*(1), 83–100.

Garre, F. G., Zwinderman, A. H., Geskus, R. B., & Sijpkens, Y. W. J. (2008). A joint latent class changepoint model to improve the prediction of time to graft failure. *Journal of the Royal Statistical Society: Series A (Statistics in Society), 171*(1), 299–308.

Ghosh, P., Ghosh, K., & Tiwari, R. C. (2011). Joint modeling of longitudinal data and informative dropout time in the presence of multiple changepoints. *Statistics in Medicine, 30*(6), 611–626.

Gould, A. L., Boye, M. E., Crowther, M. J., Ibrahim, J. G., Quartey, G., Micallef, S., & Bois, F. Y. (2015). Joint modeling of survival and longitudinal non-survival data: Current methods and issues. Report of the DIA Bayesian joint modeling working group. *Statistics in Medicine, 34*(14), 2181–2195.

Gueorguieva, R., Rosenheck, R., & Lin, H. (2012). Joint modelling of longitudinal outcome and interval-censored competing risk dropout in a schizophrenia clinical trial. *Journal of the Royal Statistical Society: Series A (Statistics in Society), 175*, 417–433.

Guo, X., & Carlin, B. P. (2004). Separate and joint modeling of longitudinal and event time data using standard computer packages. *American Statistician, 58*(1), 16–24.

Hanson, T. E., Branscum, A. J., & Johnson, W. O. (2011). Predictive comparison of joint longitudinal-survival modeling: A case study illustrating competing approaches. *Lifetime Data Analysis, 17*(1), 3–28.

Hatfield, L. A., & Carlin, B. P. (2012). Clinically relevant graphical predictions from Bayesian joint longitudinal-survival models. *Health Services and Outcomes Research Methodology, 12*(2–3), 169–181.

Hatfield, L. A., Boye, M. E., & Carlin, B. P. (2011). Joint modeling of multiple longitudinal patient-reported outcomes and survival. *Journal of Biopharmaceutical Statistics, 21*(5), 971–991.

Henderson, R., Diggle, P., & Dobson, A. (2000). Joint modelling of longitudinal measurements and event time data. *Biostatistics, 1*(4), 465–480.

Henderson, R., Diggle, P., & Dobson, A. (2002). Identification and efficacy of longitudinal markers for survival. *Biostatistics, 3*(1), 33–50.

Hogan, J. W., & Laird, N. M. (1997). Model-based approaches to analysing incomplete longitudinal and failure time data. *Statistics in Medicine, 16*(1–3), 259–272.

Hoover, D. R., Rice, J. A., Wu, C. O., & Yang, L. P. (1998). Nonparametric smoothing estimates of time-varying coefficient models with longitudinal data. *Biometrika, 85*(4), 809–822.

Hsieh, F., Tseng, Y.-K., & Wang, J.-L. (2006). Joint modeling of survival and longitudinal data: Likelihood approach revisited. *Biometrics, 62*(4), 1037–1043.

Huang, X., Stefanski, L. A., & Davidian, M. (2009). Latent-model robustness in joint models for a primary endpoint and a longitudinal process. *Biometrics, 65*(3), 719–727.

Huang, Y., Dagne, G., & Wu, L. (2011). Bayesian inference on joint models of HIV dynamics for time-to-event and longitudinal data with skewness and covariate measurement errors. *Statistics in Medicine, 30*(24), 2930–2946.

Ibrahim, J. G., Chen, M. H., & Sinha, D. (2004). Bayesian methods for joint modeling of longitudinal and survival data with applications to cancer vaccine trials. *Statistica Sinica, 14*(3), 863–883.

Ibrahim, J. G., Chu, H., & Chen, L. M. (2010). Basic concepts and methods for joint models of longitudinal and survival data. *Journal of Clinical Oncology, 28*(16), 2796–2801.

Jacqmin-Gadda, H., Commenges, D., & Dartigues, J. F. (2006). Random changepoint model for joint modeling of cognitive decline and dementia. *Biometrics, 62*(1), 254–260.

Kaplan, H., & Gurven, M. (2008). Top-down and bottom-up research in biodemography. *Demographic Research, 19*, 1587–1602.

Kurland, B. F., Johnson, L. L., Egleston, B. L., & Diehr, P. H. (2009). Longitudinal data with follow-up truncated by death: Match the analysis method to research aims. *Statistical Science, 24*(2), 211–222.

Laird, N. M., & Ware, J. H. (1982). Random-effects models for longitudinal data. *Biometrics, 38*(4), 963–974.

LaValley, M. P., & DeGruttola, V. (1996). Models for empirical Bayes estimators of longitudinal CD4 counts. *Statistics in Medicine, 15*(21–22), 2289–2305.

Law, N. J., Taylor, J. M. G., & Sandler, H. (2002). The joint modeling of a longitudinal disease progression marker and the failure time process in the presence of cure. *Biostatistics, 3*(4), 547–563.

Liang, K. Y., & Zeger, S. L. (1986). Longitudinal data analysis using generalized linear models. *Biometrika, 73*(1), 13–22.

Lillard, L., & Panis, C. W. A. (2003). *aML, multilevel multiprocess statistical software. Release 2.0*. Los Angeles: EconWare.

Lin, H. Q., McCulloch, C. E., & Mayne, S. T. (2002). Maximum likelihood estimation in the joint analysis of time-to-event and multiple longitudinal variables. *Statistics in Medicine, 21*(16), 2369–2382.

Liu, L. (2009). Joint modeling longitudinal semi-continuous data and survival, with application to longitudinal medical cost data. *Statistics in Medicine, 28*(6), 972–986.

Liu, L., Wolfe, R. A., & Kalbfleisch, J. D. (2007). A shared random effects model for censored medical costs and mortality. *Statistics in Medicine, 26*(1), 139–155.

Liu, L., Huang, X., & O'Quigley, J. (2008a). Analysis of longitudinal data in the presence of informative observational times and a dependent terminal event, with application to medical cost data. *Biometrics, 64*(3), 950–958.

Liu, L., Ma, J. Z., & O'Quigley, J. (2008b). Joint analysis of multi-level repeated measures data and survival: An application to the end stage renal disease (ESRD) data. *Statistics in Medicine, 27*(27), 5679–5691.

Lunn, D. J., Thomas, A., Best, N., & Spiegelhalter, D. (2000). WinBUGS – A Bayesian modelling framework: Concepts, structure, and extensibility. *Statistics and Computing, 10*(4), 325–337.

Martinussen, T., & Keiding, N. (1997). The Manton-Woodbury model for longitudinal data with dropouts. *Statistics in Medicine, 16*(1–3), 273–283.

McCrink, L. M., Marshall, A. H., & Cairns, K. J. (2013). Advances in joint modelling: A review of recent developments with application to the survival of end stage renal disease patients. *International Statistical Review, 81*(2), 249–269.

Murphy, T. E., Han, L., Allore, H. G., Peduzzi, P. N., Gill, T. M., & Lin, H. (2011). Treatment of death in the analysis of longitudinal studies of gerontological outcomes. *Journals of Gerontology Series A: Biological Sciences and Medical Sciences, 66*(1), 109–114.

Muthén, L.K. & Muthén, B.O. (1998–2012). *Mplus user's guide. Seventh Edition*. Los Angeles: Muthén & Muthén.

Newsom, J. T., Jones, R. N., & Hofer, S. M. (2012). *Longitudinal data analysis: A practical guide for researchers in aging, health, and social sciences*. New York: Routledge.

Pauler, D. K., & Finkelstein, D. M. (2002). Predicting time to prostate cancer recurrence based on joint models for non-linear longitudinal biomarkers and event time outcomes. *Statistics in Medicine, 21*(24), 3897–3911.

Prentice, R. L. (1982). Covariate measurement errors and parameter estimation in a failure time regression model. *Biometrika, 69*(2), 331–342.

Proust-Lima, C., & Taylor, J. M. G. (2009). Development and validation of a dynamic prognostic tool for prostate cancer recurrence using repeated measures of posttreatment PSA: A joint modeling approach. *Biostatistics, 10*(3), 535–549.

Proust-Lima, C., Sene, M., Taylor, J. M., & Jacqmin-Gadda, H. (2014). Joint latent class models for longitudinal and time-to-event data: A review. *Statistical Methods in Medical Research, 23*(1), 74–90.

Rizopoulos, D. (2010). JM: An R package for the joint modelling of longitudinal and time-to-event data. *Journal of Statistical Software, 35*(9), 1–33.

Rizopoulos, D. (2011). Dynamic predictions and prospective accuracy in joint models for longitudinal and time-to-event data. *Biometrics, 67*(3), 819–829.

Rizopoulos, D. (2012). *Joint models for longitudinal and time-to-event data with applications in R*. Boca Raton: Chapman and Hall/CRC.

Rizopoulos, D., & Ghosh, P. (2011). A Bayesian semiparametric multivariate joint model for multiple longitudinal outcomes and a time-to-event. *Statistics in Medicine, 30*(12), 1366–1380.

Rizopoulos, D., Verbeke, G., & Molenberghs, G. (2008). Shared parameter models under random effects misspecification. *Biometrika, 95*(1), 63–74.

Rizopoulos, D., Verbeke, G., & Lesaffre, E. (2009). Fully exponential Laplace approximations for the joint modelling of survival and longitudinal data. *Journal of the Royal Statistical Society, Series B (Statistical Methodology), 71*, 637–654.

Rizopoulos, D., Verbeke, G., & Molenberghs, G. (2010). Multiple-imputation-based residuals and diagnostic plots for joint models of longitudinal and survival outcomes. *Biometrics, 66*(1), 20–29.

Song, X., & Wang, C. Y. (2008). Semiparametric approaches for joint modeling of longitudinal and survival data with time-varying coefficients. *Biometrics, 64*(2), 557–566.

Song, X., Davidian, M., & Tsiatis, A. A. (2002a). A semiparametric likelihood approach to joint modeling of longitudinal and time-to-event data. *Biometrics, 58*(4), 742–753.

Song, X. A., Davidian, M., & Tsiatis, A. A. (2002b). An estimator for the proportional hazards model with multiple longitudinal covariates measured with error. *Biostatistics, 3*(4), 511–528.

Song, H., Peng, Y., & Tu, D. (2012). A new approach for joint modelling of longitudinal measurements and survival times with a cure fraction. *Canadian Journal of Statistics, 40*(2), 207–224.

Sousa, I. (2011). A review on joint modelling of longitudinal measurements and time-to-event. *Revstat-Statistical Journal, 9*(1), 57–81.

Struthers, C. A., & McLeish, D. L. (2011). A particular diffusion model for incomplete longitudinal data: Application to the multicenter AIDS cohort study. *Biostatistics, 12*(3), 493–505.

Sweeting, M. J., & Thompson, S. G. (2011). Joint modelling of longitudinal and time-to-event data with application to predicting abdominal aortic aneurysm growth and rupture. *Biometrical Journal, 53*(5), 750–763.

Swindell, W. R. (2009). Accelerated failure time models provide a useful statistical framework for aging research. *Experimental Gerontology, 44*(3), 190–200.

Tapsoba, J. D., Lee, S.-M., & Wang, C. Y. (2011a). Approximate nonparametric corrected-score method for joint modeling of survival and longitudinal data measured with error. *Biometrical Journal, 53*(4), 557–577.

Tapsoba, J. D., Lee, S.-M., & Wang, C. Y. (2011b). Joint modeling of survival time and longitudinal data with subject-specific changepoints in the covariates. *Statistics in Medicine, 30*(3), 232–249.

Taylor, J. M. G., Cumberland, W. G., & Sy, J. P. (1994). A stochastic model for analysis of longitudinal AIDS data. *Journal of the American Statistical Association, 89*(427), 727–736.

Taylor, J. M. G., Yu, M. G., & Sandler, H. M. (2005). Individualized predictions of disease progression following radiation therapy for prostate cancer. *Journal of Clinical Oncology, 23*(4), 816–825.

Troxel, A. B. (2002). Techniques for incorporating longitudinal measurements into analyses of survival data from clinical trials. *Statistical Methods in Medical Research, 11*(3), 237–245.

Tseng, Y. K., Hsieh, F. S., & Wang, J. L. (2005). Joint modelling of accelerated failure time and longitudinal data. *Biometrika, 92*(3), 587–603.

Tsiatis, A. A., & Davidian, M. (2001). A semiparametric estimator for the proportional hazards model with longitudinal covariates measured with error. *Biometrika, 88*(2), 447–458.

Tsiatis, A. A., & Davidian, M. (2004). Joint modeling of longitudinal and time-to-event data: An overview. *Statistica Sinica, 14*(3), 809–834.

Tsiatis, A. A., DeGruttola, V., & Wulfsohn, M. S. (1995). Modeling the relationship of survival to longitudinal data measured with error. Applications to survival and CD4 counts in patients with AIDS. *Journal of the American Statistical Association, 90*(429), 27–37.

Uhlenbeck, G. E., & Ornstein, L. S. (1930). On the theory of the Brownian motion. *Physical Review, 36*(5), 0823–0841.

Vasunilashorn, S., & Crimmins, E. M. (2008). Biodemography: Integrating disciplines to explain aging. In V. L. Bengtson, D. Gans, N. M. Putney, & M. Silverstein (Eds.), *Handbook of theories of aging* (pp. 63–85). New York: Springer.

Vaupel, J. W. (2010). Biodemography of human ageing. *Nature, 464*(7288), 536–542.

Vonesh, E. F., Greene, T., & Schluchter, M. D. (2006). Shared parameter models for the joint analysis of longitudinal data and event times. *Statistics in Medicine, 25*(1), 143–163.

Wachter, K. W. (2008). Biodemography comes of age. *Demographic Research, 19*, 1501–1512.

Wang, Y., & Taylor, J. M. G. (2001). Jointly modeling longitudinal and event time data with application to acquired immunodeficiency syndrome. *Journal of the American Statistical Association, 96*(455), 895–905.

Wang, P., Shen, W., & Boye, M. E. (2012). Joint modeling of longitudinal outcomes and survival using latent growth modeling approach in a mesothelioma trial. *Health Services and Outcomes Research Methodology, 12*(2–3), 182–199.

Weinstein, M., Vaupel, J. W., & Wachter, K. W. (2007). *Biosocial surveys*. Washington, DC: The National Academies Press.

Woodbury, M. A., & Manton, K. G. (1977). A random-walk model of human mortality and aging. *Theoretical Population Biology, 11*(1), 37–48.

Woodbury, M. A., Manton, K. G., & Stallard, E. (1979). Longitudinal analysis of the dynamics and risk of coronary heart disease in the Framingham study. *Biometrics, 35*(3), 575–585.

Wu, L., Hu, X. J., & Wu, H. (2008). Joint inference for nonlinear mixed-effects models and time to event at the presence of missing data. *Biostatistics, 9*(2), 308–320.

Wu, L., Liu, W., & Hu, X. J. (2010). Joint inference on HIV viral dynamics and immune suppression in presence of measurement errors. *Biometrics, 66*(2), 327–335.

Wu, L., Liu, W., Yi, G.Y., & Huang, Y. (2012). Analysis of longitudinal and survival data: Joint modeling, inference methods, and issues. *Journal of Probability and Statistics 2012*:Article ID 640153.

Wulfsohn, M. S., & Tsiatis, A. A. (1997). A joint model for survival and longitudinal data measured with error. *Biometrics, 53*(1), 330–339.

Xu, J., & Zeger, S. L. (2001a). The evaluation of multiple surrogate endpoints. *Biometrics, 57*(1), 81–87.

Xu, J., & Zeger, S. L. (2001b). Joint analysis of longitudinal data comprising repeated measures and times to events. *Journal of the Royal Statistical Society: Series C: Applied Statistics, 50*, 375–387.

Yakovlev, A. Y., & Tsodikov, A. D. (1996). *Stochastic models of tumor latency and their biostatistical applications*. New Jersey: World Scientific.

Yakovlev, A. Y., Asselain, B., Bardou, V. J., Fourquet, A., Hoang, T., Rochefordiere, A., & Tsodikov, A. D. (1993). A simple stochastic model of tumor recurrence and its application to data on premenopausal breast cancer. In B. Asselain, M. Boniface, C. Duby, C. Lopez, J. P. Masson, & J. Tranchefort (Eds.), *Biometrie et Analyse de Donnees Spatio-Temporelles, No. 12* (pp. 66–82). Rennes: Societe Francaise de Biometrie.

Yao, F. (2007). Functional principal component analysis for longitudinal and survival data. *Statistica Sinica, 17*(3), 965–983.

Yao, F. (2008). Functional approach of flexibly modelling generalized longitudinal data and survival time. *Journal of Statistical Planning and Inference, 138*(4), 995–1009.

Yashin, A. I., Manton, K. G., & Vaupel, J. W. (1985). Mortality and aging in a heterogeneous population: A stochastic process model with observed and unobserved variables. *Theoretical Population Biology, 27*(2), 154–175.

Yashin, A. I., Manton, K. G., & Stallard, E. (1986a). Dependent competing risks: A stochastic process model. *Journal of Mathematical Biology, 24*(2), 119–140.

Yashin, A. I., Manton, K. G., & Stallard, E. (1986b). Evaluating the effects of observed and unobserved diffusion processes in survival analysis of longitudinal data. *Mathematical Modelling, 7*(9–12), 1353–1363.

Yashin, A. I., Arbeev, K. G., Akushevich, I., Kulminski, A., Akushevich, L., & Ukraintseva, S. V. (2007). Stochastic model for analysis of longitudinal data on aging and mortality. *Mathematical Biosciences, 208*(2), 538–551.

Yashin, A. I., Arbeev, K. G., Akushevich, I., Arbeeva, L., Kravchenko, J., Il'yasova, D., Kulminski, A., Akushevich, L., Culminskaya, I., Wu, D., & Ukraintseva, S. V. (2010). Dynamic determinants of longevity and exceptional health. *Current Gerontology and Geriatrics Research, 2010*, 381637.

Yashin, A. I., Arbeev, K. G., Akushevich, I., Kulminski, A., Ukraintseva, S. V., Stallard, E., & Land, K. C. (2012a). The quadratic hazard model for analyzing longitudinal data on aging, health, and the life span. *Physics of Life Reviews, 9*(2), 177–188.

Yashin, A. I., Arbeev, K. G., Ukraintseva, S. V., Akushevich, I., & Kulminski, A. (2012b). Patterns of aging related changes on the way to 100: An approach to studying aging, mortality, and longevity from longitudinal data. *North American Actuarial Journal, 16*(4), 403–433.

Yashin, A.I., Arbeev, K.G., Wu, D., Arbeeva, L.S., Kulminski, A., Akushevich, I., Culminskaya, I., Stallard, E., & Ukraintseva, S. (2013). How lifespan associated genes modulate aging changes: Lessons from analysis of longitudinal data. *Frontiers in Genetics 4*:article 3.

Ye, W., Lin, X., & Taylor, J. M. G. (2008). Semiparametric modeling of longitudinal measurements and time-to-event data – A two-stage regression calibration approach. *Biometrics, 64*(4), 1238–1246.

Yu, B., & Ghosh, P. (2010). Joint modeling for cognitive trajectory and risk of dementia in the presence of death. *Biometrics, 66*(1), 294–300.

Yu, M. G., Law, N. J., Taylor, J. M. G., & Sandler, H. M. (2004). Joint longitudinal-survival-cure models and their application to prostate cancer. *Statistica Sinica, 14*(3), 835–862.

Yu, M., Taylor, J. M. G., & Sandler, H. M. (2008). Individual prediction in prostate cancer studies using a joint longitudinal survival-cure model. *Journal of the American Statistical Association, 103*(481), 178–187.

Zhang, D., Lin, X. H., Raz, J., & Sowers, M. F. (1998). Semiparametric stochastic mixed models for longitudinal data. *Journal of the American Statistical Association, 93*(442), 710–719.

Zucker, D. M., & Karr, A. F. (1990). Nonparametric survival analysis with time-dependent covariate effects: A penalized partial likelihood approach. *Annals of Statistics, 18*(1), 329–353.

Chapter 12
Stochastic Process Models of Mortality and Aging

Anatoliy I. Yashin, Konstantin G. Arbeev, Liubov S. Arbeeva, Igor Akushevich, Svetlana V. Ukraintseva, Alexander M. Kulminski, Eric Stallard, and Kenneth C. Land

12.1 Introduction

Studying biodemographic aspects of human aging, health, and longevity involves analyses of dynamic biological mechanisms dealing with regulation and manifestation of aging-related changes in biomarkers (e.g., physiological indices), and of connections of these biomarkers with morbidity and mortality risks. The individual aging-related processes are modulated by genetic factors and external disturbances. These processes and interactions, occurring in populations of individuals during the life course, form the shape of the age patterns of human population mortality rates.

Such shapes demonstrate remarkable regularities in different populations: they decline in childhood, exponentially increase in the adult ages, and tend to decelerate and even level off at the oldest-old ages (Vaupel et al. 1998). Demographers and actuaries have developed a number of parametric descriptions of mortality curves capturing all aspects of their variation with age (Gage 1991; Heligman and Pollard 1980; Mode and Busby 1982; Siler 1983). Such descriptions are not intended to explain detailed features of the age patterns but are designed to provide a good fit to the overall mortality data. In contrast, biodemographers and gerontologists aim to explain observed features of mortality curves using emerging theoretical concepts and accumulated biological information (Charlesworth 2001; Gavrilov and Gavrilova 2001; Lee 2003; Strehler and Mildvan 1960; Yashin et al. 2000, 2001a, 2002; Zheng et al. 2011). Since the chances of death are affected by internal and external stresses that challenge defense mechanisms deteriorating in aging human bodies, the shape of the age trajectories of mortality curves is likely to reflect the average pattern of such deterioration, modulated by external disturbances. To go beyond such population average patterns and make conclusions about individual aging processes by investigating the age pattern of the mortality curve requires the development of a description of this curve in terms of parameters characterizing the internal (biological) and external (environmental) processes that contribute to the shape of this curve. Such a description can be formulated using

A.I. Yashin et al., *Biodemography of Aging*, The Springer Series on Demographic Methods and Population Analysis 40, DOI 10.1007/978-94-017-7587-8_12

models of aging-related changes recorded in longitudinal data merged with corresponding data on health and survival events.

A number of statistical methods for joint modeling of longitudinal and survival (time-to-event) data have been developed during the past several decades using both frequentist and Bayesian approaches (Ibrahim et al. 2010; Sousa 2011; Tsiatis and Davidian 2004; Yu et al. 2004). Traditionally, time-to-event data are analyzed using the proportional hazards regression model. A basic approach to modeling longitudinal panel data embeds a hazards regression model in a mixed (fixed and random) effects model (Laird and Ware 1982) with the assumption that the random effects or individual-specific parameters are normally distributed (Faucett and Thomas 1996; Ibrahim et al. 2004; Wulfsohn and Tsiatis 1997; Xu and Zeger 2001a). More flexible semi-parametric approaches that do not rely on an assumed normal distribution of the random effects or individual parameters have also been developed (Brown and Ibrahim 2003; Song et al. 2002a, b; Song and Wang 2008; Tsiatis and Davidian 2001). As discussed in Chap. 11, longitudinal data have also been modeled using either an (integrated) Ornstein-Uhlenbeck or Wiener process. These approaches allow for a more flexible description of individual longitudinal dynamics and provide a better fit compared to random effects models (Henderson et al. 2000; LaValley and DeGruttola 1996; Taylor et al. 1994; Wang and Taylor 2001; Xu and Zeger 2001b; Ye et al. 2008).

To be useful as a tool for biodemographers and gerontologists who seek biological explanations for observed processes, models of longitudinal data should be based on realistic assumptions and reflect relevant knowledge accumulated in the field. An example is the shape of the risk functions. Epidemiological studies show that the conditional hazards of health and survival events considered as functions of risk factors often have U- or J-shapes (Allison et al. 1997; Boutitie et al. 2002; Kulminski et al. 2008; Kuzuya et al. 2008; Mazza et al. 2007; Okumiya et al. 1999; Protogerou et al. 2007; Troiano et al. 1996; van Uffelen et al. 2010; Witteman et al. 1994; Yashin et al. 2001b), so a model of aging-related changes should incorporate this information. In addition, risk variables, and, what is very important, their effects on the risks of corresponding health and survival events, experience aging-related changes and these can differ among individuals. Since risk variables are measured periodically in longitudinal studies of aging, health, and longevity, estimates of age trajectories of these variables as well as their effects on health and mortality risks can be studied if the total period of observation is long enough.

An important class of models for joint analyses of longitudinal and time-to-event data incorporating a stochastic process for description of longitudinal measurements uses an epidemiologically-justified assumption of a quadratic hazard (i.e., U-shaped in general and J-shaped for variables that can take values only on one side of the U-curve) considered as a function of physiological variables. Quadratic hazard models have been developed and intensively applied in studies of human longitudinal data (Manton and Yashin 2000; Woodbury and Manton 1977; Yashin 1985; Yashin and Manton 1997; Yashin et al. 1985). The prototype of a model discussed in this chapter was introduced in Woodbury and Manton (1977) where its Gaussian properties were initially characterized. Yashin (1980, 1985) investigated conditions for preserving a Gaussian distribution property of the stochastic covariates under the operation of conditional averaging, and found that the quadratic hazard model described in Woodbury and Manton (1977) satisfied these conditions. An important property of this model is that the age trajectory of the total mortality rate can be explicitly represented in terms of the first two moments of the conditional distribution of the process describing aging-related changes and influencing the conditional mortality risk. The advantage of this approach is that it allows for incorporation of new insights and ideas from research on aging into the model structure. These properties, together with flexibility in describing age trajectories of biomarkers (e.g., physiological variables) affecting conditional risk, make this model a valuable tool for studying aging, health, and longevity using longitudinal data.

Progress in the study of human aging, health, and longevity would be substantially facilitated if researchers had a tool for analyzing the wealth of available data in an integrative systemic framework having the ability to incorporate important facts, relevant research findings, and emerging theoretical concepts in the analyses of new data. Several such concepts capturing fundamental features of aging-related changes are currently emerging. They are related to: (a) the notion of allostatic load (Seeman et al. 2001), (b) aging-related changes in adaptive capacity (homeostenosis) (Hall et al. 2000; Lund et al. 2002; Rankin and Kushner 2009; Troncale 1996), (c) changes in resistance to stresses with age (Semenchenko et al. 2004; Strehler 1962; Strehler and Mildvan 1960; Ukraintseva and Yashin 2003), and (d) age dependence of physiological norms (Yashin et al. 2009, 2010).

In this chapter, we outline a mathematical model for analyzing longitudinal data on aging, health, and mortality that incorporates these four concepts of aging-related changes. We also review applications of this model to analyses of longitudinal data, and investigate its potential for performing more comprehensive analyses of such data.

An initial version of this model was suggested in Yashin et al. (2007a). Its various extensions have been developed and applied in different contexts to investigate mechanisms of aging-related changes and their connection with morbidity/mortality risks. These include: (1) analyses of age trajectories of different physiological indices (such as blood glucose, body mass index, cholesterol, diastolic blood pressure, hematocrit, pulse pressure, and pulse rate) in relation to mortality/morbidity risks (Arbeev et al. 2011; Yashin et al. 2009, 2010, 2011c); (2) studies of the aging process using "indices of cumulative deficits" (Kulminski et al. 2007; Mitnitski et al. 2001; Rockwood et al. 2005; Yashin et al. 2007b) which have proved useful for analyses of a wide spectrum of information in relation to health- and aging-related changes and better characterize the aging phenotype than chronological age (Yashin et al. 2007c); and (3) analyses of age trajectories of medical costs in relation to mortality risks (Yashin et al. 2008b). Extended versions of this model have been also used in analyses of dependent competing risks (Akushevich et al. 2011; Yashin et al. 1986a), heterogeneity in longitudinal data

(Yashin et al. 2008a), analyses of genetic effects on age trajectories of physiological indices (Arbeev et al. 2009), joint analyses of individual health histories and physiological aging (Yashin et al. 2011a), and joint analyses of data collected using different observational plans (Yashin et al. 2011b).

12.2 Models

12.2.1 General Description

To specify the algebraic form of a dynamic model describing the age trajectories of individual physiological variables and their influence on mortality risks, which have J- or U-shapes considered as a function of the risk factors, and which exploits the theory of stochastic processes, with sampling paths that can be stopped at random times (Manton et al. 1994), let Y_t (where t is age) be a k-dimensional stochastic process describing a continuously changing vector of risk factors/covariates (e.g., physiological variables), and let Z be a vector of time-independent observed covariates (e.g., a person's genetic background). The risk function or conditional hazard of death is specified in the form:

$$\mu(t, Y_t, Z) = \mu_0(t, Z) + (Y_t - f_0(t, Z))^T Q(t, Z)(Y_t - f_0(t, Z)). \tag{12.1}$$

Here $\mu_0(t,Z)$ is a background hazard characterizing the nonzero mortality rate that would remain if the vector of covariates Y_t followed the optimal trajectory (the minimum value of the risk function at each age t), $f_0(t,Z)$, "T" denotes the matrix or vector transposition, and the matrix $Q(t,Z)$ is a non-negative-definite symmetric matrix of dimension $k \times k$. We use column vectors throughout; thus, the transposition to row vectors in (12.1) is needed to produce a scalar quadratic term in the hazard rate. The one-dimensional version of (12.1) is

$$\mu(t, Y_t, Z) = \mu_0(t, Z) + \mu_1(t, Z)(Y_t - f_0(t, Z))^2, \tag{12.2}$$

where $\mu_1(t, Z)$ is a non-negative function of age, t, characterizing the effects of physiological variables on mortality risk. Changes in $\mu_1(t, Z)$ influence the U-shape of the risk function. This is important because the narrowing of the U-shape for some risk factor with age captures the age-related declines in resistance to stresses associated with changes in this factor.

The age trajectory of a physiological variable, for which the minimum value of the risk function is reached, is called the *physiological norm*. The model (12.1) specifies that this norm is a function of age. The need for a constructive definition of such a norm is widely discussed, e.g., in Yashin et al. (2010).

Temporal changes in the vector of risk factors Y_t are described by a diffusion-type stochastic differential equation:

$$dY_t = a(t,Z)(Y_t - f_1(t,Z))dt + b(t,Z)dW_t, \tag{12.3}$$

with initial value Y_0. Here W_t is specified as a k-dimensional vector Wiener process with independent components, which describes exogenous challenges affecting these covariates. The process W_t is assumed to be independent of the initial vector Y_0 and covariates Z with normally distributed components. Note that such a model preserves the Gaussian property of the distribution of a vector of physiological variables among survivors to a given age: in the case of an initial Gaussian distribution for Y_0, the distribution of Y_t among survivors is also Gaussian (Yashin 1980, 1985; Yashin et al. 1985). The initial Gaussian assumption includes the important special case of zero variance which occurs when the vector Y_0 is measured without error. Thus, our assumption of an initial Gaussian distribution and the linearity of Eq. (12.3) define the structure of the entire process Y_t. The strength of disturbances of W_t is characterized by a $k \times k$ matrix of diffusion coefficients $b(t,Z)$. In the case of a non-Gaussian initial distribution, the model can be considered as a Gaussian approximation thereto.

The trajectory of Y_t describes aging-related changes in an individual's physiological functioning in response to the complicated interplay among processes induced by the ontogenetic program, senescence, and environmental stresses. The body's response to persistent external or internal disturbances affects the age trajectories of the physiological indices, producing *allostatic adaptation. Allostasis* is the process of individual adaptation to persistent external disturbances aimed at achieving stability in key metabolic variables, through physiological or behavioral changes affecting other variables. This process is especially important in analyses of longitudinal data in which measurements of external disturbances are absent or limited. The vector-function $f_1(t,Z)$ (with the same dimension as the vector Y_t) describes a trajectory of physiological states that organisms subjected to allostasis (McEwen and Wingfield 2003) are forced to follow by the process of adaptive regulation at age t. Allostatic adaptation produces deviations from the norm in the trajectories of the process Y_t. The magnitudes of such deviations for each physiological index will be associated with components of the *allostatic load* defined as $f_1(t,Z) - f_0(t,Z)$.

Homeostatic regulation plays a fundamental role for living organisms and this regulation needs to be included in the equation describing physiological changes. The dynamic model (12.3) includes a description of negative feedback mechanisms with coefficients of homeostatic regulation given by a matrix $a(t, Z)$. According to (12.3), the age trajectory of physiological variables Y_t will tend to follow the function $f_1(t,Z)$, i.e., adapt to changes in $f_1(t,Z)$ (the absence of such negative feedback mechanism would allow the trajectories to deviate from $f_1(t,Z)$ indefinitely, which is biologically implausible). The ability to adapt depends on the absolute values of the coefficients that are components of the matrix $a(t,Z)$. Age-related changes in these coefficients characterize changes in adaptive capacity with age. Specifically, the elements of the matrix $a(t,Z)$ regulate the age trajectories of the components of the physiological state approximated by the vector Y_t, i.e., they characterize the rate of the adaptive response for any deviation of a physiological index from the state $f_1(t,Z)$ which an organism tends to follow. An important feature of aging – the *decline in adaptive capacity* – has never been measured directly in longitudinal studies of aging, health, and longevity before. The use of the matrix $a(t,Z)$ in our model allows us to evaluate this effect. For example, in a simplified one-dimensional case, when $b(t,Z)=0$, for all t, in Eq. (12.3), and constant negative $a(t,Z)=a$ for all t, the parameter a is the coefficient of negative feedback in the equation for Y_t, which keeps the trajectory Y_t close to $f_1(t,Z)$. When $f_1(t,Z) = f_1$, constant for all t and Z, the value of Y_t asymptotically approaches f_1. In the case of non-zero disturbances, the higher the absolute value of a, the closer on average Y_t is to f_1, and for any given deviation the faster Y_t tends to f_1. That is why the value $a(t,Z)$ characterizes adaptive capacity. When the absolute value of the coefficient $a(t, Z)$ declines with age, more time is needed for the trajectory of Y_t to approach $f_1(t,Z)$ at older ages compared to younger ages. The estimation of changes in adaptive capacity with age involves maximization of the likelihood function (12.6) below, in which coefficients of matrix $a(t,Z)$ are described as parametric functions of age.

The vector-function $f_0(t,Z)$ in (12.1) (or, correspondingly, the scalar function $f_0(t,Z)$ in (12.2)) is introduced to explicitly characterize age-related changes in the "optimal" physiological state corresponding to the minimum hazard at a given age. It represents the age-dependent norm for a given functional state. It may differ from $f_1(t,Z)$ since the process of allostatic adaptation does not necessarily result in the optimal physiological state.

12.2.2 Estimation Procedure

The parameters of the model described above can be estimated using the maximum likelihood method. The survival function associated with the conditional distribution of lifespan, X, is $P(X > t|Z) = \exp\left(-\int_0^t \overline{\mu}(u,Z)du\right)$, where the observed (unconditional) hazard $\overline{\mu}(u)$ has the form (Yashin and Manton 1997; Yashin et al. 1986a, b):

$$\begin{aligned}\overline{\mu}(u,Z) = \mu_0(u,Z) &+ (m(u,Z) - f_0(u,Z))^T Q(u,Z)(m(u,Z) - f_0(u,Z)) \\ &+ Tr(Q(u,Z)\gamma(u,Z)).\end{aligned} \tag{12.4}$$

Here, $Tr(\cdot)$ denotes the matrix trace operator and $m(u,Z)$ and $\gamma(u,Z)$ satisfy the following system of ordinary nonlinear differential equations:

$$\frac{dm(t,Z)}{dt} = a(t,Z)(m(t,Z) - f_1(t,Z)) - 2\gamma(t,Z)Q(t,Z)(m(t,Z) - f_0(t,Z)),$$
$$\frac{d\gamma(t,Z)}{dt} = a(t,Z)\gamma(t,Z) + \gamma(t,Z)a(t,Z)^T + b(t,Z)b(t,Z)^T - 2\gamma(t,Z)Q(t,Z)\gamma(t,Z), \tag{12.5}$$

with $m(0,Z)$ and $\gamma(0,Z)$ being the vector of means and the variance/covariance matrix of the conditional normal distribution of the initial vector Y_0, given Z. Note that in such a model the conditional distribution of Y_t among survivors is also Gaussian at any age t (Yashin 1980, 1985; Yashin et al. 1985). The mean vector and the variance/covariance matrix of this distribution at age t are given by $m(t, Z)$ and $\gamma(t, Z)$, respectively.

Let the sequence $y^i_{t^i_0}, y^i_{t^i_1}, \ldots, y^i_{t^i_{n_i}}\, \tau_i$ represent the results of measurements of the process Y_t and the life span (which may be censored) related to the ith individual. The likelihood function for N individuals is (Yashin and Manton 1997; Yashin et al. 1986a, b):

$$\prod_{i=1}^{N} \bar{\mu}^i(\tau_i, Z)^{\delta_i} \exp\left\{-\int_0^{\tau_i} \bar{\mu}^i(u, Z)du\right\} \prod_{j=0}^{n_i(\tau_i)} (2\pi)^{-\frac{k}{2}} \left|\gamma^i\left(t^i_j -, Z\right)\right|^{-\frac{1}{2}}$$
$$\times \exp\left\{-\frac{1}{2}\left(y^i_{t^i_j} - m^i\left(t^i_j -, Z\right)\right)^T \gamma^i\left(t^i_j -, Z\right)^{-1} \left(y^i_{t^i_j} - m^i\left(t^i_j -, Z\right)\right)\right\}. \tag{12.6}$$

Here, the superscript "i" denotes the ith individual, $\delta_i = (0,1)$ is a indicator that individual i was at risk (i.e., not censored) at time τ_i, $m^i(t,Z)$ and $\gamma^i(t,Z)$ satisfy Eq. (12.5) at the intervals $[t^i_0, t^i_1); [t^i_1, t^i_2); \ldots; \left[t^i_{n_i-1}, t^i_{n_i}\right); \left[t^i_{n_i}, \tau_i\right)$ with the initial conditions $y^i_{t^i_0}, y^i_{t^i_1}, \ldots, y^i_{t^i_{n_i}}$, respectively; and $\gamma^i\left(t^i_j, Z\right) = 0, \forall j$, assuming that the vectors $y^i_{t^i_0}, y^i_{t^i_1}, \ldots, y^i_{t^i_{n_i}}$ are measured without error, except for the case of randomly missing measurements at time t^i_j, in which the values of $m^i(t^i_j, Z)$ and $\gamma^i(t^i_j, Z)$ are obtained from $m^i\left(t^i_j -, Z\right)$ and $\gamma^i\left(t^i_j -, Z\right)$ by conditioning (i.e., regressing) on the observed measurements. Here $m^i\left(t^i_j -, Z\right) = \lim_{t \uparrow t^i_j} m^i(t, Z)$, and $\gamma^i\left(t^i_j -, Z\right) = \lim_{t \uparrow t^i_j} \gamma^i(t, Z)$, and $t^i_{n_i}$ is the age of the latest measurement of a functional state before death/censoring at τ_i.

The procedure of maximization of this likelihood function is computationally extensive, because it involves the solution of the systems of ordinary differential equations (ODE) (12.5) for each measurement, for each individual, and at each step of the likelihood optimization procedure. Our experience with these models is that such solutions are feasible using modern statistical and technical software, e.g., MATLAB's Optimization Toolbox and ODE solvers, or SAS OPTMODEL, implementing different optimization algorithms (such as the Newton-Raphson or trust-region methods) and the Runge-Kutta method for the ODE solution. The parameter estimates then characterize the dynamics of the stochastic process Y_t describing the trajectories of physiological aging, as well as changes in mortality risk over age.

Note that the observed values $y^i_{t^i_0}, y^i_{t^i_1}, \ldots, y^i_{t^i_{n_i}}$ are used as initial conditions for the differential Eq. (12.5) at the beginning of subsequent intervals between the observation times. Therefore, the individual trajectories of $m^i(t,Z)$ and $\gamma^i(t,Z)$ differ even for individuals having the same values of Z. Consequently, the estimates of the chances of death for individuals having different observed values of the covariates also differ.

Treating the observed values $y^i_{t^i_0}, y^i_{t^i_1}, \ldots, y^i_{t^i_{n_i}}$ as initial conditions for Eq. (12.5) provides a simple and natural way of handling missing data beyond the first examination of a longitudinal study when, conditionally on the observed covariates, the data for the missing covariates can be assumed to be missing at random. The assumption that the data are missing at random could be reasonably applied when specific measurements are missing for a given study subject at a particular examination but are present for other examinations, without a specific health-related reason being provided. This assumption could also be applied for a specific covariate that is missing for all subjects at a particular examination due to a study design decision not to collect that covariate at that examination. This source of missing data occurs frequently in long-term longitudinal studies like the Framingham Heart Study, where covariates were added, deleted, or restored over time as their predictive utility for cardiovascular disease events was established. This assumption could also be applied for a specific subject that is missing all covariates at a particular examination due to their lack of participation in the study at that examination, without a specific health-related reason being provided for such lack of participation. In this latter case, linkage to external data files such as Medicare diagnosis and billing files may be used to determine if the lack of participation was due to hospitalization, death, or another health-related reason that may not have been recorded on the study data records. Missing examinations that were scheduled shortly before the time of hospitalization or death are unlikely to be missing at random, and instead are much more likely to be missing due to health-related reasons; if this occurs, the source of missingness should be taken into account in defining the study endpoints.

We assume that the observations $y^i_{t^i_0}, y^i_{t^i_1}, \ldots, y^i_{t^i_{n_i}}$ do not contain measurement errors. This assumption is not a serious problem when the model's parameters are used to characterize the entire population of study participants. In this case, one random process describes a probability distribution function of physiological variables in the population. The individual trajectories are just sampling paths of this process, so the differences among individuals are generated by the Wiener process, values of Z, and differences in values of the physiological indices at the age of entry into the study. This rough approximation is appropriate for evaluating and predicting population characteristics (e.g., changes in distributions of aging, health, and survival indices in the population in response to changes in health care policy, modification in Medicare services, etc.).

The model can, however, be extended to the case where measurement errors are explicitly taken into account. To do so would require that the observed values $y^i_{t^i_j}$ no longer be treated as known initial conditions at time t_j^i for Eq. (12.5). Instead, computation of the values of $m^i(t_j^i,Z)$ and $\gamma^i(t_j^i,Z)$ would require specification of a measurement error sub-model with corresponding extensions to Eq. (12.6) to reflect the increased variance of $y^i_{t^i_j}$; but such extensions are not necessary for our current applications.

We also recognize that neglecting measurement errors may be a suboptimal strategy in applications dealing with "personalized" analyses, where one is more concerned about the response of individual characteristics to preventive measures or medical interventions. In this situation, individual age trajectories of physiological state have to be "tracked," i.e., different model parameters have to be used to describe the age trajectories of the functional states for different individuals. In such individualized applications, taking measurement error into account could be an important issue.

The model can be extended to handle this case using the measurement error procedures indicated above. In addition, the random process described by Eq. (12.3) can be further personalized, i.e., (ideally) each individual can be characterized by his/her own stochastic differential equation independent of the equations describing aging-related changes in the other individuals. Development of such personalized forms of the stochastic process model, with and without the measurement error procedures, represents the next logical step in applications of these models, which will lead to new opportunities for using these models in individual-level forecasting.

As is the case for any mathematical description of an empirical phenomenon, the model discussed above has a number of limitations. The Gaussian property of the conditional distribution of Y_t among survivors yields a positive probability for negative values of Y_t. Since physiological indices measured in longitudinal studies are positive, and because the model's characteristics have to be biologically interpreted, the use of appropriate constraints on the parameters of the model is necessary in the estimation procedure. In particular, (a) the distribution of Y_0 should guarantee a negligible probability of negative values; (b) the functions $f_1(t,Z)$ and $f_0(t,Z)$ should have non-negative values for each age; (c) the absolute values of feedback coefficients in the matrix $a(t,Z)$ in (12.3) should not become too small for the trajectories of Y_t to tend to $f_1(t,Z)$); (d) the background hazard $\mu_0(t,Z)$ should have non-negative values for each age and a non-decreasing age pattern; and (e) the matrix $Q(t,Z)$ should remain non-negative definite for each age. Applications of these models to the analysis of longitudinal data indicate that such constraints do not reduce the quality of the parameter estimates obtained from the proposed ML estimation procedure.

12.2.3 Simulation Studies

In a simulation experiment, we generated 100 datasets with data on life spans and a hypothetical physiological index (mimicking pulse pressure) for 5000 individuals in each dataset. We followed individuals for 56 years with 28 biennial observations of physiological indices, with ages at entry to the study uniformly distributed over the age interval [30, 60]. Individuals with life spans exceeding the age at entry plus duration of the follow-up period (56 years) were considered censored at this age. Such a data structure is similar to the Framingham Heart Study Original Cohort data (Dawber 1980; Dawber et al. 1951). We estimated a one-dimensional version of the model (Eqs. 12.2 and 12.3) with:

1. A constant diffusion coefficient, σ_1 (replacing the function b in Eq. 12.3);
2. Linear functions of age (t) for the quadratic hazard term (the function μ_1 in Eq. 12.2), the adaptive capacity term (the negative feedback coefficient a in Eq. 12.3), and the physiological "norm" (the function f_0 in Eq. 12.2):

$$\begin{aligned}\mu_1(t) &= a_{\mu 1} + b_{\mu 1}t,\\ a(t) &= a_Y + b_Y(t-30),\\ f_0(t) &= a_{f_0} + b_{f_0}(t-30);\end{aligned}$$

3. A Gompertz function for the baseline hazard: $\ln\ \mu_0(t) = \ln\ a_{\mu_0} + b_{\mu_0}(t-30)$.

For simplicity, all of these characteristics were assumed to be independent of the covariates, Z. In addition:

4. The "allostatic trajectory" term (function f_1 in Eq. 12.3) was assumed to depend on age and a dichotomous covariate Z ($Z = (0,\ 1)$; $P(Z=1) = 0.5$): $f_1(t,Z) = a_{f_1} + b_{f_1}(t-30) + \beta_{f_1}Z$..
5. The initial value of the diffusion process for Y_t, Y_0, was assumed normal, $Y_0\tilde{}N(f_1(t_0,Z),\ \sigma_0)$, where t_0 is the age at the baseline exam for the respective individual.

The results of the simulations are summarized in Table 12.1 and Fig. 12.1. The true values of the parameters are given in Table 12.1 in the row labeled **True Values**. The simulations confirm that the parameters of the models can be estimated with reasonable accuracy for a sample size of this magnitude.

Table 12.1 Means, standard deviations (St. Dev.) and minimal and maximal values of parameter estimates in 100 simulated data sets

	$\ln a_{\mu_0}$	b_{μ_0}	$a_{\mu_1} \cdot 10^4$	$b_{\mu_1} \cdot 10^5$	a_Y	$b_y \cdot 10^3$	σ_0	σ_1	a_{f_1}	b_{f_1}	a_{f_0}	b_{f_0}	β_{f_1}
Mean	−8.040	0.091	1.011	0.300	−0.200	1.000	6.003	4.998	55.001	0.200	50.043	0.099	2.994
St. Dev.	0.205	0.004	0.321	0.054	0.004	0.133	0.056	0.016	0.130	0.006	0.410	0.013	0.120
Min	−8.641	0.081	0.243	0.156	−0.209	0.618	5.872	4.952	54.697	0.188	48.975	0.066	2.639
Max	−7.621	0.102	1.946	0.434	−0.189	1.317	6.137	5.037	55.301	0.212	51.111	0.134	3.335
True values	***−8.0***	***0.09***	***1.0***	***0.3***	***−0.2***	***1.0***	***6.0***	***5.0***	***55.0***	***0.2***	***50.0***	***0.1***	***3.0***

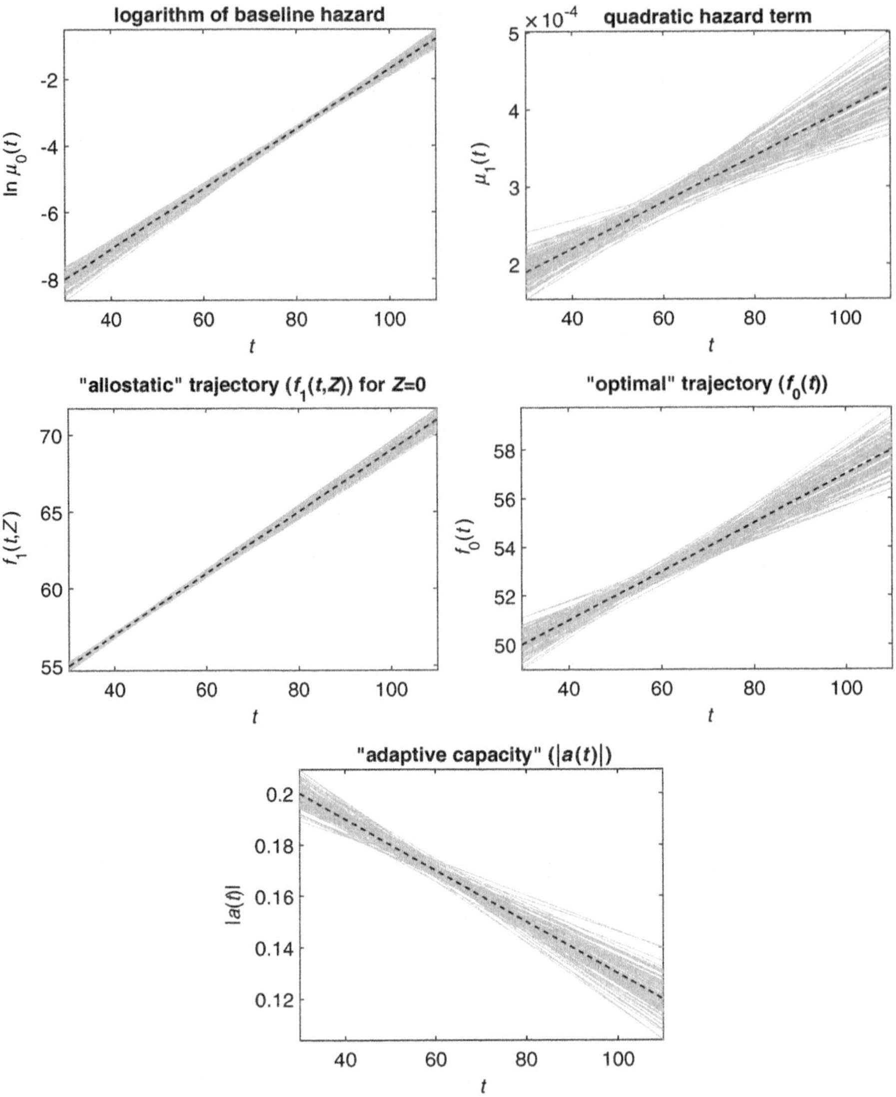

Fig. 12.1 Estimated (*solid grey lines*) and true (*dashed black lines*) trajectories in 100 simulated data sets

12.3 Discussion

12.3.1 To What Extent Can Mortality Rates Characterize Aging?

In many experimental studies of aging using populations of laboratory animals, the sensitivity of the individual aging process to external disturbances (e.g., medical interventions) or genetic manipulations is often evaluated by comparing empirical survival or mortality rate functions/curves constructed for populations in the experimental and control groups. Similarly, the slope of the logarithm of the mortality curve at the adult and old ages is often interpreted as the aging rate (Economos 1982). The limitations of such an interpretation have been discussed in a number of papers. Yashin et al. (2002b) argued that the use of such a measure of the aging rate may be misleading: changes in the slope of the mortality curve may occur for many other reasons having little to do with the aging process. For example, the slope could change as a result of changes in the heterogeneity distribution (e.g., distribution of frailty Vaupel et al. 1979; Vaupel and Yashin 1985), saving individuals' lives by providing adequate medical help in emergencies, etc. Analyzing changes in mortality rates in developed countries, Rozing and Westendorp (2008) came to the conclusion that recent progress in mortality reduction does not affect the slope of the logarithm of the mortality curves (see Yashin et al. (2002) and references therein for similar observations). Koopman et al. (2011) criticized the use of the slope of the logarithm of the mortality curve as a measure of the aging rate and proposed, instead, the use of the age derivative of the mortality rate as such a measure. Although the use of the derivative overcomes some limitations of the slope measure, it could also be seriously criticized because it does not link changes in mortality with biological changes progressing in the aging human body. Information on many such changes affecting health and survival events in humans is available in the data collected in longitudinal studies of aging, health, as well as in other sources accessible to interested researchers. That is why studies of human aging with a focus solely on the properties of the mortality curve, ignoring other relevant information accumulated in the field, often look scientifically unjustified.

12.3.2 The Strehler and Mildvan Model

Taking into account that individual chances of death depend on aging-related declines in an individual's ability to withstand the stresses of life and the random process of external disturbances, Strehler and Mildvan (1960) proposed a mortality model (SM-model) that includes two fundamental components shaping the age pattern of the mortality curve: the aging-related decline in vitality and the process of external stresses. The authors showed that each parameter of the Gompertz mortality curve is a function of both components, so it would be erroneous to interpret, for example, changes in the rate of increase in the Gompertz curve as changes in the individual aging rate. Strehler and Mildvan also showed that the two Gompertz parameters were strongly negatively correlated. They explained the origin of this correlation in terms of the parameters of the two fundamental components of their model. Yashin et al. (2012) showed that differences in Gompertz parameters for populations sharing the same environmental conditions may have a genetic background, that is, populations of individuals having different numbers of longevity genes in their genomes have different values of their respective Gompertz parameters, confirming that longevity genes are responsible for organisms' resistance to stresses.

The SM model made an important conceptual contribution to the biodemography of aging by emphasizing the importance of the two fundamental components (internal and external) in the shape of the observed mortality curve. However, this model could not be used (in its original form) in the analyses of longitudinal data to clarify additional issues important for a better understanding of the factors and mechanisms involved in regulating human aging and longevity.

12.3.3 Comparing Two Versions of the Stochastic Process Model

The need for the development of new models of aging, health, and mortality and for describing their connection with traditional demographic models was emphasized by Manton and Yashin (2000). The connection between the Gompertz model of the mortality curve with a model describing longitudinal data played an important role in better understanding of the forces and mechanisms shaping the age pattern of the demographic mortality rate; see Manton and Yashin (2000) and references therein. This connection is described by the conditional mortality rate in the form of a generalized Gompertz model:

$$\mu(t, Y_t) = \widetilde{Y}_t^T \widetilde{Q} \widetilde{Y}_t e^{\theta t}, \tag{12.7}$$

where $\widetilde{Y}_t^T = \left(1, Y_t^T\right)$ is a vector of covariates Y_t (e.g., physiological indices), t is age, $\widetilde{Q}$ is a (constant) matrix, and θ is the Gompertz exponential growth parameter (Manton and Yashin 2000). In applications of this model to longitudinal data, estimated values of the parameter θ have always been smaller than the corresponding parameter in a Gompertz model that does not include information on covariates. The reduction of the exponential growth parameter has been interpreted as an effect of measurements: the new (reduced) value of the parameter θ characterized the unexplained component of aging-related increase in the mortality rate. Versions of this model have been applied to the analyses of aging-related changes in a number of biomarkers and their connection with mortality risks.

For example, Manton et al. (1994) applied the generalized Gompertz model in (12.7) to longitudinal data from the first 18 biennial examinations of the Framingham Heart Study (FHS) and compared the results with those obtained from the first three waves of the National Long Term Care Survey (NLTCS: 1982, 1984, and 1989). The use of ten cardiovascular covariates in the FHS model reduced the θ-parameters from 9.4 % to 8.1 % for males and from 10.0 % to 8.1 % for females. The FHS study population ranged from 30 to 60 years at the start of the 34-year observation period; hence, the reduced θ-parameters implied that the FHS mortality rates would double every 8.6 years for males and 8.5 years for females if one were to hold the observed covariates constant at their age-30 values. This compares with the actual doubling times of 7.4 years for males and 6.9 years for females when the covariates were allowed to follow their "natural" trajectories.

The use of 27 ADL, IADL, and physical performance covariates in the NLTCS model reduced the θ-parameters from 8.2 % to 5.3 % for males and from 9.1 % to 4.8 % for females. The NLTCS study population was 65 years or older at each wave; hence, the reduced θ-parameters implied that the mortality rates would double every 13.1 or 14.6 years (males, females) if one were to hold the observed disability covariates constant at their age-65 values. The relative reductions in θ and the corresponding increases in the doubling times were larger for the NLTCS model, indicating that the NLTCS disability covariates explained more of the age-dependence of mortality than the FHS cardiovascular covariates explained. The differences were attributable, in part, to the age patterns of several FHS cardiovascular covariates which reached peak values near age 60, followed by declines at older ages (Yashin et al. 2011c). This contrasted markedly with the age patterns for the disability covariates which exhibited strong monotonic increases with age. Indeed, the exercise of holding the disability covariates constant at their age-65 values is purely hypothetical, given our current state of knowledge about how this might be done.

The conditional hazard rate (12.7) in the one-dimensional version of the original model can be represented as follows:

$$\mu(t, Y_t) = \left(\mu_0 + \mu_1(Y_t - c)^2\right)e^{\theta t} = \mu_0 e^{\theta t} + \mu_1 e^{\theta t}(Y_t - c)^2, \tag{12.8}$$

where μ_0, μ_1 and c are constants. The term $\left(\mu_0 + \mu_1(Y_t - c)^2\right)$ in (12.8) describes one of the parameters of the Gompertz mortality model explained by observations Y_t. Thus, Eq. (12.8) can be interpreted as providing a more detailed description of the Gompertz mortality curve widely used in demography. In another interpretation, the term $\mu_0 e^{\theta t}$ is interpreted as the "optimal" Gompertz mortality rate which would be observed in an ideal situation when $Y_t = c$. This representation clearly shows two limitations of the original version: (i) the exponential multipliers in both components of the risk function are the same; and (ii) the minimum of the second (quadratic hazard) term is reached at a constant level of the observed covariates.

In the generalized hazard model (12.1) described above, the covariates' values minimizing the risk function can change as a function of age. This is a more realistic assumption since in epidemiologic and medical practice specialists often use the notion of physiological "norm", specific to a given age. This age-dependent norm is explicitly included in the description of age trajectories of mortality risk (compare Eqs. 12.1, 12.2, and 12.8). This allows one to statistically test hypotheses about age dependence of physiological norms and verify such dependence from available data. The modified hazard model (12.1) includes the earlier version (12.8) as a particular case.

In the generalized model, the term $\mu_0(t)$ can differ from the multiplier of the quadratic hazard $\mu_1(t)$ (we omit dependence of these functions on Z to make this case comparable with (12.8)), which results in a completely new interpretation of these coefficients. The hazard rate $\mu_0(t)$ is associated with death from unmeasured factors. The risk $\mu_0(t)$ must be smaller than the total (demographic) mortality risk $\overline{\mu}(t)$ calculated in the absence of observations on risk factors. Therefore, $\mu_0(t)$ characterizes the mortality remaining after all observed covariates follow the "optimal" trajectory and its interpretation remains similar to that used in the original model. In the case of a Gompertz specification, both Gompertz parameters in $\mu_0(t)$ can be evaluated and compared with their values in $\overline{\mu}(t)$. This model allows for evaluating the gain in life expectancy when observed covariates follow normal age trajectories.

The term $\mu_1(t)$ clarifies the connections between senescence, longevity, and stress-resistance. Indeed, the increasing pattern of $\mu_1(t)$ indicates that the branches of the corresponding U-shaped risk function become steeper with increasing age, and the range of tolerable deviations of the resultant risk factor from its "optimal" value becomes narrower with age, reflecting the decline in stress resistance with age. Although many aspects of such connections have been investigated in experimental animal studies (Semenchenko et al. 2004), they have never been adequately addressed in studies of human longitudinal data. Since the decline in stress resistance is an important indicator of aging (senescence), the rate of increase in $\mu_1(t)$ (not the slope of the logarithm of the mortality curve) may characterize the rate of aging. More generally, the increasing role of senescence in mortality risk with increasing age could be captured by the widening pattern of changes in the U-shape of the relative risk, which would indicate a faster increase in $\mu_0(t)$ compared to $\mu_1(t)$. In this case the ratio $\frac{\mu_1(t)}{\mu_0(t)}$ is the declining function of age. An important methodological advance of the extended model is that it is transformed to a form wherein effects of senescence on survival, longevity, and disease development may be evaluated from longitudinal data.

Taking into account the age dependence of the functions $f_1(t,Z)$, $f_0(t,Z)$, and $a(t, Z)$ allows for testing hypotheses about factors and mechanisms affecting the dynamic properties of the age trajectories of physiological states. For example, one could test whether $f_1(t,Z)$ coincides with $f_0(t,Z)$. Differences in these functions would mean that the processes of allostatic adaptation in response to persistent external disturbances do not tend to minimize mortality risk. One can also test a hypothesis of declines in "adaptive capacity" in the aging human body and evaluate patterns of such declines. All such hypotheses can be tested using the likelihood ratio test. For example, to test the hypothesis of a decline in adaptive capacity with age, one needs to compare the model with a decline (say, a linear decline with age in $a(t,Z)$) with the model without such a decline (i.e., $a(t,Z)$ is independent of age), where all other functions (except $a(t,Z)$) are specified similarly in both models, using a likelihood ratio test.

12.3.4 Modeling Personalized Aging Changes

The use of observed fixed covariates Z in the functions $f_0(t,Z)$, $Q(t,Z)$, $a(t,Z)$, $b(t,Z)$, and $f_1(t,Z)$ makes the model more personalized. The notion of the "norm" may differ for individuals carrying different alleles or genotypes, or having different histories of events and processes experienced by an individual during the life course (e.g., diseases, environmental exposures), etc. These indicate the need for developing a more general methodology, which could incorporate individualized notions of "norms" and adaptive responses.

An important feature of the model discussed above is the preservation of the Gaussian property in the operation of conditional averaging. If the distribution of risk factors at the initial wave of observations, Y_0, is Gaussian, the distribution of Y_t among survivors is also Gaussian. This facilitates descriptions of the probability distributions of dynamic covariates in terms of the first two moments which satisfy ordinary non-linear differential equations. Note that the Gaussian distribution allows for negative values of the risk factors to occur with positive probability. Our studies show that in practice this property does not limit the analyses. The model can also be used as two-moment approximation for the age trajectories of covariates which follow non-Gaussian dynamics. In non-Gaussian cases, the model could also be extended to include higher order moments (e.g., conditional semi-invariants).

Acknowledgements The chapter was partially supported by grants, R01AG046860, P01AG043352, and P30AG034424 from the National Institute on Aging. The content is solely the responsibility of the authors and does not necessarily represent the official views of the National Institute on Aging or the National Institutes of Health.

References

Akushevich, I., Arbeev, K., Ukraintseva, S., & Yashin, A. (2011). *Theory of individual health histories and dependent competing risks*. In *Presented at JSM Proceedings, Section on Risk Analysis*.

Allison, D. B., Faith, M. S., Heo, M., & Kotler, D. P. (1997). Hypothesis concerning the U-shaped relation between body mass index and mortality. *American Journal of Epidemiology, 146*(4), 339–349.

Arbeev, K. G., Akushevich, I., Kulminski, A. M., Arbeeva, L. S., Akushevich, L., Ukraintseva, S. V., Culminskaya, I. V., & Yashin, A. I. (2009). Genetic model for longitudinal studies of aging, health, and longevity and its potential application to incomplete data. *Journal of Theoretical Biology, 258*(1), 103–111.

Arbeev, K. G., Ukraintseva, S. V., Akushevich, I., Kulminski, A. M., Arbeeva, L. S., Akushevich, L., Culminskaya, I. V., & Yashin, A. I. (2011). Age trajectories of physiological indices in relation to healthy life course. *Mechanisms of Ageing and Development, 132*(3), 93–102.

Boutitie, F., Gueyffier, F., Pocock, S., Fagard, R., & Boissel, J. P. (2002). J-shaped relationship between blood pressure and mortality in hypertensive patients: New insights from a meta-analysis of individual-patient data. *Annals of Internal Medicine, 136*(6), 438–448.

Brown, E. R., & Ibrahim, J. G. (2003). A Bayesian semiparametric joint hierarchical model for longitudinal and survival data. *Biometrics, 59*(2), 221–228.

Charlesworth, B. (2001). Patterns of age-specific means and genetic variances of mortality rates predicted by the mutation-accumulation theory of ageing. *Journal of Theoretical Biology, 210*(1), 47–65.

Dawber, T. R. (1980). *The Framingham study: The epidemiology of atherosclerotic disease*. Cambridge, MA: Harvard University Press.

Dawber, T. R., Meadors, G. F., & Moore, F. E. (1951). Epidemiological approaches to heart disease: The Framingham Study. *American Journal of Public Health, 41*(3), 279–286.

Economos, A. C. (1982). Rate of aging, rate of dying and the mechanism of mortality. *Archives of Gerontology and Geriatrics, 1*(1), 3–27.

Faucett, C. L., & Thomas, D. C. (1996). Simultaneously modelling censored survival data and repeatedly measured covariates: A Gibbs sampling approach. *Statistics in Medicine, 15*(15), 1663–1685.

Gage, T. B. (1991). Causes of death and the components of mortality: Testing the biological interpretations of a competing hazards model. *American Journal of Human Biology, 3*(3), 289–300.

Gavrilov, L. A., & Gavrilova, N. S. (2001). The reliability theory of aging and longevity. *Journal of Theoretical Biology, 213*(4), 527–545.

Hall, D. M., Xu, L., Drake, V. J., Oberley, L. W., Oberley, T. D., Moseley, P. L., & Kregel, K. C. (2000). Aging reduces adaptive capacity and stress protein expression in the liver after heat stress. *Journal of Applied Physiology, 89*(2), 749–759.

Heligman, L., & Pollard, J. H. (1980). The age pattern of mortality. *Journal of the Institute of Actuaries, 107*(1), 49–80.

Henderson, R., Diggle, P., & Dobson, A. (2000). Joint modelling of longitudinal measurements and event time data. *Biostatistics, 1*(4), 465–480.

Ibrahim, J. G., Chen, M. H., & Sinha, D. (2004). Bayesian methods for joint modeling of longitudinal and survival data with applications to cancer vaccine trials. *Statistica Sinica, 14*(3), 863–883.

Ibrahim, J. G., Chu, H., & Chen, L. M. (2010). Basic concepts and methods for joint models of longitudinal and survival data. *Journal of Clinical Oncology, 28*(16), 2796–2801.

Koopman, J. J. E., Rozing, M. P., Kramer, A., de Jager, D. J., Ansell, D., De Meester, J. M. J., Prutz, K. G., Finne, P., Heaf, J. G., Palsson, R., Kramar, R., Jager, K. J., Dekker, F. W., & Westendorp, R. G. J. (2011). Senescence rates in patients with end-stage renal disease: A critical appraisal of the Gompertz model. *Aging Cell, 10*(2), 233–238.

Kulminski, A. M., Ukraintseva, S. V., Akushevich, I. V., Arbeev, K. G., & Yashin, A. I. (2007). Cumulative index of health deficiencies as a characteristic of long life. *Journal of American Geriatrics Society, 55*(6), 935–940.

Kulminski, A. M., Arbeev, K. G., Kulminskaya, I. V., Ukraintseva, S. V., Land, K., Akushevich, I., & Yashin, A. I. (2008). Body mass index and nine-year mortality in disabled and nondisabled older U.S. individuals. *Journal of the American Geriatrics Society, 56*(1), 105–110.

Kuzuya, M., Enoki, H., Iwata, M., Hasegawa, J., & Hirakawa, Y. (2008). J-shaped relationship between resting pulse rate and all-cause mortality in community-dwelling older people with disabilities. *Journal of the American Geriatrics Society, 56*(2), 367–368.

Laird, N. M., & Ware, J. H. (1982). Random-effects models for longitudinal data. *Biometrics, 38*(4), 963–974.

LaValley, M. P., & DeGruttola, V. (1996). Models for empirical Bayes estimators of longitudinal CD4 counts. *Statistics in Medicine, 15*(21–22), 2289–2305.

Lee, R. D. (2003). Rethinking the evolutionary theory of aging: Transfers, not births, shape social species. *Proceedings of the National Academy of Sciences of the United States of America, 100*(16), 9637–9642.

Lund, J., Tedesco, P., Duke, K., Wang, J., Kim, S. K., & Johnson, T. E. (2002). Transcriptional profile of aging in C-elegans. *Current Biology, 12*(18), 1566–1573.

Manton, K. G., & Yashin, A. I. (2000). *Mechanisms of aging and mortality: A search for new paradigms* (Odense monograph on population aging no. 7). Odense: Odense University Press.

Manton, K. G., Stallard, E., Woodbury, M. A., & Dowd, J. E. (1994). Time-varying covariates in models of human mortality and aging: Multidimensional generalizations of the Gompertz. *Journals of Gerontology, 49*(4), B169–B190.

Mazza, A., Zamboni, S., Rizzato, E., Pessina, A. C., Tikhonoff, V., Schiavon, L., & Casiglia, E. (2007). Serum uric acid shows a J-shaped trend with coronary mortality in non-insulin-dependent diabetic elderly people. The CArdiovascular STudy in the ELderly (CASTEL). *Acta Diabetologica, 44*(3), 99–105.

McEwen, B. S., & Wingfield, J. C. (2003). The concept of allostasis in biology and biomedicine. *Hormones and Behavior, 43*(1), 2–15.

Mitnitski, A. B., Mogilner, A. J., & Rockwood, K. (2001). Accumulation of deficits as a proxy measure of aging. *Scientific World Journal, 1*, 323–336.

Mode, C. J., & Busby, R. C. (1982). An 8-parameter model of human mortality – the single decrement case. *Bulletin of Mathematical Biology, 44*(5), 647–659.

Okumiya, K., Matsubayashi, K., Wada, T., Fujisawa, M., Osaki, Y., Doi, Y., Yasuda, N., & Ozawa, T. (1999). A U-shaped association between home systolic blood pressure and four-year mortality in community-dwelling older men. *Journal of the American Geriatrics Society, 47*(12), 1415–1421.

Protogerou, A. D., Safar, M. E., Iaria, P., Safar, H., Le Dudal, K., Filipovsky, J., Henry, O., Ducimetiere, P., & Blacher, J. (2007). Diastolic blood pressure and mortality in the elderly with cardiovascular disease. *Hypertension, 50*(1), 172–180.

Rankin, M. M., & Kushner, J. A. (2009). Adaptive beta-cell proliferation is severely restricted with advanced age. *Diabetes, 58*(6), 1365–1372.

Rockwood, K., Song, X., MacKnight, C., Bergman, H., Hogan, D. B., McDowell, I., & Mitnitski, A. (2005). A global clinical measure of fitness and frailty in elderly people. *CMAJ, 173*(5), 489–495.

Rozing, M. P., & Westendorp, R. G. J. (2008). Parallel lines: Nothing has changed? *Aging Cell, 7*(6), 924–927.

Seeman, T. E., McEwen, B. S., Rowe, J. W., & Singer, B. H. (2001). Allostatic load as a marker of cumulative biological risk: MacArthur studies of successful aging. *Proceedings of the National Academy of Sciences of the United States of America, 98*(8), 4770–4775.

Semenchenko, G. V., Khazaeli, A. A., Curtsinger, J. W., & Yashin, A. I. (2004). Stress resistance declines with age: Analysis of data from a survival experiment with drosophila melanogaster. *Biogerontology, 5*(1), 17–30.

Siler, W. (1983). Parameters of mortality in human populations with widely varying life spans. *Statistics in Medicine, 2*(3), 373–380.

Song, X., & Wang, C. Y. (2008). Semiparametric approaches for joint modeling of longitudinal and survival data with time-varying coefficients. *Biometrics, 64*(2), 557–566.

Song, X., Davidian, M., & Tsiatis, A. A. (2002a). A semiparametric likelihood approach to joint modeling of longitudinal and time-to-event data. *Biometrics, 58*(4), 742–753.

Song, X. A., Davidian, M., & Tsiatis, A. A. (2002b). An estimator for the proportional hazards model with multiple longitudinal covariates measured with error. *Biostatistics, 3*(4), 511–528.
Sousa, I. (2011). A review on joint modelling of longitudinal measurements and time-to-event. *Revstat-Statistical Journal, 9*(1), 57–81.
Strehler, B. (1962). *Time, cells, and aging*. London: Academic.
Strehler, B. L., & Mildvan, A. S. (1960). General theory of mortality and aging. *Science, 132*(3418), 14–21.
Taylor, J. M. G., Cumberland, W. G., & Sy, J. P. (1994). A stochastic model for analysis of longitudinal AIDS data. *Journal of the American Statistical Association, 89*(427), 727–736.
Troiano, R. P., Frongillo, E. A., Sobal, J., & Levitsky, D. A. (1996). The relationship between body weight and mortality: A quantitative analysis of combined information from existing studies. *International Journal of Obesity, 20*(1), 63–75.
Troncale, J. A. (1996). The aging process: Physiologic changes and pharmacologic implications. *Postgraduate Medicine, 99*(5), 111–114, 120-122.
Tsiatis, A. A., & Davidian, M. (2001). A semiparametric estimator for the proportional hazards model with longitudinal covariates measured with error. *Biometrika, 88*(2), 447–458.
Tsiatis, A. A., & Davidian, M. (2004). Joint modeling of longitudinal and time-to-event data: An overview. *Statistica Sinica, 14*(3), 809–834.
Ukraintseva, S. V., & Yashin, A. I. (2003). Individual aging and cancer risk: How are they related? *Demographic Research, 9*(8), 163–196.
van Uffelen, J. G. Z., Berecki-Gisolf, J., Brown, W. J., & Dobson, A. J. (2010). What is a healthy body mass index for women in their seventies? Results from the Australian Longitudinal Study on Women's Health. *Journals of Gerontology Series A: Biological Sciences and Medical Sciences, 65*(8), 844–850.
Vaupel, J. W., & Yashin, A. I. (1985). Heterogeneity's ruses: Some surprising effects of selection on population dynamics. *American Statistician, 39*(3), 176–185.
Vaupel, J. W., Manton, K. G., & Stallard, E. (1979). Impact of heterogeneity in individual frailty on the dynamics of mortality. *Demography, 16*(3), 439–454.
Vaupel, J. W., Carey, J. R., Christensen, K., Johnson, T. E., Yashin, A. I., Holm, N. V., Iachine, I. A., Kannisto, V., Khazaeli, A. A., Liedo, P., Longo, V. D., Zeng, Y., Manton, K. G., & Curtsinger, J. W. (1998). Biodemographic trajectories of longevity. *Science, 280*(5365), 855–860.
Wang, Y., & Taylor, J. M. G. (2001). Jointly modeling longitudinal and event time data with application to acquired immunodeficiency syndrome. *Journal of the American Statistical Association, 96*(455), 895–905.
Witteman, J. C. M., Grobbee, D. E., Valkenburg, H. A., Vanhemert, A. M., Stijnen, T., Burger, H., & Hofman, A. (1994). J-shaped relation between change in diastolic blood pressure and progression of aortic atherosclerosis. *Lancet, 343*(8896), 504–507.
Woodbury, M. A., & Manton, K. G. (1977). A random-walk model of human mortality and aging. *Theoretical Population Biology, 11*(1), 37–48.
Wulfsohn, M. S., & Tsiatis, A. A. (1997). A joint model for survival and longitudinal data measured with error. *Biometrics, 53*(1), 330–339.
Xu, J., & Zeger, S. L. (2001a). The evaluation of multiple surrogate endpoints. *Biometrics, 57*(1), 81–87.
Xu, J., & Zeger, S. L. (2001b). Joint analysis of longitudinal data comprising repeated measures and times to events. *Journal of the Royal Statistical Society: Series C: Applied Statistics, 50*, 375–387.
Yashin, A. I. (1980). Conditional Gaussian estimation of dynamic system response on the basis of jerky observations. *Automation and Remote Control, 41*(5), 618–626.
Yashin, A. I. (1985). Dynamics in survival analysis: Conditional Gaussian property vs. Cameron-Martin formula. In N. V. Krylov, R. S. Lipster, & A. A. Novikov (Eds.), *Statistics and control of stochastic processes* (pp. 446–475). New York: Springer.

Yashin, A. I., & Manton, K. G. (1997). Effects of unobserved and partially observed covariate processes on system failure: A review of models and estimation strategies. *Statistical Science, 12*(1), 20–34.

Yashin, A. I., Manton, K. G., & Vaupel, J. W. (1985). Mortality and aging in a heterogeneous population: A stochastic process model with observed and unobserved variables. *Theoretical Population Biology, 27*(2), 154–175.

Yashin, A. I., Manton, K. G., & Stallard, E. (1986a). Dependent competing risks: A stochastic process model. *Journal of Mathematical Biology, 24*(2), 119–140.

Yashin, A. I., Manton, K. G., & Stallard, E. (1986b). Evaluating the effects of observed and unobserved diffusion processes in survival analysis of longitudinal data. *Mathematical Modelling, 7*(9–12), 1353–1363.

Yashin, A. I., Iachine, I. A., & Begun, A. S. (2000). Mortality modeling: A review. *Mathematical Population Studies, 8*(4), 305–332.

Yashin, A. I., Begun, A. S., Boiko, S. I., Ukraintseva, S. V., & Oeppen, J. (2001a). The new trends in survival improvement require a revision of traditional gerontological concepts. *Experimental Gerontology, 37*(1), 157–167.

Yashin, A. I., Ukraintseva, S. V., De Benedictis, G., Anisimov, V. N., Butov, A. A., Arbeev, K., Jdanov, D. A., Boiko, S. I., Begun, A. S., Bonafe, M., & Franceschi, C. (2001b). Have the oldest old adults ever been frail in the past? A hypothesis that explains modern trends in survival. *Journals of Gerontology Series A: Biological Sciences and Medical Sciences, 56*(10), B432–B442.

Yashin, A. I., Begun, A. S., Boiko, S. I., Ukraintseva, S. V., & Oeppen, J. (2002). New age patterns of survival improvement in Sweden: Do they characterize changes in individual aging? *Mechanisms of Ageing and Development, 123*(6), 637–647.

Yashin, A. I., Arbeev, K. G., Akushevich, I., Kulminski, A., Akushevich, L., & Ukraintseva, S. V. (2007a). Stochastic model for analysis of longitudinal data on aging and mortality. *Mathematical Biosciences, 208*(2), 538–551.

Yashin, A. I., Arbeev, K. G., Kulminski, A., Akushevich, I., Akushevich, L., & Ukraintseva, S. V. (2007b). Cumulative index of elderly disorders and its dynamic contribution to mortality and longevity. *Rejuvenation Research, 10*(1), 75–86.

Yashin, A. I., Arbeev, K. G., Kulminski, A., Akushevich, I., Akushevich, L., & Ukraintseva, S. V. (2007c). Health decline, aging and mortality: How are they related? *Biogerontology, 8*(3), 291–302.

Yashin, A. I., Arbeev, K. G., Akushevich, I., Kulminski, A., Akushevich, L., & Ukraintseva, S. V. (2008a). Model of hidden heterogeneity in longitudinal data. *Theoretical Population Biology, 73*(1), 1–10.

Yashin, A. I., Arbeev, K. G., Kulminski, A., Akushevich, I., Akushevich, L., & Ukraintseva, S. V. (2008b). What age trajectories of cumulative deficits and medical costs tell us about individual aging and mortality risk: Findings from the NLTCS-Medicare data. *Mechanisms of Ageing and Development, 129*(4), 191–200.

Yashin, A. I., Ukraintseva, S. V., Arbeev, K. G., Akushevich, I., Arbeeva, L. S., & Kulminski, A. M. (2009). Maintaining physiological state for exceptional survival: What is the normal level of blood glucose and does it change with age? *Mechanisms of Ageing and Development, 130*(9), 611–618.

Yashin, A. I., Arbeev, K. G., Akushevich, I., Ukraintseva, S. V., Kulminski, A., Arbeeva, L. S., & Culminskaya, I. (2010). Exceptional survivors have lower age trajectories of blood glucose: Lessons from longitudinal data. *Biogerontology, 11*(3), 257–265.

Yashin, A. I., Akushevich, I., Arbeev, K. G., Kulminski, A., & Ukraintseva, S. (2011a). Joint analysis of health histories, physiological states, and survival. *Mathematical Population Studies, 18*(4), 207–233.

Yashin, A. I., Akushevich, I., Arbeev, K. G., Kulminski, A., & Ukraintseva, S. V. (2011b). New approach for analyzing longitudinal data on health, physiological state, and survival collected using different observational plans. In *Presented at JSM proceedings, section on government statistics*.

Yashin, A. I., Arbeev, K. G., Ukraintseva, S. V., Akushevich, I., & Kulminski, A. (2011c). Patterns of aging related changes on the way to 100: An approach to studying aging, mortality, and longevity from longitudinal data. In *2011 living to 100 monograph. Society of Actuaries Monograph M-LI11-1*. Schaumburg: Society of Actuaries.

Yashin, A. I., Wu, D., Arbeev, K. G., Stallard, E., Land, K. C., & Ukraintseva, S. V. (2012). How genes influence life span: The biodemography of human survival. *Rejuvenation Research, 15*(4), 374–380.

Ye, W., Lin, X., & Taylor, J. M. G. (2008). Semiparametric modeling of longitudinal measurements and time-to-event data – a two-stage regression calibration approach. *Biometrics, 64*(4), 1238–1246.

Yu, M. G., Law, N. J., Taylor, J. M. G., & Sandler, H. M. (2004). Joint longitudinal-survival-cure models and their application to prostate cancer. *Statistica Sinica, 14*(3), 835–862.

Zheng, H., Yang, Y., & Land, K. C. (2011). Heterogeneity in the Strehler-Mildvan general theory of mortality and aging. *Demography, 48*(1), 267–290.

Chapter 13
The Latent Class Stochastic Process Model for Evaluation of Hidden Heterogeneity in Longitudinal Data

Konstantin G. Arbeev, Kenneth C. Land, and Anatoliy I. Yashin

13.1 Introduction

Various approaches to statistical model building and data analysis that incorporate unobserved heterogeneity are ubiquitous in different scientific disciplines. Unobserved heterogeneity in models of health and survival outcomes can arise because there may be relevant risk factors affecting an outcome of interest that are either unknown or not measured in the data. Frailty models introduce the concept of unobserved heterogeneity in survival analysis for time-to-event data. In demographic applications, the term "frailty" first appeared in a seminal paper by Vaupel et al. (1979) and since then a considerable body of literature with various extensions and applications of frailty models has been generated. The breadth and "heterogeneity" of this literature can be seen in recent books devoted to frailty models (Duchateau and Janssen 2008; Hanagal 2011; Wienke 2010), which have extensive (but not exclusively overlapping) lists of references.

As discussed in more detail in Chap. 11, biodemography integrates biological knowledge and methods with traditional demographic analyses. One particularly valuable feature of biodemographic approaches is that they can incorporate longitudinal observations of physiological variables (biomarkers) to elucidate the impact of aging on health and longevity. Longitudinal data provide an additional source of heterogeneity that can contribute to differences in risks of time-to-event outcomes. Individual age trajectories of biomarkers can differ due to various observed as well as unobserved (and unknown) factors and such individual differences propagate to differences in risks of related time-to-event outcomes such as the onset of a disease or death. In this chapter, we briefly review recent biostatistical approaches to deal with such heterogeneity. As in Chap. 11, we focus on approaches that model both time-to-event and longitudinal data. This excludes methods focusing on analyses of longitudinal data alone where events are generally treated as a nuisance factor to be

A.I. Yashin et al., *Biodemography of Aging*, The Springer Series on Demographic Methods and Population Analysis 40, DOI 10.1007/978-94-017-7587-8_13

adjusted for and approaches that do not include time-to-event information (e.g., onset of a disease) but include, for example, binary indicators such as prevalence of a disease.

13.2 Approaches to the Incorporation of Hidden Heterogeneity in Analyses of Longitudinal and Time-to-Event Data

The joint analysis of longitudinal and time-to-event data is the realm of a special area of biostatistics named "joint models for longitudinal and time-to-event data" or simply "joint models"; see Chap. 11. Such models can also be viewed as extensions of traditional frailty models in survival analysis, see, e.g., the brief discussion in Chapter 7.3 of the book on frailty models by Duchateau and Janssen (2008). A standard approach in joint modeling is to include the "true" (unobserved) value of a longitudinal outcome represented by a linear mixed-effects model as a covariate in the time-to-event sub-model (typically, the Cox proportional hazards model) (Faucett and Thomas 1996; Wulfsohn and Tsiatis 1997). In this case, the random intercept from the longitudinal sub-model that enters the time-to-event sub-model can be thought of as the (log-normal) frailty term.

Some papers on joint modeling specifically introduce an additional random variable in the time-to-event sub-model referred to as "frailty." This term accommodates any heterogeneity in the risk of events that is not explained by the shared random effects in the longitudinal and time-to-event sub-models. For example, Henderson et al. (2000) (see also Guo and Carlin 2004) model the joint distribution of longitudinal measurements and events via an unobserved (latent) zero-mean bivariate Gaussian process, with correlation between the two components of the process inducing dependence between the longitudinal data and event times. Specification of a component of the Gaussian process in the time-to-event sub-model (which is a Cox proportional hazards model) contains a Gaussian random variable representing a (log-normal) frailty term which is assumed independent of the random variables in the longitudinal sub-model. Lin et al. (2002a) suggested a joint model with a mixed-effects model for the multivariate longitudinal data and a proportional hazards model with a gamma-distributed frailty term in the time-to-event sub-model. Ratcliffe et al. (2004) developed a joint model with frailty for analysis of data with clustering (e.g., hospitals or enrollment sites, or any other type of grouping such that ignoring this clustering is thought to be a possible source of bias in statistical inference). The longitudinal data are modeled using a random effects model with subject- and cluster-level random effects. The survival data are modeled using a Cox proportional hazards model with the same cluster-level random effect acting multiplicatively on the baseline hazard. Thus, this cluster-level random effect links the two types of data (survival and longitudinal), and it represents frailty in the context of frailty models in survival analysis. This two-level (subject- and cluster-level) modeling of random effects adds flexibility and efficiency in modeling data where such "informative clustering" occurs. Liu et al. (2008) extended the joint model by Ratcliffe et al. (2004) assuming that each individual's survival depends not only on random effects at the cluster (i.e., medical center) level, but also on the individual-level random effects. The model is implemented in the freely available software aML (Lillard and Panis 2003) and uses Gaussian quadrature for numerical calculation of integrals over normally distributed random effects which is important for practical applications as it improves the computation time. Yet the formulation of the model in the multi-level settings makes it very comprehensive and suitable for applications in different circumstances where such multi-level repeated measurements and dependent failure times arise. Ko (2010) extended the joint models with a zero-mean Gaussian process described by Henderson et al. (2000, 2002), incorporating a frailty term in the survival sub-model. In contrast to Ratcliffe et al. (2004), Ko (2010) specified a gamma-distributed frailty acting multiplicatively on the hazard in line with the gamma-frailty models extensively used in demographic applications since their introduction to demography by Vaupel et al. (1979).

Another area of application of joint frailty models is analysis of competing risks (Elashoff et al. 2007, 2008; Hu et al. 2009; Huang et al. 2010, 2011; Li et al. 2009). Such models can also accommodate informative censoring, treating it as a competing risk for the event of interest. Competing risks can be modeled using a mixture sub-model (Larson and Dinse 1985; Ng and McLachlan 2003), as in Elashoff et al. (2007), or adopting a cause-specific hazards model (Prentice et al. 1978) with frailty. The mixture model approach produces estimates of the effects of risk factors on the marginal probabilities of occurrence of different risks. The approach with a cause-specific hazards frailty sub-model allows one to account for correlated competing risks and dependent censoring. In the latter approach, it is assumed that the joint distribution of random effects from the longitudinal and cause-specific hazards sub-models is multivariate normal. These random effects ("frailties") in the survival sub-model induce a correlation between different failure types. Simulation studies in Elashoff et al. (2007) showed that the efficiency of estimates of frailty in the time-to-event sub-model improves when information on the longitudinal outcome is included in the model (provided that the two outcomes are correlated and this correlation is modeled correctly). Simulations also showed that this joint model has more power in statistical tests for effects of factors on the time-to-event endpoint compared to separate analyses using the time-to-event data alone, which is valuable in practical applications.

Approaches that incorporate heterogeneity in populations through random variables with continuous distributions (as in the standard joint models and their extensions cited above) assume that the risks of events and longitudinal trajectories follow similar patterns for all individuals in a population (e.g., that biomarkers change linearly with age for all individuals). Although such homogeneity in patterns can be justifiable for some applications, generally this is a rather strict assumption not applicable in all circumstances. A population under study may consist of subpopulations with distinct patterns of longitudinal trajectories of biomarkers that can also have different effects on the time-to-event outcome in each subpopulation. When such subpopulations can be defined on the base of observed covariate(s), one can perform stratified analyses applying different models for each subpopulation. However, observed covariates may not capture the entire heterogeneity in the population in which case it may be useful to conceive of the population as consisting of *latent* subpopulations defined by unobserved characteristics. Special methodological approaches are necessary to accommodate such hidden heterogeneity. Within the joint modeling framework, a special class of models, joint latent class models, was developed to account for such heterogeneity in a population (Lin et al. 2002b; Proust-Lima et al. 2009; Proust-Lima and Taylor 2009). A recent review of joint latent class models can be found in Proust-Lima et al. (2014).

The joint latent class model has three components. First, it is assumed that a population consists of a fixed number of (latent) subpopulations. The latent class indicator represents the latent class membership and the probability of belonging to the latent class is specified by a multinomial logistic regression function of observed covariates. It is assumed that individuals from different latent classes have different patterns of longitudinal trajectories of biomarkers and different risks of event. The key assumption of the model is conditional independence of the biomarker and the time-to-events given the latent classes. Then the class-specific models for the longitudinal and time-to-event outcomes constitute the second and third component of the model thus completing its specification. The longitudinal sub-model is typically modeled in the mixed-effects context and the time-to-event component can be represented by a Cox proportional hazards model (Proust-Lima et al. 2009; Proust-Lima and Taylor 2009) or the gamma-frailty model (Lin et al. 2002b). Computationally, joint latent class models are more feasible than standard joint models with shared random effects because the former do not require numerical integration in the likelihood (although there can still be many challenges related to fitting such models, see Discussion). The R package *lcmm* (Proust-Lima et al. 2012) can be used to fit joint latent class models (see also Chapter 5.4 in Rizopoulos 2012). One particular area of applications of joint latent class models is to develop dynamic prognostic tools that can be updated according to the observed values of the longitudinal outcome (Commenges et al. 2012; Garre et al. 2008; Proust-Lima et al. 2014; Proust-Lima and Taylor 2009).

The stochastic process model (Yashin et al. 2007, 2012) represents a special class of models for joint analysis of longitudinal and time-to-event data that is especially relevant for biodemographic applications as discussed in Chaps. 11 and 12. This model allows for indirect estimation of hidden components of aging that are manifested in individual age trajectories of physiological variables measured for participants of a longitudinal study, which helps to advance our understanding of the impact of processes related to aging on health and longevity. Such hidden components include adaptive capacity, resistance to stresses, physiological norms, and effects of allostatic adaptation which are known from the literature to be characteristics of the processes of aging. The original stochastic process model (Yashin et al. 2007) assumes that all such characteristics follow similar patterns in all individuals in a population. However, as noted above, a population may consist of latent subpopulations with distinct patterns of longitudinal trajectories with different effects on the time-to-event outcome in each such subpopulation. The presence of such heterogeneity is a realistic scenario which cannot be simply ignored in statistical analyses of longitudinal data. For example, carriers of some alleles or genotypes can have distinct patterns of aging-related characteristics. If the corresponding genetic marker is not available in the data, then there is no way to evaluate the true characteristics from the data unless we can incorporate such hidden heterogeneity into the model. If this latent structure in a population is ignored, then the results can be biased and erroneous conclusions can be made. For example, suppose that carriers of a (unobserved in the data) "longevity" allele have a survival advantage relative to non-carriers of the allele because carriers have, say, no age-related decline in adaptive capacity or resistance to stress whereas these effects are prominent in non-carriers of such an allele. Then, if the proportion of carriers of this "longevity" allele in a population is small, estimates of these characteristics in a population will be dominated by those of non-carriers so that conclusions will be made about the presence of an aging-related decline in adaptive capacity and stress resistance in the general population without realizing that the conclusions are wrong for both carriers and non-carriers. For carriers, the analysis will fail to identify the absence of declines in these characteristics with age and for non-carriers the corresponding true values will be underestimated.

The extension of the stochastic process model (Yashin et al. 2007) to accommodate such hidden heterogeneity was suggested in Yashin et al. (2008). In this chapter, we present a version of this extended model, which we call the "latent class stochastic process model," modified to include dependence of the probability of the latent class membership as well as other components of the model on observed covariates. The model was briefly introduced in Arbeev et al. (2014). Here we elaborate it in more detail and present the likelihood estimation procedure and simulation studies to illustrate the approach.

13.3 The Latent Class Stochastic Process Model

13.3.1 Specification of the Model

Consider a population consisting of a finite number G of latent subpopulations or latent classes. One example could be that these subpopulations represent carriers of alleles/genotypes at some gene or single nucleotide polymorphism (SNP) when the corresponding genetic information is not available in the data (note that if such genetic data are available for some subsample of a longitudinal study, then this leads to the methods presented in Chap. 14). Denote by K the random indicator variable identifying the latent class membership for an individual, that is, $K = g$ if an individual belongs to the class $g = 1 \ldots G$ (e.g., he/she has the unobserved allele/genotype g). Then we can specify the probabilities of the latent class membership,

p_g, conditional on observed covariates. Following conventional specifications in joint latent class models (Lin et al. 2002b; Proust-Lima et al. 2009, 2014; Proust-Lima and Taylor 2009), we represent this probability using a multinomial logistic regression:

$$p_g = P(K = g \mid X) = \frac{e^{\beta_{0g} + \beta_{1g}^T X}}{1 + \sum_{c=1}^{G-1} e^{\beta_{0c} + \beta_{1c}^T X}}, \tag{13.1}$$

for $g = 1 \ldots G - 1$, and

$$p_G = P(K = G \mid X) = 1 - \sum_{g=1}^{G-1} P(K = g \mid X) = \frac{1}{1 + \sum_{g=1}^{G-1} e^{\beta_{0g} + \beta_{1g}^T X}}. \tag{13.2}$$

Here β_{0g} is the intercept for the latent class g, β_{1g} is the vector of class-specific regression parameters, and X is the corresponding vector of time-independent covariates ("T" denotes transposition; here and below we will use column vectors if not stated otherwise).

For each latent class g, we then specify a stochastic differential equation for the age dynamics of an M-dimensional vector of biomarkers, Y_t (t is age), similar to the original stochastic process model by Yashin et al. (2007):

$$dY_t = a(t, g, X)(Y_t - f_1(t, g, X))dt + B(t, g, X)dW_t, \tag{13.3}$$

with initial condition Y_{t_0}. We omit the dependence of Y_t on the latent class g and covariates X for conciseness of notation. Here W_t is an M-dimensional vector Wiener process independent of the vector of initial values Y_{t_0} and the latent class indicator. It describes unobserved disturbances affecting the trajectory of biomarkers, and it incorporates stochasticity into the model. The strength of such disturbances is characterized by the $M \times M$ matrix of diffusion coefficients $B(t, g, X)$. The vector-function $f_1(t, g, X)$ (with the same dimension as Y_t) introduces the notion of allostasis into the model as a representation of the age trajectories of biomarkers that organisms are forced to follow by the process of allostatic adaptation. The $M \times M$ matrix $a(t, g, X)$, the negative feedback coefficient in Eq. (13.3), describes the adaptive (homeostatic) capacity in an aging organism. The elements of this matrix correspond to the rate of adaptive response to any deviation of trajectories Y_t from the trajectories $f_1(t, g, X)$ "prescribed" by the processes of allostatic adaptation.

The time-to-event sub-model specifies the latent class-specific expressions for the hazard rates conditional on the vector of biomarkers Y_t and the vector of observed covariates X:

$$\mu(t|Y_t, g, X) = \mu_0(t, g, X) + (Y_t - f_0(t, g, X))^T Q(t, g, X)(Y_t - f_0(t, g, X)). \quad (13.4)$$

Here $\mu_0(t, g, X)$ is the baseline hazard characterizing the risk that would remain if the vector Y_t followed the trajectory $f_0(t, g, X)$ and $Q(t, g, X)$ is a non-negative-definite symmetric $M \times M$ matrix. The M-dimensional vector-function $f_0(t, g, X)$ introduces the concept of an age-dependent physiological norm into the model—it corresponds to the values of the biomarkers that minimize the risk at each age t. The matrix $Q(t, g, X)$ in the quadratic hazard term can be associated with the decline in resistance to stresses with age, as discussed in Yashin et al. (2007, 2012) and Arbeev et al. (2011).

The model in Yashin et al. (2008) is formulated for two latent classes with the latent class membership probability and other components of the model not dependent on the observed covariates. The model presented in this chapter thus naturally generalizes the model in Yashin et al. (2008). More details on the biological interpretation of different components of the stochastic process model can be found in Chap. 12 and in Yashin et al. (2007, 2012). The dependence of all components in the specification of the longitudinal and time-to-event sub-models on the latent class indicator g allows the corresponding aging-related mechanisms to work differently in the different latent subpopulations (for example, in the carriers of some alleles/genotypes). In practice, one can estimate the restricted models with some parameters in the different classes equated in order to test hypotheses about the differences of the respective characteristics in the latent subpopulations using the likelihood ratio test.

13.3.2 Likelihood Estimation Procedure

For simplicity of presentation and following Yashin et al. (2008), we describe the likelihood estimation procedure for the model with two latent classes $g = 1, 2$. Generalizations for more latent classes are straightforward.

Let $\widetilde{Y}_0^t = (Y_{t_0}, Y_{t_1}, \ldots Y_{t_i})$, $t_i \leq t < \tau$, where τ is a random stoppage time (denoting age at death/censoring) of the process Y_t and Yt_i is the observation of this process at age t_i. Denote by $\pi(t|X) = P(K = 1|\widetilde{Y}_0^t, \tau > t, X)$ the conditional probability that an individual belongs to latent class 1, given that he/she has a vector of longitudinal measurements $\widetilde{Y}_0^t$, survived until age t, and has a vector of observed covariates X. The evolution of $\pi(t|X)$ starts at the initial age t_0 and continues first until age t_1. The age dynamics of $\pi(t|X)$ at the interval follows the nonlinear differential equation (Yashin 1985):

$$\frac{d\pi(t|X)}{dt} = \pi(t|X)(\overline{\overline{\mu}}(t|X) - \overline{\mu}(t,1,X)), \tag{13.5}$$

with initial condition (see Eq 13.1):

$$\pi(t_0|X) = p_1 = P(K=1|X) = \frac{e^{\beta_0+\beta_1^T X}}{1+e^{\beta_0+\beta_1^T X}}. \tag{13.6}$$

Here $\overline{\overline{\mu}}(t|X)$ is expressed through $\pi\left(t\middle|X\right)$ and $\overline{\mu}(t,g,X)$, $g=1, 2$, as follows:

$$\overline{\overline{\mu}}(t|X) = \pi(t|X)\overline{\mu}(t,1,X) + (1-\pi(t|X))\overline{\mu}(t,2,X) \tag{13.7}$$

and $\overline{\mu}(t,g,X)$ is given by:

$$\begin{aligned}\overline{\mu}(t,g,X) = \mu_0(t,g,X) + (m(t,g,X) - f_0(t,g,X))^T Q(t,g,X) \\ \times (m(t,g,X) - f_0(t,g,X)) + Tr(Q(t,g,X)\gamma(t,g,X)),\end{aligned} \tag{13.8}$$

where $Tr(\cdot)$ is the matrix trace operator and $m(t,g,X)$, $\gamma(t,g,X)$ denote the mean and variance/covariance matrix of the conditional distribution $P\left(Y_t \le y\middle|K=g, \tau > t, X\right)$ ($Y_t \le y$ means this inequality holds for each component of the vector) satisfying the ordinary differential equations for the age interval $t_0 \le t < t_1$ (see the formulae for the model without latent classes in Yashin and Manton (1997)):

$$\begin{aligned}\frac{dm(t,g,X)}{dt} &= a(t,g,X)(m(t,g,X) - f_1(t,g,X)) - 2\gamma(t,g,X)Q(t,g,X) \\ &\quad \times (m(t,g,X) - f_0(t,g,X)),\end{aligned} \tag{13.9}$$

$$\begin{aligned}\frac{d\gamma(t,g,X)}{dt} = \; & a(t,g,X)\gamma(t,g,X) + \gamma(t,g,X)a(t,g,X)^T \\ & + B(t,g,X)B(t,g,X)^T - 2\gamma(t,g,X)Q(t,g,X)\gamma(t,g,X),\end{aligned} \tag{13.10}$$

with initial conditions $m(t_0,g,X)$, $\gamma(t_0,g,X)$ representing the mean and the variance/covariance matrix of the conditional normal distribution of the initial vector Y_0 in the latent class g, given X. Parameters in the specifications of these quantities need to be estimated along with all other parameters of the model.

The above expressions describe the dynamics of $\pi\left(t\middle|X\right)$ for the interval $t_0 \le t < t_1$. At age $t = t_1$, the process Y_t is observed and this new observation yields new information about the latent class membership that modifies $\pi\left(t\middle|X\right)$. The relationship between $\pi\left(t_1\middle|X\right)$ and $\pi(t_1-|X) = \lim_{t\uparrow t_1}\pi(t|X)$ is given by the Bayes rule (for conciseness, we present these equations for the one-dimensional case):

$$\pi(t_1|X) = \frac{\pi(t_1-|X)\sqrt{\gamma(t_1-,2,X)}V(t_1,1,X)}{\pi(t_1-|X)\sqrt{\gamma(t_1-,2,X)}V(t_1,1,X)+(1-\pi(t_1-|X))\sqrt{\gamma(t_1-,1,X)}V(t_1,2,X)}, \tag{13.11}$$

where $V(t,g,X) = \exp\left\{-(Y_t - m(t-,g,X))^2/2\gamma(t-,g,X)\right\}$.

The value $\pi(t_1|X)$ is an initial condition for $\pi(t|X)$ that evolves according to Eq. (13.5) for the interval $t_1 \le t < t_2$. This process continues for all subsequent age intervals until the last one before the death/censoring time. The expressions for the age interval $t_i \le t < t_{i+1}$ are as in Eq. (13.11) with t_1 replaced by t_i and the dynamics of $m(t, g, X)$ and $\gamma(t, g, X)$ at this interval are given by Eqs. (13.9) and (13.10) with initial conditions $m(t_i, g, X) = Y_{t_i}$, and $\gamma(t_i, g, X) = 0$.

At $t = \tau$, we have

$$P\left(K=1|\widetilde{Y}_0^\tau, \tau=t, X\right) = \pi(\tau-|X)\frac{\overline{\mu}(\tau,1,X)}{\overline{\overline{\mu}}(\tau|X)}, \tag{13.12}$$

where $\pi(\tau-|X) = \lim_{t_i\uparrow t}\pi(t_i|X)|_{t=\tau}$. Let $\widetilde{\pi}(t|X) = P(K=1|\widetilde{Y}_0^t, Z_t, X)$, where $Z_t = I(\tau \le t)$. Then the trajectory of $\widetilde{\pi}(t|X)$ during the interval $t_i \le t \le \tau$ can be described by the stochastic differential equation with one jump:

$$d\widetilde{\pi}(t|X) = \widetilde{\pi}(t-|X)\left(\frac{\mu(t,1,X)}{\overline{\overline{\mu}}(t|X)} - 1\right)\left(dZ_t - \overline{\overline{\mu}}(t|X)dt\right). \tag{13.13}$$

Here $\overline{\overline{\mu}}(t|X) = \widetilde{\pi}(t|X)\overline{\mu}(t,1,X) + (1-\widetilde{\pi}(t|X))\overline{\mu}(t,2,X)$ (note also that $\widetilde{\pi}(t|X)I(\tau > t) = \pi(t|X)I(\tau > t)$).

The likelihood function is the product of two terms:

$$L = L_Y L_\tau, \tag{13.14}$$

where

$$L_Y = \prod_{j=1}^{N}\prod_{i=0}^{n(j)}\left(\frac{\pi_j(t_i^j-|X_j)}{\sqrt{2\pi\gamma_j(t_i^j-,1,X_j)}}e^{-(y_j(t_i^j)-m_j(t_i^j-,1,X_j))^2/2\gamma_j(t_i^j-,1,X_j)} + \frac{\left(1-\pi_j(t_i^j-|X_j)\right)}{\sqrt{2\pi\gamma_j(t_i^j-,2,X_j)}}e^{-(y_j(t_i^j)-m_j(t_i^j-,2,X_j))^2/2\gamma_j(t_i^j-,2,X_j)}\right) \tag{13.15}$$

and

$$L_\tau = \prod_{j=1}^{N} \overline{\overline{\mu}}_j(\tau_j|X_j)^{\delta_j} e^{-\int_0^{\tau_j} \overline{\overline{\mu}}_j(u|X_j)du}. \quad (13.16)$$

Here the subscript j refers to the characteristics for the jth individual, N is the number of individuals, δ_j is the at-risk indicator ($\delta_j = 1$ if jth individual died at age τ_j and $\delta_j = 0$ if he/she was censored at that age), $n(j)$ is the number of measurements of the process Y_t for the jth individual (with measurement at age t_i^j denoted by $y_j(t_i^j)$).

13.4 Simulation Studies

13.4.1 Simulation Study for Latent Class Stochastic Process Model

We performed a simulation study to illustrate the approach using the discrete-time one-dimensional version of the model (13.1–13.16). The following specifications of the model's components were used for two latent classes, $g = 1, 2$:

1. Gompertz baseline hazards: $\mu_0(t, g, X) = \ln \mu_0(t, g) = \ln a_{\mu_0}(g) + b_{\mu_0}(g)(t - t_{\min})$, where $t_{\min} = 30$;
2. Linear functions of age for multipliers in the quadratic hazard terms: $Q(t, g, X) = Q(t, g) = a_Q(g) + b_Q(g)t$;
3. Linear functions of age for the "optimal trajectories": $f_0(t, g, X) = f_0(t, g) = a_{f_0}(g) + b_{f_0}(g)(t - t_{\min})$;
4. Linear functions of age for feedback coefficients in (13.3) ("adaptive capacities"): $a(t, g, X) = a(t, g) = a_Y(g) + b_Y(g)(t - t_{\min})$, where $a_Y(g) < 0$ and $b_Y(g) \geq 0$;
5. Linear functions of age for the "mean allostatic trajectories": $f_1(t, g, X) = f_1(t, g) = a_{f_1}(g) + b_{f_1}(g)(t - t_{\min})$;
6. Constant diffusion coefficients: $B(t, g, X) = \sigma_1(g)$;
7. Normally distributed initial values of the process Y_t with means $f_1(t_0^j, g, X)$ (where t_0^j is age at the first exam for the jth individual) and variances $\sigma_0^2(g)$; and
8. Probability of the latent class 1 membership: $p_1 = e^{\beta_0 + \beta_1 X_1 + \beta_2 X_2} / (1 + e^{\beta_0 + \beta_1 X_1 + \beta_2 X_2})$, where $X_1 = 0, 1$ with $P(X_1 = 1) = 0.5$, and X_2 has standard normal distribution.

Thus, the parameters to be estimated in this model are: $a_{\mu_0}(g)$, $b_{\mu_0}(g)$, $a_Q(g)$, $b_Q(g)$, $a_Y(g)$, $b_Y(g)$, $a_{f_1}(g)$, $b_{f_1}(g)$, $a_{f_0}(g)$, $b_{f_0}(g)$, $\sigma_0(g)$, $\sigma_1(g)$, β_0, β_1, and β_2.

To make the simulations more realistic, we generated a structure resembling that of the Framingham Heart Study data (Dawber et al. 1951) with parameters selected to produce reasonable mortality rates close to those observed in modern human populations. Pulse pressure was used as a prototype for the process Y_t. Age at entry into the study was simulated as a discrete random variable uniformly distributed over the interval [30, 60]. The interval between observations of Y_t is 2 years. The number of observations (exams) is 30. We simulated 100 data sets with 5200 individuals in each, and estimated the likelihood function of the model (13.1–13.16) for each data set. Table 13.1 shows mean values, standard deviations, and minimal and maximal values of the estimated parameters in these 100 data sets. Figure 13.1 displays estimated trajectories of different components of the model (logarithms of baseline hazard, multipliers in the quadratic hazard terms, mean allostatic trajectories, optimal trajectories, and adaptive capacity) in two latent classes for 100 simulated data sets. The results show that the model correctly evaluates all model components for two latent classes. The next section illustrates what happens if we ignore the latent structure in the data and apply the original stochastic process model (Yashin et al. 2007) to these data sets.

13.4.2 *Simulation Study for Stochastic Process Model That Ignores Latent Classes*

We estimated the parameters of the original stochastic process model (Yashin et al. 2007) which does not take into account the latent structure of the simulated data. This model is essentially given by Eqs. (13.3) and (13.4) without dependence on the latent class g:

$$dY_t = a(t,X)(Y_t - f_1(t,X))dt + B(t,X)dW_t, \tag{13.17}$$

$$\mu(t|Y_t,X) = \mu_0(t,X) + (Y_t - f_0(t,X))\ ^TQ(t,X)(Y_t - f_0(t,X)). \tag{13.18}$$

We estimated a one-dimensional version of this model, applying it to the same simulated data from Sect. 13.4.1. We used specifications of the model's components similar to those described in Sect. 13.4.1, but without dependence on the latent class g. Thus, the estimated parameters in this case are: a_{μ_0}, b_{μ_0}, a_Q, b_Q, a_Y, b_Y, a_{f_1}, a_{f_0}, b_{f_0}, σ_0, and σ_1 (note that we do not have parameters β_0, β_1, and β_2 here because we do not model the probability of the latent class membership).

The results of these estimates are shown in Table 13.1 (see the "***No LC***" panel of the table) and Fig. 13.2. It is clear from the table that this model, which ignores the latent structure of the data, produces parameter estimates that deviate from the true values in the two latent classes. Correspondingly, the resulting "population"

Table 13.1 Simulation study for latent class stochastic process model: Means, standard deviations (STD), minimal (MIN) and maximal (MAX) values of parameter estimates for latent class stochastic process model in 100 simulated data sets

	$\ln a_{\mu_0}$	b_{μ_0}	$a_Q \cdot 10^4$	$b_Q \cdot 10^4$	a_Y	$b_Y \cdot 10^3$	σ_0	σ_1	a_{f_1}	b_{f_1}	a_{f_0}	b_{f_0}	β_0	β_1	β_2
$g=1$															
MEAN	−7.50	0.110	0.92	0.51	−0.201	1.01	5.00	4.00	55.00	0.299	49.92	0.302	1.01	−0.20	0.30
STD	0.17	0.004	0.63	0.12	0.007	0.24	0.07	0.02	0.17	0.006	0.46	0.018	0.12	0.10	0.05
MIN	−7.90	0.102	-0.31	0.20	−0.218	0.23	4.82	3.93	54.61	0.288	48.78	0.236	0.70	−0.46	0.16
MAX	−7.11	0.120	2.52	0.74	−0.176	1.58	5.23	4.05	55.32	0.315	51.45	0.350	1.32	0.01	0.43
TRUE	***−7.5***	***0.11***	***1.0***	***0.5***	***−0.20***	***1.0***	***5.0***	***4.0***	***55.0***	***0.3***	***50.0***	***0.3***	***1.0***	***−0.2***	***0.3***
$g=2$															
MEAN	−8.56	0.111	0.48	0.11	−0.251	1.00	5. 00	4.00	49.97	0.300	45.00	0.295			
STD	0.32	0.006	0.77	0.13	0.010	0.24	0.12	0.02	0.26	0.007	2.16	0.068			
MIN	−9.57	0.099	−1.22	−0.18	−0.276	0.14	4.69	3.95	49.43	0.286	38.25	0.124			
MAX	−7.85	0.128	2.15	0.41	−0.218	1.64	5.24	4.05	50.52	0.321	48.79	0.482			
TRUE	***−8.5***	**0.11**	***0.5***	***0.1***	***−0.25***	***1.0***	***5.0***	***4.0***	***50.0***	***0.3***	***45.0***	***0.3***			
No LC															
MEAN	−7.16	0.091	1.12	0.26	−0.186	0.47	5.50	4.05	53.94	0.272	49.63	0.233			
STD	0.10	0.002	0.33	0.06	0.004	0.14	0.05	0.01	0.10	0.004	0.31	0.012			
MIN	−7.37	0.084	0.30	0.10	−0.196	0.16	5.37	4.02	53.64	0.264	48.85	0.199			
MAX	−6.86	0.095	1.95	0.41	−0.177	0.81	5.65	4.07	54.18	0.284	50.42	0.258			

Notes:

1. Dependence of parameters on the latent class (g) is omitted for conciseness
2. Parameters. a_Q, b_Q . and b_Y are rescaled for better visibility
3. Values for the model without latent classes ("***No LC***") estimating these same data sets are given for comparison

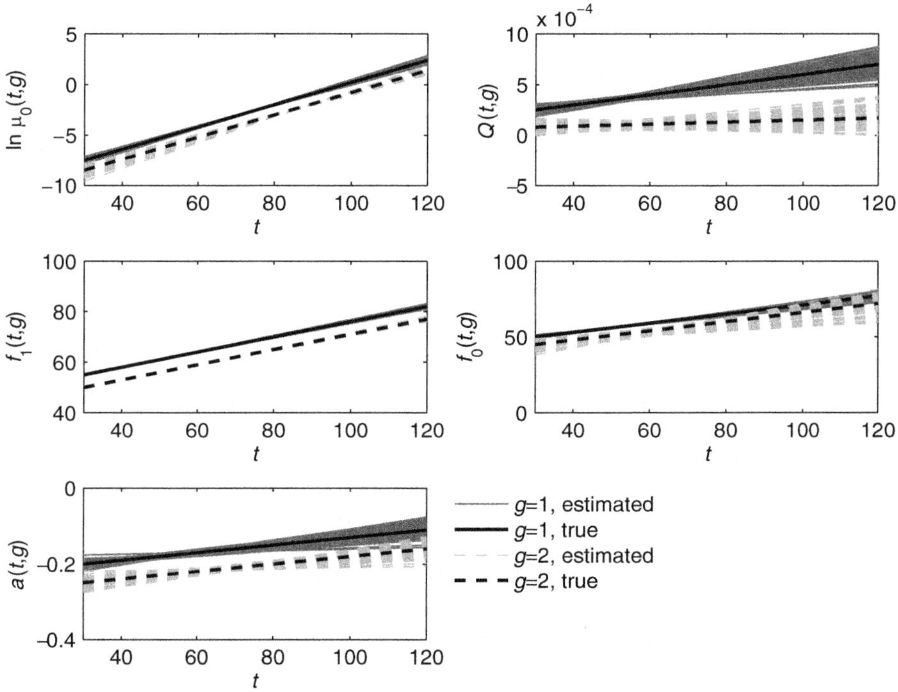

Fig. 13.1 Simulation study for latent class stochastic process model: estimated trajectories of logarithms of baseline hazard ($\ln \mu_0(t, g)$), multipliers in the quadratic hazard terms ($Q(t, g)$), mean allostatic trajectories ($f_1(t, g)$), optimal trajectories ($f_0(t, g)$) and adaptive capacity ($a(t, g)$) in two latent classes ($g = 1$ and $g = 2$) for 100 simulated data sets. The true trajectories in two latent classes are denoted by *thick lines*

trajectories of the logarithm of the baseline hazard ($\ln \mu_0(t, X)$), multipliers in the quadratic hazard terms ($Q(t, X)$), mean allostatic trajectories ($f_1(t, X)$), optimal trajectories ($f_0(t, X)$), and adaptive capacity ($a(t, X)$) deviate from the actual trajectories in both latent classes (see Fig. 13.2). Thus, ignoring latent classes can lead to incorrect estimates and wrong conclusions. For example, using the estimates of the model (13.17–13.18) leads to the inference that the estimated "population" trajectory ($f_0(t, X)$) is a universal optimal trajectory for all individuals. However, such an "optimal" trajectory will actually be not optimal for any individual in terms of minimizing mortality risk: if an individual belongs to the latent class g and his/her trajectory Y_t follows the function ($f_0(t, X)$) then his/her risk of death will be higher than it would be had he/she followed the truly optimal trajectory $f_0(t, g, X)$ (the risk of death is $\mu_0(t, g, X) + (f_0(t, X) - f_0(t, g, X))^T Q(t, g, X)\,(f_0(t, X) - f_0(t, g, X))$ in the former case, whereas in the latter case the risk reduces to the baseline level $\mu_0(t, g, X)$).

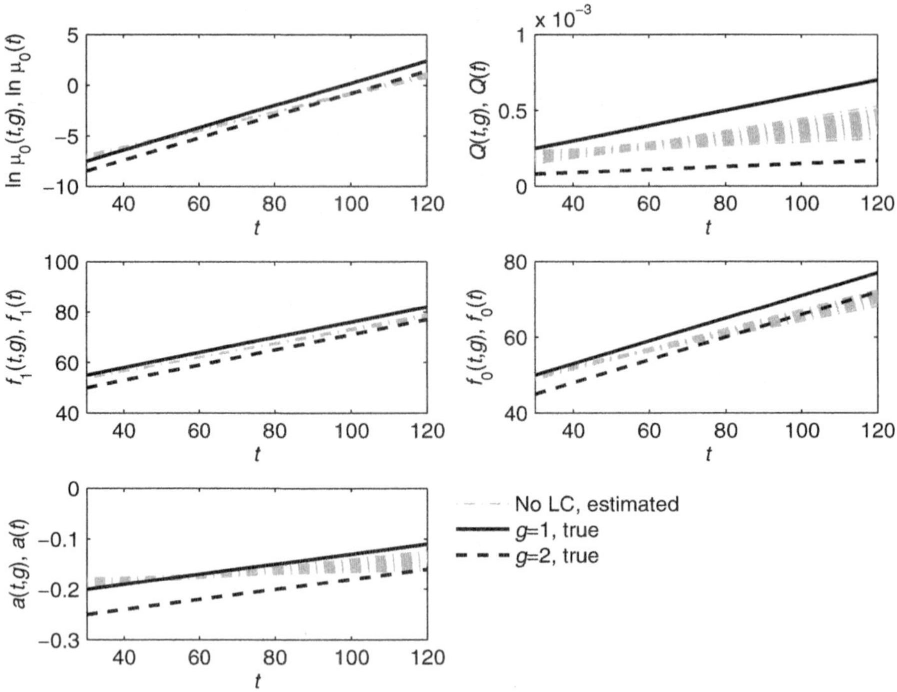

Fig. 13.2 Simulation study for stochastic process model that ignores latent classes: estimated trajectories of logarithms of baseline hazard ($\ln \mu_0(t)$), multipliers in the quadratic hazard terms ($Q(t)$), mean allostatic trajectories ($f_l(t)$) optimal trajectories ($f_0(t)$) and adaptive capacity ($a(t)$) for 100 simulated data sets containing latent classes (same as in Fig. 13.1). "No LC" denotes no latent classes. The corresponding true trajectories in two latent classes $g = 1$ and $g = 2$ are denoted by *thick lines*

13.5 Discussion and Conclusion

In this chapter, we have described approaches for dealing with unobserved heterogeneity in joint analyses of longitudinal and time-to-event outcomes. Special attention has been paid to a specific class of such approaches that accommodates hidden heterogeneity in the population due to the presence of latent subpopulations with distinct longitudinal patterns with different relations to the risk of an event. We also presented a latent class stochastic process model that takes into account such hidden heterogeneity and allows for indirect estimation of hidden components of aging that are manifested in individual age trajectories of physiological variables measured in participants of a longitudinal study. These hidden components of aging and their impact on the risk of events can be evaluated for latent subpopulations. This can help to unravel hidden effects in the data that otherwise can remain masked if a model that ignores this structures is applied to such data.

Several computational challenges should be mentioned here as a caution concerning the practical implementation of the latent class stochastic process model. These challenges are similar to those encountered by the joint latent class models that have been discussed in the literature on such models (and, generally, in the mixture models literature). Since the joint latent class models and latent class stochastic process model have many characteristics in common, and as our practical experience with our model suggests, careful attention to these matters is necessary. *First*, it is well known in mixture models that the likelihood functions may have local maxima. Therefore, calculating the estimation algorithm from different initial values is a safeguard measure for ensuring convergence to the global maximum, although at the price of additional computation time, and defining good initial values becomes practically important in latent class models (Han 2009). In our experience, parameter estimates obtained from fitting the data using the original stochastic process model generally provide a good starting point for initial values for the latent class stochastic process model. *Second*, the number of latent classes for a given set of data generally is not known prior to estimation of latent class models. Therefore, it is necessary to estimate models with different numbers of latent classes and select the model with the number of classes that gives the best fit to the data. The Bayesian information criterion (BIC) is recommended for selecting the optimal number of latent classes (Proust-Lima et al. 2014; Proust-Lima and Taylor 2009). *Third*, the conditional independence assumption (i.e., that the longitudinal biomarker and the time-to-events are assumed to be conditionally independent given the latent classes) is crucial in the formulation of latent class models and derivation of the likelihood functions. This assumption is difficult to test in practice because the latent classes are unobserved. However, several statistical approaches for evaluating the conditional independence assumption have been suggested in the literature on joint latent class models (Jacqmin-Gadda et al. 2010; Lin et al. 2004; Proust-Lima et al. 2009). *Fourth*, in the latent class stochastic process model, similarly to joint latent class models, permutations of the latent class parameters between latent classes produces the same likelihood. However, this does not cause problems in the estimation procedure, because it is based on maximum likelihood estimation. The only inconvenience arises in simulation studies where one must always check whether the latent classes were estimated in the same order in all datasets to report the correct average values of parameters in the simulated datasets.

With all these considerations taken into account, the latent class stochastic process model nevertheless provides a useful tool for dealing with unobserved heterogeneity in joint analyses of longitudinal and time-to-event outcomes and taking into account hidden components of aging in their joint influence on health and longevity. This approach is also helpful for sensitivity analyses in applications of the original stochastic process model. We recommend starting the analyses with the original stochastic process model and estimating the model ignoring possible hidden heterogeneity in the population. Then the latent class stochastic process model can be applied to test hypotheses about the presence of hidden heterogeneity in the data in order to appropriately adjust the conclusions if a latent structure is revealed. Such an approach can be implemented not only with the original model described in Chap. 12 but also with its extensions, for example, with the genetic stochastic process model described in Chap. 14.

Acknowledgements This chapter was partly supported by the National Institute on Aging of the National Institutes of Health under Award Numbers R01AG030198, R01AG032319, R01AG030612, R01AG046860, P01AG043352, and P30AG034424. The content is solely the responsibility of the authors and does not necessarily represent the official views of the National Institutes of Health.

References

Arbeev, K. G., Ukraintseva, S. V., Akushevich, I., Kulminski, A. M., Arbeeva, L. S., Akushevich, L., Culminskaya, I. V., & Yashin, A. I. (2011). Age trajectories of physiological indices in relation to healthy life course. *Mechanisms of Ageing and Development, 132*(3), 93–102.

Arbeev, K. G., Akushevich, I., Kulminski, A. M., Ukraintseva, S., & Yashin, A. I. (2014). Joint analyses of longitudinal and time-to-event data in research on aging: Implications for predicting health and survival. *Frontiers in Public Health, 2*, article 228.

Commenges, D., Liquet, B., & Proust-Lima, C. (2012). Choice of prognostic estimators in joint models by estimating differences of expected conditional Kullback-Leibler risks. *Biometrics, 68*(2), 380–387.

Dawber, T. R., Meadors, G. F., & Moore, F. E. (1951). Epidemiological approaches to heart disease: The Framingham Study. *American Journal of Public Health, 41*(3), 279–286.

Duchateau, L., & Janssen, P. (2008). *The frailty model*. New York: Springer.

Elashoff, R. M., Li, G., & Li, N. (2007). An approach to joint analysis of longitudinal measurements and competing risks failure time data. *Statistics in Medicine, 26*(14), 2813–2835.

Elashoff, R. M., Li, G., & Li, N. (2008). A joint model for longitudinal measurements and survival data in the presence of multiple failure types. *Biometrics, 64*(3), 762–771.

Faucett, C. L., & Thomas, D. C. (1996). Simultaneously modelling censored survival data and repeatedly measured covariates: A Gibbs sampling approach. *Statistics in Medicine, 15*(15), 1663–1685.

Garre, F. G., Zwinderman, A. H., Geskus, R. B., & Sijpkens, Y. W. J. (2008). A joint latent class changepoint model to improve the prediction of time to graft failure. *Journal of the Royal Statistical Society, Series A (Statistics in Society), 171*(1), 299–308.

Guo, X., & Carlin, B. P. (2004). Separate and joint modeling of longitudinal and event time data using standard computer packages. *American Statistician, 58*(1), 16–24.

Han, J. (2009). Starting values for EM estimation of latent class joint model. *Communications in Statistics-Simulation and Computation, 38*(7), 1519–1534.

Hanagal, D. D. (2011). *Modeling survival data using frailty models*. Boca Raton: Chapman & Hall/CRC.

Henderson, R., Diggle, P., & Dobson, A. (2000). Joint modelling of longitudinal measurements and event time data. *Biostatistics, 1*(4), 465–480.

Henderson, R., Diggle, P., & Dobson, A. (2002). Identification and efficacy of longitudinal markers for survival. *Biostatistics, 3*(1), 33–50.

Hu, W., Li, G., & Li, N. (2009). A Bayesian approach to joint analysis of longitudinal measurements and competing risks failure time data. *Statistics in Medicine, 28*(11), 1601–1619.

Huang, X., Li, G., & Elashoff, R. M. (2010). A joint model of longitudinal and competing risks survival data with heterogeneous random effects and outlying longitudinal measurements. *Statistics and Its Interface, 3*(2), 185–195.

Huang, X., Li, G., Elashoff, R. M., & Pan, J. (2011). A general joint model for longitudinal measurements and competing risks survival data with heterogeneous random effects. *Lifetime Data Analysis, 17*(1), 80–100.

Jacqmin-Gadda, H., Proust-Lima, C., Taylor, J. M. G., & Commenges, D. (2010). Score test for conditional independence between longitudinal outcome and time to event given the classes in the joint latent class model. *Biometrics, 66*(1), 11–19.

Ko, F.-S. (2010). Using frailty models to identify the longitudinal biomarkers in survival analysis. *Communications in Statistics Theory and Methods, 39*(18), 3222–3237.

Larson, M. G., & Dinse, G. E. (1985). A mixture model for the regression analysis of competing risks data. *Journal of the Royal Statistical Society, Series C (Applied Statistics), 34*(3), 201–211.

Li, N., Elashoff, R. M., & Li, G. (2009). Robust joint modeling of longitudinal measurements and competing risks failure time data. *Biometrical Journal, 51*(1), 19–30.

Lillard, L., & Panis, C. W. A. (2003). *aML, multilevel multiprocess statistical software. Release 2.0*. Los Angeles: EconWare.

Lin, H. Q., McCulloch, C. E., & Mayne, S. T. (2002a). Maximum likelihood estimation in the joint analysis of time-to-event and multiple longitudinal variables. *Statistics in Medicine, 21*(16), 2369–2382.

Lin, H. Q., Turnbull, B. W., McCulloch, C. E., & Slate, E. H. (2002b). Latent class models for joint analysis of longitudinal biomarker and event process data: Application to longitudinal prostate-specific antigen readings and prostate cancer. *Journal of the American Statistical Association, 97*(457), 53–65.

Lin, H. Q., McCulloch, C. E., & Rosenheck, R. A. (2004). Latent pattern mixture models for informative intermittent missing data in longitudinal studies. *Biometrics, 60*(2), 295–305.

Liu, L., Ma, J. Z., & O'Quigley, J. (2008). Joint analysis of multi-level repeated measures data and survival: An application to the end stage renal disease (ESRD) data. *Statistics in Medicine, 27*(27), 5679–5691.

Ng, S. K., & McLachlan, G. J. (2003). An EM-based semi-parametric mixture model approach to the regression analysis of competing-risks data. *Statistics in Medicine, 22*(7), 1097–1111.

Prentice, R. L., Kalbfleisch, J. D., Peterson, A. V., Flournoy, N., Farewell, V. T., & Breslow, N. E. (1978). Analysis of failure times in presence of competing risks. *Biometrics, 34*(4), 541–554.

Proust-Lima, C., & Taylor, J. M. G. (2009). Development and validation of a dynamic prognostic tool for prostate cancer recurrence using repeated measures of posttreatment PSA: A joint modeling approach. *Biostatistics, 10*(3), 535–549.

Proust-Lima, C., Joly, P., Dartigues, J.-F., & Jacqmin-Gadda, H. (2009). Joint modelling of multivariate longitudinal outcomes and a time-to-event: A nonlinear latent class approach. *Computational Statistics and Data Analysis, 53*(4), 1142–1154.

Proust-Lima, C., Diakite, A., & Liquet, B. (2012). *lcmm: Estimation of latent class mixed models, joint latent class mixed models and mixed models for curvilinear outcomes*. R package, version 1.5.8, http://cran.r-project.org/web/packages/lcmm/index.html

Proust-Lima, C., Sene, M., Taylor, J. M., & Jacqmin-Gadda, H. (2014). Joint latent class models for longitudinal and time-to-event data: A review. *Statistical Methods in Medical Research, 23*(1), 74–90.

Ratcliffe, S. J., Guo, W. S., & Ten Have, T. R. (2004). Joint modeling of longitudinal and survival data via a common frailty. *Biometrics, 60*(4), 892–899.

Rizopoulos, D. (2012). *Joint models for longitudinal and time-to-event data with applications in R*. Boca Raton: Chapman & Hall/CRC.

Vaupel, J. W., Manton, K. G., & Stallard, E. (1979). Impact of heterogeneity in individual frailty on the dynamics of mortality. *Demography, 16*(3), 439–454.

Wienke, A. (2010). *Frailty models in survival analysis*. Boca Raton: Chapman & Hall/CRC.

Wulfsohn, M. S., & Tsiatis, A. A. (1997). A joint model for survival and longitudinal data measured with error. *Biometrics, 53*(1), 330–339.

Yashin, A. I. (1985). Dynamics in survival analysis: Conditional Gaussian property vs. Cameron-Martin formula. In N. V. Krylov, R. S. Lipster, & A. A. Novikov (Eds.), *Statistics and control of stochastic processes* (pp. 446–475). New York: Springer.

Yashin, A. I., & Manton, K. G. (1997). Effects of unobserved and partially observed covariate processes on system failure: A review of models and estimation strategies. *Statistical Science, 12*(1), 20–34.

Yashin, A. I., Arbeev, K. G., Akushevich, I., Kulminski, A., Akushevich, L., & Ukraintseva, S. V. (2007). Stochastic model for analysis of longitudinal data on aging and mortality. *Mathematical Biosciences, 208*(2), 538–551.

Yashin, A. I., Arbeev, K. G., Akushevich, I., Kulminski, A., Akushevich, L., & Ukraintseva, S. V. (2008). Model of hidden heterogeneity in longitudinal data. *Theoretical Population Biology, 73*(1), 1–10.

Yashin, A. I., Arbeev, K. G., Akushevich, I., Kulminski, A., Ukraintseva, S. V., Stallard, E., & Land, K. C. (2012). The quadratic hazard model for analyzing longitudinal data on aging, health, and the life span. *Physics of Life Reviews, 9*(2), 177–188.

Chapter 14
How Biodemographic Approaches Can Improve Statistical Power in Genetic Analyses of Longitudinal Data on Aging, Health, and Longevity

Konstantin G. Arbeev and Anatoliy I. Yashin

14.1 Introduction

The modern era of revolutionary advances in genetics provides great opportunities and challenges for the field of biodemography. The imperative to integrate the principles of genetics and genomics into biodemography is so evident that this problem will continue to be at the forefront of demographic analysis for decades into the future (Carey 2008; Wachter 2008). The importance of "genetic biodemography" will continue to grow in the coming years, because many studies that have collected data on biomarkers will include (or already have included) genetic information. The ongoing incorporation of genetic information into longitudinal studies is considered potentially "the most revolutionary element of the addition of biological data in large-scale surveys" (Suzman 2010) and such studies will "increasingly provide analyses of the interactions of genetic, biological, social, economic, and demographic characteristics" (Crimmins et al. 2010).

To realize the full potential of such rich data, special attention should be paid to approaches to the analysis of the diverse information contained therein. Consider, for example, the case in which the research objective is the evaluation of genetic effects on some time-to-event outcome, e.g., risk of death or onset of a disease. Comparison of the age patterns of incidence or mortality rates for carriers of different alleles/genotypes can contribute to an understanding of the role of genetic factors in survival or the development of aging-associated diseases. To simplify, let us set aside for a moment research questions involving longitudinal measurements of biomarkers (which require special consideration, see Chap. 11) and assume that only socioeconomic and demographic covariates are involved in analyses. Traditional methods for estimation of the effect of genetic markers in such cases can be enhanced if we complement them with a demographic approach that takes into account the demographic structure of the population under study. Specifically, when genetic data are included in longitudinal studies of aging, we have several relevant sources of information for analyses of genetic influences on lifespan

A.I. Yashin et al., *Biodemography of Aging*, The Springer Series on Demographic Methods and Population Analysis 40, DOI 10.1007/978-94-017-7587-8_14

(or onset of diseases), in addition to the genetic data themselves and socioeconomic and demographic covariates.

The first is follow-up data on the outcome of interest (e.g., mortality). Second, genetic data are usually collected in longitudinal studies from participants at different ages. This provides information on the age structure of the population at the time of biospecimen collection. Along with follow-up data, the population age structure also contains information about the effect of genetic variants on lifespan, and the full potential of the data is underused when this information is ignored in analyses, especially when genotyping is performed at advanced ages with noticeable attrition due to mortality. Indeed, in order to be genotyped, an individual has to survive until the age at biospecimen collection. Hence, if the proportion of carriers of some genetic variant increases with age (here we mean the age at biospecimen collection) then this variant should favor longevity. This implies that we can associate genetic variants with lifespan even without follow-up data by using the "gene frequency" method (Yashin et al. 1999, 2000). We can expect therefore that, if we use both follow-up data and data on population age structure, this will provide us with more accurate estimates of parameters and additional power compared to the use of follow-up data alone. Such data can be analyzed jointly using appropriate methods (Arbeev et al. 2011b; Yashin et al. 2007b).

The third source of information in longitudinal studies of aging that is relevant for genetic analyses stems from the history of incorporation of genetic information into such studies. While in some modern longitudinal studies the genetic data can be collected at the baseline, it is a common situation that many older long-established longitudinal studies commenced before genetic data collection began. Hence, in such studies genetic data are available only for a sub-sample of participants of the longitudinal study (i.e., for those who survived until the time of biospecimen collection). It is also possible that genetic data were collected only for a sub-sample of participants due to, for example, budgetary restrictions. However, in such cases information on the outcome of interest (e.g., follow-up on mortality) can be available for all (genotyped and non-genotyped) participants of the longitudinal study. This information should not be neglected in genetic analyses because it provides an additional source for increasing power and improving the accuracy of the estimates. Indeed, the group of non-genotyped individuals is a mixture of carriers/non-carriers of the same alleles/genotypes collected in the genetic data and a similar age-specific form of the mortality rate can be assumed for the entire sample. Therefore, this information can be appropriately combined in the likelihood function with information for genotyped individuals (see Arbeev et al. 2011b).

Incorporation of genetic information in studies that collect longitudinal measurements of biomarkers along with follow-up data offers new opportunities for analyses of genetic influences on aging, health, and longevity. As discussed in Chap. 11, joint analysis of time-to-event outcomes and longitudinal measurements of biomarkers requires special methodological considerations to take into account measurement errors and biological variability of biomarkers and to avoid biased estimates. Joint models provide a setting for performing such analyses involving genetic data, longitudinal measurements, and follow-up data. Joint models, in particular, permit evaluating the effect of a covariate (i.e., a genetic marker in these applications) on both the longitudinal trajectories and times-to-events, so that it is possible to distinguish genetic effects on these two outcomes. Joint models provide more efficient estimates of the effect of a covariate (such as a genetic marker) on the time-to-event outcome in the case in which there is also an effect of the covariate on the longitudinal trajectory of a biomarker. This means that analyses of longitudinal and time-to-event data in joint models may require smaller sample sizes to achieve comparable statistical power with analyses based on time-to-event data alone (Chen et al. 2011).

As mentioned above, participants of a longitudinal study for whom genetic information was not collected, but for whom other outcomes (longitudinal measurements of biomarkers, follow-up data, and possibly other relevant covariates) are available, provide an additional source for increasing the accuracy and statistical power in analyses of genetic effects on longitudinal and time-to-event outcomes. However, following similar arguments as in Chap. 11, longitudinal measurements of biomarkers cannot be simply incorporated as time-dependent covariates in joint analyses of genotyped and non-genotyped participants of longitudinal studies (Arbeev et al. 2011b). An approach for jointly analyzing longitudinal measurements of biomarkers and time-to-event outcomes for genotyped and non-genotyped participants of longitudinal studies has been developed recently within the framework of the stochastic process model (SPM) of aging discussed in detail in Chap. 12 (see Arbeev et al. 2009, 2012). Such a model, named the "genetic stochastic process model," or the "genetic SPM," is especially relevant in the context of biodemographic research. Biodemography of aging aims at integrating demographic and biological theory and methods to advance our understanding of the impact of processes related to aging on health and longevity. Genetic biodemography aims at elucidating genetic components in such processes and their influence on health and longevity. Therefore, the particular advantages of the genetic SPM for biodemographic applications are that it is based on biological theory, it incorporates several essential mechanisms of aging-related changes in organisms, and it allows for evaluating genetic effects on such characteristics and their influence on mortality or onset of a disease. Such "hidden components" of aging-related changes incorporated into this model include: adaptive capacity, resistance to stresses, physiological norms, and effects of allostatic adaptation (for more details, see Chap. 12). As is known from the literature, all these variables play important roles in aging processes. Therefore, their inclusion in the model is crucial for a better understanding of regulatory mechanisms driving observed aging-related changes in physiological variables and their influence on risks of death or getting a disease, as well as for evaluating the genetic component in such processes. However, relevant variables associated with such "hidden components of aging" are typically not directly measured in longitudinal data and, hence, they cannot be directly estimated from the data using, for example, joint models. The genetic SPM thus provides a useful approach for working with such "hidden components of aging" indirectly. Importantly, it also provides an additional possibility for improving the power of genetic analyses by joint analysis of data for genotyped and non-genotyped sub-samples of the study (Arbeev et al. 2009).

The remainder of this chapter is organized as follows. Section 14.2 presents results of simulation studies of the longitudinal genetic-demographic model (Arbeev et al. 2011b) which illustrate that inclusion of information on ages at biospecimen collection in addition to follow-up data improves power in analyses of genetic effects on mortality or morbidity risks (see also Yashin et al. 2013b). Section 14.3 describes simulation studies of the genetic SPM (Arbeev et al. 2009) which show the increase in statistical power of joint analyses of genotyped and non-genotyped participants of a longitudinal study compared to analyses of genotyped participants alone in different scenarios to test relevant biologically-based hypotheses. Section 14.4 discusses the results and possible generalizations of the approaches.

14.2 Simulation Studies of the Longitudinal Genetic-Demographic Model

The longitudinal genetic-demographic model (or the genetic-demographic model for longitudinal data) is described in Arbeev et al. (2011b). The full model combines three sources of information in the likelihood function: (1) follow-up data on survival (or, generally, on some time-to-event) for genotyped individuals; (2) (cross-sectional) information on ages at biospecimen collection for genotyped individuals; and (3) follow-up data on survival for non-genotyped individuals. In the simulation study presented in this section, we utilize only the first two sources. Of course, follow-up information for non-genotyped individuals provides an additional source for improving the power of genetic analyses but this simulation study illustrates that, even for the studies where genetic data are collected for all participants, the use of information on ages at biospecimen collection still makes a difference for the power of genetic analyses.

Let x_k^0, $k = 1,\ldots, K$, be the ages at baseline (entry to the study) of individuals from the genotyped subsample of the data and let x_{m,x_k^0}, $m = 1,\ldots, M_k$, be their ages at the time of biospecimen collection. Denote by $N\left(x_{m,x_k^0}\right) = N_1\left(x_{m,x_k^0}\right) + N_0\left(x_{m,x_k^0}\right)$ the number of individuals in the genotyped subsample who were aged x_{m,x_k^0} at the time of biospecimen collection and aged x_k^0 at baseline. Here $N_g\left(x_{m,x_k^0}\right)$ denotes the number of non-carriers ($g = 0$) and carriers ($g = 1$) of a specific allele/genotype. Let τ denote the life span (it may be censored). Denote by $\mu(x|G = g)$ the hazard rate for carriers/non-carriers and by $\pi\left(x_{m,x_k^0}|x_k^0\right) = P\left(G = 1|\tau > x_{m,x_k^0}, x_k^0\right)$ the proportion of carriers at age x_{m,x_k^0} given that the individuals were aged x_k^0 at baseline. Denote by $S_g(x) = P(\tau > x|G = g)$ the survival functions for carriers/non-carriers and by $P_1 = P(G = 1)$ the initial proportion (at birth) of carriers of the allele/genotype in a population, which is assumed here to be the same for different birth cohorts represented in the study. The total (population) survival function is then $S(x) = P_1 S_1(x) + (1 - P_1) S_0(x)$. Conditional survival functions for the individuals aged x_k^0 at the baseline are $S_g(x|x_k{}^0) = P(\tau > x|G = g, x_k{}^0)$. The hazard rates for carriers/non-carriers can be of any parametric form, e.g., the Gompertz curves, as in our simulations presented below. The proportions $\pi\left(x_{m,x_k^0}|x_k^0\right)$ are:

$$\pi\left(x_{m,x_k^0}|x_k^0\right) = \frac{P(G = 1|x_k^0) S_1\left(x_{m,x_k^0}|x_k^0\right)}{P(G = 1|x_k^0) S_1\left(x_{m,x_k^0}|x_k^0\right) + (1 - P(G = 1|x_k^0)) S_0\left(x_{m,x_k^0}|x_k^0\right)}, \tag{14.1}$$

where $P(G = 1|x_k{}^0) = P_1 S_1(x_k{}^0)/S(x_k{}^0)$.

The likelihood function of the data on the ages at biospecimen collection (L_A) and the likelihood function of the follow-up data (L_{FU}) are (Arbeev et al. 2011b):

$$L_A \sim \prod_{k=1}^{K} \prod_{m=1}^{M_k} \pi\left(x_{m,x_k^0}|x_k^0\right)^{N_1\left(x_{m,x_k^0}\right)} \left(1 - \pi\left(x_{m,x_k^0}|x_k^0\right)\right)^{N_0\left(x_{m,x_k^0}\right)} \tag{14.2}$$

and

$$L_{FU} \sim \prod_{k=1}^{K} \prod_{m=1}^{M_k} \prod_{g=0}^{1} \prod_{i=1}^{N_g\left(x_{m,x_k^0}\right)} \mu(\tau_i|G = g)^{\delta_i} S_g\left(\tau_i|x_{m,x_k^0}\right), \tag{14.3}$$

where δ_i is an at-risk indicator ($\delta_i = 1$ if τ is a death time; $\delta_i = 0$ if τ is a censoring time). The total likelihood function of the data relevant for genetic analyses of the genotyped subsample is the product of these two likelihood functions:

$$L_{FU+A} \sim L_{FU} L_A. \tag{14.4}$$

In our simulation studies, we compared two methods of estimating parameters of the allele- or genotype-specific hazard rates: (1) a method that uses only follow-up data, i.e., the likelihood function L_{FU} (14.3); and (2) a method that uses both data on the ages at biospecimen collection and follow-up data, i.e., the likelihood function L_{FU+A} (14.4).

We assumed that carriers and non-carriers of the allele in a population have mortality rates $\mu(x|G) = \mu_0(x)e^{\gamma G}$, where the variable G denotes carriers ($G = 1$) or non-carriers ($G = 0$), the baseline mortality $\mu_0(x)$ is the Gompertz function, i.e., $\ln \mu_0(x) = \ln a + bx$, with $\ln a = -10.0$ and $b = 0.09$, and the proportion of carriers at birth $P_1 = 0.25$. We varied the parameter γ from -0.5 to 0.5 with the interval 0.05 to simulate scenarios with different effect sizes.

We generated a "general population" of 10,000,000 individuals, assigning the genetic status (i.e., variable G) to individuals in accordance with the initial proportion P_1. Then we generated life spans for all individuals from the corresponding probability distributions (i.e., those corresponding to the hazard $\mu_0(x)e^{\gamma G}$ for carriers and $\mu_0(x)$ for non-carriers, with the parameters defined above). Then we assigned a hypothetical "age at entry" into the study to each individual in the population generated as a discrete random variable uniformly distributed over the interval 40–100 years. We assumed that individuals were genotyped at the baseline, i.e., their age at biospecimen collection coincided with age at entry. We collected a sample of 4500 individuals whose life spans exceeded their hypothetical "age at entry." We considered two scenarios: a short follow-up period (6 years) and a long follow-up period (60 years). Individuals with simulated life spans exceeding "age at entry" plus the follow-up period were considered censored at that age in the scenario. In the scenario with a long follow-up period almost all individuals experienced the event under study, whereas in the scenario with a short follow-up period a substantial proportion of individuals are censored. This procedure was repeated 1000 times (in each scenario with a different γ and follow-up period) to generate 1000 datasets whose parameters were estimated using the likelihoods (14.3) and (14.4).

Figure 14.1a illustrates the empirical power (i.e., the proportion of datasets in which the null hypothesis H_0: $\gamma = 0$ was rejected at $\alpha = 0.05$) in the scenario with a short follow-up for different effect sizes (i.e., values of the parameter γ). We also fitted these empirical values with the power curves of a one-sample Z-test of the mean and found the values of the standard deviations that produced the best fit to the empirical power curves for each method (0.059 for "FU + A" (14.4) and 0.088 for "FU" (14.3)). Figure 14.1b shows the level of the test (shown as $-\log_{10}(\alpha)$ for better visibility) that yields power $w = 0.8$, as a function of the effect size in both methods (the curves were calculated using the above mentioned values of standard deviations). Figure 14.1c, d display similar quantities for the scenario with a long follow-up period.

Figure 14.1a, b show that, in the case of a short follow-up period, the use of information on ages at biospecimen collection in addition to follow-up data substantially increases the power compared to the traditional approach that uses the follow-up data alone. For example, Fig. 14.1b shows that for effect size $\gamma = 0.3$ the p-value decreases approximately from 10^{-2} to 10^{-5} and for effect size $\gamma = 0.4$ the p-value drops approximately from 10^{-4} to 10^{-9}. This means that many genetic variants which would not reach genome-wide significance in genome-wide association studies (GWAS) using the traditional approach of analyzing the follow-up data alone could become highly significant if the data on ages at biospecimen collection were also used. Figure 14.1c, d reveal that this effect diminishes for a long follow-up period. In the case of a long follow-up period, information from the long follow-up makes a more substantial contribution compared to information hidden in the distributions of ages at biospecimen collection. Conversely, in the

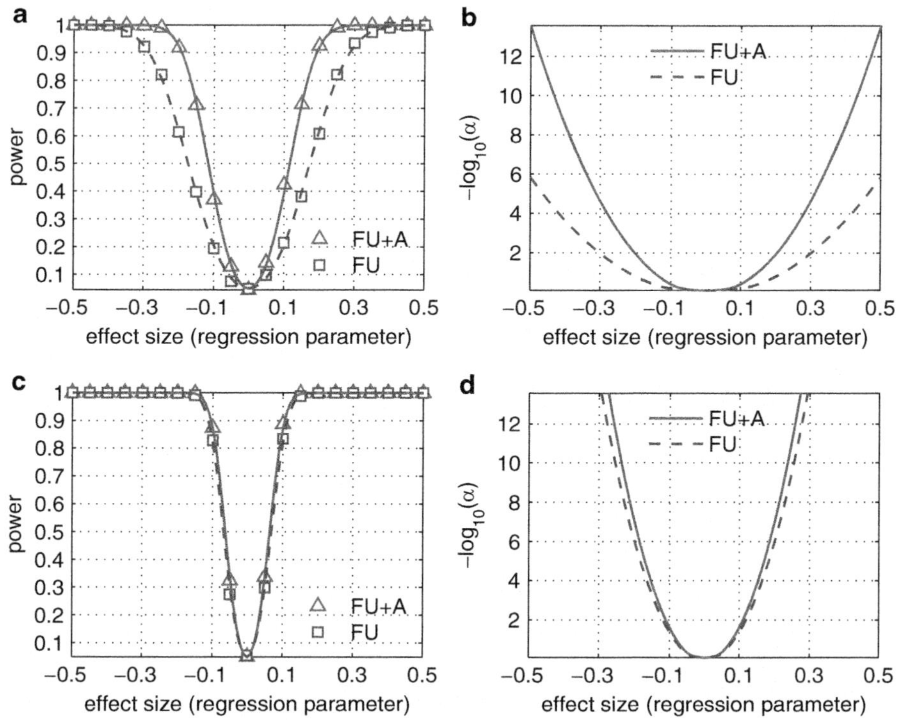

Fig. 14.1 Simulation studies in longitudinal genetic-demographic model: (**a**) Power in two methods (with follow-up only, "FU", and follow-up and ages at biospecimen collection, "FU+A") for different effect sizes (i.e., values of the regression parameter γ) and $\alpha = 0.05$ in the scenario with a short follow-up period (6 years). The lines denote the fit of the empirical curves by the power curves of a one-sample Z-test of the mean (the standard deviations that produced the best fit are 0.059 for "FU+A" and 0.088 for "FU"). (**b**) The level of the test (shown as $-\log_{10}(\alpha)$ for better visibility) that yields power $w = 0.8$, as a function of the effect size in both methods (the curves are calculated using the above mentioned values of standard deviations) in the scenario with a short follow-up period (6 years). (**c**) Is same as (**a**) but for a long follow-up period (60 years). The standard deviations that produced the best fit are 0.032 for "FU+A" and 0.035 for "FU." (**d**) Is same as (**b**) but for a long follow-up period (60 years)

case of a short follow-up period, the distribution of the ages at biospecimen collection plays a more important role in differentiating the allele- or genotype-specific survival patterns compared to the follow-up data (for the case of a substantial proportion of censored individuals, as in our simulations).

Our simulations thus illustrate that the use of information on ages at biospecimen collection may have important implications for GWAS of longevity or onset of diseases for studies with short follow-up periods (which are the majority of datasets currently available).

14.3 Simulation Studies in Genetic Stochastic Process Model

The genetic stochastic process model was originally developed in Arbeev et al. (2009). In Arbeev et al. (2014) we briefly introduced its version modified to include the dependence of the model's components on a vector of observed (time-independent) covariates available at baseline. In this section, we elaborate this modification in more detail and also describe simulation studies to illustrate the increase in the power of joint analyses of genotyped and non-genotyped participants in a longitudinal study compared to analyses of only genotyped participants in different scenarios to test relevant biologically-based hypotheses.

Let g, $g = 1,\ldots, G$, denote the presence of allele/genotype g in the genome of an individual. We can specify the probabilities of having this allele/genotype, p_g, conditional on some vector of time-independent covariates X. One possibility, for example, is to specify this probability using a multinomial logistic regression:

$$p_g = \frac{e^{\beta_{0g}+\beta_{1g}^T X}}{1+\sum_{c=1}^{G-1} e^{\beta_{0c}+\beta_{1c}^T X}}, \tag{14.5}$$

for $g = 1,\ldots, G–1$, and

$$p_G = \frac{1}{1+\sum_{g=1}^{G-1} e^{\beta_{0g}+\beta_{1g}^T X}}. \tag{14.6}$$

Here "T" denotes transposition (we will use column vectors if not stated otherwise).

Let Y_t (where t is age) be the stochastic process representing the age dynamics of an M-dimensional vector of biomarkers in carriers of allele/genotype g with the following stochastic differential equation:

$$dY_t = a(t,g,X)(Y_t - f_1(t,g,X))dt + B(t,g,X)dW_t, \tag{14.7}$$

with initial condition Y_{t_0}. Here W_t is an M-dimensional vector Wiener process independent of the vector of initial values Y_{t_0} which represents unobserved disturbances affecting the trajectory of biomarkers. The strength of such disturbances is characterized by the $M \times M$ matrix of diffusion coefficients $B(t,g,X)$. The vector-function $f_1(t,g,X)$ (having the same dimension as Y_t) introduces the notion of allostasis into the model representing the age trajectories of biomarkers that organisms are forced to follow by the process of allostatic adaptation (see detailed description of the meaning of different components of the stochastic process model in Chap. 12 and Arbeev et al. (2009)). The negative feedback coefficient in Eq. (14.7), the $M \times M$ matrix $a(t, g, X)$, describes the adaptive (homeostatic) capacity in an aging organism. The elements of this matrix correspond to the rate of adaptive response to any deviation of trajectories Y_t from the trajectories $f_1(t, g, X)$.

The hazard rates for carriers of allele/genotype g conditional on the vector of biomarkers Y_t and the vector of observed covariates X are given as:

$$\mu(t|Y_t, g, X) = \mu_0(t, g, X) + (Y_t - f_0(t, g, X))^T Q(t, g, X)(Y_t - f_0(t, g, X)). \quad (14.8)$$

Here $\mu_0(t, g, X)$ is the baseline hazard for carriers of allele/genotype g characterizing the risk that would remain if the vector Y_t followed the trajectory $f_0(t, g, X)$, and $Q(t, g, X)$ is a non-negative-definite symmetric $M \times M$ matrix. The M-dimensional vector-function $f_0(t, g, X)$ introduces the concept of an age-dependent physiological norm into the model which corresponds to the values of biomarkers which minimize the risk at each age for carriers of allele/genotype g. The matrix $Q(t, g, X)$ in the quadratic hazard term can be associated with the decline in resistance to stresses with age, as discussed in Yashin et al. (2007a, 2012) and Arbeev et al. (2011a).

Although the model (14.5), (14.6), (14.7) and (14.8) looks similar to the latent class stochastic process model presented in Chap. 13, there is one important distinction: in the model presented in this chapter information on g (i.e., the presence of some allele/genotype) is assumed to be available for genotyped participants in the longitudinal study, whereas information on the latent class g is not available for any individual in the model in Chap. 13 (hence its name). Therefore, the likelihood estimation procedure is different from that presented in Chap. 13. The likelihood function for the model (14.5), (14.6), (14.7) and (14.8) is a straightforward modification of the likelihood for the original model in Chap. 12 (see Arbeev et al. (2009)) and is not presented here. Note that the likelihood function contains parts for the genotyped and non-genotyped sub-samples and that both parts contain the same parameters of the model. Hence, the use of available information from the non-genotyped participants (i.e., the longitudinal measurements of biomarkers and time-to-event data) provides an opportunity for increasing the statistical power compared to analyses based on the genotyped sample alone. The advantage of the genetic stochastic process model is that it has different components which represent specific biological concepts and aging-related mechanisms for which the respective parameters have clear biological interpretations. Dependence of the model's components on variable g allows for formulating and testing different hypotheses about the genetic effect of the alleles/genotypes on aging-related characteristics (such as stress resistance, adaptive capacity, age-dependent physiological norms, etc.). Below we present the results of a simulation study that compares the power for tests of several such hypotheses using two approaches: (1) using only information from the genotyped participants; and (2) in joint analyses of genotyped and non-genotyped individuals.

We used the following specifications of the model's components in the simulations:

Table 14.1 Simulation studies of the genetic stochastic process model: Parameters used to generate data (parameters used to define the null hypotheses to be tested in each simulation are ***highlighted***)

Simulation	G	Baseline hazard ($\mu_0(t, g, X)$)			Quadr. hazard ($Q(t, g, X)$)		Adaptive capacity ($a(t, g, X)$)		Mean allostatic trajectory ($f_1(t, g, X)$)		Physiological norm ($f_0(t, g, X)$)		Other parameters		
		$\ln a^g_{\mu_0}$	$b^g_{\mu_0}$	β^g_X	a^g_Q	b^g_Q	a^g_Y	b^g_Y	$a^g_{f_1}$	$b^g_{f_1}$	$a^g_{f_0}$	$b^g_{f_0}$	σ^g_0	σ^g_1	p_1
1	1	***−9.0***	***0.080***		0.5	0.1	−0.25	1.0	45.0	0.20	45.0	0.1	5.0	4.0	0.25
	2	***−8.5***	***0.082***		0.3	0.1	−0.20	1.0	50.0	0.25	40.0	0.1	5.0	4.0	
2	1	−9.0	0.080	***−0.014***	0.5	0.1	−0.25	1.0	45.0	0.20	45.0	0.1	5.0	4.0	0.25
	2	−8.5	0.082	***−0.014***	0.3	0.1	−0.20	1.0	50.0	0.25	40.0	0.1	5.0	4.0	
3	1	−9.0	0.080		***0.5***	***0.1***	−0.25	1.0	45.0	0.20	45.0	0.1	5.0	4.0	0.25
	2	−8.5	0.082		***0.5***	***0.4***	−0.20	1.0	50.0	0.25	40.0	0.1	5.0	4.0	
4	1	−9.0	0.080		0.5	0.1	***−0.22***	***1.0***	45.0	0.20	45.0	0.1	5.0	4.0	0.25
	2	−8.5	0.082		0.3	0.1	***−0.20***	***1.0***	50.0	0.25	40.0	0.1	5.0	4.0	
5	1	−9.0	0.080		0.5	0.1	−0.25	1.0	***45.0***	***0.20***	45.0	0.1	5.0	4.0	0.25
	2	−8.5	0.082		0.3	0.1	−0.20	1.0	***46.0***	***0.20***	40.0	0.1	5.0	4.0	
6	1	−9.0	0.080		0.5	0.1	−0.25	1.0	45.0	0.20	***50.0***	***0.1***	5.0	4.0	0.25
	2	−8.5	0.082		0.3	0.1	−0.20	1.0	50.0	0.25	***40.0***	***0.1***	5.0	4.0	

Notes:
(1) Some parameters are rescaled for better visibility in the table: a^g_Q is multiplied by 10^4; b^g_Q is multiplied by 10^5; b^g_Y is multiplied by 10^3

1. Gompertz baseline hazards: $\ln \mu_0(t, g, X) = \ln a^g_{\mu_0} + b^g_{\mu_0} t + \beta^g_X X$, where $g = 1$, 2 for carriers and non-carriers of a hypothetical allele (genotype), $X = c - c_0$, c is year of birth (cohort), $c_0 = 1890$, in simulation #2 (see Table 14.1) and $\ln \mu_0(t, g, X) = \ln a^g_{\mu_0} + b^g_{\mu_0} t$ in the other simulations;
2. Linear functions for the multipliers in the quadratic hazard: $Q(t, g, X) = a^g_Q + b^g_Q t$;
3. Linear functions for the mean allostatic trajectories: $f_1(t, g, X) = a^g_{f_1} + b^g_{f_1} t$;
4. Linear functions for physiological norms: $f_0(t, g, X) = a^g_{f_0} + b^g_{f_0} t$;
5. Linear functions for the negative feedback coefficient in (14.7) representing the adaptive capacity of an organism: $a(t, g, X) = a^g_Y + b^g_Y t$, with $a^g_Y \le 0$ and $b^g_Y \ge 0$;
6. Constant diffusion coefficients: $B(t, g, X) = \sigma^g_1$;
7. Normally distributed initial values of the process Y_t with means $f_1(t_0^j, g, X)$ (where t_0^j is age at the first exam for the jth individual) and standard deviations ${\sigma_0}^g$; and
8. Initial probability of carrying the allele/genotype (p_1) is independent of covariates X.

The values of the parameters were chosen to provide realistic samples resembling real data on mortality in the Framingham Original Cohort data (Dawber et al. 1951) and with longitudinal dynamics Y_t mimicking pulse pressure. Table 14.1 contains a summary of the parameters used in the simulation studies.

We performed six simulation studies for testing different biological hypotheses of genetic effects on aging-related characteristics (see the "**Null Hypothesis**" and "**Interpretation of Null Hypothesis**" columns in Table 14.2). In each scenario, we simulated 100 datasets with data on age at death/censoring and the longitudinal dynamics of Y_t for 2500 individuals followed-up for 60 years with ages at baseline uniformly distributed over the interval 30–60 years and with 30 biennial exams measuring Y_t. Year of birth c for simulation #2 was defined as 1950 minus age at baseline. We assumed that 500 individuals were genotyped and genetic data were not available for the rest of the sample. Power was estimated as the proportion of datasets in which a null hypothesis was rejected at the 0.05 significance level by the likelihood ratio test (see Table 14.2). For these purposes, we estimated original/unrestricted models and restricted models that assume that the parameters highlighted in Table 14.1 are equal for carriers and non-carriers (simulation #2 assumes the restriction $\beta^g_X = 0$). The column "**Gen. Only**" in Table 14.2 corresponds to the likelihood that used only information from the genotyped participants and column "**Gen. + Non-Gen.**" displays the power for the likelihood with joint analyses of the genotyped and non-genotyped individuals. The table shows that joint analysis of the genotyped and non-genotyped individuals allows for a substantial increase in the power compared to analyses based on information from the genotyped participants alone, thus making it possible to detect genetic effects on aging-related characteristics that would remain non-significant in analyses of the genotyped subsample.

Table 14.2 Simulation studies of the genetic stochastic process model: Power (for $\alpha = \mathbf{0.05}$ and effect sizes defined by the parameters from Table 14.1) for estimation of the likelihood only using data on genotyped individuals (column "**Gen. Only**") and data on both genotyped and non-genotyped individuals (column "**Gen. + Non-Gen.**")

Simulation	Null hypothesis	Interpretation of null hypothesis	Power	
			Gen. only	Gen. + Non-gen
1	$\mu_0(t, g, X) = \mu_0(t, X)$	No genetic effect on baseline hazard	0.42	0.85
2	$\mu_0(t, g, X) = \mu_0(t, g)$	No cohort changes in baseline hazard	0.25	0.89
3	$Q(t, g, X) = Q(t, X)$	No genetic effect on stress resistance	0.41	0.90
4	$a(t, g, X) = a(t, X)$	No genetic effect on adaptive capacity	0.40	0.82
5	$f_1(t, g, X) = f_1(t, X)$	No genetic effect on mean allostatic trajectory	0.71	0.89
6	$f_0(t, g, X) = f_0(t, X)$	No genetic effect on physiological norm	0.44	0.91

14.4 Discussion

In this chapter, we presented different approaches that can be applied in genetic biodemography to work with data available in modern longitudinal studies of aging, health, and longevity that collect genetic information in addition to follow-up data on events and longitudinal measurements of biomarkers.

The longitudinal genetic-demographic model described in Sect. 14.2 (see also Arbeev et al. 2011b) provides a method for enhancing genetic analyses of time-to-event outcomes from longitudinal data combining several sources of information: follow-up data on the outcome of interest (e.g., mortality) for genotyped individuals, information on the age structure of the population at the time of biospecimen collection, and follow-up data on the outcomes for non-genotyped participants. Such joint analyses of genotyped and non-genotyped individuals can result in substantial improvements in statistical power and accuracy of estimates compared to analyses of the genotyped subsample alone if the proportion of non-genotyped participants is large. Situations in which genetic information cannot be collected for all participants of longitudinal studies are not uncommon. They can arise for several reasons: (1) the longitudinal study may have started some time before genotyping was added to the study design so that some initially participating individuals dropped out of the study (i.e., died or were lost to follow-up) by the time of genetic data collection; (2) budget constraints prohibit obtaining genetic information for the entire sample; (3) some participants refuse to provide samples for genetic analyses. Nevertheless, even when genotyped individuals constitute a majority of the sample or the entire sample, application of such an approach is still beneficial in terms of estimation accuracy and power because it takes into account the population structure at the time of biospecimen collection which has additional information on genetic effects on the risk of death complementing the follow-up data (Yashin et al. 2013b). Clearly, any statistical model is just an approximation of reality and the use of even the most advanced models does not replace the need to collect large-scale genetic data in longitudinal studies. The genetic-demographic model presented in Sect. 14.2 and in Arbeev et al. (2011b) uses parametric specifications of allele- or genotype-specific survival functions. More flexible specifications, such as semiparametric and non-parametric models or methods that correct for unobserved heterogeneity effects, can be formulated and estimated (see Yashin et al. 1999).

The genetic stochastic process model presented in Sect. 14.3 adds a new dimension to genetic biodemographic analyses, combining information on longitudinal measurements of biomarkers available for participants of a longitudinal study with follow-up data and genetic information. Such joint analyses of different sources of information collected in both genotyped and non-genotyped individuals allow for more efficient use of the research potential of longitudinal data which otherwise remains underused when only genotyped individuals or only subsets of available information (e.g., only follow-up data on genotyped individuals) are involved in analyses. Similar to the longitudinal genetic-demographic model presented in Sect. 14.2, the benefits of combining data on genotyped and non-genotyped individuals in the genetic SPM come from the presence of common parameters describing characteristics of the model for genotyped and non-genotyped subsamples of the data. This takes into account the knowledge that the non-genotyped subsample is a mixture of carriers and non-carriers of the same alleles or genotypes represented in the genotyped subsample and applies the ideas of heterogeneity analyses (Vaupel and Yashin 1985). When the non-genotyped subsample is substantially larger than the genotyped subsample, these joint analyses can lead to a noticeable increase in the power of statistical estimates of genetic parameters compared to estimates based only on information from the genotyped subsample. This approach is applicable not only to genetic data but to any discrete time-independent variable that is observed only for a subsample of individuals in a longitudinal study.

The genetic stochastic process model enhances biodemographic analyses by allowing for hidden components of aging (such as age-specific physiological norms, allostasis and allostatic load, decline in adaptive capacity, and stress resistance with age) that are typically not directly measured in longitudinal data and, hence, can be estimated only indirectly. Different components and mechanisms characterizing the same process of aging should be mutually dependent and work in concert. Therefore, unification of such concepts in a comprehensive model of aging is an important step forward in the development of a systemic methodology in aging research. As in the original stochastic process model, the genetic SPM allows working with several mechanisms of aging-related changes under the overarching framework of one statistical model. In addition, the genetic SPM evaluates genetic effects on such mechanisms, thus providing deeper insights into genetic determinants of the processes of aging affecting mortality and morbidity risks. It permits one to address new questions in biodemographic analyses concerning genetic influences on aging-related changes in humans, questions that cannot be studied using conventional approaches, for example, joint models (see Chap. 11) or standard demographic methods. Simulations in Sect. 14.3 provided several examples of hypotheses that can be tested using the genetic SPM and illustrated the differences in statistical power resulting from the addition of information on non-genotyped individuals to the analyses.

Several practical considerations should be mentioned about applications of the genetic SPM to real data. As with any parametric model, the genetic SPM relies on the description of its components as specific parametric functions. Although the basic components of the model (such as the quadratic shape of the hazard, physiological norm, average allostatic trajectory, negative feedback coefficient) are all based on solid biological theories that justify their presence in the model, the specific parametric forms of these components are unknown and may be hard to justify biologically. Moreover, the parametric forms of these components generally cannot be empirically evaluated, because they model hidden components of the aging process not directly associated with any measurable variables in the data (one exception might be the baseline hazard rate which, with some degree of confidence, can be assumed to have the same shape as the hazard rate in the total population, e.g., Gompertz, Weibull, gamma-Gompertz, or gamma-Weibull baseline rates can be chosen depending on the application). Therefore, it is advisable to perform sensitivity analyses with different parametric specifications of the components of the model, e.g., linear, quadratic, or higher order polynomial functions, and select the best fitting model using formal criteria such as the likelihood ratio test for nested models or the Akaike Information Criterion for non-nested models.

In brief, the specific types of genetic influences on the hidden components of aging are not known a priori. Thus, versions of the model with different types of genetic influences should be tested in applications. For example, dominant, recessive, or additive form of action of the minor allele on the outcome characteristics can be investigated. Similarly, joint analyses of two or more genetic markers might be of interest in applications. The genetic SPM can be straightforwardly extended to work with multiple genetic markers. However, this results in a larger number of parameters and a smaller number of individuals in different groups that can reduce the reliability of estimates.

The computational burden should always be taken into account in practical implementations of statistical methods, especially in large-scale problems involving studies with large sample sizes and/or extensive amounts of genetic data. For example, genome-wide association studies (GWAS) data are collected in different longitudinal studies that can contain millions of single nucleotide polymorphisms (SNPs) for thousands of participants (see the dbGaP website, http://www.ncbi.nlm.nih.gov/gap?db=gap). For such data, the computational burden of parameter estimation in the genetic SPM suggests that its routine application to each SNP in the dataset is not feasible for modern computers, especially in high dimensional cases. At the present time, a more relevant application of this model is to work with a much smaller set of SNPs pre-selected according to some criterion (Yashin et al. 2013a). The likelihood estimation procedure in the longitudinal genetic-demographic model is considerably faster and, therefore, it is suitable for large-scale applications. Our experience with the version of the model by Arbeev et al. (2011b) indicates that estimation of GWAS data on thousands of individuals and more than a hundred thousand SNPs can be performed in a reasonable time. Nevertheless, both the genetic SPM and longitudinal genetic-demographic models can be used in studies with candidate genes or SNPs to investigate their connections with mortality risk and risks of diseases and to evaluate genetic contributions to hidden components of aging that affect these risks.

Several further generalizations of the methods to evaluate genetic influences on hidden components of aging can be considered. As discussed in Chap. 13 and Yashin et al. (2008), ignoring hidden heterogeneity in a population due to the presence of latent subpopulations defined by some unobserved characteristics can lead to erroneous conclusions concerning biological regularities of aging-related processes estimated by the stochastic process model. The same, of course, is true for the genetic SPM. Therefore, the generalization of the genetic SPM to include latent classes can be useful for sensitivity analyses to test the presence of hidden heterogeneity that can affect the results of the genetic SPM.

Another direction for possible extension of the genetic SPM is the "individualization" of longitudinal trajectories. In its present form, all individuals in the model have the same ("population") parameters of the adaptive capacity and the allostatic trajectory. The parameters of these components can be specified as random variables or realizations of some stochastic process to describe individual patterns of adaptive capacity and the allostatic load. Since such additional random effects and the "original" random process (i.e., the Wiener process W_t in the equation for the dynamics of the longitudinal biomarker Y_t (14.7)) may "compete" for the same correlation structure in the longitudinal data, the feasibility of such an approach needs careful investigation. See also relevant discussions of the use of complicated random effects structures vs. the use of stochastic processes in the joint models literature (Rizopoulos 2012; Tsiatis and Davidian 2004).

Investigation of genetic effects on hidden components of aging and their relation to risks of death and onset of diseases can also be performed in the framework of extended versions of the stochastic process model aimed at analyses of dependent competing risks (Akushevich et al. 2011; Manton et al. 1992; Yashin et al. 1986), using longitudinal data on individual health histories and mortality (Yashin et al. 2011a), which may be collected using different observational plans (Yashin et al. 2011b). Such analyzes would allow addressing many new problems that cannot be investigated using standard approaches. For example, the role of genetic factors in competing risks of death can be detected without the traditional assumption of independent risks for different causes of death, how genes affect hidden mechanisms of aging manifested in the longitudinal dynamics of physiological variables can be investigated, and their relation to these dependent competing risks can be explored. The introduction of jumping components describing health states in the model allows for comprehensive analyses of genetic effects on both fast changes in health status and slower changes in the physiological state of an organism associated with aging processes, and their effects on mortality. This can help in uncovering pre-disease physiological pathways and differences in aging-related characteristics among carriers of different alleles or genotypes. Generalized stochastic process models with jumping components are discussed in Chaps. 15 and 16.

Acknowledgements This chapter was partly supported by the National Institute on Aging of the National Institutes of Health under Award Numbers R01AG030612, R01AG046860, P01AG043352, and P30AG034424. The content is solely the responsibility of the authors and does not necessarily represent the official views of the National Institutes of Health.

References

Akushevich, I., Arbeev, K., Ukraintseva, S., & Yashin, A. (2011). Theory of individual health histories and dependent competing risks. In *JSM proceedings, section on risk analysis* (pp. 5385–5399).

Arbeev, K. G., Akushevich, I., Kulminski, A. M., Arbeeva, L. S., Akushevich, L., Ukraintseva, S. V., Culminskaya, I. V., & Yashin, A. I. (2009). Genetic model for longitudinal studies of aging, health, and longevity and its potential application to incomplete data. *Journal of Theoretical Biology, 258*, 103–111.

Arbeev, K. G., Ukraintseva, S. V., Akushevich, I., Kulminski, A. M., Arbeeva, L. S., Akushevich, L., Culminskaya, I. V., & Yashin, A. I. (2011a). Age trajectories of physiological indices in relation to healthy life course. *Mechanisms of Ageing and Development, 132*, 93–102.

Arbeev, K. G., Ukraintseva, S. V., Arbeeva, L. S., Akushevich, I., Kulminski, A. M., & Yashin, A. I. (2011b). Evaluation of genotype-specific survival using joint analysis of genetic and non-genetic subsamples of longitudinal data. *Biogerontology, 12*, 157–166.

Arbeev, K. G., Ukraintseva, S. V., Kulminski, A. M., Akushevich, I., Arbeeva, L. S., Culminskaya, I. V., Wu, D., & Yashin, A. I. (2012). Effect of the APOE polymorphism and Age trajectories of physiological variables on mortality: Application of genetic stochastic process model of aging. *Scientifica, 2012*, 568628.

Arbeev, K. G., Akushevich, I., Kulminski, A. M., Ukraintseva, S., & Yashin, A. I. (2014). Joint analyses of longitudinal and time-to-event data in research on aging: Implications for predicting health and survival. *Frontiers in Public Health, 2*, 228.

Carey, J. R. (2008). Biodemography: Research prospects and directions. *Demographic Research, 19*, 1749–1757.

Chen, L. M., Ibrahim, J. G., & Chu, H. (2011). Sample size and power determination in joint modeling of longitudinal and survival data. *Statistics in Medicine, 30*, 2295–2309.

Crimmins, E., Kim, J. K., & Vasunilashorn, S. (2010). Biodemography: New approaches to understanding trends and differences in population health and mortality. *Demography, 47*, S41–S64.

Dawber, T. R., Meadors, G. F., & Moore, F. E. (1951). Epidemiological approaches to heart disease: The Framingham study. *American Journal of Public Health, 41*, 279–286.

Manton, K. G., Stallard, E., & Singer, B. (1992). Projecting the future size and health status of the United States elderly population. *International Journal of Forecasting, 8*, 433–458.

Rizopoulos, D. (2012). Joint models for longitudinal and time-to-event data with applications in R. Boca Raton: Chapman and Hall/CRC.

Suzman, R. (2010). Prologue: Research on the demography and economics of aging. *Demography, 47*, S1–S4.

Tsiatis, A. A., & Davidian, M. (2004). Joint modeling of longitudinal and time-to-event data: An overview. *Statistica Sinica, 14*, 809–834.

Vaupel, J. W., & Yashin, A. I. (1985). Heterogeneity's ruses: Some surprising effects of selection on population dynamics. *American Statistician, 39*, 176–185.

Wachter, K. W. (2008). Biodemography comes of age. *Demographic Research, 19*, 1501–1512.

Yashin, A. I., Manton, K. G., & Stallard, E. (1986). Dependent competing risks: A stochastic process model. *Journal of Mathematical Biology, 24*, 119–140.

Yashin, A. I., De Benedictis, G., Vaupel, J. W., Tan, Q., Andreev, K. F., Iachine, I. A., Bonafe, M., Deluca, M., Valensin, S., Carotenuto, L., & Franceschi, C. (1999). Genes, demography, and life span: The contribution of demographic data in genetic studies on aging and longevity. *American Journal of Human Genetics, 65*, 1178–1193.

Yashin, A. I., De Benedictis, G., Vaupel, J. W., Tan, Q., Andreev, K. F., Iachine, I. A., Bonafe, M., Valensin, S., De Luca, M., Carotenuto, L., & Franceschi, C. (2000). Genes and longevity: Lessons from studies of centenarians. *Journals of Gerontology. Series A, Biological Sciences and Medical Sciences, 55*, B319–B328.

Yashin, A. I., Arbeev, K. G., Akushevich, I., Kulminski, A., Akushevich, L., & Ukraintseva, S. V. (2007a). Stochastic model for analysis of longitudinal data on aging and mortality. *Mathematical Biosciences, 208*, 538–551.

Yashin, A. I., Arbeev, K. G., & Ukraintseva, S. V. (2007b). The accuracy of statistical estimates in genetic studies of aging can be significantly improved. *Biogerontology, 8*, 243–255.

Yashin, A. I., Arbeev, K. G., Akushevich, I., Kulminski, A., Akushevich, L., & Ukraintseva, S. V. (2008). Model of hidden heterogeneity in longitudinal data. *Theoretical Population Biology, 73*, 1–10.

Yashin, A. I., Akushevich, I., Arbeev, K. G., Kulminski, A., & Ukraintseva, S. (2011a). Joint analysis of health histories, physiological states, and survival. *Mathematical Population Studies, 18*, 207–233.

Yashin, A. I., Akushevich, I., Arbeev, K. G., Kulminski, A., & Ukraintseva, S. V. (2011b). New approach for analyzing longitudinal data on health, physiological state, and survival collected using different observational plans. In *JSM proceedings, section on government statistics* (pp. 5336–5350).

Yashin, A. I., Arbeev, K. G., Akushevich, I., Kulminski, A., Ukraintseva, S. V., Stallard, E., & Land, K. C. (2012). The quadratic hazard model for analyzing longitudinal data on aging, health, and the life span. *Physics of Life Reviews, 9*, 177–188.

Yashin, A. I., Arbeev, K. G., Wu, D., Arbeeva, L. S., Kulminski, A., Akushevich, I., Culminskaya, I., Stallard, E., & Ukraintseva, S. (2013a). How lifespan associated genes modulate aging changes: Lessons from analysis of longitudinal data. *Frontiers in Genetics, 4*, 3.

Yashin, A. I., Arbeev, K. G., Wu, D., Arbeeva, L. S., Kulminski, A. M., Akushevich, I., Culminskaya, I., Stallard, E., & Ukraintseva, S. (2013b). How the quality of GWAS of human lifespan and health span can be improved. *Frontiers in Genetics, 4*, 125.

Chapter 15
Integrative Mortality Models with Parameters That Have Biological Interpretations

Anatoliy I. Yashin, Igor Akushevich, Konstantin G. Arbeev, Alexander M. Kulminski, and Svetlana V. Ukraintseva

15.1 Introduction

Mortality rates are important characteristics of life span distributions that integrate the influences of many external and internal factors affecting individuals during their life course. These include the ontogenetic program, individual aging processes, exposure to external (environmental) and internal (biological) factors, and changes in health status, as well as effects of compensatory adaptation to damages and changes induced by all these processes. Various parametric models of human mortality rates are used in analyses of survival data in demographic and epidemiological applications, experimental studies of aging and longevity using laboratory animals, etc.

Despite an existing tradition of interpreting differences in the shapes or parameters of the mortality rates (survival functions) resulting from the effects of exposure to different conditions or other interventions in terms of characteristics of individual aging, this practice has to be used with care. This is because such characteristics are difficult to interpret in terms of properties of external and internal processes affecting the chances of death. An important question then is: What kind of mortality model has to be developed to obtain parameters that are biologically interpretable? The purpose of this chapter is to describe an approach to mortality modeling that represents mortality rates in terms of parameters of physiological changes and declining health status accompanying the process of aging in humans. In contrast to traditional demographic and actuarial models dealing with mortality data, the proposed model is appropriate for analyses of longitudinal data on aging, health, and longevity. We use a diffusion-type continuous-time stochastic process for describing the evolution of physiological states over the life course, and a finite-state continuous-time process for describing changes in health status during this period. We derive equations for the resulting mortality models, and approximate changes in physiological states by a Gaussian process conditional on health status.

A.I. Yashin et al., *Biodemography of Aging*, The Springer Series on Demographic Methods and Population Analysis 40, DOI 10.1007/978-94-017-7587-8_15

These equations will be used in Chap. 16 in the joint analyses of data collected using different observational plans.

Could a demographic mortality model be developed whose parameters can characterize aging-related declines in health status and physiological/biological functioning? To address this question, we specify the conditional mortality rate as a function of health status and physiological/biological state, describe age-related changes in health and physiological variables, and perform averaging with respect to all unobserved characteristics. A traditional (demographic) description of changes in individual health/survival status is performed using a continuous-time random Markov process with a finite number of states, and age-dependent transition intensity functions (transitions rates). Transitions to the absorbing state are associated with death, and the corresponding transition intensity is a mortality rate. Although such a description characterizes connections between health and mortality, it does not allow for studying factors and mechanisms involved in the aging-related health decline. Numerous epidemiological studies provide compelling evidence that health transition rates are influenced by a number of factors. Some of them are fixed at the time of birth (e.g., genetic background). Others experience stochastic changes over the life course. Examples include variables describing physiological states, behavioral, or socio-economic factors, etc. The presence of such randomly changing influential factors violates the Markov assumption, and makes the description of aging-related changes in health status more complicated.

The age dynamics of influential factors (e.g., physiological variables) in connection with mortality risks has been described using a stochastic process model of human mortality and aging (Woodbury and Manton 1977; Yashin 1985; Yashin and Manton 1997). Recent extensions of this model have been used in analyses of longitudinal data on aging, health, and longevity, collected in the Framingham Heart Study (Akushevich et al. 2005; Arbeev et al. 2009, 2011; Yashin et al. 2008, 2007b). This model and its extensions are described in terms of a Markov stochastic process satisfying a diffusion-type stochastic differential equation. The stochastic process is stopped at random times associated with individuals' deaths. The quadratic hazard assumption about the form of the conditional mortality function, given covariates values and certain regularity conditions, guarantees the Gaussian property of the conditional distribution of the covariates value at any given age. This yielded a description of the aging-related changes in terms of the first two moments of a multidimensional Gaussian distribution. When an individual's health status is taken into account, the coefficients of the stochastic differential equations become dependent on values of the jumping process. This dependence violates the Markov assumption and renders the conditional Gaussian property invalid. So the description of this (continuously changing) component of aging-related changes in the body also becomes more complicated.

Since studying age trajectories of physiological states in connection with changes in health status and mortality would provide more realistic scenarios for analyses of available longitudinal data, it would be a good idea to find an appropriate mathematical description of the joint evolution of these interdependent processes in aging organisms. For this purpose, we propose a comprehensive model of human aging, health, and mortality in which the Markov assumption is fulfilled by a two-component stochastic process consisting of jumping and continuously changing processes. The jumping component is used to describe relatively fast changes in health status occurring at random times, and the continuous component describes relatively slow stochastic age-related changes of individual physiological states.

15.2 Conditional Risk of Death and Demographic Mortality Rate

Let $\mu_{\theta_t}(Y_t, Z, G, t)$ be the risk of death conditional on the values of the stochastic processes θ_t and Y_t and random variables (observed covariates) Z, G, and age t. Let T be a non-negative random variable describing the life span. In our applications, the processes θ_t and Y_t describe health status and physiological state, respectively, and the variables Z and G correspond to static (age independent) covariates and fixed genetic factors, respectively. Let $\overline{\overline{\mu}}(t)$ be the demographic mortality rate in the population cohort whose individuals experience the influences of dynamic θ_t, Y_t, and static factors Z, G, and age t. The correspondence between $\mu_{\theta_t}(Y_t, Z, G, t)$ and $\overline{\overline{\mu}}(t)$ extends the relationship well known in analyses of heterogeneous populations, see Vaupel et al. (1979), Vaupel and Yashin (1985), and Yashin and Manton (1997), among others:

$$\overline{\overline{\mu}}(t) = E\big(\mu_{\theta_t}(Y_t, Z, G, t)|T > t\big). \tag{15.1}$$

Here the symbol E denotes the operation of mathematical expectation and $E\big(\mu_{\theta_t}(Y_t, Z, G, t)|T > t\big)$ is the result of this operation applied to $\mu_{\theta_t}(Y_t, Z, G, t)$ conditional on the event $\{T > t\}$ (life span is larger than t). Such conditioning means that the average mortality rate at age t in the population cohort is defined for only those individuals who survived to this age. Practical implementation of Eq. (15.1) involves three steps. First, the stochastic processes θ_t and Y_t, the random variables Z and G, and the function $\mu_{\theta_t}(Y_t, Z, G, t)$ have to be described. At this step, available information about aging, health, and longevity accumulated in the research field and relevant to the research problems of the study can be incorporated into the mortality model. Second, a parametric description of the demographic mortality rate $\overline{\overline{\mu}}(t)$ has to be obtained by performing the conditional averaging in (15.1). This procedure depends on the specification of the components θ_t and Y_t of the stochastic process, random variables Z and G, as well as the function $\mu_{\theta_t}(Y_t, Z, G, t)$, and may require the use of approximation techniques. Third, the model parameters are estimated using available data and statistical hypotheses about strength of the specified relationships and the effects of corresponding variables on risks of disease and death are tested. The parameter estimation procedure is performed using maximization of the likelihood of the observed data. The procedure allows for testing statistical hypotheses about parameter values using likelihood ratio tests.

The product of the first two steps is a representation of the mortality model (15.1) in terms of parameters characterizing the stochastic processes θ_t and Y_t, the random variables Z and G, as well as the dependence of the conditional hazard rate μ_{θ_t} (Y_t, Z, G, t) on these factors and age t. When these processes and variables describe each person's health status, physiological state, and genetic and non-genetic risk factors, then the parameters of the model have a biological interpretation. The possibility of such an interpretation opens a unique opportunity for obtaining new insights into mechanisms connecting health and survival outcomes with aging-related changes in biomarkers. It also allows for further integration of the research findings obtained in related disciplines. The third step allows one to make conclusions about dynamic aspects of aging-related changes, the contribution of each component of the model to risks of disease and death, and other key issues related to the regulation of aging, health, and longevity from available data.

15.3 Description of the Processes θ_t and Y_t and Their Connections to t

In addition to the demographic mortality rate, interest centers on estimating parameters of mortality rates conditional on observed covariates Z and G:

$$\overline{\overline{\mu}}(Z, G, t) = E\big(\mu_{\theta_t}(Y_t, Z, G, t)|T > t, Z, G\big). \tag{15.2}$$

In this case, one has to further specify the processes θ_t (health status) and Y_t (physiological states) conditional on Z (fixed covariates) and G (fixed genetic factors). Let $\theta_t, t \geq 0$ be the finite-state (jumping) continuous-time stochastic process (i.e., $\theta_t \in \{1,2,\ldots,M\}$, where M is the number of states), and let $Y_t, t \geq 0$ be a K-dimensional stochastic process with continuous components, where K is the number of physiological states that are monitored. We assume that Y_t satisfies a stochastic differential equation with coefficients depending on θ_t:

$$dY_t = A_{\theta_t}(Y_t, Z, G, t)dt + B_{\theta_t}(Z, G, t)dW_t, \quad Y_{t_0}. \tag{15.3}$$

Here $A_{\theta_t}(Y_t, Z, G, t)$ is a K-dimensional vector function, $B_{\theta_t}(Z, G, t)$ is a matrix of corresponding dimension, Y_{t_0} is a random vector of initial conditions, and W_t is a p-vector Wiener process with independent components that is independent of the initial value, Y_{t_0}.

The finite-state continuous-time process $\theta_t, t \geq 0$, describing jumping changes in health/well-being status is characterized by a conditional transition intensity matrix (from state k to state r) with elements:

$$\lambda_{kr}(Y_t,\theta_t,Z,G,t),k,r=1,2,\ldots M;\ \text{and}\ \lambda_{kk}(Y_t,\theta_t,Z,G,t)= -\sum\nolimits_{r=1,y\neq k}^{M}\lambda_{kr}(Y_t,\theta_t,Z,G,t) \tag{15.4}$$

with initial probabilities $P(\theta_{t_0}=j|Z,G,T>t_0),j=1,2,\ldots M$.

Let T be a non-negative random variable describing life span. Its distribution characterizes the variability in life span among individuals in human cohorts observed in longitudinal data. An individual's death at time T means that the trajectories of θ_t and Y_t are stopped at time T. The conditional distribution of T given trajectories of θ_u, Y_u,$0\leq u\leq t$, as well as values of Z, G, is completely characterized by the conditional hazard (mortality) rate $\mu_{\theta_t}(Y_t,Z,G,t)$. The triple θ_t, Y_t, T describes the joint evolution of individual health/survival status and physiological variables over age during a person's life course.

The use of stochastic differential equations for random continuously changing covariates has been studied intensively in the analysis of longitudinal data, see Arbeev et al. (2009) and Yashin et al. (2007a, 2008) and references therein. Such a description is convenient since it captures the feedback mechanism typical of biological systems reflecting regular aging-related changes and takes into account the presence of random noise affecting individual trajectories. It also captures the dynamic connections between aging-related changes in health and physiological states, which are important in many applications.

15.4 Evolution of the Conditional Distribution of θ_t and Y_t Among Those Who Survived to Age t

To calculate $\overline{\overline{\mu}}(t)$ in (15.1) one needs $f(y,j|t)=\frac{\partial}{\partial y}P(Y_t\leq y,\theta_t=j|Z,G,T>t)$ which is the joint conditional probability density function (p.d.f.), with respect to Y_t, and the probability with respect to θ_t, given $\{T>t\}$, Z,G and the functional form of the conditional mortality rate $\mu_{\theta_t}(Y_t,Z,G,t)$. Using standard Bayesian arguments similar to that used in Yashin et al. (1985, 1995), the following partial differential equation for $f(y,j,|t)$ can be derived:

$$\frac{d}{dt}f(y,j|t)=\sum_{i=1}^{M}\lambda_{ij}(y,t)f(y,i|t)-\frac{\partial}{\partial y}\left(A_j(y,t)f(y,j|t)\right)+\frac{1}{2}\frac{\partial^2}{\partial y^2}\left(B\circ B_j(t)f(y,j|t)\right)$$
$$+f(y,j|t)(\overline{\overline{\mu}}(t)-\mu_i(y,t)),\quad f(y,j|t_0). \tag{15.5}$$

For simplicity, we omit the dependence of coefficients in (15.5) on the variables Z, G. Here the functions $A_j(y,t)$ are defined in (15.1) and $B\circ B_j(t)=B_j(t)B_j(t)^*$, where the symbol * denotes transposition, and the transition intensities $\lambda_{kr}(y,t)$, $k,r=1,2,\ldots,M$ are defined by (15.4). Since $f(y,j\,|t)$ multiplies $\overline{\overline{\mu}}(t)$ in (15.5), and $\overline{\overline{\mu}}(t)$ is the result of integration of $\mu_{\theta_t}(Y_t,Z,G,t)$ with respect to $f(y,j\,|t)$ the relationship (15.5) is a nonlinear partial integral-differential equation with respect to $f(y, j\,|t)$. Note that the mortality rate $\bar{\bar{\mu}}(t)$ in (15.5) can also be represented as follows:

$$\bar{\bar{\mu}}(t) = \sum_{j=1}^{M} \bar{\mu}_j(t)\pi_j(t) \tag{15.6}$$

where $\pi_j(t) = P(\theta_t = j|T > t, Z, G)$, and

$$\bar{\mu}_j(t) = E\left(\mu_{\theta_t}(Y_t, t)|\theta_t = j, Z, G, T > t)\right). \tag{15.7}$$

To calculate (15.6) and (15.7), one needs $\pi_j(t) = P(\theta_t = j|Z, G, T > t)$ and the conditional p.d.f. $f(y|j, t) = \partial P(Y_t \leq y|\theta_t = j, Z, G, T > t)/\partial y$ for each $t \geq 0$. An equation for $\pi_j(t)$ can be derived by integrating $f(y,j|t)$ in (15.5) with respect to y:

$$d\pi_j(t)/dt = \sum\nolimits_{k=1}^{M} \bar{\lambda}_{jg}(t)\pi_k(t) + \pi_j(t)\left(\bar{\bar{\mu}}(t) - \bar{\mu}_j(t)\right), \ \pi_j(t_0) \quad j = 1, 2, \ldots M. \tag{15.8}$$

Here $\bar{\bar{\mu}}(t)$ and $\bar{\mu}_j(t)$ are given by (15.6) and (15.7), and $\bar{\lambda}_{ij}(t)$ is defined as follows:

$$\bar{\lambda}_{ij}(t) = E\left(\lambda_{ij}(Y_t, t)|\theta_t = i, Z, G, T > t\right) = \int_{R^K} \lambda_{ij}(y, t) f(y|i, t)dy. \tag{15.9}$$

Integration in (15.7) and (15.9) requires $f(y|j, t), j = 1, 2, \ldots, M$. The equations for this conditional p.d.f. follow from Bayes' rule and Eqs. (15.5) and (15.8):

$$\begin{aligned} \frac{\partial}{\partial t} f(y|j, t) = & \sum_{i=1}^{N} \left(\lambda_{ij}(y, t) f(y|i, t) - \bar{\lambda}_{ij}(t), f(y|j, t)\right) \frac{\pi_i(t)}{\pi_j(t)} \\ & - \frac{\partial}{\partial y}\left(A_j(y, t) f(y|j, t)\right) + \frac{1}{2}\frac{\partial^2}{\partial y^2}\left(B_j(t) f(y|j, t)\right) \\ & + f(y|j, t)\left(\bar{\mu}_j(t) - \mu_j(y, t)\right), \quad f(y|j, t_0). \end{aligned} \tag{15.10}$$

Note that (15.8) and (15.10) constitute a system of nonlinear (partial and ordinary) differential equations that must be solved together. Calculation of $\bar{\bar{\mu}}(Z, G, t)$ and $\bar{\mu}_j(Z, G, t)$ in (15.8) and (15.10) will require a parametric description of $\mu_{\theta_t}(Y_t, Z, G, t)$. A special form of such a description—the quadratic hazard—is used in a Gaussian approximation as follows.

15.5 Gaussian Approximation

To solve Eqs. (15.8) and (15.10), the functional forms for the conditional mortality rate $\mu_{\theta_t}(Y_t, t)$ in (15.1) and the elements of the conditional transition intensities matrix $\lambda_{kr}(Y_t,t)$ in (15.9), as well as the coefficients, $A_{\theta_t}(y, t)$ and $B_{\theta_t}(t)$ in (15.5) and (15.10) have to be specified, and the integrations have to be performed to obtain $\overline{\overline{\mu}}(t), \overline{\mu}_j(t)$ and $\overline{\lambda}_{ij}(t)$. It is convenient and epidemiologically justified to describe the conditional hazard functions as quadratic forms of Y_t:

$$\lambda_{kr}(Y_t, t) = \lambda_{0\,kr}(t) + (Y_t - g_{0\,k}(t))^* \Lambda_{kr}(t)(Y_t - g_{0k}(t)) \tag{15.11}$$

$$\mu_{\theta_t}(Y_t, t) = \mu_{0\theta_t}(t) + \left(Y_t - f_{0\theta_t}(t)\right)^* Q_{\theta_t}(t)\left(Y_t - f_{0\theta_t}(t)\right). \tag{15.12}$$

Here $\Lambda_{kr}(t)$ and $Q_j(t)$ are symmetric non-negative-definite $K \times K$ matrices, $g_{0k}(t)$ and $f_{0\theta_t}(t)$ are K-vector functions, and $\lambda_{0kr}(t)$ and $\mu_{0r}(t)$ are parametric functions of t for $k, r, j = 1, 2,\ldots,M$; $t \geq t_0$. It is convenient to modify Eq. (15.1) to explicitly describe the mechanism of physiological regulation in the presence of external disturbances. This mechanism can be described in terms of linear stochastic differential equations with feedback loops:

$$dY_t = a_{\theta_t}(t)\left(Y_t - f_{1\theta_t}(t)\right)dt + b_{\theta_t}(t)dW_t, \quad Y_{t_0}. \tag{15.13}$$

Here $a_{\theta_t}(t)$ and $b_{\theta_t}(t)$ are matrices of appropriate dimensions, Y_{t_0} is a random vector of initial conditions, and W_t is a vector Wiener process with independent components, which is independent of the initial value, Y_{t_0}. The components of the vector function $f_{1\theta_t}(t)$ characterize the effects of allostatic adaptation on the physiological state (Yashin et al. 2007a, 2012). Equation (15.13) includes negative feedback loops, which reflect basic regularities of an organisms' biological functioning. The strength of feedback regulation depends on the absolute values of the elements of the matrix $a_{\theta_t}(t)$. The dependence of these elements on θ_t indicates that strength of the feedback regulation at age t may depend on an individual's health status at this age.

Conditions (15.12) and (15.13) together with the assumptions about normality of the distribution for Y_{t_0} and the absence of the jumping process yield a Gaussian conditional probability distribution of the process Y_t among survivors to age t (Yashin and Manton 1997; Yashin et al. 1985). The presence of the jumping process, θ_t, affecting the coefficients of Eq. (15.13) for Y_t, and hence its age dynamics, violates the Gaussian property of this distribution. However, the quadratic forms for the conditional transition intensity functions (15.11) and mortality risks (15.12), as well as the linear structure of (15.13), suggest the possibility of a Gaussian approximation of the conditional p.d.f. $f(y|j, t) = \partial P(Y_t \leq y|\theta_t = j, Z, G, T > t)/\partial y$.

The conditional mortality risk and conditional transition intensity functions given $\{T > t\}, \{\theta_t = j\}$, Z, G, can be represented as follows:

$$\overline{\mu}_j(t) = \mu_{0j}(t) + \left(m_j(t) - f_{0j}(t)\right)^* Q_j(t)\left(m_j(t) - f_{0j}(t)\right) + Tr\left(Q_j(t)\gamma_j(t)\right) \quad (15.14)$$

$$\overline{\lambda}_{jk}(t) = \lambda_{0jk}(t) + \left(m_j(t) - g_{0j}(t)\right)^* \Lambda_{jk}(t)\left(m_j(t) - g_{0j}(t)\right) + Tr\left(\Lambda_{jk}(t)\gamma_j(t)\right) \quad (15.15)$$

where

$$m_j(t) = E(Y_t|\theta_t = j, Z, G, T > t))$$

and

$$\gamma_j(t) = E\Big(\left(Y_t - m_j(t)\right) - \left(Y_t - m_j(t)\right)^* |\theta_t = j, Z, G, T > t\Big).$$

These conditional moments satisfy the following ordinary differential equations:

$$\frac{dm_j(t)}{dt} = \sum_i \frac{\pi_i(t)}{\pi_j(t)}\left[m_{ij}(t)\overline{\lambda}_{ij}(t) - 2\gamma_i(t)\Lambda_{ij}(t)\hat{g}_{0i}(t)\right] - a_j(t) \times \hat{f}_{1j}(t) + 2y_j(t)Q_j(t)\hat{f}_{0j}(t), \quad (15.16)$$

$$\frac{d\gamma_j(t)}{dt} = \sum_i \frac{\pi_i(t)}{\pi_j(t)}\Big[\left(\gamma_i(t) - \gamma_j(t) + m_{ij}(t)\cdot m^*_{ij}(t)\right)\overline{\lambda}_{ij}(t) + 2\left(\gamma_i(t)\Lambda_{ij}(t)\gamma_i(t)\right. - \gamma_i(t)\Lambda_{ij}(t)\hat{g}_{0i}(t)\cdot m^*_{ij}(t) - m_{ij}(t)\cdot \hat{g}^*_{0i}(t)\Lambda_{ij}(t)\gamma_i(t))\Big] + a_j(t)\gamma_i(t) + \gamma_j(t)a^*_j(t) + B_j(t) - 2\gamma_j(t)Q_i(t)\gamma_j(t) \quad (15.17)$$

Here $\overline{\lambda}_{ij}(t)$ is given by (15.16), $m_{ij}(t) = m_i(t) - m_j(t)$, and the "hat" variables are defined as $\hat{f}_{0j}(t) = f_{0j}(t) - m_j(t), \hat{f}_{1j}(t) = f_{1j}(t) - m_j(t)$, $\hat{g}_i(t) = g_i(t) - m_i(t)$.

15.6 Conclusion

Researchers in experimental studies of aging often use mortality curves or survival functions for comparing the effects of external exposures or genetic manipulations on the aging process and life span. Although such a practice is efficient for detecting effects of interventions on survival, it does not allow for addressing more sophisticated research questions about biological mechanisms regulating changes in survival distributions in response to such interventions. This is because observing changes in mortality and survival has limited utility for understanding the biological machinery of aging and disease development (Yashin et al. 2002). In this chapter, we showed that, to better understand how people lose health and functional capacities during the aging process and how these changes influence survival outcomes, integrative mortality models can be developed, the parameters of which have a biological interpretation in terms of aging-related declines in health status and changes in physiological or other indices affecting health and survival outcomes.

These integrative mortality models have more sophisticated mathematical descriptions than those conventionally used in demography, actuarial science, and biostatistics. These descriptions include stochastic differential equations of aging-related changes in biomarkers, and their effects on health and survival across the life course. Mathematically, such descriptions involve stochastic processes with continuous and jumping sampling paths. A corresponding parametric representation of the all-cause mortality rate involves the operation of conditional mathematical expectation of the mortality risk given the event $\{T > t\}$, where T is the death time and t is current age. Such conditioning indicates that averaging has to be performed among only those individuals who survived to age t. The models include partial differential/integral equations for conditional probability distributions of the values of biomarkers and health status at a given age among survivors.

Although integrative mortality models look more complex and cumbersome compared to demographic models, their research potential exceeds that of demographic models. The models derived above open up opportunities for investigating how people age by analyzing changes in biological variables that take place in aging human bodies, how these changes influence health, and how aging and health affect survival outcomes. The proper interpretation of modeling results requires identification of the model's parameters using appropriate longitudinal data.

Many longitudinal datasets are now available to researchers. These datasets are collected by distinct research groups, using different study designs, different (also possibly overlapping) sets of biomarkers, and different time intervals between subsequent examinations. They also may use different health characteristics. In other words, each dataset is collected in accordance with the unique observational plan specific to a given study, and different datasets may have distinct structures because of that. The integrative mortality models developed in this chapter have important features that make them promising tools in analyses of fundamental problems of aging for which large amounts of comprehensive information are required. These models allow for joint analyses of several dataset collected using different observational plans. In the next chapter, we explain how these analyses can be performed. The use of the Gaussian approximation allows for substantial facilitation of numerical calculations involved in maximization of the likelihood function of the data.

Acknowledgements The research reported in this chapter was supported by the National Institute on Aging grants R01AG027019, R01AG030612, R01AG030198, 1R01AG046860, and P01AG043352. The content is solely the responsibility of the authors and does not necessarily represent the official views of the National Institute on Aging or the National Institutes of Health. The Framingham Heart Study (FHS) is conducted and supported by the National Heart, Lung and Blood Institute (NHLBI) in collaboration with the FHS Investigators. This chapter was prepared using a limited access dataset obtained from the NHLBI and does not necessarily reflect the opinions or views of the FHS or the NHLBI.

References

Akushevich, I., Kulminski, A., & Manton, K. (2005). Life tables with covariates: Dynamic model for nonlinear analysis of longitudinal data. *Mathematical Population Studies, 12*(2), 51–80.

Arbeev, K. G., Akushevich, I., Kulminski, A. M., Arbeeva, L. S., Akushevich, L., Ukraintseva, S. V., Culminskaya, I. V., & Yashin, A. I. (2009). Genetic model for longitudinal studies of aging, health, and longevity and its potential application to incomplete data. *Journal of Theoretical Biology, 258*(1), 103–111.

Arbeev, K. G., Ukraintseva, S. V., Arbeeva, L. S., Akushevich, I., Kulminski, A. M., & Yashin, A. I. (2011). Evaluation of genotype-specific survival using joint analysis of genetic and non-genetic subsamples of longitudinal data. *Biogerontology, 12*(2), 157–166.

Vaupel, J. W., & Yashin, A. I. (1985). Heterogeneity's ruses: Some surprising effects of selection on population dynamics. *American Statistician, 39*(3), 176–185.

Vaupel, J. W., Manton, K. G., & Stallard, E. (1979). The impact of heterogeneity in individual frailty on the dynamics of mortality. *Demography, 16*(3), 439–454.

Woodbury, M. A., & Manton, K. G. (1977). A random-walk model of human mortality and aging. *Theoretical Population Biology, 11*(1), 37–48.

Yashin, A. I. (1985). Dynamics in survival analysis: Conditional Gaussian property vs. Cameron-Martin formula. In N. V. Krylov, R. S. Lipster, & A. A. Novikov (Eds.), *Statistics and control of stochastic processes* (pp. 446–475). New York: Springer.

Yashin, A. I., & Manton, K. G. (1997). Effects of unobserved and partially observed covariate processes on system failure: A review of models and estimation strategies. *Statistical Science, 12*(1), 20–34.

Yashin, A. I., Manton, K. G., & Vaupel, J. W. (1985). Mortality and aging in a heterogeneous population: A stochastic process model with observed and unobserved variables. *Theoretical Population Biology, 27*(2), 154–175.

Yashin, A. I., Manton, K. G., Woodbury, M. A., & Stallard, E. (1995). The effects of health histories on stochastic process models of aging and mortality. *Journal of Mathematical Biology, 34*(1), 1–16.

Yashin, A. I., Ukraintseva, S. V., Boiko, S. I., & Arbeev, K. G. (2002). Individual aging and mortality rate: How are they related? *Social Biology, 49*(3–4), 206–217.

Yashin, A. I., Arbeev, K. G., Akushevich, I., Kulminski, A., Akushevich, L., & Ukraintseva, S. V. (2007a). Stochastic model for analysis of longitudinal data on aging and mortality. *Mathematical Biosciences, 208*(2), 538–551.

Yashin, A. I., Arbeev, K. G., Kulminski, A., Akushevich, I., Akushevich, L., & Ukraintseva, S. V. (2007b). Health decline, aging and mortality: How are they related? *Biogerontology, 8*(3), 291–302.

Yashin, A. I., Arbeev, K. G., Akushevich, I., Kulminski, A., Akushevich, L., & Ukraintseva, S. V. (2008). Model of hidden heterogeneity in longitudinal data. *Theoretical Population Biology, 73*(1), 1–10.

Yashin, A. I., Arbeev, K. G., Akushevich, I., Kulminski, A., Ukraintseva, S. V., Stallard, E., & Land, K. C. (2012). The quadratic hazard model for analyzing longitudinal data on aging, health, and the life span. *Physics of Life Reviews, 9*(2), 177–188.

Chapter 16
Integrative Mortality Models for the Study of Aging, Health, and Longevity: Benefits of Combining Data

Anatoliy I. Yashin, Igor Akushevich, Konstantin G. Arbeev, Alexander M. Kulminski, and Svetlana V. Ukraintseva

16.1 Introduction

In a number of longitudinal studies, individual health and physiological/biological variables are repeatedly measured for a relatively large number of study subjects. These data capture aging-related changes in biomarkers as well as health and survival outcomes that take place during these individuals' life courses. Such data have good potential for investigating properties of dynamic mechanisms involved in the regulation of aging-related changes, as well as in evaluating roles of genetic and non-genetic factors affecting them.

Despite the relatively large sample sizes of each longitudinal dataset, the number of study subjects is often not enough to guarantee either high quality statistical estimates of dynamic characteristics in multidimensional models or of tests of statistical hypotheses related to fundamental research questions on causes and mechanisms of aging and disease development. Partly, these desirable goals depend on the complexity of the model, number of model parameters, prevalence of diseases under study, etc. Often it happens that measurements of some important variables or health outcomes that are omitted in one dataset were measured in another dataset. In such cases, combining data would be a promising alternative for comprehensive analyses of mechanisms of aging-related changes, health decline, and life span. These analyses can be performed within the framework of a comprehensive model of human aging, health, and mortality. In this chapter, we describe a method of statistical modeling for joint analyses of longitudinal data on aging, health, and longevity collected using different observational plans. The method is based on the mathematical model described in Chap. 15. Observational plans corresponding to each dataset play a crucial role in specifying the likelihood functions of observed components of the data. The results of our analyses indicate that parameters of both continuous and jumping components of the model can be identified from the combined data, and jointly estimated using the method of maximum likelihood.

A.I. Yashin et al., *Biodemography of Aging*, The Springer Series on Demographic Methods and Population Analysis 40, DOI 10.1007/978-94-017-7587-8_16

16.2 Observational Plans and Combining Data

16.2.1 Likelihood Function of Life Span Data

The concept of an observational plan is used to characterize differences in the structures of data in different datasets. The simplest observational plan relevant to the mortality model described in Chap. 15 includes measurements of values of variables Z, G, and life spans of study subjects with no observation of components θ_t and Y_t. Let $T_1, T_2, \ldots, T_N$, be life span data (possibly censored), $Z_1, Z_2, \ldots, Z_N$ be data on non-genetic covariates, and $G_1, G_2, \ldots, G_N$ be genetic data on N individuals. Let $\overline{\overline{\mu}}(t)$ be a parametrically specified conditional mortality rate given Z and G. Then the likelihood function of the life span data conditional on Z and G is:

$$L(T_1, T_2, \ldots T_N) = \prod_{i=1}^{N} \overline{\overline{\mu_i}}(T_i)^{\delta_i} \exp\left\{ - \int_0^{T_i} \overline{\overline{\mu}}_i(u)du \right\} \tag{16.1}$$

where δ_i is a censoring variable (i.e. an at-risk indicator): $\delta_i = 1$ if T_i is the life span of the i-th individual and $\delta_i = 0$ if the lifespan of the i-th individual is censored at age T_i. The likelihood function (16.1) must be maximized with respect to the parameters describing the mortality risk $\overline{\overline{\mu}}_i(t) = \overline{\overline{\mu}}_i(Z_i, G_i, t)$. As these parameters are involved in the characterization of the process θ_t, Y_t and variables Z, G, their interpretation has biological and physiological meaning. As in Chap. 15, we omit writing the variables Z and G in coefficients of the corresponding equations for brevity. Note that the difference of $\overline{\overline{\mu}}_i(t) = \overline{\overline{\mu}}_i(Z_i, G_i, t)$, from a parametric demographic mortality model or epidemiologic model of mortality risk is that, in modeling $\overline{\overline{\mu}}_i(t) = \overline{\overline{\mu}}_i(Z_i, G_i, t)$ we assume that it is generated by the processes θ_t and Y_t in accordance with Eqs. 15.11, 15.12, and 15.13 in Chap. 15. These processes are not observed in this observational plan. To calculate mortality models $\overline{\overline{\mu}}_i(t)$ in (16.1), a conditional averaging procedure given $\{T > t\}, Z, G$, ((15.2), Chap. 15) has to be performed. In most cases, such a procedure cannot be done analytically, and hence $\overline{\overline{\mu}}_i(t)$ usually is not represented as an explicit function of the model parameters. Therefore, maximization of the likelihood function (16.1) involves intensive computations that include solving non-linear partial differential or ordinary differential equations at each step of the likelihood maximization procedure. Such calculations can be performed using modern optimization software packages.

An important feature of the integrative mortality models described in Chap. 15 is that the data on life spans alone are usually not enough to identify all model parameters. The problem could be resolved if data from longitudinal studies of aging, health and longevity were used in the analyses. Since many datasets currently used for aging studies in humans were initially designed to address issues related to specific chronic diseases, and were collected by different research groups, they often measure different biological variables, record different health outcomes, and use different timings for making repeated observations of biomarkers. The likelihood functions of the data, the parameter estimation algorithms, as well as parameter estimates, will reflect these differences in the observational plans. At first, it may not seem possible to combine the data collected according to different observational plans. It turns out, however, that the use of comprehensive mortality models allows such analyses to be performed. Here we describe the likelihood functions of the data corresponding to several different observational plans and show how these data can be jointly analyzed to estimate the model parameters.

16.2.2 Longitudinal Data on Physiological Variables: Health Changes Are Not Observed: Observational Plan #1

In a number of longitudinal studies of aging and longevity, the values of physiological variables are measured repeatedly during the life course. The age trajectories of physiological variables for each individual are described by (15.3) of Chap. 15. The parameters of this equation characterize the dynamics of aging-related changes in a population or in a group of study participants. Individual variations in age trajectories of these variables are assumed to be generated by differences in the initial conditions and by a random Wiener process. The longitudinal data on physiological states consist of the results of measurements of individual physiological variables in a group of individuals at a series of subsequent time (age) points during the individuals' life courses. These data play a key role in estimating the model parameters using the likelihood maximization procedure. In such a description, the model parameters characterize the population under study.

To form the various likelihood functions, assume that the continuously changing variables are measured at age points $t_0, t_1, t_2, \ldots, t_n; t_n \leq T$. Let $\widetilde{Y}_0^t = Y_{t_0}, Y_{t_1}, Y_{t_2}, \ldots, Y_{t_n}; t_n \leq T$ be a random vector of observations of the process Y_t at these age points. It follows that $\widetilde{Y}_0^{t_k-} = \widetilde{Y}_0^{t_{k-1}}$ and $\widetilde{Y}_0^t = \widetilde{Y}_0^{t_k}$, if $t_k \leq t < t_{k+1}$. Here $t_k- = \lim_{u \uparrow t_{k+1}} t_u$. Denote by

$$\widetilde{\pi}_j(t) = P\left(\theta_t = j \middle| \widetilde{Y}_0^t, Z, G, T > t\right) \tag{16.2}$$

the conditional probability of having health/well-being status j, given $\widetilde{Y}_0^t, Z, G, \{T > t\}$. Let

$$\widetilde{f}(y|j,t) = \frac{\partial}{\partial y} P\left(Y_t \leq y \middle| \widetilde{Y}_0^t, \theta_t = j, Z, G, T > t\right) \tag{16.3}$$

be the conditional probability density function of Y_t. The evolution of $\widetilde{\pi}_j(t)$ and $\widetilde{f}(y|j,t)$ starts at age t_0, and continues at the intervals

$t_0 \le t < t_1; t_1 \le t < t_2; \ldots; t_{n-1} \le t_n; t < T$. At each such interval, these functions satisfy Eqs. (15.8) and (15.10) of Chap. 15, respectively.

An important property of the age trajectories of $\widetilde{\pi}_j(t)$ and $\widetilde{f}(y|j,t)$ is that they both experience jumps at the observation times $t_1, t_2, \ldots t_n; t_n \le T$. The values of these functions immediately after the jumps follow from standard Bayes' arguments:

$$\widetilde{\pi}_j(t_k) = \widetilde{\pi}_j(t_k-)\frac{\widetilde{f}(Y_{t_k}|j,t_k-)}{\sum_{r=1}^{M}\widetilde{\pi}_r(t_k-)\widetilde{f}(Y_{t_k}|r,t_k-)}; \widetilde{f}(y|j,t_k) = \delta(y - Y_{t_k}), \tag{16.4}$$

respectively. Here $\widetilde{\pi}_j(t_k-) = \lim_{t\uparrow t_k}\widetilde{\pi}_j(t)$ and $\widetilde{f}(Y_{t_k}|j,t_k-) = \frac{\partial}{\partial y}P\left(Y_t \le y|\widetilde{Y}_0^{t-},\theta_t = j, T > t.\right)_{t=t_k, y=Y_{t_k}}$, and $\delta(y - Y_{t_k})$ is the delta-function. Thus $\widetilde{\pi}_j(t)$ and $\widetilde{f}(y|j,t)$ are solutions of Eqs. (15.8) and (15.10) of Chap. 15 at the interval $[t_{k-1}, t_k)$ with $\overline{\lambda}_{kj}(t), \overline{\mu}_i(t), \overline{\mu}_i(t)$ replaced by $\widetilde{\overline{\lambda}}_{kj}\left(\widetilde{Y}_0^t, t\right), \widetilde{\overline{\mu}}_{kj}\left(\widetilde{Y}_0^t, t\right), \widetilde{\mu}_j\left(\widetilde{Y}_0^t, t\right)$, respectively. Here

$$\widetilde{\overline{\lambda}}_{kj}\left(\widetilde{Y}_0^t, t\right) = E\left(\lambda_{kj}(Y_t, t)|\widetilde{Y}_0^t, \theta_t = k, Z, G, T > t\right), \tag{16.5}$$

$$\widetilde{\overline{\mu}}\left(\widetilde{Y}_0^t, t\right) = \sum_{j=1}^{M}\widetilde{\mu}_j\left(\widetilde{Y}_0^t, t\right)\widetilde{\pi}_j(t), \tag{16.6}$$

and

$$\widetilde{\mu}_j\left(\widetilde{Y}_0^t, t\right) = E\left(\mu(\theta_t, Y_t, t)|\widetilde{Y}_0^t, \theta_t = j, Z, G, T > t\right). \tag{16.7}$$

The values (16.4) serve as initial conditions for $\widetilde{\pi}_j(t)$ and $\widetilde{f}(y|j,t)$ satisfying Eqs. (15.8) and (15.10) of Chap. 15 at the intervals $[t_k, t_{k+1})$, $k = 0,1,2,\ldots$.

This notation allows us to form likelihood functions for the data collected in such examinations. They include data on the sequences of discrete-time measurements of continuously changing component (e.g., physiological state) for each study participant, plus survival data. The component of the likelihood function dealing with discrete-time measurements involves the conditional p.d.f.: $\widetilde{\phi}(y|t) = \partial P\left(Y_t \le y|\widetilde{Y}_0^t, Z, G, T > t\right)/\partial y$, which may be represented as $\widetilde{\phi}(y|t) = \sum_{r=1}^{M}\widetilde{f}(y|r,t)\widetilde{\pi}_r(t)$. The second component describes survival data and involves the conditional mortality rate $\widetilde{\mu}_j\left(\widetilde{Y}_0^t, t\right)$. Thus, the component of the likelihood function for the *i-th* individual having measurements $y^i_{t^i_1}, y^i_{t^i_2}, \ldots y^i_{t^i_{n(i)}}$, T_i is:

$$L_i\left(y^i_{t^i_1}, y^i_{t^i_2}, \ldots y^i_{t^i_{n(i)}}, T_i\right) =$$
$$\widetilde{\phi}\left(y^i_{t^i_{n(i)}}\middle|t^i_{t^i_{n(i)}}-\right)\widetilde{\phi}\left(y^i_{t^i_{n(i)-1}}\middle|t^i_{t^i_{n(i)-1}}-\right)\ldots\ldots\widetilde{\phi}\left(y_{t^i_1}\middle|t^i_{t^i_1}-\right)\widetilde{\overline{\mu}}^i(T_i)^{\delta_i}\exp\left\{-\int_0^{T_i}\widetilde{\overline{\mu}}^i(u)du\right\}. \tag{16.8}$$

Here $\widetilde{\phi}\left(y_{t^i_k}|t^i_k-\right) = \sum_{r=1}^{M}\widetilde{f}\left(y_{t^i_k}|r, t^i_k-\right)\pi_r\left(t^i_k-\right)$, and δ_i is the at-risk indicator for the *i-th* individual. Maximization of the likelihood function (16.8) requires solving the modified Eqs. (15.8) and (15.10) from Chap. 15 for different values of intermediate parameters at each step of the likelihood maximization procedure. Solving a system of non-linear partial differential equations at each iteration step may be computationally extensive. To reduce the computational load, a Gaussian approximation of the conditional probability density function $\widetilde{f}(y|j,t)$ could be used. Such an approximation allows one to replace numerical solutions of the partial differential equations by numerical solutions of the ordinary differential equations for the first two moments of this distribution.

16.2.3 Gaussian Approximation of the Model of Physiological Variables

The use of the Gaussian approximation described in Chap. 15 transforms the likelihood function (16.8) into (16.9). In this case, the likelihood becomes a function of parameters determining the dynamic properties of Eqs. (15.8), (15.16), and (15.17) of Chap. 15:

$$L_i\left(y^i_{t^i_0}, y^i_{t^i_1}, \ldots, y^i_{t^i_{n_i}}, T_i\right) = \widetilde{\overline{\mu}}^i\left(y_0^{T_i-}, T_i\right)^{\delta_i}\exp\left\{-\int_0^{T_i}\widetilde{\overline{\mu}}^i\left(\widetilde{y}_0^{iu}, u\right)du\right\}$$
$$\times\prod_{j=0}^{n_i(i)}\left(\sum_{k=1}^{M}\widetilde{\pi}^i_k\left(t^i_j-\right)\left(2\pi\left|\widetilde{\gamma}^i_k\left(t^i_{t^i_j}-\right)\right|\right)\right)^{-\frac{k}{2}}$$
$$\exp\left\{-\frac{1}{2}\left(y^i_{t^i_j}-\widetilde{m}^i_k\left(t^i_j-\right)\right)^*\widetilde{\gamma}^i_k\left(t^i_{t^i_j}-\right)^{-1}\left(y^i_{t^i_j}-\widetilde{m}^i_k\left(t^i_j-\right)\right)\right\} \tag{16.9}$$

where $\widetilde{\overline{\mu}}\left(\widetilde{Y}^t_0, t\right)$ is defined by (16.6), and $\widetilde{\mu}_j\left(\widetilde{Y}^t_0, t\right)$ is represented as:

$$\widetilde{\mu}_j\left(\widetilde{Y}_0^t, t\right) = \mu_{0j}(t) + \left(\widetilde{m}_j(t) - f_j(t)\right)^* Q_j(t)\left(\widetilde{m}_j(t) - f_j(t)\right) + Tr\left(Q_j(t)\widetilde{\gamma}_j(t)\right) \quad (16.10)$$

with

$$\widetilde{m}_k(t) = E\left(Y_t \middle| \widetilde{Y}_0^t, \theta_j = k, Z, G, T > t\right);$$

$$\widetilde{\gamma}_k(t) = E\left((Y_t - \widetilde{m}_k(t))^* (Y_t - \widetilde{m}_k(t)) \middle| \widetilde{Y}_0^t, \theta_j = k, Z, G, T > t\right). \quad (16.11)$$

The transition intensities $\widetilde{\lambda}_{kj}\left(\widetilde{Y}_0^t, t\right)$ in 15.8, Chap. 15 for $\widetilde{\pi}_j(t)$ are:

$$\widetilde{\lambda}_{kj}\left(\widetilde{Y}_0^t, t\right) = \lambda_{0kj}(t) + (\widetilde{m}_k(t) - g_k(t))^* \Lambda_{kj}(t)(\widetilde{m}_k(t) - g_k(t) - g_k(t)) + Tr\left(\Lambda_{kj}(t)\widetilde{\gamma}_k(t)\right). \quad (16.12)$$

Above, δ_i is an at-risk indicator, K is the dimension of the vector Y_t, $\widetilde{m}_k^i(t)$ and $\widetilde{\gamma}_k^i(t)$ satisfy Eqs. (15.16) and (15.17) of Chap. 15 at the intervals $\left[t_0^i, t_0^i\right)$; $\left[t_1^i, t_2^i\right); \ldots;$ $\left[t_{n_i-1}^i, t_{n_i}^i\right)$; $\left[t_{n_i}^i, T_i\right)$ with the initial conditions $y_{t_0^i}^i, y_{t_1^i}^i, \ldots, y_{t_{n(i)}^i}^i$, for $\widetilde{m}_k^i\left(t_j^i\right) = y_{t_j^i}^i$, and $\widetilde{\gamma}_k^i\left(t_j^i-\right) = y_{t_j^i}^i$, respectively, $\widetilde{m}_k^i\left(t_j^i-\right) = \lim_{t\uparrow t_j^i}\widetilde{m}_k^i(t)$, and $\widetilde{\gamma}_k^i\left(t_j^i-\right) = \lim_{t\uparrow t_j^i}\widetilde{\gamma}_k^i(t)$, and $t_{n(i)}^i$ is the age of the latest measurement of the physiological index before death at T_i for the ith individual. Note that using the index i in $\widetilde{\pi}_k^i(t)$, $\widetilde{m}_k^i(t)$, and $\widetilde{\gamma}_k^i(t)$ in these equations is necessary because the values of these estimates depend on the individual histories of the process Y_t observed in discrete times.

The conditions (16.4) can now be represented in the form:

$$\widetilde{\pi}_j(t_i) = \widetilde{\pi}_j(t_i-)\frac{\left(2\pi\left|\widetilde{\gamma}_j(t_i-)\right|\right)^{-\frac{K}{2}}\exp\left\{-\frac{1}{2}\left(Y_{t_i} - \widetilde{m}_j(t_i-)\right)^*\widetilde{\gamma}_j^{-1}(t_i-)\left(Y_{t_i} - \widetilde{m}_j(t_i-)\right)\right\}}{\sum_{k=1}^{M}\widetilde{\pi}_k(t_i-)\left(2\pi\left|\widetilde{\gamma}_k(t_i-)\right|\right)^{-\frac{K}{2}}\exp\left\{-\frac{1}{2}\left(Y_{t_i} - \widetilde{m}_k(t_i-)\right)^*\widetilde{\gamma}_k^{-1}(t_i-)\left(Y_{t_i} - \widetilde{m}_k(t_i-)\right)\right\}}, \quad (16.13)$$

$\widetilde{m}_j(t_i) = Y_{t_i}$ and $\widetilde{y}_j(t_i) = 0$.

The dynamics of $\widetilde{\pi}_j(t)$ follow Eq. (15.8) of Chap. 15 with $\widetilde{\lambda}_{kj}\left(\widetilde{Y}_0^t, t\right)$ used instead of $\overline{\lambda}_{kj}(t)$ at the intervals $t_0 \le t < t_1; t_1 \le t < t_2; \ldots; t_{n-1} \le t < t_n; t < T$. The initial values of $\widetilde{\pi}_j(t)$ at the beginning of the ith interval $[t_i \le t < t_{i+1})$ are given by the relationship which involves values of $\widetilde{\pi}_j(t_i-)$, $\widetilde{m}_j(t_i-)$, and $\widetilde{\gamma}_j(t_i-)$ which are the solutions of Eqs. (15.8), 15.16, and (15.17) of Chap. 15 at the end of the interval $[t_{i-1\leq}t < t_i)$.

16.2.4 Data on Health Transitions Without Measurements of the Physiological State: Observational Plan #2

Let $\widehat{f}(y,t) = \frac{\partial}{\partial y}P\left(Y_t \leq y \middle| \theta_0^t, Z, G, T > t\right)$, with $\tau_1, \tau_2, \ldots, \tau_m$, denoting ages at which changes in an individual's health status took place (i.e. times of jumps of the process θ_t). Then for the conditional probability density function of Y_t, given age trajectories of health history θ_0^t, variables Z, G and $\{T > t\}$, the following equation is operative:

$$\frac{\partial}{\partial t}\widehat{f}(y,t) = -\frac{\partial}{\partial t}\left(A_{\theta_t}(y,t)\,\widehat{f}(y,t)\right) + \frac{1}{2}\frac{\partial^2}{\partial y^2}\left(B_{\theta_t}(t)\,\widehat{f}(y,t)\right) + \widehat{f}(y,t)$$
$$\times\left(\sum_{k=1,\,k\neq\theta_{t-}}^{M}\widehat{\lambda}_{\theta_{t-},k}(t) - \sum_{k=1,\,k\neq\theta_{t-}}^{M}\lambda_{\theta_{t-},k}(y,t)\right) + \widehat{f}(y,t)\left(\widehat{\mu}_{\theta_{t-}}(t)\right.$$
$$\left. - \mu_{\theta_{t-}}(y,t)\right) \tag{16.14}$$

The presence of two selection terms on the right side of Eq. (16.14) is easy to understand because the probability density function $\widehat{f}(y,t) = \frac{\partial}{\partial y}P\left(Y_t \leq y \middle| \theta_0^t, Z, G, \tau_1 > t, T > t\right)$ is conditional on the events that neither the first transition of the process θ_t, nor death, happened before age t. Similarly, the function $\widehat{f}(y,t) = \frac{\partial}{\partial y}P\left(Y_t \leq y \middle| \theta_0^t, Z, G, \tau_k > t, T > t\right)$ is conditional on neither the k-th transition of the process θ_t, nor death, happening before age t. This equation has to be solved at the intervals $[t_0, \tau_1), [\tau_1, \tau_2), [\tau_2, \tau_3), \ldots, [\tau_m, T)$, i.e., between subsequent jumps of the process θ_t. To avoid multiple hierarchical indexing, we will use notation $\theta_t \equiv \theta(t)$. The initial conditions at the beginning of each interval $[\tau_0, \tau_1)$, $[\tau_1, \tau_2), [\tau_2, \tau_3), \ldots, [\tau_m, T)$ are

$$\widehat{f}\left(y, \tau_p\right) = \widehat{f}\left(y, \tau_p-\right)\frac{\lambda_{\theta(\tau_p-),\theta(\tau_p)}\left(y, \tau_p-\right)}{\widehat{\lambda}_{\theta(\tau_p-),\theta(\tau_p)}\left(y, \tau_p-\right)}. \tag{16.15}$$

Here $\widehat{f}\left(y, \tau_p-\right) = \frac{\partial}{\partial y}P\left(Y_{t_p} \leq y \middle| \theta_0^{\tau_{p-1}}, Z, G, T > \tau_p\right)$ is the solution of Eq. (16.14) at the interval $\left[\tau_{p-1}, \tau_p\right)$ at the time just before the *p-th* jump of the process θ_t at time τ_p:

$$\widehat{\lambda}_{\theta(\tau_p-),\theta(\tau_p)}(\tau_p) = E\left(\lambda_{\theta(\tau_p-),x}(Y_{\tau_p},\tau_p)\middle|\theta_0^{\tau_p-},Z,G,T>\tau_p\right)\Big|_{x=\theta(\tau_p)} \tag{16.16}$$

$$\widehat{\mu}_{\theta(t)}(t) = E\left(\mu_{\theta(t)}(Y_t,t)\middle|\theta_0^t,Z,G,T>t\right) \tag{16.17}$$

16.2.5 The Likelihood of the Data on Health Transitions

The component of the likelihood corresponding to the data on ages (times) of change in the health status (age at onset of diseases) for the *i-th* individual with *m* (*i*) changes in health status occurring at the time points $\tau_1^i, \tau_2^i, \ldots, \tau_{m(i)}^i$ and death (censoring) at age T_i is:

$$\begin{aligned}\widehat{L}_i\left(\theta^i(\tau_1^i),\theta^i(\tau_2^i),\ldots\theta^i\left(\tau^i_{m(i)}\right),Y_i\right) &= p(\theta^i(t_0))\\ \times\prod_{p=1}^{m(i)}\widehat{\lambda}_{\theta^i(\tau_p^i-),\theta^i(\tau_p^i)}\left(t_p^i\right)&\exp\left\{-\int_{t_{p-1}^i}^{t_p^i}\left(\sum_{k=1,\,k\neq\theta^i(t-)}^{M}\widehat{\lambda}_{\theta^i(t-),k}(t)++\widehat{\mu}_{\theta^i(t-)}(t)\right)dt\right\}\\ \times\widehat{\mu}_{\theta^i\left(\tau^i_{m(i)}\right)}(T_i)^{\delta_i}&\exp\left\{-\int_{\tau^i_{m(i)}}^{T_i}\left(\sum_{k=1,\,k\neq\theta^i(t-)}^{M}\widehat{\lambda}_{\theta^i(t-),k}(t)+\widehat{\mu}_{\theta^i(t-)}(t)\right)dt\right\}\end{aligned} \tag{16.18}$$

Here $p(\theta^i(t_0))$ is the initial distribution of health status, and $\theta(\tau_1^i-)=\theta(t_0)$ by definition.

16.2.6 Gaussian Approximation of the Model with Health Transitions

Since health transitions are observed, the equations for the first two moments are:

$$\frac{d\widehat{m}_j(t)}{dt} = -a(t)\hat{f}_{1j}(t) + \sum_{k\neq j}2\widehat{\gamma}_j(t)\Lambda_{jk}(t)\hat{g}_j(t) + 2\gamma_j(t)\hat{f}_j(t), \tag{16.19}$$

$$\frac{d\gamma_j(t)}{dt} = a_j(t)\widehat{\gamma}_j(t) + \widehat{\gamma}_j(t)a_j^*(t) + B_j(t) - \sum_{k\neq j} 2\widehat{\gamma}_j(t)\Lambda_{jk}(t)\widehat{\gamma}_j(t) - 2\widehat{\gamma}_j(t)Q_j(t)\widehat{\gamma}_j(t) \tag{16.20}$$

Here we use the index j to indicate dependence of these moments on the process θ_t, when $\theta_{t-} = j$. Strictly speaking, these moments depend on the entire trajectory of θ_t at the interval $[t_0, t)$. These equations have to be solved at the intervals $[\tau_1, \tau_2)$, $[\tau_2, \tau_3), \ldots, [\tau_m, T)$, i.e., between subsequent jumps of the process θ_t. When $\theta(\tau_p-) = k$ and $\theta^i(\tau_p-) = j$, and $\lambda_{kj}(Y_t, Z, G, t)$ is as described by Eq. (15.15) of Chap. 15, we have for the initial values $\widehat{m}_j(\tau_p)$ and $\widehat{\gamma}_j(\tau_p)$:

$$\widehat{m}_j(\tau_p) = \widehat{m}_k(\tau_p-) - \frac{2\widehat{\gamma}_k(\tau_p-)\Lambda_{kj}(\tau_p)\hat{g}_{0k}(\tau_p-)}{\widehat{\lambda}_{kj}(\tau_p-)} \tag{16.21}$$

$$\widehat{\gamma}_j(\tau_p) = \widehat{\gamma}_k(\tau_p-) - \frac{2\widehat{\gamma}_k(\tau_p-)\Lambda_{kj}(\tau_p)\widehat{\gamma}_k(\tau_p-)}{\widehat{\lambda}_{kj}(\tau_p-)} \tag{16.22}$$

with

$$\widehat{g}_{0k}(\tau_p-) = g_{0k}(\tau_p) - \widehat{m}_k(\tau_p-) \tag{16.23}$$

and

$$\widehat{\lambda}_{kj}(t) = \lambda_{0kj}(t) + \left(\widehat{m}_k(t) - g_{0k}(t)\right)^* \Lambda_{kj}(t)\left(\widehat{m}_k(t) - g_{0k}(t)\right) + Tr\left(\Lambda_{kj}(t)\widehat{\gamma}_k(t)\right) \tag{16.24}$$

16.2.7 Discrete Time Observations of the Physiological State and Health Transitions: Observational Plan #3

Let us assume that the physiological state is repeatedly measured at a sequence of discrete times, and all health transitions are also observed. In this case, the changes in information about the *i-th* individual take place at times of discrete observation of the process $Y_t : y^i_{t^i_1}, y^i_{t^i_2}, \ldots, y^i_{t^i_{n(i)}}$ or the values of the jumping process θ_t right after the jumps: $\theta^i(\tau^i_1), \theta^i(\tau^i_2), \ldots, \theta^i(\tau^i_{m(i)})$. Here $\tau^i_{m(i)}$ is the last observation of the health transition, and $t^i_{n(i)}$ is the last measurement of the physiological state for individual

$i, \tau^i_{m(i)} < T_i$ and $t^i_{n(i)} < T_i$. Each individual is characterized by the ordered sequence of ages at which different measurements took place. For example, the sequence (t_1, $t_2, \tau_1, t_3, \tau_2, \tau_3, t_4, \ldots, \tau_m$) indicates that the first health transition happened between the second and third measurements of the physiological state, the second and third health transitions occurred between the third and fourth physiological measurements, etc. Such sequences could be different for different study participants. Note that in our model $P(\tau_k = t_r) = 0$ for any k and r.

Let $\widehat{\widehat{f}}(y,t) = \frac{\partial}{\partial y} P\left(Y_t \leq y \middle| \widetilde{Y}^t_0, \theta^t_0, T > t\right)$ be the conditional probability density function of Y_t given observations $\widetilde{Y}^t_0, \theta^t_0$ and $\{T > t\}$. This function satisfies Eq. (15.10) of Chap. 15 with $\widehat{\lambda}_{\theta_{t-},k}(t)$ and $\widehat{\mu}_{\theta_{t-}}(t)$ replaced by $\widehat{\widehat{\lambda}}_{\theta_{t-},k}(t)$ and $\widehat{\widehat{\mu}}_{\theta_{t-}}(t)$ where

$$\widehat{\widehat{\lambda}}_{\theta_{t-},k}(t) = E\left(\lambda_{\theta_{t-},k}(Y_t,t) \middle| \widetilde{Y}^t_0, \theta^{t-}_0, T > t\right) \tag{16.25}$$

and

$$\widehat{\widehat{\mu}}_{\theta_{t-}}(t) = E\left(\mu_{\theta_{t-}}(Y_t,t) \middle| \widetilde{Y}^t_0, \theta^{t-}_0, T > t\right) \tag{16.26}$$

at each interval resulting from combining and ordering $t_1, t_2, \ldots, t_n$ and $\tau_1, \tau_2, \ldots, \tau_m$. If an interval starts with t_k, then modified forms of Eq. (15.10, Chap. 15) for $\widehat{\widehat{f}}(y,t)$ have to be used with the initial condition:

$$\widehat{\widehat{f}}(y_{t_k}, t_k) = \delta(y - y_{t_k}).$$

If an interval starts with τ_p, then these equations have to be used with initial conditions:

$$\widehat{\widehat{f}}(y, \tau_p) = \widehat{\widehat{f}}(y, \tau_p-) \frac{\lambda_{\theta(\tau_p-),\theta(\tau_p)}(y, \tau_p)}{\widehat{\widehat{\lambda}}_{\theta(\tau_p-),\theta(\tau_p)}(\tau_p)} \tag{16.27}$$

where

$$\widehat{\widehat{\lambda}}_{\theta(\tau_p-),\theta(\tau_p)}(\tau_p) = E\left(\lambda_{\theta(\tau_p-),x}(Y_{\tau_p}, \tau_p) \middle| \widetilde{Y}^{\tau_p}_0, \theta^{\tau_p-}_0, T > \tau_p\right)\Big|_{x=\theta(\tau_p)}. \tag{16.28}$$

An example of the likelihood function for the *i-th* individual with observations occurring at the sequence of times $t^i_1, t^i_2, \tau^i_1, t^i_3, \tau^i_2, \tau^i_3, \ldots, \tau^i_{m(i)}, t^i_{n(i)}, T^i$ is:

$$\widehat{\widehat{L}}_i\left(y^i_{t^i_1}, y^i_{t^i_2}, \ldots, y^i_{t^i_{n(i)}}, \theta^i(\tau^i_1), \theta^i(\tau^i_2), \ldots, \theta^i\left(\tau^i_{m(i)}\right), T_i\right)$$

$$= \widehat{\widehat{f}}\left(y_{t^i_1}, t^i_1-\right)\widehat{\widehat{f}}\left(y_{t^i_2}, t^i_2-\right)\widehat{\widehat{f}}\left(y_{\tau^i_1}, \tau^i_1-\right)\widehat{\widehat{f}}\left(y_{t^i_3}, t^i_3-\right)\widehat{\widehat{f}}\left(y_{\tau^i_2}, \tau^i_2-\right)$$

$$\widehat{\widehat{f}}\left(y_{\tau^i_3}, \tau^i_3-\right)\ldots\widehat{\widehat{f}}\left(y_{s^i}, s^i-\right) \times p\left(\theta^i\left(t^i_0\right)\right)\prod_{p=1}^{m(i)}\widehat{\widehat{\lambda}}_{\theta^i\left(\tau^i_p-\right),\theta^i\left(\tau^i_p\right)}\left(\tau^i_p\right)\widehat{\widehat{\mu}}_{\theta^i(T_i)}(T_i)^{\delta_i}$$

$$\exp\left\{-\int_{t^i_0}^{T_i}\left(\sum_{k=1,\,k\neq\theta^i(t-)}^{M}\widehat{\widehat{\lambda}}_{\theta^i(t-),k}(t) + \widehat{\widehat{\mu}}_{\theta^i(t-)}(t)\right)dt\right\}. \tag{16.29}$$

Here $s^i = \max\left\{t^i_{n(i)}, \tau^i_{m(i)}\right\}$, $\cdot\ \widehat{\widehat{f}}\left(y, t^i_k-\right) = \frac{\partial}{\partial y}P\left(Y_{t^i_k} \leq y \middle| Y_0^{t^i_k-}\theta_0^{t^i_k-}, T > t^i_k\right)$ is the solution of the modified forms of equations Eq. (15.10, Chap. 15) either at the interval $[\tau_p, t^i_k)$, or at the interval $\left[t^i_{k-1}, t^i_k\right)$, assuming that these intervals do not contain other observations; and $\widehat{\widehat{f}}\left(y, \tau^i_k-\right) = \frac{\partial}{\partial y}P\left(Y_{t^i_k} \leq y \middle| Y_0^{t^i_k-}\theta_0^{t^i_k-}, T > t^i_k\right)$ is the solution of the modified forms of equations Eq. (15.10, Chap. 15) either at the interval $[\tau_{k-1}, \tau_k)$, or at the interval $\left[t^i_{k-1}, t^i_k\right)$, also assuming that these intervals do not contain other observations.

16.2.8 *Gaussian Approximation of the Model of Longitudinal Data on Physiological Variables and Health Transitions*

The Gaussian approximation facilitates analyses of combined data. Let $\widehat{m}(t) = E\left(Y_t \middle| \widetilde{Y}^t_0, \theta^t_0, T > t\right)$ and $\widehat{\widehat{\gamma}}(t) = E\left(\left(Y_t - \widehat{m}(t)\right)\left(Y_t - \widehat{m}(t)\right)^* \middle| \widetilde{Y}^t_0, \theta^t_0, T > T\right)$ be the first two moments of the conditional probability density function $\widehat{\widehat{f}}(y, t) = \frac{\partial}{\partial y}P\left(Y_t \leq y \middle| \widetilde{Y}^t_0\theta^t_0, T > t\right)$. These moments satisfy Eqs. (15.16) and (15.17) of Chap. 15 at each interval resulting from combining and ordering the observation times t_1, $t_2, \ldots t_n$ and times of health transitions $\tau_1, \tau_2, \ldots, \tau_m$.

If an interval starts with t_k, then equations Eq. (15.16, Chap. 15) for $\widehat{m}(t)$ starts with the initial condition $\widehat{m}(t_k) = y_{t_k}$ and, for $\widehat{\gamma}(t)$, with the condition $\widehat{\gamma}(t_k) = 0$. If an interval starts with τ_p, then equations Eq. (15.16, Chap. 15) starts with the initial condition (16.21) with $\widehat{m}(\tau_p-)$ and $\widehat{\gamma}(\tau_p-)$ replaced by $\widehat{m}(\tau_p-)$ and $\widehat{\gamma}(\tau_p-)$.

The likelihood function of the data for the *i-th* individual when both physiological variables and health transitions are measured is:

$$\widehat{L}_i^G\left(y_{t_1^i}^i, y_{t_2^i}^i, \ldots y_{t_{n(i)}^i}^i, \theta^i(\tau_1^i), \theta^i(\tau_2^i), \ldots, \theta^i\left(\tau_{(m)i}^i\right), T_i\right) = \prod_{j=1}^{n(i)+m(i)}\left(\left(2\pi\left|\widehat{\gamma}^i\left(u_{t_j^i}^i-\right)\right|\right)^{-\frac{K}{2}}\right.$$
$$\exp\left\{-\frac{1}{2}\left(y_{u_j^i}^i - \widehat{m}^i\left(u_j^i\right)\right)^* \widehat{\gamma}_k^i\left(u_{t_j^i}^i-\right)^{-1}\left(y_{t_j^i}^i - \widehat{m}^i\left(u_j^i-\right)\right)\right)$$
$$\times p\left(\theta^i(t^i)\right)\prod_{p=1}^{m(i)}\widehat{\lambda}_{\theta^i(\tau_p^i-),\theta^i(\tau_p^i)}\left(\tau_p^i\right)\widehat{\mu}_{\theta^i(T^i)}(T_i)^{\delta_i}$$
$$\exp\left\{-\int_{t_0^i}^{T_i}\left(\sum_{k=1,\,k\neq\theta^i(t-)}^{M}\widehat{\lambda}_{\theta^i(t-),k}(t) + \widehat{\mu}_{\theta^i(t-)}(t)\right)dt\right\}. \tag{16.30}$$

Here $u_k^i, k = 1, 2, \ldots n(i) + m(i)$ is the element of an ordered sequence combined from $t_1^i, t_2^i, \ldots, t_{n(i)}^i$ and $\tau_1^i, \tau_2^i, \ldots, \tau_{m(i)}^i$. If an interval starts with t_k^i, then Eqs. (15.16) and (15.17) of Chap. 15 start with the initial conditions $m_{t_k^i} = y_{t_k^i}, \gamma_{t_k^i} = 0$. If an interval starts with τ_p^i, then Eqs. (15.16) and (15.17) of Chap. 15 start with the initial conditions (16.21) and (16.22):

$$\widehat{\mu}(t) = \mu_0(t) + \left(\widehat{m}(t) - f_j(t)\right)^* Q_j(t)\left(\widehat{m}(t) - f_j(t)\right) + Tr\left(Q_j(t)\,\widehat{\gamma}(t)\right). \tag{16.31}$$

The transition intensities $\widehat{\lambda}_{kj}(t)$ are:

$$\widehat{\lambda}_{kj}(t) = \lambda_{0kj}(t) + \left(\widehat{m}(t) - g_k(t)\right)^* \Lambda_{kj}(t)\left(\widehat{m}(t) - g_k(t)\right)$$
$$+ Tr\left(\Lambda_{kj}(t)\,\widehat{\gamma}(t)\right). \tag{16.32}$$

The likelihood function (16.30) can be used for combining data, as shown below.

16.3 A Simulation Study

When the mortality model is specified in terms of the processes θ_t and Y_t and random variables G and Z, its parameters can be identified from the data only if the observational plan is detailed enough. For example, using one of Observational Plans #1 or #2 alone is not enough to identify the model parameters. However, when the parameters of the health transition process θ_t are known (say, from some other study), the use of longitudinal data provided by the Observational Plan #1 allows one to evaluate the parameters of the biomarker process Y_t. Similarly, when the parameters associated with the biomarker process Y_t are known (from some other study), the use of longitudinal data provided by the Observational Plan #2 allows one to evaluate the parameters of the health transition process θ_t. This situation suggests that if one has two datasets for the two groups of individuals, one collected using Observational Plan #1 and another collected using Observational Plan #2, then combining these two datasets and performing their analyses jointly may result in identifying the parameters of a comprehensive model. To confirm the feasibility of such identification, we performed the following simulation/estimation experiment.

16.3.1 The Model with Repeated Measurements of a Physiological Variable and Changes in Health State: Observational Plan #3

First, we estimated parameters of a model in which each individual could be characterized by a continuously changing physiological index Y_t, whose dynamics are described by the stochastic differential equations Eq. (15.13, Chap. 15), and possible transitions from healthy ("*H*"), to unhealthy (diseased "*D*") states. In this model, the process Y_t represented diastolic blood pressure will hypertension used as the unhealthy state. We assumed: (i) time independence of parameters describing the age dynamics of physiological variables in each of two states, i.e., a_H,a_D,b_H,$b_D f_{1H}$ and f_{1D}, (ii) a Gompertz type function for $\mu_{0i}(t) = \mu_{0i}\exp(\alpha_i t)$, $i = H, D$, and for $\lambda_{0HD}(t) = \lambda_{0HD}\exp(\alpha_{HD}t)$, and (iii) time independence of other parameters describing transition probabilities, i.e., Q_H, Q_D, f_H, f_D, Λ_{HD} and g_{HD}.

The mortality rates from healthy and unhealthy states and transition intensity functions (Eqs. (15.12) and (15.13), Chap. 15) are specified as:

$$\begin{aligned}\bar{\mu}_H(t) &= \mu_{0H}(t) + (m_H(t) - f_H(t))^2 Q_H(t) + Q_H(t)\gamma_H(t), \\ \bar{\mu}_D(t) &= \mu_{0D}(t) + (m_D(t) - f_D(t))^2 Q_D(t) + Q_D(t)\gamma_D(t),\end{aligned} \qquad (16.33)$$

$$\bar{\lambda}_{HD}(t) = \lambda_{0HD}(t) + (m_H(t) - g_H(t))^2 \Lambda_{HD}(t) + \Lambda_{HD}(t)\gamma_H(t). \qquad (16.34)$$

The differential equations for the conditional moments become:

$$\frac{dm_H}{dt} = 2\gamma_H \Lambda_{HD} \hat{g}_H - a_H \hat{f}_{1H} + 2\gamma_H Q_H \hat{f}_H, \tag{16.35}$$

$$\frac{dm_D}{dt} = \frac{\pi_H}{\pi}\left[m_{HD}\overline{\lambda}_{HD} - 2\gamma_H \Lambda_{HD} \hat{g}_H\right] - a_D \hat{f}_{1D} + 2\gamma_D Q_D \hat{f}_D, \tag{16.36}$$

$$\frac{d\gamma_H}{dt} = -2\gamma_H^2 \Lambda_{HD} + 2a_H\gamma_H + b_H^2 - 2\gamma_H^2 Q_H. \tag{16.37}$$

$$\frac{d\gamma_D}{dt} = \frac{\pi_H}{\pi_D}\left[\left(\gamma_H - \gamma_D + m_{HD}^2\right)\overline{\lambda}_{HD} + 2\left(\gamma_H^2 \Lambda_{HD} - 2\gamma_H \Lambda_{HD} \hat{g}_H \cdot m_{HD}\right)\right] + 2a_D\gamma_D + b_D^2 - 2\gamma_D^2 Q_D. \tag{16.38}$$

Finally, the equations for the conditional probabilities of being in the healthy and unhealthy states are:

$$\begin{aligned} \frac{d\pi_H}{dt} &= -\pi_H\overline{\lambda}_{HD} + \pi_H\left(\pi_H\overline{\mu}_H + \pi_D\overline{\mu}_D - \overline{\mu}_H\right), \\ \frac{d\pi_D}{dt} &= \pi_H\overline{\lambda}_{HD} + \pi_D\left(\pi_H\overline{\mu}_H + \pi_D\overline{\mu}_D - \overline{\mu}_D\right). \end{aligned} \tag{16.39}$$

For simplicity, we omitted the dependence on t in the variables in the right hand side of these equations. The initial values for $\widehat{m}\left(\tau_p\right)$ and $\widehat{\gamma}\left(\tau_p\right)$ for the case of a time interval starting by a jump are

$$\frac{\widehat{m}\left(\tau_p\right) = \widehat{m}\left(\tau_p-\right) - 22\,\widehat{\gamma}\left(\tau_p-\right)\Lambda_{HD}\hat{g}_j\left(\tau_p-\right)}{\dfrac{\widehat{\lambda}_{HD}\left(\tau_p-\right) \text{ and } \widehat{\gamma}\left(\tau_p\right) = \widehat{\gamma}\left(\tau_p-\right) + 2\widehat{\gamma}^2\left(\tau_p-\right)\Lambda_{HD}}{\widehat{\lambda}_{HD}\left(\tau_p-\right).}} \tag{16.40}$$

The purpose of these analyses is to get a set of realistic model parameters appropriate for use in data simulation. Several parameters estimates were rounded to simplify the description. The following set of parameters was finally used: $a_H = -0.05$, $a_D = -0.03$, $b_H^2 = 10$, $b_D^2 = 15$, $f_{1H} = 80$, $f_{1D} = 90$, μ_{0H} =0.00002, $\alpha_H = 0.08$, $\mu_{0D} = 0.002$, $\alpha_D = 0.045$, $Q_H = 0.00001$, $Q_H = 0.00007$, $\Lambda_{HD} = 0.00005$, $\lambda_{0HD} = 0.00005$, $\alpha_{HD} = 0.065$, $f_{0H} = 80$, $f_{0D} = 80$, and $g_H = 72$. Here f_{1H}, $f_{1D} f_{0H} f_{0D}, g_H, b_H$, and b_D are measured in the units of the selected covariate (e.g., mmHg for blood pressure), a_H and a_D are dimensionless, $\mu_{0H}, \alpha_H, \mu_{0D}, \alpha_D, \lambda_{0HD}$, and α_{HD} are in units of year^{-1}, and, finally, Q_H, Q_D and Λ_{HD} are in units of year^{-1} multiplied by the reciprocal of the covariate unit squared.

To test the quality of the simulation procedure, we simulated data on ten cohorts using these parameters. The starting age for individuals in these cohorts was

50 years with 25 % of the individuals initially unhealthy. The starting values for the covariate (diastolic blood pressure) were 80 mmHg for healthy and 85 mmHg for unhealthy individuals. Mortality and transitions to the unhealthy state were simulated for each month. Information about the vital status, the health state, and the covariate value was recorded annually. Then reconstructed characteristics of the simulated cohorts were compared with the theoretical two-stage model, which can be specified for these two stages as described below.

Figure 16.1 illustrates the quality of the data simulation. It compares the theoretical curves, corresponding to the equations derived above and calculated with the parameters used in the data simulation procedure, and the age patterns of the respective characteristics in the simulated cohorts empirically averaged over 5000 individuals.

It can be seen from this figure that the simulated data are in good correspondence with the theoretical model. The results of the simulation studies are similar to those presented in Yashin et al. (2011) where $f_{1D} = 85$ was used for simulation. The plots are qualitatively similar, which shows the stability of the model predictions.

Next, we use the simulated data and the estimation scheme based on maximization of the likelihood (16.30) (Observational Plan #3) to estimate the model parameters again. The likelihood function for the case of two discrete states can be represented as a product of terms whose functional form depends on the types of observational events forming the respective age intervals. These events are associated with: (i) measurements of physiological state, (ii) changes in health status, and (iii) transitions to death/censoring. There are five types of such interval-specific contributions shown in Table 16.1 below.

Measurement → measurement	$\widehat{f}_c(y_{t_k}, t_{k-1}, -) S_c(t_{k-1}, t_k)$ for $c = H, D$
Measurement → jumping (HD)	$\widehat{\lambda}_{HD}(\tau) S_H(t_k, \tau)$
Measurement → death/censoring	$\widehat{\mu}_k(T)^{\delta_i} S_k(t_k, T)$ for $k = H, D$
Jumping → measurement	$\widehat{f}_D(y_{t_k}, \tau, t_k -) S_D(\tau, t_k)$
Jumping → death	$\widehat{\mu}_D(T)^{\delta_i} S_D(\tau, T)$

where $\widehat{f}_G(y_{t_k}, u, t_k -) = \left(2\pi \left|\widehat{\gamma}(t_k -)\right|\right)^{-\frac{1}{2}} \exp\left\{-\frac{1}{2}\left(y_{t_k} - \widehat{m}(t_k -)\right)^* \widehat{\gamma}(t_k -)^{-1}\right.$ $\left.\left(y_{t_k} - \widehat{m}(t_k -)\right)\right\}$, where $\widehat{m}(t)$ and $\widehat{\gamma}(t)$, obtained as solutions of (16.35 and 16.36) and (16.37 and 16.38) with initial condition (16.40) if $u = \tau$, and initial condition (16.13) if $u = t_k$; where

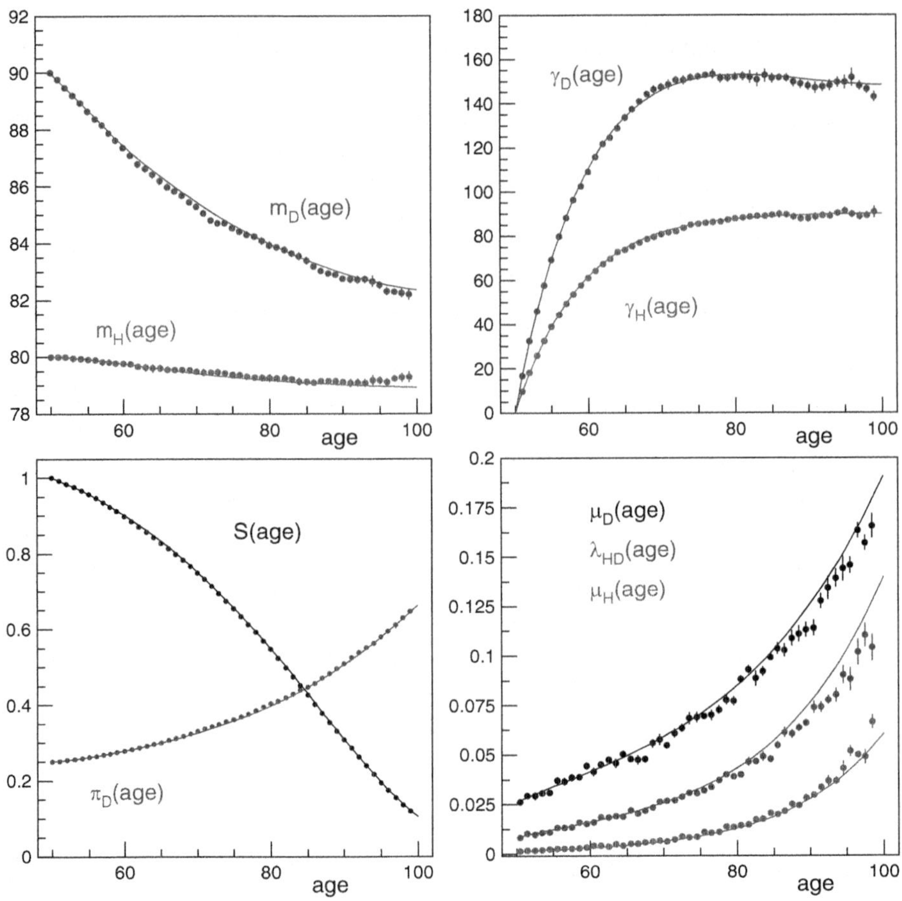

Fig. 16.1 Age patterns of dynamic characteristics in the simulation analysis. *Solid lines* represent the "true" age trajectories constructed by solving the differential equations with model parameters used in the data simulation. The *dots* with error bars represent the age trajectories of the mean values of these characteristics evaluated from empirical analyses of longitudinal data on aging, health, and mortality from 10 simulated cohorts with 5000 individuals each. The following characteristics are presented: (1) *Top left panel*: First moments of the hypothetical variable describing the physiological state of the simulated cohort of healthy and unhealthy individuals, $m_H(t)$ and $m_D(t)$. (2) *Top right panel*: Age trajectories of second central moments of the probability distributions of the hypothetical variable describing the physiological state for the simulated cohort of healthy and unhealthy individuals, $\gamma_H(t)$ and $\gamma_D(t)$. (3) *Bottom left panel*: Age patterns of the survival function and prevalence of unhealthy individuals in the simulated cohort. (4) *Bottom right panel*. Age patterns of mortality rates for the healthy and unhealthy states and the transition rate from the healthy to unhealthy states

Table 16.1 Results of the simulation experiment for Observational Plan #3 with ten datasets

		True	Mean	SD	SE
a_H		**−0.05**	−0.051	0.002	0.001
a_D		**−0.03**	−0.03	0.002	0.001
b_H	[C]	**10**	9.97	0.06	0.02
b_D	[C]	**15**	15	0.07	0.02
f_{1H}	[C]	**80**	80.1	0.25	0.08
f_{1D}	[C]	**85**	84.8	0.55	0.18
$\mu_{0H} \times 10^5$	Year^{-1}	**2**	1.96	0.33	0.1
$\mu_{0D} \times 10^5$	Year^{-1}	**200**	201.3	24.3	7.7
α_H	Year^{-1}	**0.08**	0.082	0.002	0.001
α_D	Year^{-1}	**0.045**	0.045	0	0
$Q_H \times 10^5$	$\text{Year}^{-1}[\text{C}]^{-2}$	**1**	1.23	0.21	0.07
$Q_D \times 10^5$	$\text{Year}^{-1}[\text{C}]^{-2}$	**7**	7.14	0.5	0.16
$\lambda_{0HD} \times 10^5$	Year^{-1}	**20**	23.5	4.77	1.51
α_{HD}	Year^{-1}	**0.065**	0.063	0.002	0.001
$\Lambda_{HD} \times 10^5$	$\text{Year}^{-1}[\text{C}]^{-2}$	**5**	4.66	0.62	0.2
f_H	[C]	**80**	79.3	1.3	0.4
f_D	[C]	**80**	80	0.96	0.3
g_H	[C]	**72**	71.7	1.01	0.32

[C] denotes dimensionality of a covariate

$$S_H(t_1,t_2) = \exp\left\{-\int_{t_1}^{t_2}\left(\widehat{\lambda}_{HD}(t) + \widehat{\mu}_H(t)\right)dt\right\} \text{ and } S_D(t_1,t_2) = \exp\left\{-\int_{t_1}^{t_2}\left(\widehat{\mu}_D(t)\right)dt\right\}.$$

Figure 16.2 presents the characteristics of the simulated cohorts calculated using the estimated parameter values and compares them with the trajectories obtained using the original data.

Some bias in the empirical and reconstructed parameters is likely to be due to "doublings" of the subsequent events (healthy→unhealthy transition and death) occurring at the same time interval that are interpreted in Observational Plan #3 as death from the healthy state.

16.3.2 Combining Data with Observational Plans #1 and #2

To confirm the benefits from such analyses, we considered two subsets of simulated datasets. One subset deals with data collected according to Observational Plan #1, i.e., these data represent a cohort of individuals for whom the measurements of physiological variables have been performed during each individual's life course.

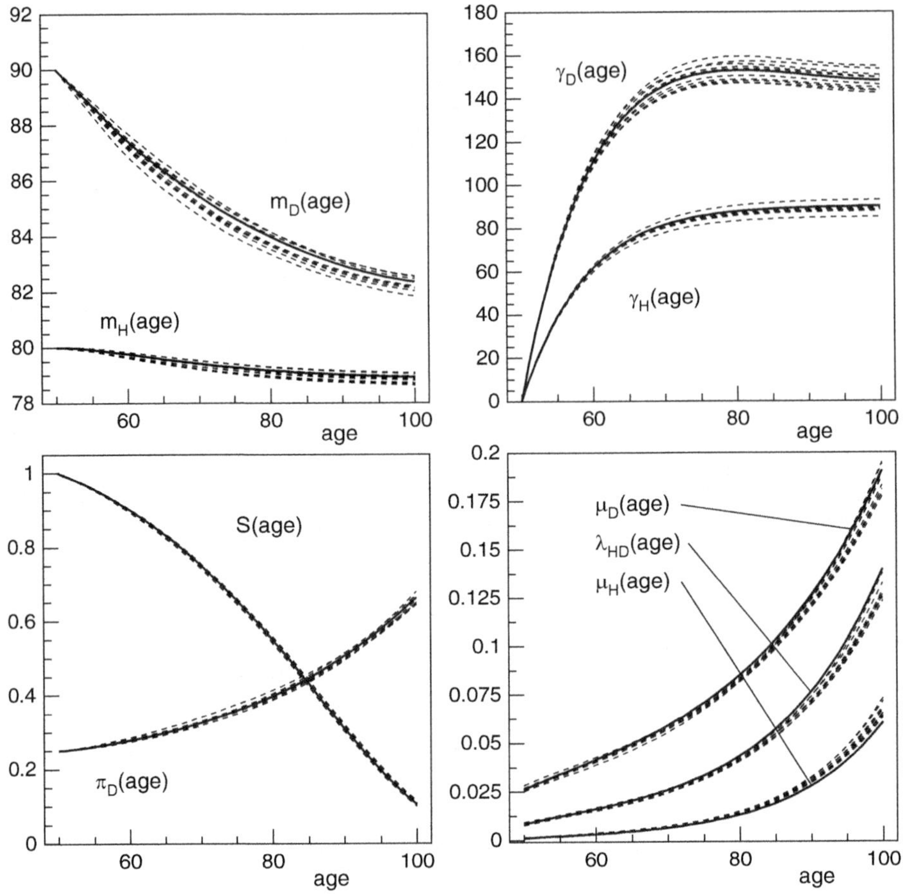

Fig. 16.2 Age patterns of physiological and life history characteristics calculated from the theoretical model (*solid lines*), and using parameter estimates obtained by maximization of the likelihood for observational plan #3 for ten simulated cohorts with 5000 individuals each (*dashed lines*). The same characteristics as in Fig. 16.1 are presented in the four panels

Another dataset consists of records of health transitions for another group of individuals who do not have records of measurements of physiological indices. We used the estimation scheme based on the likelihood (16.30) in which each component of the likelihood represents the corresponding dataset. Note that the use of data collected in Observational Plan #1 alone does not allow for evaluating all parameters characterizing health transitions and physiological age trajectories. Similarly, having data collected in Observational Plan #2 alone does not allow for evaluating all parameters characterizing changes in physiological age trajectories. However, maximization of the joint likelihood function for the two sub-cohorts of individuals allows for reliable estimation of model parameters for both components of the model.

Table 16.2 The results of the five simulation experiments of the 18-parameter model for joint analyses of data with Observational Plans #1 and #2 (1000 persons for Plan 1 and 10,000 persons for Plan 2)

		True	1	2	3	4	5
a_H		**−0.05**	−0.056	−0.050	−0.053	−0.051	−0.044
a_D		**−0.03**	−0.026	−0.049	−0.034	−0.034	−0.055
b_H	[C]	**10**	10.0	10.1	10.0	10.2	9.8
b_D	[C]	**15**	14.9	14.9	14.6	15.0	15.7
f_{1H}	[C]	**80**	81.9	81.3	82.4	78.5	79.4
f_{1D}	[C]	**85**	76.1	86.3	82.6	83.5	86.2
$\mu_{1H} \times 10^5$	Year^{-1}	**2**	3.2	0.7	2.8	4.2	2.1
α_H	Year^{-1}	**0.08**	0.076	0.091	0.076	0.071	0.078
$Q_H \times 10^5$	$\text{Year}^{-1}[\text{C}]^{-2}$	**1**	<0.1	3.1	0.4	<0.1	1.4
$\mu_{0D} \times 10^5$	Year^{-1}	**200**	188.9	213.0	156.5	205.4	185.7
α_D	Year^{-1}	**0.045**	0.045	0.044	0.047	0.045	0.047
$Q_D \times 10^5$	$\text{Year}^{-1}[\text{C}]^{-2}$	**7**	8.0	10.1	10.6	9.5	5.0
$\lambda_{0HD} \times 10^5$	Year^{-1}	**20**	11.1	15.9	8.1	28.5	19.9
α_{HD}	Year^{-1}	**0.065**	0.071	0.067	0.074	0.061	0.065
$\Lambda_{HD} \times 10^5$	$\text{Year}^{-1}[\text{C}]^{-2}$	**5**	6.2	4.5	6.6	2.9	5.2
f_H	[C]	**80**	78.6	78.8	75.2	80.0	79.9
f_D	[C]	**80**	78.9	83.5	79.7	82.9	80.0
g_H	[C]	**72**	72.3	70.2	71.7	69.9	71.8

[C] denotes dimensionality of a covariate

The approach described above opens remarkable opportunities for joint analyses of incompatible data collected using different observational plans. In this section, we implement this idea in the joint analysis of data corresponding to Observational Plans 1 and 2. An important property of the model is that it can be used for constructing likelihood functions of the data for each of the plans (see Eqs. (16.9) and (16.18). These functions depend on the same parameters and represent different data on independent individuals; therefore, the likelihood of combined data is the product of these two likelihood functions. Table 16.2 presents results of such parameter estimates for five simulation experiments.

The parameter estimation procedure based on maximization of this likelihood improves the accuracy of estimates for parameters which could be estimated from one of the two datasets. However, the main advantage of such joint analyses of several datasets is the possibility of estimating new parameters which could not be estimated by analyzing the two datasets separately. This means that combining incompatible data in joint analyses allows for addressing new research questions and obtaining new findings, which will contribute to a better understanding of the nature of the processes under study.

16.4 Discussion and Conclusion

The results of recent studies of aging and longevity in laboratory animals motivate studying the possibility of postponing or slowing down aging processes in humans. The beneficial consequences of such postponement for health and survival have been widely discussed in the literature (Blagosklonny 2012; Butler et al. 2004; Cevenini et al. 2013; Hefti and Bales 2006; Kirkland and Peterson 2009; Kristjuhan 2012; Le Bourg 2009; Olshansky et al. 2007; Rattan 2008; Vijg and Wei 1995; Yang et al. 2012; Yashin 2009). To study the connections between human aging and chronic conditions, data collected in longitudinal studies of aging, health, and longevity can be used. A good example of such a longitudinal dataset is the Framingham Heart Study. The data on the original cohort of this study contain the results of biennial examinations of physiological states, biological indicators, and ages at onset of a number of chronic conditions, such as cardiovascular disease, cancer, and diabetes, performed during more than 60 years of follow-up.

Progress in understanding aging processes depends to a large extent on the possibility of evaluating properties of complicated mechanisms that link aging and development of chronic diseases. The quality of such an evaluation, as well the possibility of addressing fundamental research questions about aging, health, and longevity, depend on the amount of information that can be used in the analyses. For these reasons, the joint analyses of large amount of data describing various aspects of aging-related changes and disease development looks promising. Such analyses require appropriate mathematical description of aging mechanisms and their connections to available data, as well as efficient statistical methods for estimating parameters of the corresponding processes from the data. The results of this chapter indicate that comprehensive analyses of available longitudinal data on human aging, health, and longevity can be successfully performed.

The use of a diffusion-type continuous-time stochastic process for describing the evolution of physiological states over the life course allows us to take recent findings in the area of aging into account, and incorporate them into the model of aging-related changes in physiological states. The use of a finite-state continuous-time stochastic process with transition rates depending on current values of physiological variables for describing changes in health status during this period captures important connections between aging-related changes, morbidity, and mortality risks. We derived equations for integrative mortality models, and approximate changes in physiological state in terms of the first two moments of a conditional Gaussian process, given the health state. The simulations showed that the model parameters can be successfully estimated from the data.

Different longitudinal studies of aging, health, and longevity have different designs, use different systems of measurements (observational plans), and measure different biological variables and health outcomes. In this chapter, we showed (Table 16.2) that data with different observational plans describing the same basic phenomena (aging-related changes) can be jointly analyzed successfully. Although such analyses are computationally extensive, the reward is clear: more sophisticated research questions can be addressed in joint analyses of combined data. This is because separate analyses of each dataset are not able to estimate all links among aging-related changes and health and survival outcomes. This means that separate analyses can be performed only with simplified models of aging, health, and longevity which have lower scientific value.

In sum, this chapter has illustrated how the study of aging, health, and longevity using comprehensive mortality models may require the use several longitudinal datasets to estimate important features of the aging mechanisms. The parameters of such models can be interpreted in terms of aging-related declines in physiological functions and deterioration in health/well-being status. The parameter estimation procedures become computationally more extensive, compared to those used in traditional analyses of survival or demographic data. The analyses, however, can be successfully performed using currently available computational equipment and software. The parameters of the proposed model can also be identified in the joint analyses of two or more incomplete datasets. The reward for these efforts is a better understanding of how changes, developing in the aging human body, affect the risks of diseases and survival. The approach described above can be extended to evaluate properties of individualized dynamic mechanisms involved in the regulation of aging-related changes in each study participant. These types of analyses will contribute to scientific knowledge promoting the development of personalized preventive and treatment strategies.

Acknowledgements The research reported in this chapter was supported by the National Institute on Aging grants R01AG027019, R01AG030612, R01AG030198, 1R01AG046860, and P01AG043352. The content is solely the responsibility of the authors and does not necessarily represent the official views of the National Institute on Aging or the National Institutes of Health. The Framingham Heart Study (FHS) is conducted and supported by the National Heart, Lung and Blood Institute (NHLBI) in collaboration with the FHS Investigators. This chapter was prepared using a limited access dataset obtained from the NHLBI and does not necessarily reflect the opinions or views of the FHS or the NHLBI.

References

Blagosklonny, M. V. (2012). Rapalogs in cancer prevention: Anti-aging or anticancer? *Cancer Biology and Therapy, 13*(14), 1349–1354.

Butler, R. N., Warner, H. R., Williams, T. F., Austad, S. N., Brody, J. A., Campisi, J., Cerami, A., Cohen, G., Cristofalo, V. J., Drachman, D. A., Finch, C. E., Fridovich, I., Harley, C. B., Havlik, R. J., Martin, G. M., Miller, R. A., Olshansky, S. J., Pereira-Smith, O. M., Smith, J. R., Sprott, R. L., West, M. D., Wilmoth, J. R., & Wright, W. E. (2004). The aging factor in health and disease: The promise of basic research on aging. *Aging Clinical and Experimental Research, 16*(2), 104–111; discussion 111–102.

Cevenini, E., Monti, D., & Franceschi, C. (2013). Inflamm-ageing. *Current Opinion in Clinical Nutrition and Metabolic Care, 16*(1), 14–20.

Hefti, F. F., & Bales, R. (2006). Regulatory issues in aging pharmacology. *Aging Cell, 5*(1), 3–8.

Kirkland, J. L., & Peterson, C. (2009). Healthspan, translation, and new outcomes for animal studies of aging. *Journals of Gerontology. Series A, Biological Sciences and Medical Sciences, 64*(2), 209–212.

Kristjuhan, U. (2012). Postponing aging and prolonging life expectancy with the knowledge-based economy. *Rejuvenation Research, 15*(2), 132–133.

Le Bourg, E. (2009). Hormesis, aging and longevity. *Biochimica et Biophysica Acta, 1790*(10), 1030–1039.

Olshansky, S. J., Perry, D., Miller, R. A., & Butler, R. N. (2007). Pursuing the longevity dividend: Scientific goals for an aging world. *Annals of the New York Academy of Sciences, 1114*, 11–13.

Rattan, S. I. (2008). Principles and practice of hormetic treatment of aging and age-related diseases. *Human and Experimental Toxicology, 27*(2), 151–154.

Vijg, J., & Wei, J. Y. (1995). Understanding the biology of aging: The key to prevention and therapy. *Journal of the American Geriatrics Society, 43*(4), 426–434.

Yang, Y., Li, T., & Nielsen, M. E. (2012). Aging and cancer mortality: Dynamics of change and sex differences. *Experimental Gerontology, 47*(9), 695–705.

Yashin, A. I. (2009). Hormesis against aging and diseases: Using properties of biological adaptation for health and survival improvement. *Dose-Response: A Publication of International Hormesis Society, 8*(1), 41–47.

Yashin, A. I., Akushevich, I., Arbeev, K. G., Kulminski, A., & Ukraintseva, S. (2011). Joint analysis of health histories, physiological states, and survival. *Mathematical Population Studies, 18*(4), 207–233.

Chapter 17
Analysis of the Natural History of Dementia Using Longitudinal Grade of Membership Models

Eric Stallard and Frank A. Sloan

17.1 Introduction

Continuing increases in longevity have long been expected to produce dramatic increases in the incidence and prevalence of mental disorders and associated chronic diseases and disabilities (Kramer 1980). This expectation is now a reality for Alzheimer's disease and related dementias (Alzheimer's Association 2016); the resulting challenge for biodemographers is to extend their analytic repertoire to adequately describe the complex multidimensional long-term progressive array of cognitive, functional, behavioral, and clinical characteristics of dementia cases to facilitate the analysis of longitudinal panel data on these subpopulations.

This chapter presents a promising solution to this challenge based on a longitudinal form of the Grade of Membership (GoM) model (Woodbury et al. 1993; Kinosian et al. 2000; Stallard 2007; Stallard et al. 2010; Razlighi et al. 2014). The presentation of the model is self-contained—providing sufficient mathematical details that the reader can fully understand the concepts and methodology—and the application is realistic—addressing the complexities of the dementia process and the difficulties in such analysis. The specific formulation of the dementia model presented in this chapter has been independently validated for Alzheimer's disease (a subset the dementia syndrome) in Stallard et al. (2010) and Razlighi et al. (2014).

Dementia is the primary cause of cognitive impairment and subsequent deterioration in functioning among older persons, leading to long-term care (LTC), nursing home admission, and death. The disease is progressive, generally fatal, and characterized by substantial heterogeneity in etiology, symptomatology at diagnosis, and rates of progression post-diagnosis. The absence of well-defined biomarkers complicates the diagnosis and contributes further to the apparent heterogeneity of cases due to differences in the manifestations of the disease at the time of diagnosis (Grossberg and Desai 2003).

Much of the literature on the progression of dementia is based on samples of patients selected through contacts with medical providers or recruited into clinical

A.I. Yashin et al., *Biodemography of Aging*, The Springer Series on Demographic Methods and Population Analysis 40, DOI 10.1007/978-94-017-7587-8_17

trials after meeting specified protocols (Kinosian et al. 2000). Alzheimer's disease (AD), the primary component of dementia, has been described as exhibiting three stages—mild, moderate, and severe—with each stage lasting 2–3 years and the overall process lasting 5–8 years, on average, but with substantial individual variation in overall duration ranging from 2 to 20 years (Grossberg and Desai 2003).

Stern et al. (1996) used a growth curve model to show that declines in cognitive and physical functioning due to AD were nonlinear and multidimensional with an estimated overall average duration of 15–18 years, with initial losses in cognition preceding losses of independence in instrumental activities-of-daily-living (IADLs) and basic activities-of-daily-living (ADLs), on average, by 5 and 7 years, respectively. The process was nonlinear because the rates of loss for each of the three types of functioning (cognition, IADL, and ADL) differed over time. The process was multidimensional because the three types of functioning exhibited different patterns of time-dependent rates of loss. The following patterns were found:

- Loss of cognitive functioning was slow for the first 5 years, then rapid for years 5–15, beyond which it was slow again.
- IADL loss started at 5 years, was initially rapid, and was effectively finished at 12 years.
- ADL loss started later, at 7 years, increased more slowly and for a longer time, and was effectively finished at 17 years.

The above results indicated that improved descriptions of the onset and progression of dementia from the first manifestation of cognitive impairment to death would require (1) multiple indicators of cognitive and functional impairment in longitudinally followed cohorts with at least 15 years of follow-up and (2) multivariate models of changes in those indicators capable of representing health state transitions that underlie the losses in the various types of functioning.

When analyzing survey data with large numbers of questionnaire items, researchers are increasingly relying on the Grade of Membership (GoM) model—a flexible nonparametric model designed to analyze large numbers of categorical variables with a minimum number of assumptions (e.g., Berkman et al. 1989; Wachter 1999; Portrait et al. 2001; Seplaki et al. 2006; Wieland et al. 2013). The GoM model readily deals with substantial amounts of "missing data", a common and pervasive problem in longitudinal data due to death or early withdrawal from the study, to the use of different questionnaires for different types of respondents (e.g., one questionnaire for community residents, another for institutional residents), and to the presence of different "skip patterns" depending on the answers to specific questions. This can be done without strong distributional assumptions such as joint multivariate normality of observed and unobserved variables. The GoM model is a latent state model that produces, as a direct outcome of its estimation algorithm, vectors of "intensity" or "severity" scores for each individual person (i.e., study subject) that can be interpreted as parsimonious multivariate summaries of each individual's array of measurements. With longitudinal data on general population samples, these intensity scores (which are more generally referred to as "GoM scores") can be tracked over time and used to describe the natural history of the targeted disease process.

The GoM model has been applied to the analysis of longitudinal data with repeated time-varying measures by treating each observation of the same person as if it were that of a new person, i.e., as a statistically independent observation (Manton et al. 1991, 1992). Under this approach, the resulting estimated GoM scores are used to form a new longitudinal dataset which can be analyzed separately in an independent longitudinal model. A limitation of this approach is that the estimation of the GoM scores neither explicitly recognizes nor exploits the information contained in the assumed form of GoM-score changes represented in the independent longitudinal model. This limitation produces a loss of efficiency in estimation.

We present an optimized longitudinal form of the GoM model for the analysis of initial and subsequent patterns of cognitive and functional impairment in longitudinal cohort data in which each follow-up observation of the same person is linked to all prior observations through a set of transition matrices that use the initial GoM scores for that person to generate all subsequent values of those scores. Duration-specific rates of transition from more to less healthy states within the GoM model are consistent with reports of the rates of progression of dementia described in the literature (e.g., Stern et al. 1996; Stallard et al. 2010). The linkage of each follow-up vector of GoM scores to the immediately prior vector of GoM scores through the corresponding transition matrix generalizes the growth curve model used by Stern et al. (1996) in which the change in each test score over a follow-up interval is assumed to be a function of the test score at the start of that interval. Our generalization allows more complex forms of transition from one observation to the next and allows the transitions to change with the time since onset of the disease.

Earlier versions of this model were employed in analyzing the natural history of AD (Kinosian et al. 2000, 2004) and in analyzing the age-dependence of disability and mortality in the general population (Stallard 2007). The current version tests the assumption that dementia is a complex irreversible multidimensional process governed by a latent three-dimensional bounded state-space process with upper-triangular transition matrices. Maximum likelihood parameter estimation is based on a modified Newton-Raphson algorithm that meets the Kuhn-Tucker conditions (Kuhn and Tucker 1951).

The model was estimated using longitudinal panel data from the National Long Term Care Survey (NLTCS) of 1984, 1989, 1994, and 1999. All respondents included in the model were determined to have been cognitively intact prior to the interview at which dementia was first detected. Longitudinal follow-up extended up to 15 years for those with onset detected in 1984, with shorter follow-up for those with onset detected in 1989 and 1994, and no follow-up for those with onset detected in 1999. All cases meeting the criteria for recent onset of dementia had at least one interview; 33 % had at least two interviews; 7 % had three or four interviews; and 1 % had four interviews. Data from all completed interviews were used in model estimation.

Cognitive measures included individual items in the Short Portable Mental Status Questionnaire (SPMSQ; Pfeiffer 1975) and the Mini-Mental State Exam (MMSE; Folstein et al. 1975); and self-reported and physician-reported diagnostic codes for AD and related dementias.

Functional performance measures included individual items from standard scales of basic activities of daily living (ADLs; Katz and Akpom 1976), instrumental activities of daily living (IADLs; Lawton and Brody 1969), and physical performance (Nagi 1976).

The remainder of the chapter contains four sections:

- Section 17.2 (Methods) describes the longitudinal GoM model and the constrained Newton-Raphson estimation procedures.
- Section 17.3 (Data) describes the NLTCS and related administrative files from Medicare used to select cases with recent onset of dementia.
- Section 17.4 (Results) presents and interprets the parameter estimates.
- Section 17.5 (Discussion) considers the implications of the analyses and potential future applications of the model.

The chapter describes the specification and estimation of the longitudinal GoM model and provides a substantively meaningful application to the natural history of dementia from the time of initial clinical symptoms to death. The methodology and the application are mutually reinforcing and both are necessary to motivate the model for the study of individual level processes based on generalizations of standard biodemographic methods. The substantive results indicated that the natural history of dementia was highly variable over individuals with respect to their cognitive and physical functioning at onset and the subsequent rates of loss of such functioning. Estimates of life expectancy ranged from 2 to 12 years, depending on cognitive and physical functioning at onset. This range is so large that the use of population averages may be highly misleading when used as prognosticators for subgroups of dementia cases with similar characteristics at onset. The results showed that the dementia population was continuously distributed on the derived GoM-score dimensions indicating that there was no evidence for the existence of clusters of homogeneous subgroups that could be modeled using more traditional demographic procedures—consistent with the findings in Stallard et al. (2010), based on similar analyses of AD in the Predictors Study (Stern et al. 1996).

17.2 Methods

17.2.1 Model

The basic form of the GoM model as a fuzzy-set representation of disease states based on discrete clinical variables (i.e., categorical data) was introduced in Woodbury and Clive (1974). Modifications to the basic model that facilitated estimation were presented in Woodbury et al. (1978). Manton et al. (1994) presented a comprehensive review of statistical properties and applications of the GoM model. Wachter (1999) provided an alternative derivation of the GoM model based on concepts of dimensionality reduction similar to those used in Principal Component Analysis (PCA).

Applications of the GoM model to longitudinal data were introduced in (1) Manton et al. (1991)—using an approach in which the transition matrices were estimated separately from the basic GoM model—and (2) Woodbury et al. (1993) and Kinosian et al. (2000)—using an approach in which the transition matrices were estimated simultaneously with the basic GoM model using either maximum likelihood procedures based on the fixed-point iteration algorithm or approximations to such maximum likelihood procedures based on the faster Newton-Raphson iteration algorithm. Limitations of the separate estimation approach were discussed in Stallard (2007) and will not be considered further herein. For the simultaneous estimation approach, the primary limitation was that the Newton-Raphson iteration algorithms could be applied to the approximating likelihood function, but not to the exact likelihood function. This chapter introduces new estimation procedures for the exact likelihood function based on three linked sets of Newton-Raphson iteration algorithms that satisfy the Kuhn-Tucker conditions for convergence and that, depending on the specific estimation problem, are one to three orders of magnitude faster than existing procedures based on fixed-point iteration algorithms.

The basic GoM model is specified in terms of a K-element vector of "GoM-scores," $\mathbf{g}_i$, defined for each individual i in a study population of size N, where $i = 1, \ldots, N$. Each GoM-vector (of GoM-scores) must satisfy "convexity constraints" such that all K elements are nonnegative and their sum is exactly equal to 1. The convexity constraints combine a "summation constraint", i.e., $\sum_k g_{ik} = 1$, with K "boundary constraints", i.e., $g_{ik} \geq 0,\ k = 1, \ldots, K$. Given the summation constraint, it follows that the GoM-vectors generate a $(K-1)$-dimensional space with all points falling within or on the boundaries of a regular $(K-1)$-simplex with K vertices. Geometrically, these objects are a point for $K = 1$, a line for $K = 2$, an equilateral triangle for $K = 3$, a regular tetrahedron for $K = 4$, etc.

The longitudinal GoM model is specified by adding a time subscript t to the basic GoM vector $\mathbf{g}_i$, yielding $\mathbf{g}_{it}$, with initial values $\mathbf{g}_i \equiv \mathbf{g}_{i0}$, by convention, which are updated using a sequence of transition matrices, $\mathbf{U}_t$, $t = 0, \ldots, T-1$, governing the changes in $\mathbf{g}_{it}$ from time 0 to time T. It follows that:

$$\mathbf{g}_{it}' = \mathbf{g}_{i0}' \left\{ \prod_{r=0}^{t-1} \mathbf{U}_r \right\} = \mathbf{g}_i' \mathbf{V}_{0,t} \tag{17.1}$$

where $\mathbf{g}_{it}'$ is the transpose of $\mathbf{g}_{it}$ and, in general,

$$\mathbf{V}_{s,t} = \prod_{r=s}^{t-1} \mathbf{U}_r. \tag{17.2}$$

$\mathbf{V}_{0,t}$ is the cumulative transition matrix from time 0 to time t. To preserve the convexity constraints on the GoM-vectors over time, each row of each $\mathbf{U}$-matrix must satisfy convexity constraints such that all K elements are nonnegative (a boundary constraint) and their sum is exactly equal to 1 (a summation constraint).

We refer to the GoM vector $\mathbf{g}_{it}$ as the vector of "adjusted" GoM-scores to distinguish it from the basic "unadjusted" GoM vector $\mathbf{g}_i$. This distinction becomes important when considering the impact of the $\mathbf{V}$-matrices on the GoM-score distribution at time $t > 0$ (e.g., see Figs. 17.2a, 17.2b, 17.2c, and 17.2d below).

The GoM-scores are unobserved latent state-space variables hypothesized to explain the associations among observed measurements (Woodbury and Clive 1974; Wachter 1999; Stallard et al. 2010). Precise connections between GoM-scores and observed measurements are as follows.

Let $\mathbf{x}_{it}$ denote the J-element vector of responses to J categorical variables obtained for individual i at time t. The elements of $\mathbf{x}_{it}$ are indexed by j, $j = 1, \ldots, J$, and are denoted x_{ijt}. Because the variables are categorical, it is convenient to index the response outcomes by l, $l = 1, \ldots, R_j$, where R_j is the number of response outcomes for variable j. The responses are further characterized by defining a vector of auxiliary variables, $\mathbf{y}_{it}$, where the elements of $\mathbf{y}_{it}$ are binary coded using the following convention:

$$y_{ijlt} = \begin{cases} 1, & \text{if } x_{ijt} = l \\ 0, & \text{if } x_{ijt} \neq l. \end{cases} \tag{17.3}$$

The second condition includes the case where x_{ijt} is missing at random or missing because of planned survey skip patterns including the case where a person is lost to follow-up due to withdrawal or mortality. Informative missing data can be represented by coding appropriate supplementary indicator variables.

The fundamental equation of the longitudinal GoM model is a bilinear form that describes the probability of each observed outcome m, $m = 1, \ldots, M$, in terms of a K-element vector of probabilities, denoted $\boldsymbol{\lambda}_m$, in which each element is associated with one of K latent states or pure types:

$$\text{Prob}\left(y_{ijlt} = 1\right) \equiv p_{ijlt} = \mathbf{g}_i' \left\{ \prod_{r=0}^{t-1} \mathbf{U}_r \right\} \boldsymbol{\lambda}_{mjl} = \mathbf{g}_i' \mathbf{V}_{0,t} \boldsymbol{\lambda}_{mjl}, \tag{17.4}$$

where the subscript t indexes time and the subscript m is a known function of the combination (j, l) of variable and outcome indexes, which implies that $M = \sum_j R_j$.

For each variable j, there are $R_j \boldsymbol{\lambda}$-vectors which can be arranged to form the columns of a $K \times R_j$ matrix $\boldsymbol{\Lambda}_j$; to conform to the convexity constraints on the GoM-vectors, each row of $\boldsymbol{\Lambda}_j$ must satisfy convexity constraints such that all R_j

elements are nonnegative (a boundary constraint) and their sum is exactly equal to 1 (a summation constraint).

The characterization of GoM as a fuzzy-set model is based on the bilinear form of the probability of each observed outcome as a weighted average of the pure-type probabilities for that outcome, using the same set of GoM scores as weights for all outcomes for a given individual. To the extent that the pure types can be described as discrete classes in a Latent Class Analysis (LCA; Lanza et al. 2007), the "fuzziness" introduced by GoM represents the capacity of the model to describe individuals who are not exactly like any of the pure types, or equivalently, who do not unambiguously fall into any of the standard categories used to classify the progression of the disease (Eisdorfer et al. 1992).

Fuzziness is an important aspect of GoM. It allows one to generalize a discrete latent class process to a fuzzy-set latent membership process with intermediate state-space locations that are specified as abstractions from a more complex multidimensional process. If the fuzzy-set generalization is in fact unnecessary, then the model will revert to the standard form of a discrete latent class process. For the analysis in this chapter, where 17 % of the cases were exactly like one of the four pure types, and 83 % were not (see Figs. 17.1a, 17.1b, 17.1c, and 17.1d), the fuzzy-set generalization was essential to the specification of an adequate model.

The fuzziness in GoM induces the close connections with PCA noted by Wachter (1999). One remaining difference between PCA and GoM is that the PCA scores are unrestricted whereas the GoM scores are restricted by the convexity constraints. This difference is less restrictive than commonly recognized: the PCA

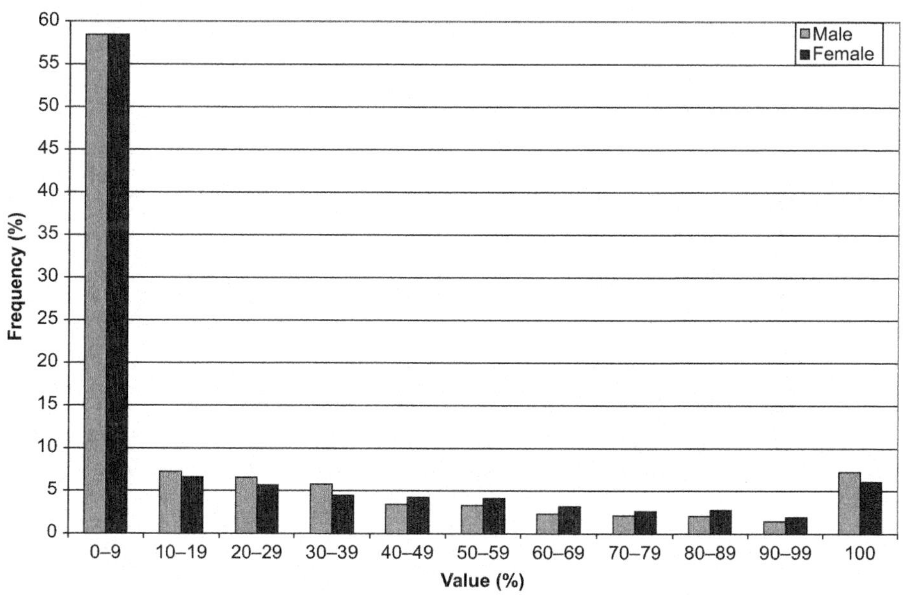

Fig. 17.1a Distribution of GoM scores for pure type I, by sex

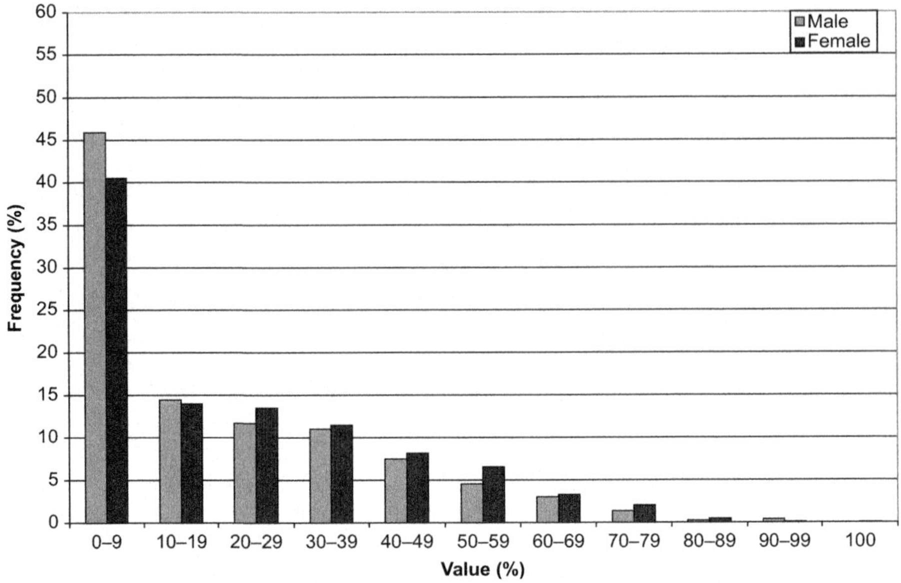

Fig. 17.1b Distribution of GoM scores for pure type II, by sex

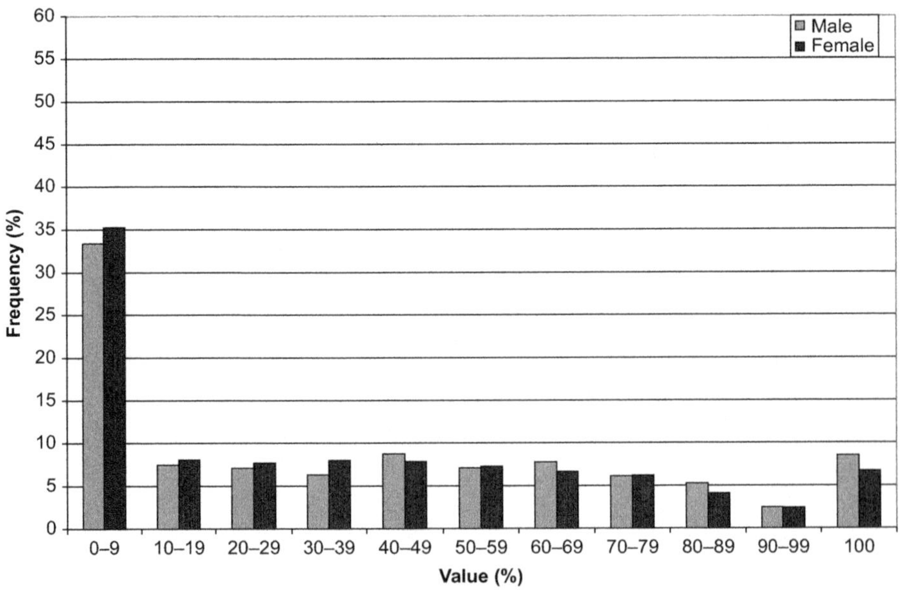

Fig. 17.1c Distribution of GoM scores for pure type III, by sex

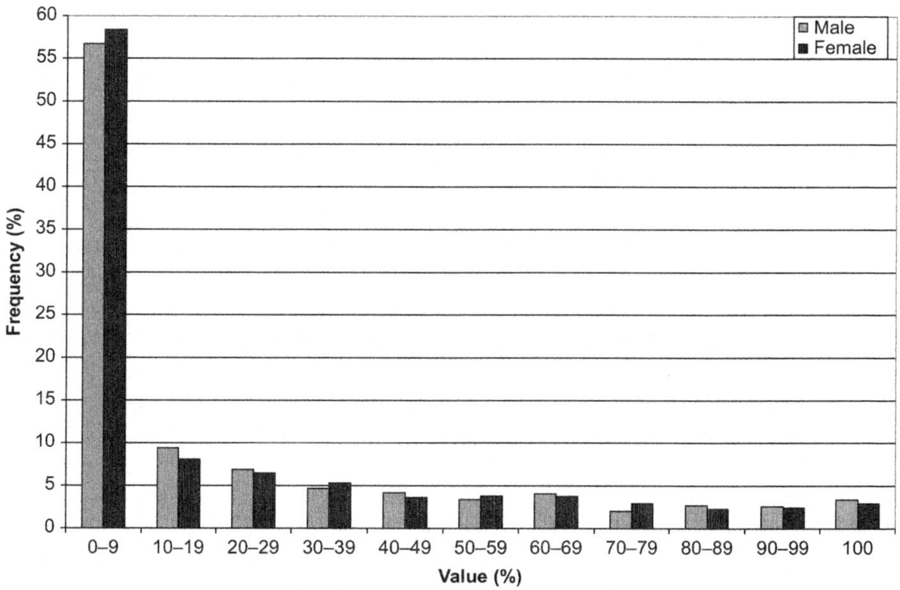

Fig. 17.1d Distribution of GoM scores for pure type IV, by sex

scores can be transformed to meet the convexity constraints using a suitably defined homeomorphic mapping from the Euclidian space $\mathbf{R}^{(K-1)}$ used in PCA onto the interior of the $(K-1)$-simplex $\mathbf{S}^{(K-1)}$ used in GoM. The mapping will be homeomorphic if it is bijective (i.e., one-to-one and onto), continuous, and open.

Such a mapping from PCA to GoM will yield GoM scores that are interior to the $(K-1)$-simplex, with extreme PCA scores mapping to locations arbitrarily close to the simplex boundaries. For maximum likelihood estimation of the GoM model, the directly estimated GoM scores for these extreme cases are allowed to fall on the boundaries, even though the "true" GoM scores are near, but not precisely on, the boundaries. As noted by Birch (1964), the possibility that some maximum likelihood estimates may lie on the boundaries presents no special problems for the asymptotic theory as long as the true values are in the interior of the parameter space. The large sample properties of longitudinal GoM are considered in Sect. 17.2.9.

17.2.2 Likelihood

The likelihood is the product over i, j, l, and t of the set $\{\text{Prob}(y_{ijlt} = 1)\}$:

$$L = \prod_i \prod_j \prod_l \prod_t \left(\mathbf{g}_i' \left\{ \prod_{r=0}^{t-1} \mathbf{U}_r \right\} \boldsymbol{\lambda}_{m_{jl}} \right)^{y_{ijlt}}. \tag{17.5}$$

The product form of the likelihood expresses the assumption of statistical independence of the $i = 1, \ldots, N$ sample persons over the $t = 0, \ldots, T$ observation times and the $j = 1, \ldots, J$ variables. The product over the $l = 1, \ldots, R_j$ response levels reflects the binary coding of $\mathbf{y}_{it}$, not an independence assumption.

The assumption that different individuals are statistically independent is standard in biodemographic analysis. The assumption that different variables are statistically independent is the defining assumption of the GoM model. That is, the GoM-score vectors are precisely the vectors of latent variables required to generate conditional distributions of the observed variables that are statistically independent. By treating the GoM-scores as unknown parameters, we can ensure that the fitted model best matches this conditional independence assumption (Wachter 1999).

The assumption that repeated observations of the same individuals are statistically independent over time was implicit in prior applications of the GoM model to longitudinal data analysis (Manton et al. 1991, 1994; Woodbury et al. 1993). We assume that the **U**-matrices are precisely the set of transition matrices required to transform the GoM-score vectors so that the resulting conditional distributions of the observed variables are statistically independent over time. By treating both the GoM-score vectors and the **U**-matrices as unknown parameter sets, we can ensure that the fitted model best matches the independence assumption over variables and time.

The product form of the likelihood over the set of J variables readily accommodates the coding of survey skip patterns and missing responses. Skip patterns occur when the interrogatories for given questions depend on specific responses to prior questions. For example, the SPMSQ and MMSE questions were asked only to self-respondents in the NLTCS. If the relevant screening variable was coded to indicate a proxy interview, then all SPMSQ and MMSE questions would be coded as missing (i.e., by setting $y_{ijlt} = 0$, for all values of l associated with the relevant indexes j and t). Similarly, because all NLTCS surveys used either the SPMSQ or MMSE, but not both, the complementary questions were coded as missing. More fundamentally, if the person was not alive at a given survey or if the T-year follow-up survey had not yet been conducted, all of the survey responses would be coded as missing. Occasionally, a respondent's record failed to provide an answer to a question that should have been provided. These cases were coded as missing in the current analysis.

The effect of coding variable responses as missing (by setting $y_{ijlt} = 0$) is to replace the corresponding terms in the likelihood with the value 1. This is equivalent to integrating over the conditional distributions of missing responses, given the vectors of GoM scores, $\mathbf{g}_{it}$, assuming that the missing responses are missing at random (MAR; see Rubin 1976). This is fully consistent with Orchard and Woodbury's (1971, p. 699) missing information principle under which "the missing data tells you nothing."

17.2.3 Log-Likelihood

Parameter estimation is facilitated by maximizing the natural logarithm of the likelihood function, rather than the likelihood function itself. The log-likelihood is:

$$\ln L = \sum_i \sum_j \sum_l \sum_t y_{ijlt} \ln p_{ijlt}, \tag{17.6}$$

where p_{ijlt} is the predicted probability that $y_{ijlt} = 1$.

To establish notation for the derivative expressions, we introduce three equivalent factorizations of p_{ijlt} to isolate the g-parameters and the λ-parameters as follows:

$$p_{ijlt} = \left(\mathbf{g}_i^{'} \mathbf{V}_{0,t}\right) \boldsymbol{\lambda}_{m_{jl}} = \mathbf{g}_{it}^{'} \boldsymbol{\lambda}_{m_{jl}} \tag{17.7}$$

$$p_{ijlt} = \mathbf{g}_i^{'} \left(\mathbf{V}_{0,t} \boldsymbol{\lambda}_{m_{jl}}\right) = \mathbf{g}_i^{'} \boldsymbol{\lambda}_{m_{jl}t} \tag{17.8}$$

$$p_{ijlt} = \left(\mathbf{g}_i^{'} \mathbf{V}_{0,s}\right) \left(\mathbf{V}_{s+1,t} \boldsymbol{\lambda}_{m_{jl}}\right) = \mathbf{g}_{is}^{'} \boldsymbol{\lambda}_{m_{jl}(s+1)t} \tag{17.9}$$

where

$$\boldsymbol{\lambda}_{m_{jl}t} = \mathbf{V}_{0,t} \boldsymbol{\lambda}_{m_{jl}} \tag{17.10}$$

$$\boldsymbol{\lambda}_{m_{jl}(s+1)t} = \mathbf{V}_{s+1,t} \boldsymbol{\lambda}_{m_{jl}}. \tag{17.11}$$

Thus, time indexes can be attached to the $\boldsymbol{\lambda}$-vectors as well as to the $\mathbf{g}$-vectors. The subscript m_{jl} will be reserved for use with K-element $\boldsymbol{\lambda}$-vectors, which may have 0, 1, or 2 time indexes as indicated above.

We refer to the $\boldsymbol{\lambda}$-vector $\boldsymbol{\lambda}_{m_{jl}t}$ as the vector of "adjusted" λ-parameters to distinguish it from the basic "unadjusted" $\boldsymbol{\lambda}$-vector $\boldsymbol{\lambda}_{m_{jl}}$ This distinction becomes important when considering the impact of the $\mathbf{V}$-matrices on the $\boldsymbol{\lambda}$-vectors at time $t > 0$ (e.g., see Figs. 17.3a, 17.3b, 17.3c, and 17.3d below).

17.2.4 Derivatives of Log-Likelihood

We need the following first- and second-order derivatives of the log-likelihood function:

$$\frac{\partial \ln L}{\partial g_{ik}} = \sum_j \sum_l \sum_t y_{ijlt} \lambda_{kjlt} / p_{ijlt} \tag{17.12}$$

$$\frac{\partial \ln L}{\partial \lambda_{kjl}} = \sum_i \sum_t y_{ijlt} g_{ikt} / p_{ijlt} \tag{17.13}$$

$$\frac{\partial \ln L}{\partial u_{kct}} = \sum_i \sum_j \sum_l \sum_{s=t+1} y_{ijls} g_{ikt} \lambda_{cjl(t+1)s} / p_{ijls} \tag{17.14}$$

$$\frac{\partial^2 \ln L}{\partial g_{ik} \partial g_{ik'}} = -\sum_j \sum_l \sum_t y_{ijlt} \lambda_{kjlt} \lambda_{k'jlt} / p_{ijlt}^2 \tag{17.15}$$

$$\frac{\partial^2 \ln L}{\partial \lambda_{kjl} \partial \lambda_{k'jl'}} = \begin{cases} -\sum_i \sum_t y_{ijlt} g_{ikt} g_{ik't} / p_{ijlt}^2 & \text{if} \quad l = l' \\ 0 & \text{if} \quad l \neq l' \end{cases} \tag{17.16}$$

$$\frac{\partial^2 \ln L}{\partial u_{kct} \partial u_{k'c't}} = -\sum_i \sum_j \sum_l \sum_{s=t+1} y_{ijls} g_{ikt} g_{ik't} \lambda_{cjl(t+1)s} \lambda_{c'jl(t+1)s} / p_{ijls}^2. \tag{17.17}$$

17.2.5 Constrained Log-Likelihood

Under Lagrange's method, the summation constraints on the g-, λ-, and u-parameters can be represented by adding an appropriately defined set of "side conditions" (additive terms) to the log-likelihood function. The specification of the side conditions must be such that the additional terms sum to zero when the summation constraints are satisfied. Such a constrained log-likelihood can be defined as follows:

$$\ln L^* = \ln L + \sum_i \rho_i \left(1 - \sum_k g_{ik}\right) + \sum_k \sum_j \theta_{kj} \left(1 - \sum_l \lambda_{kjl}\right) + \sum_k \sum_t \varphi_{kt} \left(1 - \sum_c u_{kct}\right), \tag{17.18}$$

where ρ_i, θ_{kj}, and φ_{kt} are "Lagrange multipliers" (i.e., unknown constants). To avoid confusion, we refer to ln L as the unconstrained log-likelihood in the following.

17.2.6 Kuhn-Tucker Conditions

Kuhn and Tucker (1951: Lemmas 1 and 2) provided necessary and sufficient conditions for maximum likelihood estimation under the convexity constraints of the GoM model:

1. Each parameter must be zero or positive (a boundary constraint).
2. The derivative of the constrained log-likelihood with respect to each Lagrange multiplier must be zero (equivalent to the summation constraint).
3. The derivative of the constrained log-likelihood with respect to each parameter must be zero or negative.
4. The product of each parameter and the derivative of the constrained log-likelihood with respect to that parameter must be zero (ensuring that only parameters on the boundary can have negative derivatives).
5. The log-likelihood function must be concave.

Conditions 1–4 are necessary; Conditions 1–5 are sufficient. Conditions 1 and 2 restate the convexity constraints. Conditions 2–4 require consideration of the first-order derivatives. Condition 5 requires consideration of the second-order derivatives.

If Conditions 1–5 are met for all parameters, then the entire set of estimates may constitute a global maximum likelihood solution. If Conditions 1–4 are met and Condition 5 is met for the subset of all parameters with 0-valued derivatives (i.e., excluding one or more parameters with negative derivatives), then the entire set of estimates constitutes a local maximum likelihood solution. The global maximum likelihood solution can be found by solving for all local maxima and selecting the one with the largest value of the log-likelihood function.

17.2.7 Derivatives of Constrained Log-Likelihood

The first-order derivatives of the constrained and unconstrained log-likelihood functions are related as follows:

$$\frac{\partial \ln L^*}{\partial g_{ik}} = \frac{\partial \ln L}{\partial g_{ik}} - \rho_i \tag{17.19}$$

$$\frac{\partial \ln L^*}{\partial \lambda_{kjl}} = \frac{\partial \ln L}{\partial \lambda_{kjl}} - \theta_{kj} \tag{17.20}$$

$$\frac{\partial \ln L^*}{\partial u_{kct}} = \frac{\partial \ln L}{\partial u_{kct}} - \varphi_{kt}. \tag{17.21}$$

The second-order derivatives of the constrained and unconstrained log-likelihood functions with respect to the g-, λ-, and u-parameters, respectively, are identical. It follows from Kuhn-Tucker Conditions 2 and 4 that the Lagrange multipliers are weighted averages of the derivatives of the unconstrained log-likelihood function with weights equal to the associated parameter values:

$$\rho_i = \sum_k g_{ik} \frac{\partial \ln L}{\partial g_{ik}} = \sum_j \sum_l \sum_t y_{ijlt} \tag{17.22}$$

$$\theta_{kj} = \sum_l \lambda_{kjl} \frac{\partial \ln L}{\partial \lambda_{kjl}} \tag{17.23}$$

$$\varphi_{kt} = \sum_c u_{kct} \frac{\partial \ln L}{\partial u_{kct}}. \tag{17.24}$$

It follows from Kuhn-Tucker Condition 3 that the derivatives of the unconstrained log-likelihood in each weighted average corresponding to 0-valued derivatives of the constrained log-likelihood are equal.

Kuhn-Tucker Condition 4 requires that:

$$g_{ik} \frac{\partial \ln L^*}{\partial g_{ik}} = g_{ik} \frac{\partial \ln L}{\partial g_{ik}} - g_{ik}\rho_i = 0, \tag{17.25}$$

with similar expressions for the λ- and u-parameters. Although this equation does not allow one to solve for the value of each g-parameter, it does allow one to express each g-parameter as a function of itself and the λ- and u-parameters:

$$g_{ik} = g_{ik} \times \frac{\sum_j \sum_l \sum_t y_{ijlt} \lambda_{kjlt} / p_{ijlt}}{\sum_j \sum_l \sum_t y_{ijlt}}. \tag{17.26}$$

Except for the addition of the time dimension, this expression matches Eq. (3.11) in Woodbury and Clive (1974). The numerator and denominator of the factor to the right of the "×" are the first and second terms, respectively, of the derivative of the constrained log-likelihood with respect to g_{ik} in Eq. (17.19). The second term in Eq. (17.19) is preceded by a minus sign: hence, when the derivative is zero the factor is 1; when the derivative is positive the factor is greater than 1, and, when negative, less than 1. Moreover, g_{ik} is functionally independent of $g_{i'k'}$, when $i' \neq i$.

Similarly, one obtains

$$\lambda_{kjl} = \lambda_{kjl} \times \frac{\sum_i \sum_t y_{ijlt} g_{ikt} / p_{ijlt}}{\sum_{l'} \sum_i \sum_t y_{ijl't} g_{ikt} \lambda_{kjl't} / p_{ijl't}}, \tag{17.27}$$

which corrects and adds a time dimension to Eq. (4.6) in Woodbury and Clive (1974; see Eq. (1.22) in Erosheva 2002); and

$$u_{kct} = u_{kct} \times \frac{\sum_i \sum_j \sum_l \sum_{s=t+1} y_{ijls} g_{ikt} \lambda_{cjl(t+1)s} / p_{ijls}}{\sum_{c'} \sum_i \sum_j \sum_l \sum_{s=t+1} y_{ijls} g_{ikt} u_{kc't} \lambda_{c'jl(t+1)s} / p_{ijls}}; \tag{17.28}$$

λ_{kjl} is functionally independent of $\lambda_{k'j'l'}$, when $j' \neq j$ or $k' \neq k$, and u_{kct} is functionally independent of $u_{k'c't}$ when $k' \neq k$.

Woodbury and Clive (1974) used their Eqs. (3.11) and (4.6) as the basis of a fixed-point iteration procedure operating independently and sequentially on each set of g- and λ-parameters. Each set was conditionally optimized given the current values of the other parameters. Once a given set was optimized, the focus shifted to the other set which was then conditionally optimized given the latest values of the given set. This process was repeated as long as the parameters continued to change.

At each iteration, the fixed-point update equations yield parameter estimates that satisfy the convexity constraints, and hence Kuhn-Tucker Conditions 1 and 2. Infinitesimally small positive values are treated as true zeroes at convergence which allows the solution to satisfy Kuhn-Tucker Condition 4. At convergence, the fixed-point factors are less than or equal to 1, which satisfies Kuhn-Tucker Condition 3. Kuhn-Tucker Condition 5, however, is not explicitly considered.

The most significant limitation of the fixed-point iteration procedure is that it has great difficulty moving infinitesimally small positive values away from 0 when the fixed-point factors are close to 1 (Stallard 2007). This can lead to problems where Kuhn-Tucker Conditions 3 and 4 are not satisfied but the iterations are unable to improve the log-likelihood to reach even a local maximum. Use of a Newton-Raphson procedure bypasses these problems by using additive rather than multiplicative updates.[1] An added benefit is the relatively much faster convergence of the Newton-Raphson procedure in cases where both procedures lead to the same solution.

17.2.8 *Constrained Newton-Raphson Procedures*

Maximum likelihood estimation can be conducted via three linked sets of Newton-Raphson procedures operating independently and sequentially on **g**-vectors, **Λ**-matrices, and **U**-matrices. Each vector or matrix of parameters is conditionally optimized given the current values of the other parameters. Once a given set of

---

[1] Woodbury et al. (1978) implemented a set of Newton-Raphson procedures for the cross-sectional GoM model but found it necessary to change the equation for p_{ijl} (without a time index) to the form:

$$p_{ijl} = \mathbf{g}_i'\boldsymbol{\lambda}_{m_{jl}} / \sum_{l'} \mathbf{g}_i'\boldsymbol{\lambda}_{m_{jl'}}$$

which allowed the g- and λ-parameters to be estimated without Lagrange side conditions by removing the summation constraints on the λ-parameters.

Although this modification facilitated the estimation of g- and λ-parameters, it constituted a fundamental change in the parametric specification of the model in which the λ-parameters were no longer interpretable as probabilities. An alternative approach is to modify the Newton-Raphson procedure to be consistent with all of the convexity constraints.

parameters is optimized, the focus shifts to the next vector or matrix which is then conditionally optimized given the latest values of the other parameters. This process is repeated until Kuhn-Tucker Conditions 1–4 are met and Condition 5 is met for the subset of all parameters with 0-valued derivatives. We guard against local maximum likelihood solutions by (1) re-estimating each model with multiple sets of "start values" for the parameter estimates and (2) using multiple sets of perturbations on the tentatively accepted global solutions to confirm that each perturbation returns to the same solution.

We implemented three linked sets of Newton-Raphson procedures where each vector of K g-parameters was estimated separately over person i, $i = 1, \ldots, N$; each matrix of $K \times R_j \lambda$-parameters was estimated separately over variable j, $j = 1, \ldots, J$; and each matrix of $K \times K$ u-parameters was estimated separately over time t, $t = 0, \ldots, T-1$. The procedures are essentially identical across the different parameter sets. Thus, we present the general form of the algorithm, indicating special conditions as needed.

Let $\mathbf{b}$ denote a Q-element vector of parameters with $Q = K$, $K \times R_j$, $K \times K$, or fewer elements depending on whether the iterations involve the g-, λ-, or u-parameters, and whether one or more of these parameters are excluded from the current iteration due to the boundary constraints; let $\mathbf{d}$ denote the corresponding Q-element vector of first-order derivatives of the constrained log-likelihood function and $\mathbf{H}$ the corresponding $Q \times Q$ (Hessian) matrix of second-order derivatives multiplied by -1. Kuhn-Tucker Condition 5 requires that $\mathbf{H}$ be positive definite; this can be verified during the computation of $\mathbf{H}^{-1}$.

The standard form of the Newton-Raphson update is:

$$\mathbf{b} \leftarrow \mathbf{b} + \alpha \mathbf{H}^{-1}\mathbf{d}, \tag{17.29}$$

where $\alpha \leq 1$ is an iteration control parameter that is typically set to 1 when the log-likelihood function is quadratic. One may set $\alpha < 1$ if the log-likelihood is not quadratic or if the parameters are bounded, such as in the GoM model. One can maintain the boundary constraints in the GoM model by setting $\alpha = \min(1, \alpha_1, \ldots, \alpha_Q)$, where

$$\alpha_n = \begin{cases} \dfrac{1 - b_n}{\Delta b_n} & \text{if} \quad \Delta b_n > 0 \\ \dfrac{-b_n}{\Delta b_n} & \text{if} \quad \Delta b_n < 0 \end{cases} \tag{17.30}$$

and Δb_n is the nth element of the Newton-Raphson change vector, $\mathbf{\Delta b}$, where

$$\mathbf{\Delta b} = \mathbf{H}^{-1}\mathbf{d}. \tag{17.31}$$

In this case, a boundary has been reached for parameter b_n if $\alpha_n = \alpha < 1$ and $b_n = 0$ or 1 after the Newton-Raphson update. The corresponding derivative, d_n, is evaluated in each subsequent iteration to determine if the log-likelihood will increase if the parameter moves away from the boundary. If so, the parameter is included in the subsequent update equation; if not, the parameter is held fixed at its boundary value and excluded from the calculation of the subsequent update. If the parameter is excluded, the number of parameters in the subsequent iteration decreases from Q to $Q-1$. To simplify notation, we let Q be the number of parameters included in the current iteration, recognizing that Q might change from one iteration to the next.

In contrast to the fixed-point update equations which always satisfy the summation constraints, the Newton-Raphson update equations generally do not satisfy them, nor is it intuitively obvious how one might impose such constraints on them.

To solve this problem, we first consider the case where the initial estimates of the parameters satisfy the convexity constraints. Hence the summation constraints will be satisfied on each subsequent iteration if the Newton-Raphson change vectors are constrained to sum to zero. To motivate our treatment of these constraints, we observe that the Lagrange multipliers appear as offsets in the derivatives of the constrained log-likelihood in Eqs. (17.19), (17.20), and (17.21), and hence in the Newton-Raphson change vectors in Eqs. (17.29) and (17.31). Given that these offsets are generally unknown, we further constrain the Newton-Raphson change vectors by introducing appropriately specified additional offsets, as follows:

$$\mathbf{\Delta b}^{*} = \mathbf{H}^{-1}(\mathbf{d} - \mathbf{W}\boldsymbol{\delta}), \tag{17.32}$$

where $\mathbf{W}$ is a $Q \times C$ design matrix with as many rows as there are parameters within the current set of g-, λ-, or u-parameters and as many columns as there are summation constraints; and $\boldsymbol{\delta}$ is a C-element vector containing "correction" terms (defined below) for the current estimates of the Lagrange multipliers. For the g-parameters, $C = 1$; for the λ- and u-parameters, $C = K$.

The elements of $\mathbf{W}$ indicate whether ($w_{nc} = 1$) or not ($w_{nc} = 0$) parameter b_n is part of constraint c. Thus, the columns of $\mathbf{W}$ are in a one-to-one correspondence with the Lagrange multipliers in the constrained log-likelihood. For the g-parameters with $C = 1$, $\boldsymbol{\delta}$ is a scalar. Kuhn-Tucker Condition 3 implies that $\boldsymbol{\delta}$ will converge to zero as convergence of the parameters is achieved, which explains why $\boldsymbol{\delta}$ is described as a "correction" vector.

It follows from the convexity of the current set of parameter estimates that $\mathbf{W}'\mathbf{b} = \mathbf{1}$, a C-element vector of 1's. To maintain the summation constraint at the next iteration, we require that $\mathbf{W}'\mathbf{\Delta b}^{*} = \mathbf{0}$, a C-element vector of 0's. This condition is satisfied by defining $\boldsymbol{\delta}$ as:

$$\boldsymbol{\delta} = \left(\mathbf{W}'\mathbf{H}^{-1}\mathbf{W}\right)^{-1}\mathbf{W}'\mathbf{H}^{-1}\mathbf{d}. \tag{17.33}$$

Moreover, it follows immediately that $\boldsymbol{\delta} \to \mathbf{0}$ as $\mathbf{d} \to \mathbf{0}$, as required. Thus, the modified Newton-Raphson change vectors satisfy the summation constraints, and hence Kuhn-Tucker Condition 2, at each iteration.

To better understand the mathematical properties of the modification in Eq. (17.32), consider a $Q\times Q$ orthogonal projection matrix $\mathbf{P}$ which projects any Q-element vector onto the subspace spanned by the columns of $\mathbf{W}$, where

$$\mathbf{P} = \mathbf{W}\left(\mathbf{W}'\mathbf{H}^{-1}\mathbf{W}\right)^{-1}\mathbf{W}'\mathbf{H}^{-1}, \tag{17.34}$$

with properties $\mathbf{P}^2 = \mathbf{P}$, $\mathbf{H}^{-1}\mathbf{P} = \mathbf{P}'\mathbf{H}^{-1}$ and $\mathbf{W}'(\mathbf{I} - \mathbf{P}') = \mathbf{0}$, a $C\times Q$ matrix of 0's. Using these properties, we can rewrite Eq. (17.32) as follows:

$$\begin{aligned} \Delta\mathbf{b}^* &= \mathbf{H}^{-1}(\mathbf{d} - \mathbf{W}\boldsymbol{\delta}) \\ &= \mathbf{H}^{-1}\left(\mathbf{I} - \mathbf{W}\left[\mathbf{W}'\mathbf{H}^{-1}\mathbf{W}\right]^{-1}\mathbf{W}'\mathbf{H}^{-1}\right)\mathbf{d} \\ &= \mathbf{H}^{-1}(\mathbf{I} - \mathbf{P})\mathbf{d} \\ &= (\mathbf{I} - \mathbf{P}')\mathbf{H}^{-1}\mathbf{d} \\ &= (\mathbf{I} - \mathbf{P}')\Delta\mathbf{b}, \end{aligned} \tag{17.35}$$

from which we immediately obtain $\mathbf{W}'\Delta\mathbf{b}^* = \mathbf{0}$, as required. This latter condition implies that the modified change vector ($\Delta\mathbf{b}^*$) is an orthogonal projection of the standard change vector ($\Delta\mathbf{b}$) onto the nullspace of $\mathbf{W}'$, a subspace of dimension $Q - C$. It follows that the modified change vector is the unique vector within the nullspace of $\mathbf{W}'$ that is closest to (i.e. has the highest correlation with) the standard change vector.

The boundary constraints in Kuhn-Tucker Condition 1 can be imposed via the α-multipliers defined in Eq. (17.30), yielding the final constrained update equation:

$$\mathbf{b} \leftarrow \mathbf{b} + \alpha(\mathbf{I} - \mathbf{P}')\mathbf{H}^{-1}\mathbf{d}. \tag{17.36}$$

Zero-valued parameters are held fixed at their boundary values in a given iteration if they have negative derivatives; they are excluded from the current update equation. Parameters with values of 1 are necessarily associated with a set of $K - 1$ or $R_j - 1$ other 0-valued parameters; they are excluded only when all of the 0-valued parameters are excluded.

Thus, the modified Newton-Raphson procedure satisfies the convexity constraints without resort to ad-hoc adjustments. This allows the overall search to switch between sets of parameters prior to convergence, which is attained only when Kuhn-Tucker Conditions 1–4 are met and Condition 5 is met for the subset of all parameters with 0-valued derivatives.

17.2.9 Consistency and Asymptotic Normality

Bradley and Gart (1962, Theorem 2) provided general conditions for the consistency and asymptotic normality of maximum likelihood estimators that are applicable to the longitudinal GoM model in Eq. (17.5) if two additional assumptions are met: [2]

1. The GoM transition matrices (**U**-matrices) can be represented by a finite number of parameters; and
2. The true values of the g-, λ-, and u-parameters lie in the interior of the respective parameter spaces.

Asymptotic theory considers the behavior of a model as the number of measurements is increased to infinity. Bradley and Gart's (1962) conditions allow the infinite increase in the number of measurements to occur along the temporal dimension of the longitudinal GoM model. The frequency of measurement is increased without limit by scheduling the longitudinal observations closer and closer together.

Assumption 1 ensures that the total number of parameters is finite for fixed N and J. The restriction to a finite number of u-parameters is implemented by assuming that the transition matrices for measurement intervals below some threshold length are fractional powers of the transition matrices for the threshold interval, i.e., the transition intensity matrices are assumed to be piecewise constant. This is reasonable for the proposed biodemographic applications.

Assumption 2 is required to avoid technical difficulties that arise when the true values of one or more parameters fall on the boundaries of the parameter space. Given that the admissible parameter values can be arbitrarily close to the boundaries of that space, this does not represent a serious restriction. The parameter estimates can fall on the boundaries of the parameter space; when this happens, however, up to half of each approximating normal distribution will be replaced by a point mass. Assumption 2 ensures that such occurrences will become increasingly rare as the number of longitudinal observations increases.

Our applications to date have been for problems in which the ratio of number of data points per parameter was in the range 20–80, with the lower end of the range typical of NLTCS applications (Stallard 2007 and this chapter) and the upper end typical for Predictors 1 and 2 applications (Stallard et al. 2010). Asymptotic theory indicates that higher values of this ratio are preferable. Further investigation is needed to determine for practical applications how large (or small) these ratios need to be for the large sample approximations to be reasonably accurate.

Comment 1 Our appeal to Bradley and Gart (1962, Theorem 2) in combination with Assumptions 1 and 2 effectively responds to Haberman's (1995) comment that

---

[2] Bradley and Gart's (1962) proof of uniqueness of the consistent estimator in their Theorem 2 (iii) follows that of Chanda's (1954) Theorem 2 and, hence, suffers from the deficiency noted by Tarone and Gruenhage (1975) who provided a corrected proof in their Theorem 2′. This means that Bradley and Gart's (1962) result on uniqueness can be proven following Tarone and Gruenhage's (1975) Theorem 2′, but not Chanda's (1954) Theorem 2. Our appeal to Bradley and Gart's (1962) Theorem 2 assumes that Tarone and Gruenhage's (1975) correction has been made, even though Bradley and Gart did not actually do it.

existing asymptotic theory does not apply to GoM because the number of g-parameters becomes increasingly large as N becomes large. In the case of longitudinal GoM, N is fixed so the number of g-parameters is likewise fixed. N may become arbitrarily large and this will accelerate the asymptotic convergence of the estimators of the λ- and u-parameters. Nonetheless, we do not assume that N will become infinite. Instead, the required infinite increase in the number of data points occurs along the temporal dimension of the model. Assumption 1 is employed to avoid having an infinite number of u-parameters. With this assumption, the model becomes a finite parameter model and this allows Bradley and Gart (1962, Theorem 2) to resolve the critical issue that the observations in Eq. (17.5) are not identically distributed, or even approximately so. Indeed, identification of the differences in these distributions over individuals and times of measurement is the focus of the longitudinal GoM model; the ability to deal with population and process heterogeneity is what makes the model attractive for biodemographic analysis.

Comment 2 The standard proof of consistency of the general (i.e., cross-sectional) GoM model was provided by Tolley and Manton (1992) for fixed J and increasing N using a marginal likelihood formulation in which, using an assumed mixing distribution for the GoM scores, the g-parameters were integrated out of the final expression, which yielded a consistent estimator of the λ-parameters. While this formulation yielded the desired consistency result for the marginal GoM likelihood, it did not prove that the conditional GoM likelihood (i.e., the expression inside the integral obtained by removing the time index from Eq. (17.5)) was likewise consistent—as noted by Haberman (1995).

We comment on this limitation in the Appendix, where: (1) we present a new synthesis indicating that the large sample behavior of the conditional GoM model is not consistent except, possibly, for the special case that N and J both go to infinity; (2) we describe a generalization of conditional GoM in which the empirical GoM-score mixing distribution is used to create an empirical marginal likelihood with the same set of λ- and g-parameters; (3) we conjecture that the empirical marginal GoM estimator is optimal in the sense of minimizing the Kullback-Leibler divergence (i.e., relative entropy; a distance measure) between the estimated and the true model (Kullback and Leibler 1951); and (4) we observe that the conditional GoM likelihood is approximately proportional to the empirical marginal GoM likelihood and that the accuracy of the approximation increases with increasing J; hence it is also conjectured to be optimal in the sense of minimizing (approximately) the Kullback-Leibler divergence. Further investigation is needed: (1) to prove/disprove our conjecture; and (2) to determine for practical applications how large (or small) J needs to be for the conditional GoM approximation to be acceptable for given sizes of N and configurations of the empirical GoM-score mixing distribution.

We note that the concepts of relative entropy and Kullback-Leibler divergence were fundamental to Akaike's (1973, 1974) development of the AIC information criterion which is widely used for model selection in practical applications (e.g., see below).

17.2.10 Model Testing

Model testing is based on likelihood ratio chi-squared statistics using the conventional chi-squared approximation (Wilks 1938) for the difference in the logarithms of the likelihood functions for testing pairs of nested models. The log-likelihood for the longitudinal GoM model with $K \geq 1$ is:

$$\ln L = \sum_i \sum_j \sum_l \sum_t y_{ijlt} \ln p_{ijlt}, \tag{17.37}$$

where p_{ijlt} is the predicted probability that $y_{ijlt} = 1$. This model contains the baseline or null model (Model 0) with $K = 1$ as a special case in which the likelihood simplifies to the standard multinomial form for J independent variables, with log likelihood:

$$\ln L^{(0)} = \sum_i \sum_j \sum_l \sum_t y_{ijlt} \ln \mu_{jl}, \tag{17.38}$$

which is maximized by setting the μ-parameters equal to the observed proportions for each response l within each variable j. With these specifications, the Wilks chi-squared statistic is

$$\chi^2(L) = -2 \times \left[\ln L^{(0)} - \ln L\right]. \tag{17.39}$$

The number of degrees of freedom (d.f.) in each model is equal to the number of parameters minus the number of constraints. The d.f. for the Wilks chi-squared statistic is equal to the difference in d.f. for the two models being compared.

More generally, longitudinal GoM models with different K-values may be compared using Akaike's (1973, 1974) information criterion (AIC), defined for each likelihood function L as

$$\text{AIC}_L = -2 \times \ln L + 2 \times \text{d.f.}(L), \tag{17.40}$$

where d.f.(L) denotes the total number of unconstrained parameters in the corresponding model, with the "best" model among a set of alternatives being the one having the smallest value of AIC_L. The difference in values of AIC_L for the null model and the model with K pure types can be written as

$$\begin{aligned} \Delta\text{AIC} &= \text{AIC}_{L^{(0)}} - \text{AIC}_L \\ &= \chi^2(L) - 2 \times \left[\text{d.f.}(L) - \text{d.f.}\left(L^{(0)}\right)\right] \\ &= \chi^2(L) - 2 \times \text{d.f.}(\chi^2(L)), \end{aligned} \tag{17.41}$$

showing that ΔAIC is a chi-squared statistic reduced by twice its d.f. If ΔAIC is positive, the null model is rejected. For d.f. >7, the AIC test is more conservative than the Wilks test.

An even more conservative alternative is provided by the Bayesian information criterion (BIC; Schwarz 1978), defined for each likelihood function L as

$$\text{BIC}_L = -2 \times \ln L + \text{d.f.}(L) \times \ln N^*, \tag{17.42}$$

where $\ln N^*$ is the weighted arithmetic mean of $\ln N_r$, r is an index that distinguishes each subset of g-, λ-, or u-parameters in Eqs. (17.12), (17.13), and (17.14), N_r is the number of responses for the rth parameter subset, and each such subset comprises only those g-, λ-, or u-parameters that share the same summation constraint. The weights are the number of unconstrained parameters within each subset of g-, λ-, and u-parameters (Stallard et al. 2010).[3] N^* is interpreted as the effective sample size for the given model (Raftery 1995). The "best" model among a specified set of alternatives is the one having the smallest value of BIC_L. The difference in values of BIC_L for the null model and the model with K pure types can be written as

$$\begin{aligned} \Delta\text{BIC} &= \text{BIC}_{L^{(0)}} - \text{BIC}_L \\ &= \chi^2(L) - \left[\text{d.f.}(L) - \text{d.f.}\left(L^{(0)}\right)\right] \times \ln N^* \\ &= \chi^2(L) - \text{d.f.}(\chi^2(L)) \times \ln N^*, \end{aligned} \tag{17.43}$$

showing that ΔBIC is a chi-squared statistic reduced by its d.f. times the logarithm of N^*. If ΔBIC is positive, the null model is rejected. For $N^* \geq 8$, the BIC test is more conservative than the AIC test, which will almost always be the case.

When a model with $K > 1$ pure types is accepted over the null model, one then needs to consider the significance of the different variables. The Bayesian information criterion approach also provides a suitable metric for rank ordering the variables according to their significance in the model.

BIC_j is defined for variable j as

$$\begin{aligned} \text{BIC}_j = -2 \times \sum_l \sum_i \sum_t y_{ijlt} \ln p_{ijlt} + K\left(R_j - 1\right) \\ \times \ln\left(\sum_l \sum_i \sum_t y_{ijlt}\right), \end{aligned} \tag{17.44}$$

where the second term is the product of the d.f.(L) (for the λ-parameters) and the logarithm of the (actual) sample size. The difference in BIC_j's for the null model and the model with K pure types is

[3] N^* was denoted as N^{**} and the corresponding BIC measure as BIC2 in Stallard et al. (2010).

$$\Delta\text{BIC}_j = 2 \times \sum_l \sum_i \sum_t y_{ijlt} \ln\left(p_{ijlt}/\mu_{jl}\right) - (K-1)\left(R_j - 1\right) \times \ln\left(\sum_l \sum_i \sum_t y_{ijlt}\right). \tag{17.45}$$

The first term is the component of the Wilks chi-squared approximation associated with variable j. The second term "adjusts" the chi-squared statistic for the d.f. and sample size, conditional on the g- and u-parameters which are now assumed to be held fixed. The sample-size adjustment provided by BIC was judged important for our application because the sample sizes for variables introduced at the later waves of the NLTCS were substantially smaller than for variables measured at all waves.

When $\Delta\text{BIC}_j > 0$, the model with K sets of λ_j-parameters is "more probable" than the null model with one set (i.e., for $K = 1$, the marginal probabilities). ΔBIC_j values in the ranges 0–2, 2–6, 6–10, and 10–14, and 14+ are considered "weak," "positive," "strong," "very strong," and "conclusive" evidence, respectively, in favor of the model with K sets of λ_j-parameters (Kass and Raftery 1995; Raftery 1995). Moreover, it follows from the Bayesian decision rule (Schwarz 1978) that the ΔBIC_j's can be rank-ordered across the J variables such that the largest ΔBIC_j identifies the variable j for which the non-null model is the most probable. This is the most significant or most informative variable. Similar interpretations apply to the other rank-ordered variables. Thus, ΔBIC_j provides a metric that controls for differences in d.f. and sample size, assuming that the addition or deletion of variable j would not impact the estimates of the g- and u-parameters for the remaining $J - 1$ variables, a reasonable assumption with $J = 94$ variables in our application.

17.3 Data

17.3.1 National Long Term Care Survey

The primary data source was the National Long Term Care Survey (NLTCS; 1982, 1984, 1989, 1994, and 1999) which contains both longitudinal and cross-sectional data on a nationally representative sample of 41,947 U.S. elderly persons aged 65 years or older at some point during 1982–1999, with 17,286–22,139 age-eligible survivors at each of the five waves of whom 3112–5552 were classified as disabled with 1036–1946 in institutional residence. A sixth round of the survey, fielded in late 2004 with mortality follow-up through 2009, will be used in future work in this area.

The initial 1982 sampling frame for the NLTCS was a list of aged persons enrolled in Medicare on April 1, 1982. This sampling frame was extended in subsequent years to include about 5000 new Medicare enrollees reaching age

65 between surveys to replace deaths occurring since the prior survey and to ensure that each survey sample was cross-sectionally representative of the entire elderly population aged 65+. The response rates were excellent for all five waves of the survey (approximately 95 %: Manton et al. 1997; Manton and Gu 2001).

The sampling frame covered both institutionalized and non-institutionalized persons. Institutionalized persons received face-to-face institutional interviews except in 1982. A telephone screener interview targeted non-institutionalized disabled persons for further study using face-to-face community interviews. The screener interview was designed to minimize the risk of false negative assessments of ADL and IADL impairments.

A person "screened-in" as disabled if he or she had a problem performing at least one of seven ADLs or one of eight IADLs without the help of another person or special equipment, where the expected duration of the problem was 3 months or longer; except for outdoor mobility, the IADL problems were counted only if they were due to a disability or health problem. A person also screened-in as disabled if he or she was institutionalized in a long-term care facility at the time of the survey.

Once a person screened-in for any round of the NLTCS, that person was designated for automatic follow-up in all subsequent rounds of the NLTCS, receiving the detailed face-to-face community or institutional interviews, as appropriate.

The NLTCS established special procedures to sample nondisabled persons. Altogether, a total of 4207 distinct persons received the detailed community interview who were classified as nondisabled at least one time during the 1984, 1989, 1994, and 1999 surveys.

The NLTCS detailed interviews provided data on age, sex, race, residence type (community vs. institutional, 1982–1999; and assisted living, 1999), cognitive status (SPMSQ 1982–1994, MMSE 1999; see Lee et al. 1998), and seven ADLs. The community-resident questionnaire provided additional data on 9 IADLs, 30 major medical conditions, 7 physical performance items (Nagi 1976), subjective health status, aberrant behaviors, number and relationship (to respondent) of caregivers, and caregiver hours/days and type of activity for which help was provided. Height, weight, alcohol and cigarette use, and exercise questions were added to the community-resident questionnaire in 1994 and 1999. The institutional-resident questionnaire provided limited additional data on one IADL and, in 1999, three major medical conditions.

Four characteristics of the NLTCS made it well-suited for use in the current analysis.

First, all respondent records were linked to Medicare vital statistics and beneficiary claims data for calendar years 1982 and later, with ongoing periodic updating to allow tracking of mortality, Medicare claims, and enrollment/disenrollment in a managed care plan (e.g., a health maintenance organization [HMO] or similar arrangement with fixed monthly costs). The Medicare claims data records contained information on dates and costs of service, types of providers, and ICD-9-CM diagnosis and procedure codes.

Second, the sampling frame provided large sample sizes at ages 85 and older—a population for which it is often difficult to sample effectively because of its small relative size. The NLTCS included at least 2400 people aged 85+ and at least 825 people aged 90+ in each survey; the 1994 and 1999 surveys included additional oversampling of the population aged 95+ (537 respondents in 1994; 598 in 1999).

Third, the same core set of disability and medical condition questions was asked in each round of the NLTCS using essentially the same field procedures executed by the same survey organization (U.S. Census Bureau). The stability of the methods and procedures minimized bias in disability trend estimates and other parameters (Freedman et al. 2002). This stability also made these data suitable for longitudinal analysis using models such as longitudinal GoM that required repeated measurements of the same variables on the same persons.

Fourth, the inclusion of 4207 functionally nondisabled persons in the detailed interviews allowed these data to be used to analyze both cognitive and functional impairments. Although the two types of impairment are correlated, our analytic subsample of 3290 cases was not restricted to persons with functional impairment; it also included 433 persons (13 %) with no impairment in ADLs or IADLs at onset of dementia.

17.3.2 Sample Selection

Identification of respondents with recent onset of dementia within the NLTCS was accomplished using modifications of the two-stage procedure for identifying AD cases developed in Kinosian et al. (2000, 2004), as shown in Table 17.1:

I. Two sets of "inclusion criteria" were used to establish a sample of respondents with any form of dementia, either (A) in the NLTCS interviews or (B) in the linked Medicare diagnostic files; and

II. Five sets of "exclusion criteria" were used to reject respondents meeting the inclusion criteria because the cases were not clearly dementia (A–C) or not clearly of recent onset (D, E).

The inclusion criteria employed a broad set of indicators of dementia (Taylor et al. 2002; Kinosian et al. 2004).

The exclusion criteria retained cases that had forms of dementia other than AD. Exclusion criteria A and B excluded respondents with mental retardation and Parkinson's disease at any time during the NLTCS follow-up, but did not exclude respondents with arteriosclerosis, atherosclerosis, stroke, other cerebrovascular disease, or certain ICD-9-CM non-AD dementia codes (290.4, 291.2, 294.0, and 294.1). Exclusion criterion C was designed to ensure that respondents with a documented lack of deterioration would be excluded from the sample. Exclusion criteria D and E rejected dementia cases whose onset was not clearly determined to be recent.

Table 17.1 Inclusion and exclusion criteria used to identify older persons with recent onset of dementia

I. Inclusion criteria
A. National Long Term Care Survey
1. Cognitive impairment as assessed by the Short Portable Mental Status Questionnaire (SPMSQ) or the Mini-Mental State Exam (MMSE):
SPMSQ criteria are 3+ mistakes for individuals without a high school education or 2+ mistakes for those with a high school education
MMSE criteria are 9+ mistakes for individuals without a high school education or 7+ mistakes for those with a high school education
2. Alzheimer's disease by caregiver report
3. Senility by caregiver report
B. Medicare Part A and B claims (at least two claims with any of the following ICD-9-CM codes; timing is based on the first such code)
1. Alzheimer's disease (331.0)
2. Uncomplicated senile dementia (290.0)
3. Presenile dementia (290.1)
4. Senile dementia with delusional or depressive features (290.2)
5. Senile dementia with delirium (290.3)
6. Unspecified senile psychotic condition (290.9)
7. Other chronic organic brain syndromes (chronic) (294.8)
8. Pick's disease (331.1)
9. Senile degeneration of brain (331.2)
10. Cerebral degeneration in diseases classified elsewhere (331.7)
11. Senility without mention of psychosis (797)
II. Exclusion criteria
A. National Long Term Care Survey
1. Mental retardation
2. Parkinson's disease
B. Medicare part A and B claims (at least two claims with any of the following ICD-9-CM codes)
1. Mental retardation (317–319)
2. Parkinson's disease (332)
C. Cognitively intact
Reject case if at the time of first inclusion any concurrent or future application of SPMSQ or MMSE indicates no cognitive impairment using the NLTCS inclusion criteria listed in item I.A.1
D. Unsure if case is new inclusion
Reject case if the included person does not have a prior NLTCS survey in which he or she did not meet the inclusion criteria
E. Institutional resident at time of first inclusion
Reject case if institutionalized during the NLTCS survey given at the time of first inclusion (or at the next NLTCS survey if the inclusion is based on Medicare diagnoses provided between NLTCS surveys)

Kinosian et al. (2000, 2004) attempted to minimize false positive AD-selections by use of their exclusion criteria. The narrower set of exclusion criteria in Table 17.1 identified 1526 additional cases of non-AD dementia, all of whom met the inclusion criteria, increasing our sample size by 87 % (from 1764 to 3290 cases).

Table 17.2 displays the results of the case sampling using the criteria in Table 17.1. The number of retained cases ranged from 756 to 887 per year, with a total of 3290 cases over the four surveys, representing a weighted population count of 4.7 million incident cases. The sources of the inclusions were as follows: 58 % were selected based on information in the NLTCS alone; 14 % were selected based on information in the Medicare files alone; and 28 % were selected based on information that was corroborated in both files. Among the latter group, however, 74 % met the NLTCS criteria prior to or in the same year as meeting the Medicare criteria (labeled "NLTCS+Medicare"), suggesting that the NLTCS criteria may detect milder forms of dementia than the Medicare criteria. Consistent with these results, Pressley et al. (2003) recommended use of multiple data sources for case identification because each source uniquely identified substantial numbers of cases that could not be identified from the other sources while no single data source identified all cases.

Prior to 1991, Medicare diagnostic codes were reported on Part A (Hospital Insurance) but not on Part B (Supplementary Medical Insurance) claim records. Taylor et al. (2004) reported marked increases in AD prevalence during 1991–1999 using linked NLTCS-Medicare data. These two observations may explain the increase in Medicare inclusions between 1984 and 1994. Table 17.2 shows that the number of inclusions based on Medicare alone, but not the total number of Medicare inclusions (i.e., using both NLTCS and Medicare data), increased from 1994 to 1999.

The case counts in Table 17.2 reflect recent onset of dementia. These cases were a subset of all cases of onset of dementia occurring within each 5-year period. The

Table 17.2 Distribution of dementia by year of onset and source of information meeting inclusion criteria

		Sources and order of information				
	Onset year	NLTCS alone	Medicare alone	NLTCS +Medicare	Medicare +NLTCS	Total
Unweighted count	1984	659	–	141	–	805
	1989	475	27	231	23	756
	1994	394	145	238	110	887
	1999	385	289	66	102	842
Total unweighted count		1913	–	676	–	3290
Weighted Count	1984	768,860	–	175,623	–	950,734
	1989	640,906	42,462	316,412	31,764	1,031,544
	1994	588,550	220,365	349,034	168,171	1,326,120
	1999	634,883	468,959	108,344	157,726	1,369,911
Total weighted count		2,633,199	–	949,413	–	4,678,310

Note: "–" denotes cells suppressed due to Medicare data-suppression rules for small sample sizes

difference was substantial. Among those who met all criteria in Table 17.1, 2.2 million weighted cases (1384 unweighted cases) were alive in 1999; if exclusion criteria D and E (recent onset) were deleted from Table 17.1, then we would have estimated that 3.3 million weighted cases (2260 unweighted cases) were alive in 1999. Thus, about one-third of dementia cases were dropped from our analysis due to exclusion criteria D and E.

17.4 Results

17.4.1 Model Selection

Longitudinal GoM models were independently estimated for 1033 males and 2257 females with recent onset of dementia using the criteria in Table 17.1 to identify such cases in the NLTCS. Ninety-four covariates were included in the analysis; however, each case provided a maximum of 84 covariates per wave, not 94, because the SPMSQ was replaced with the MMSE in 1999. The total over all waves for all respondents was 81,774 data items for males (averaging 79.2 per case) and 194,758 data items for females (averaging 86.3 per case). Survival status was assessed for 5-year periods following each wave, ending with the survival assessment in 2004, during which 892 male deaths and 1832 female deaths were recorded (representing 86 % and 81 %, respectively, of the total sample of 1033 males and 2257 females).

The 94 covariates included 7 ADLs, 9 IADLs, 7 physical performance measures, vision, 28 medical conditions (after dropping mental retardation and Parkinson's disease), institutionalization (after initial wave), subjective health status, cognitive functioning (10 SPMSQ items or 11 MMSE items supplemented with two memory tests, proxy interview, and four behaviors), alcohol use, cigarette use, race, height, three exercise measures, three measures of body mass index (BMI = ratio of weight [in kilograms] to height-squared [in meters^2]; at interview, age 50, and 1 year prior to interview), and two variables encoding 5-year survival (one based on GoM scores at the start of the interval, the other based on GoM scores at the end of the interval). Survey skip patterns and modest amounts of missing data were coded as described in Sect. 17.2.2.

Three models were estimated separately by sex as shown in Table 17.3; the designations M0, M1, M2, F0, F1, and F2 identify combinations of sex (M and F) and model number. Model 0, the null model, assumed that each sex was homogeneous with no differences between cases or over time. These assumptions implied that $K = 1$ and $p_{ijlt} = \mu_{jl}$, the marginal probability for response l to variable j, for every pair of indexes i and t.

Model 1, the primary model, assumed that each sex was described by a longitudinal GoM model with $K = 4$ pure types corresponding to the extreme states of a

Table 17.3 Test statistics for null model and two alternatives

Sex and model	*K*	-*L*	*N*	*J*	*M*	df GoM scores	U	Lambda	Total df	Chi-squared	df (chi-sq)	Ratio	AIC	BIC
Males														
M0	1	50,797.46	1033	94	277	0	0	183	183	–	–	–	101,960.92	102,757.60
M1	4	39,750.25	1033	94	277	3099	18	732	3849	22,094.41	3666	6.03	**87,198.51**	**97,462.18**
M2	4	39,744.66	1033	94	277	3099	36	732	3867	11.18	18	0.62	87,223.33	97,590.32
*N**						70.1	2298.9	574.5	106.3					
ln(*N**)						4.25	7.74	6.35	4.67					
Females														
F0	1	122,639.28	2257	94	277	0	0	183	183	–	–	–	245,644.57	246,603.11
F1	4	96,709.91	2257	94	277	6771	18	732	7521	51,858.75	7338	7.07	**208,461.82**	**228,015.08**
F2	4	96,709.91	2257	94	277	6771	36	732	7539	0.00	18	0.00	208,497.82	228,177.64
*N**						73.9	8357.4	1391.2	99.5					
ln(*N**)						4.30	9.03	7.24	4.60					
Combined sexes														
M0 & F0	1	173,436.74	3290	188	554	0	0	366	366	–	–	–	347,605.49	349,360.71
M1 & F1	4	136,460.16	3290	188	554	9870	36	1464	11,370	73,953	11,004	6.72	**295,660.33**	**325,477.26**
M2 & F2	4	136,454.57	3290	188	554	9870	72	1464	11,406	11.18	36	0.31	295,721.14	325,767.96
Between sexes														
M1 <– F1	4	42,260.69	1033	94	277	3099	0	0	3099	5020.88	750	6.69	90,719.39	97,693.02
F1 <– M1	4	102,007.20	2257	94	277	6771	0	0	6771	10,594.57	750	14.13	217,556.39	233,148.93

Note: Boldface fonts denote the best (i.e., minimum) values of AIC or BIC. For both sexes, the best AIC and BIC values are for Model 1. The *N** values shown by parameter type are constant over the three models. The *N** values under the header *Total df* are for Model 1; the corresponding values for Model 2 are 107.9 and 100.5, respectively, for males and females

latent irreversible three-dimensional process that was sufficiently rich to represent individual differences in measures of ADL, IADL, and cognitive functioning, both at onset and during the subsequent progression of the disease.

These assumptions, implying that the 4×4 **U**-matrices are upper-triangular in form, were based on prior research showing that the clinical course of AD has at least three irreversible stages—mild, moderate, and severe—with substantial individual variability in the durations of each stage (Eisdorfer et al. 1992; Stern et al. 1996; Grossberg and Desai 2003) and on prior GoM analyses showing that $K = 4$ was the best value for the sex-specific AD models (Fillenbaum and Woodbury 1998; Kinosian et al. 2004; Stallard et al. 2010). Stallard et al. (2010) tested sex-specific models for AD cases in the Predictors Study data with $K = 1, 2, 3$ and rejected them in favor of $K = 4$ using AIC, BIC, and related measures. Kinosian et al. (2004) tested sex-specific models for the AD subset of the NLTCS data with $K = 4$ vs. $K = 5$; the models with $K = 4$ were rejected by the Wilks chi-squared test but were favored by the AIC and BIC measures. Fillenbaum and Woodbury (1998) tested combined-sex GoM models with $K = 1, \ldots, 7$ for AD cases using clinical measures obtained at entry to the CERAD study (i.e., with one observation per case). Based on comparisons of models with K vs. $K + 1$, the Wilks test indicated that $K = 6$ was the smallest acceptable K-value. Examination of the λ-parameters showed that two of the six pure types were predominantly male and four were predominantly female.

Thus, given the close connection between AD and other dementias, these prior analyses supported our selection of $K = 4$ as the best value for the sex-specific dementia models. This selection maximizes the comparability of the current dementia analysis with prior and ongoing AD analyses. A more comprehensive investigation into the use of alternative K-values will be a topic for future research.

Model 2 differed from Model 1 in that the irreversibility assumption was relaxed, allowing the **U**-matrices to have an unrestricted form. Model 2 was included primarily to confirm our assumption that dementia, as defined in Table 17.1, was irreversible.

For both sexes, the Wilks chi-squared test statistics were highly statistically significant for Model 1 vs. Model 0 and statistically non-significant for Model 2 vs. Model 1, indicating that Model 1 was the best. For both sexes, Model 1 had the smallest value of the three AIC measures, indicating that Model 1 was also best under Akaike's (1974) decision rule, with identical conclusions reached using the BIC measures.

Differences between the sexes were tested by interchanging the u- and λ-parameters with the g-parameters re-estimated to match the new sets of u- and λ-parameters. The Wilks chi-squared statistics (each with 750 d.f.) were highly statistically significant, indicating that the u- and λ-parameters were significantly different between the sexes. The results were confirmed by the AIC and BIC measures.

17.4.2 Model Description

In describing Model 1, we present the λ-parameters first, followed by the u- and g-parameters, an ordering that corresponds to describing the characteristics of the pure types, their changes over time, and the distribution of the population across the pure types. Nonetheless, the parameters were jointly estimated and must be interpreted as a linked set.

Because the number of λ-parameters in Model 1 was large (i.e., $4 \times 277 = 1108$ for each sex), we used the Bayesian information criterion (ΔBIC_j) to rank the variables according to their significance in the model (Schwarz 1978) and we grouped the variables into 19 sets (11 of which contained 2 or more logically related variable). The results are displayed in Table 17.4 separately by sex using the average values of ΔBIC_j per variable as representative values for each group.

The top six and bottom five groups were the same for males and females, but there were several differences in the order as well as in the number of groups with $\Delta\text{BIC}_j > 0$ indicating support for differential effects across the four pure types for 14 groups for males and 15 groups for females. Among these 14 ΔBIC_j's for males, BMI's ΔBIC_j provided positive evidence, race's ΔBIC_j, strong evidence, SPMSQ's ΔBIC_j, very strong evidence, and the 11 top ranked ΔBIC_j's conclusive evidence of differential effects. For females, all 15 ΔBIC_j provided conclusive evidence of differential effects.

For both sexes, ADLs were the first group and IADLs the second. The 5-year survival variable was third for males and fifth for females. The proxy interview variable was fourth and the physical performance variables were fifth, for males. For females the same groups were one rank higher. Subjective health status was sixth for both sexes. LTC institutionalization was eighth for females and ninth for males. Race was 7th for females and 13th for males.

Because race was fixed, however, its impact on the longitudinal GoM model was more difficult to interpret than for the remaining variables which clearly change over time. One resolution would stratify the model by race and sex, rather than just by sex. An alternative would modify the likelihood function to use only the initial observation of race (and other fixed variables). We are assessing these options in our ongoing work.

Perhaps the most surprising finding in Table 17.4 was the relatively low ranking of the cognitive variables SPMSQ, MMSE, and memory. However, these items were asked only of self-respondents, not proxy respondents. Proxy respondents were asked if the sample person currently had senility or AD. Among the medical conditions, senility's ΔBIC_j (not shown) was 890.90 for females and 329.34 for males, which would be ranked sixth for both sexes in Table 17.4; and AD's ΔBIC_j (not shown) was 373.75 for females and 116.60 for males, which would be ranked eighth for females and seventh for males in Table 17.4. Moreover, the proxy interview variable was ranked third for females and fourth for males in Table 17.4. The proxy interview variable and the senility and AD variables were more significant and informative than the SPMSQ, MMSE, and memory variables.

Table 17.4 Rank-ordering of variables by BIC values

Sex and rank	Variable or variable group	Variables (*J*)	Response outcomes (*L*)	Responses (*N*)	Chi-squared	df	df*ln (*N**)	ΔBIC	ΔBIC/*J*
Males									
1	ADL	7	40	8981	5846.08	99	708.54	5137.54	733.93
2	IADL	9	18	10,954	6587.67	27	191.81	6395.85	710.65
3	5-Year survival status	2	4	2698	1269.20	6	43.24	1225.96	612.98
4	Respondent is proxy	1	2	1143	548.60	3	21.12	527.47	527.47
5	Physical performance	7	28	7341	3709.55	63	438.16	3271.39	467.34
6	Subjective health status	1	4	953	298.86	9	61.74	237.13	237.13
7	See well enough to read newspaper	1	2	1056	90.22	3	20.89	69.33	69.33
8	Exercise	3	15	1502	431.47	36	223.77	207.70	69.23
9	Residence type: institutional vs. non-institutional	1	2	260	80.09	3	16.68	63.40	63.40
10	Medical conditions	28	56	29,686	1933.41	84	585.08	1348.34	48.15
11	Behavior	4	9	4218	230.78	15	104.41	126.37	31.59
12	SPMSQ	10	20	6236	327.26	30	192.96	134.30	13.43
13	Race	1	3	2382	56.54	6	46.65	9.88	9.88
14	Body mass index	3	12	1399	174.23	27	165.87	8.36	2.79
15	Alcohol use	1	3	522	36.81	6	37.55	−0.74	−0.74
16	MMSE	11	41	1342	412.40	90	432.36	−19.96	−1.81
17	Cigarette use	1	3	526	17.18	6	37.59	−20.41	−20.41
18	Memory	2	10	69	21.52	24	84.89	−63.37	−31.69
19	Height	1	5	506	22.54	12	74.72	−52.18	−52.18
	Total	94	277	81,774	22,094.41	549	3488.04	18,606.38	197.94
Females									
1	ADL	7	39	21,644	15,350.09	96	771.51	14,578.58	2082.65
2	IADL	9	18	25,156	13,592.04	27	214.26	13,377.78	1486.42

3	Respondent is proxy	1	2	2822	1430.19	3	23.84	1406.35	1406.35
4	Physical performance	7	28	17,128	8430.59	63	491.53	7939.06	1134.15
5	5-year survival status	2	4	6610	2051.15	6	48.62	2002.53	1001.27
6	Subjective health status	1	4	2273	624.92	9	69.56	555.36	555.36
7	Race	1	3	5562	486.62	6	51.74	434.88	434.88
8	Residence type: institutional vs. non-institutional	1	2	862	354.33	3	20.28	334.06	334.06
9	Exercise	3	15	3715	971.76	36	256.37	715.39	238.46
10	See well enough to read newspaper	1	2	2487	254.06	3	23.46	230.61	230.61
11	Medical conditions	28	56	69,591	4917.84	84	656.67	4261.17	152.18
12	SPMSQ	10	20	15,863	1289.18	30	220.94	1068.24	106.82
13	Body mass index	3	12	3286	499.29	27	188.91	310.39	103.46
14	Behavior	4	9	9931	457.05	15	117.26	339.79	84.95
15	MMSE	11	41	3839	1010.70	90	526.96	483.75	43.98
16	Alcohol use	1	3	1290	58.48	6	42.97	15.51	15.51
17	Cigarette use	1	3	1293	17.75	6	42.99	−25.24	−25.24
18	Memory	2	11	174	39.09	27	120.37	−81.28	−40.64
19	Height	1	5	1232	23.60	12	85.40	−61.79	−61.79
	Total	94	277	194,758	51,858.75	549	3973.63	47,885.12	509.42

Note: *df* is the sum of the degrees of freedom over the indicated group of variables; $\ln(N^*)$ is the corresponding *df*-weighted arithmetic mean of $\ln(N)$ over the same group of variables

Table 17.5 displays the sex-specific λ-parameters and summary measures for the 19 groups of variables in the same order as shown in Table 17.4.

Berkman et al. (1989) noted that the patterns of sex-specific λ-parameters can be used to characterize each pure type based on the relative sizes of ratios of individual λ-parameters to the corresponding marginal probabilities. The λ-parameters and summary measures in Table 17.5 were coded such that large values generally reflected "less healthy" responses; boldface fonts indicate values that are greater than the corresponding marginal values.

Type IV exceeded the marginal values on 11 of the top 12 ranked variables/variable-groups, the exception being for race for females. This was consistent with the expectation that Type IV would represent severe cognitive impairment with significant impact on ADLs, IADLs, physical performance, health status, institutionalization, and survival. At the other extreme, Type I exceeded the marginal values only for variables ranked 13th or lower, which were the least significant in the model.

Based on Stallard et al. (2010), Types II and III were expected to provide a more diverse set of outcomes for mild and moderate manifestations of dementia than would be possible by restricting these manifestations to be intermediate between Types I and IV. This was confirmed in Table 17.5 where Type II exceeded the marginal values on 6–7 (M–F) of the top 12 ranked variables/variable-groups, with Type III below the marginal values on 11 of the top 12, the exception being the 5-year death rate for females (75 %). For males, the 5-year death rate was high (50 %), but not above the marginal death rate (66 %). The 5-year death rates for Type II were 54 % and 46 %, respectively, for males and females. Thus, Type III appears "healthier" than Type II for both sexes; for females, however, Type III had a higher death rate. In addition, Type III had more rapid progression to Type IV (see below).

Type II had higher than average BMI levels, with the probability of exceeding 28 kg/m^2 being 70 % for males and 65 % for females. The probability of exceeding 32 kg/m^2 was 26 % for males and 30 % for females. At age 50, the corresponding probability was 24 % for males and 48 % for females. The Type II probability of self-reported obesity (or being medically overweight; one of the 28 medical conditions—not shown) was 37 % for males and 54 % for females, which more closely corresponded to the age-50 BMI values than to the BMI values at the time of the survey.

Table 17.6 displays the 5-year transition and cumulative transition matrices $\mathbf{U}_t$ and $\mathbf{V}_{0,t}$. All of the matrices are upper triangular. The u-transitions from Type I were predominantly to Type III, an exception being for males transitioning to Type II at 5–10 years after onset of dementia. The u-transitions from Type II were predominantly to Type IV. The diagonals of the $\mathbf{V}$-matrices show that the persistency rates in Type I were 37 % for males and 7 % for females 15 years after onset of dementia. The corresponding persistency rates in Type II were 20 % and 85 %, respectively, and 0 % (both sexes) in Type III. Thus, the healthier set of initial symptoms for Type III deteriorated rapidly; all had progressed to Type IV within 5 years.

Table 17.5 λ-parameters and summary measures for rank-ordered variables, by sex

Sex and rank	Variable or variable group	Responses (N)	Pure type I	II	III	IV	Observed	Predicted	Difference
Males									
1	ADL: 0–7 with standby/active personal help	8981	0.00	0.89	0.00	**6.20**	1.68	1.74	−0.06
1	ADL: % of maximum	8981	0.00	12.78	0.00	**88.61**	23.98	24.84	−0.86
2	IADL: 0–9 impairments	10,954	0.00	**5.30**	0.18	**9.00**	3.35	3.13	0.22
2	IADL: % of maximum	10,954	0.02	**58.89**	2.01	**100.00**	37.20	34.79	2.41
3	5-Year survival status: % dead	1349	0.00	53.96	50.00	**100.00**	66.12	69.12	−3.00
4	Proxy – % proxy interview	1143	11.74	0.00	0.18	**100.00**	34.30	30.78	3.52
5	Physical performance: 0–7 very difficult/can't do	7341	0.00	**3.26**	0.08	**4.49**	1.82	1.84	−0.02
5	Physical performance: % of maximum	7341	0.00	**46.64**	1.12	**64.10**	26.00	26.28	−0.28
6	Subjective health status: % fair/poor	953	20.61	**100.00**	27.58	**77.12**	53.31	53.61	−0.30
7	Vision: % can't read newspaper	1056	12.21	**48.62**	13.40	**54.23**	31.06	31.12	−0.06
8	Exercise: % none	496	7.62	**51.81**	9.54	**71.73**	34.07	34.09	−0.02
9	LTC institution: % resident in NH	260	0.00	0.00	0.00	**66.47**	26.15	27.17	−1.01
10	Medical conditions: 0–28	29,686	2.37	**9.53**	1.49	**5.77**	4.50	4.48	0.02
10	Medical conditions: % of maximum	29,686	8.48	**34.05**	5.32	**20.59**	16.06	15.99	0.07
11	Behavior: 0–4 problems	4218	0.31	0.48	0.16	**1.16**	0.52	0.52	0.00
11	Behavior: % of maximum	4218	7.82	12.05	3.98	**29.08**	13.09	12.99	0.10
12	SPMSQ: 0–10 errors	6236	3.29	1.95	3.61	**7.36**	3.68	3.68	−0.01
12	SPMSQ: % of maximum	6236	32.93	19.54	36.14	**73.58**	36.77	36.82	−0.05
13	Race: % non-white	2382	**28.49**	**22.29**	0.69	12.45	14.23	14.09	0.14
13	Race: % black	2382	**27.88**	**21.32**	0.00	9.94	12.72	12.57	0.15
14	Body mass index: % currently ≥ 28 kg/m^2	498	12.11	**69.66**	0.00	4.36	16.87	16.81	0.05
14	Body mass index: % currently ≥ 32 kg/m^2	498	0.00	**25.92**	0.00	0.00	4.82	4.77	0.05

(continued)

Table 17.5 (continued)

Sex and rank	Variable or variable group	Responses (N)	Pure type I	II	III	IV	Observed	Predicted	Difference
14	Body mass index: % $\geq$ 32 kg/m^2 at age 50	432	0.00	**23.75**	3.15	5.78	7.18	7.10	0.08
15	Alcohol use: % current users	522	**39.32**	1.26	**18.69**	5.39	16.67	16.45	0.22
16	MMSE: 0–30 errors	1342	**11.14**	7.53	5.36	**19.27**	9.21	9.42	−0.21
16	MMSE: % of maximum	1342	**37.13**	25.11	17.86	**64.22**	30.69	31.39	−0.69
17	Cigarette use: % current users	526	1.56	**19.05**	**15.20**	4.21	9.70	9.76	−0.07
18	Memory: 0–12 DWR errors	34	9.99	**11.35**	**10.92**	**12.00**	10.91	10.90	0.01
19	Height: % < 68 in.	506	22.81	**51.88**	33.70	**36.03**	35.18	35.16	0.02
Females									
1	ADL: 0–7 with standby/active personal help	21,644	0.00	0.73	0.03	**6.39**	1.97	1.97	0.00
1	ADL: % of maximum	21,644	0.00	10.38	0.46	**91.27**	28.19	28.14	0.05
2	IADL: 0–9 impairments	25,156	0.00	**3.97**	0.90	**8.80**	3.46	3.24	0.22
2	IADL: % of maximum	25,156	0.00	**44.16**	10.01	**97.81**	38.42	35.97	2.45
3	Proxy – % proxy interview	2822	0.00	0.00	0.00	**96.95**	30.72	29.90	0.83
4	Physical performance: 0–7 very difficult/can't do	17,128	0.06	**3.75**	0.37	**4.88**	2.27	2.26	0.01
4	Physical performance: % of maximum	17,128	0.90	**53.57**	5.23	**69.78**	32.46	32.30	0.16
5	5-year survival status: % dead	3305	0.00	45.87	**75.16**	**87.74**	55.43	56.91	−1.48
6	Subjective health status: % fair/poor	2273	23.21	**99.98**	16.54	**68.88**	50.46	50.89	−0.43
7	Race: % non-white	5562	0.16	**52.39**	0.98	1.12	12.12	11.62	0.49
7	Race: % black	5562	0.00	**52.39**	0.00	0.00	11.42	10.92	0.50
8	LTC institution: % resident in NH	862	0.00	0.00	0.00	**78.94**	35.96	36.91	−0.95
9	Exercise: % none	1212	23.62	**48.29**	0.00	**91.38**	39.85	40.93	−1.08
10	Vision: % can't read newspaper	2487	8.41	**41.27**	10.79	**57.07**	29.55	29.67	−0.11
11	Medical conditions: 0–28	69,591	1.90	**9.88**	1.38	**5.12**	4.55	4.55	0.00
11	Medical conditions: % of maximum	69,591	6.79	**35.30**	4.94	**18.29**	16.26	16.26	0.01

12	SPMSQ: 0–10 errors	15,863	3.04	2.30	3.88	**7.73**	3.88	3.90	−0.01
12	SPMSQ: % of maximum	15,863	30.40	22.99	38.81	**77.27**	38.85	38.98	−0.13
13	Body mass index: % currently $\geq$ 28 kg/m^2	1179	16.14	**64.93**	6.06	4.08	20.95	20.72	0.23
13	Body mass index: % currently $\geq$ 32 kg/m^2	1179	5.52	**30.48**	0.00	0.00	8.06	7.93	0.13
13	Body mass index: % $\geq$ 32 kg/m^2 at age 50	1000	0.75	**47.78**	0.00	4.53	12.40	11.92	0.48
14	Behavior: 0–4 problems	9931	0.15	**0.44**	0.15	**0.96**	0.43	0.43	0.00
14	Behavior: % of maximum	9931	3.75	**10.94**	3.72	**24.07**	10.77	10.74	0.03
15	MMSE: 0–30 errors	3839	**11.00**	4.30	7.67	**16.64**	9.28	9.37	−0.08
15	MMSE: % of maximum	3839	**36.68**	14.33	25.57	**55.47**	30.95	31.23	−0.28
16	Alcohol use: % current users	1290	**19.94**	0.00	**17.18**	4.26	10.47	10.55	−0.08
17	Cigarette use: % current users	1293	**6.62**	1.39	**10.99**	3.04	5.72	5.72	0.00
18	Memory: 0–12 DWR errors	85	10.34	10.27	**11.56**	9.66	10.78	10.78	0.00
19	Height: % < 64 in.	1232	53.30	52.50	**69.96**	**67.61**	61.69	61.69	0.00

Note: Boldface fonts denote values greater than the observed marginal value

Table 17.6 V and U matrices, by sex

Sex and time since onset	V-matrices Pure type	I	II	III	IV	Time interval & sample size (N)	U-matrices Pure type	I	II	III	IV
Males											
0	I	1.0000	0.0000	0.0000	0.0000	0–5 ($N = 1349$)	I	0.5274	0.0000	0.4726	0.0000
	II	0.0000	1.0000	0.0000	0.0000		II	0.0000	0.8763	0.0000	0.1237
	III	0.0000	0.0000	1.0000	0.0000		III	0.0000	0.0000	0.0000	1.0000
	IV	0.0000	0.0000	0.0000	1.0000		IV	0.0000	0.0000	0.0000	1.0000
5	I	0.5274	0.0000	0.4726	0.0000	5–10 ($N = 316$)	I	0.7022	0.2978	0.0000	0.0000
	II	0.0000	0.8763	0.0000	0.1237		II	0.0000	0.2308	0.0000	0.7692
	III	0.0000	0.0000	0.0000	1.0000		III	0.0000	0.0000	0.6944	0.3056
	IV	0.0000	0.0000	0.0000	1.0000		IV	0.0000	0.0000	0.0000	1.0000
10	I	0.3704	0.1571	0.3281	0.1444	10–15 ($N = 48$)	I	1.0000	0.0000	0.0000	0.0000
	II	0.0000	0.2022	0.0000	0.7978		II	0.0000	1.0000	0.0000	0.0000
	III	0.0000	0.0000	0.0000	1.0000		III	0.0000	0.0000	0.7810	0.2190
	IV	0.0000	0.0000	0.0000	1.0000		IV	0.0000	0.0000	0.0000	1.0000
15	I	0.3704	0.1571	0.2563	0.2163						
	II	0.0000	0.2022	0.0000	0.7978						
	III	0.0000	0.0000	0.0000	1.0000						
	IV	0.0000	0.0000	0.0000	1.0000						
Females											
0	I	1.0000	0.0000	0.0000	0.0000	0–5 ($N = 3305$)	I	0.5054	0.0000	0.4946	0.0000
	II	0.0000	1.0000	0.0000	0.0000		II	0.0000	0.9745	0.0085	0.0170
	III	0.0000	0.0000	1.0000	0.0000		III	0.0000	0.0000	0.0000	1.0000
	IV	0.0000	0.0000	0.0000	1.0000		IV	0.0000	0.0000	0.0000	1.0000

5	I	0.5054	0.0000	0.4946	0.0000	5–10 ($N = 1048$)	I	0.5772	0.0000	0.4228	0.0000
	II	0.0000	0.9745	0.0085	0.0170		II	0.0000	0.8740	0.0000	0.1260
	III	0.0000	0.0000	0.0000	1.0000		III	0.0000	0.0000	0.3364	0.6636
	IV	0.0000	0.0000	0.0000	1.0000		IV	0.0000	0.0000	0.0000	1.0000
10	I	0.2917	0.0000	0.3801	0.3282	10–15 ($N = 216$)	I	0.2386	0.0756	0.6859	0.0000
	II	0.0000	0.8516	0.0029	0.1455		II	0.0000	1.0000	0.0000	0.0000
	III	0.0000	0.0000	0.0000	1.0000		III	0.0000	0.0000	0.6923	0.3077
	IV	0.0000	0.0000	0.0000	1.0000		IV	0.0000	0.0000	0.0000	1.0000
15	I	0.0696	0.0220	0.4632	0.4452						
	II	0.0000	0.8516	0.0020	0.1464						
	III	0.0000	0.0000	0.0000	1.0000						
	IV	0.0000	0.0000	0.0000	1.0000						

Figures 17.1a, 17.1b, 17.1c, and 17.1d display the univariate GoM-score distributions for the four pure types for both sexes separately at the time of onset of dementia. Other than the spikes at 0, the distributions for Types I and III were relatively uniform. In contrast, the frequencies for Type II declined with increasing GoM scores, dropping below 1 % above GoM-score values of 80 %; the frequencies for Type IV declined with increasing GoM scores up to 40 %, after which they were relatively uniform. For each of the four pure types, the GoM scores were distributed such that only 0–9 % of cases had GoM scores of 100 %, with the 0 %-frequencies occurring for Type II. Considering all pure types together, we found that 19 % of males and 16 % of females had some GoM score equal to 100 %. Thus, 81 % of males and 84 % of females had partial membership in at least two pure types.

Figures 17.2a, 17.2b, 17.2c, and 17.2d display the mean GoM scores for the four pure types for both sexes separately, for survivors at each of four durations (i.e., 0, 5, 10, and 15 years) after onset of dementia, in a 100 %-stacked-line format with and without adjustments for the changes over time since onset of dementia. The means of the unadjusted GoM scores exhibited the following patterns:

- Type I means increased from 22 % initially to 74 % of the total mean GoM score (100 %) for males at 15 years duration. The corresponding increases for females were from 23 % to 55 %.
- Type II means were relatively stable for males in the range 17–19 % at all durations and increased for females from 21 % to 29 % at 15 years duration.
- Type III means declined for both sexes over 15 years, from 38 % to 1 % for males and from 35 % to 10 % for females.

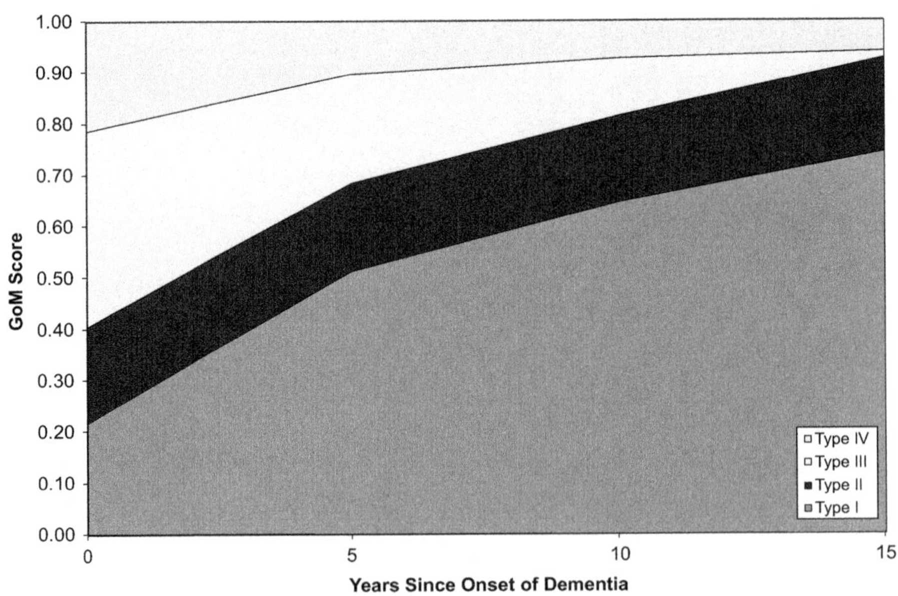

Fig. 17.2a Unadjusted GoM score distribution, male survivors

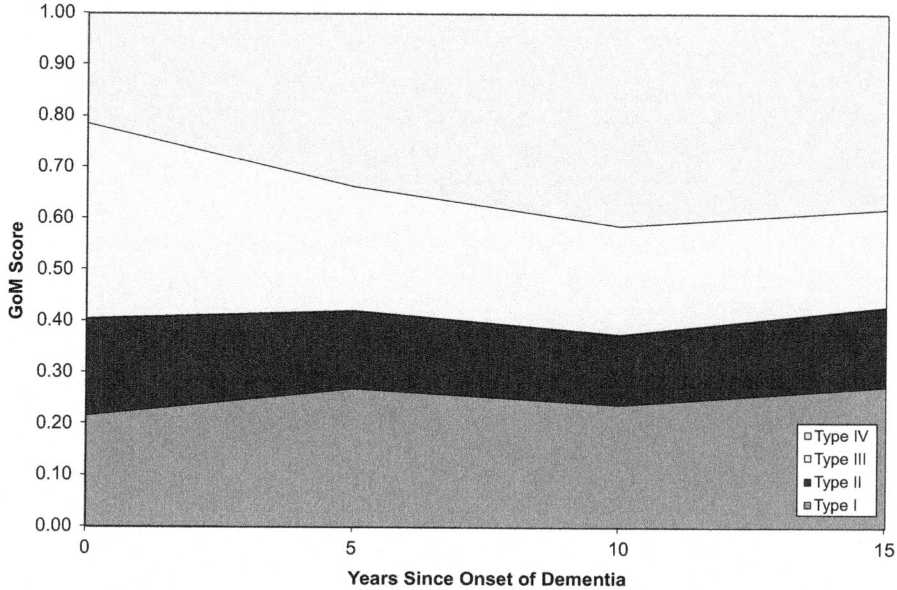

Fig. 17.2b Adjusted GoM score distribution, male survivors

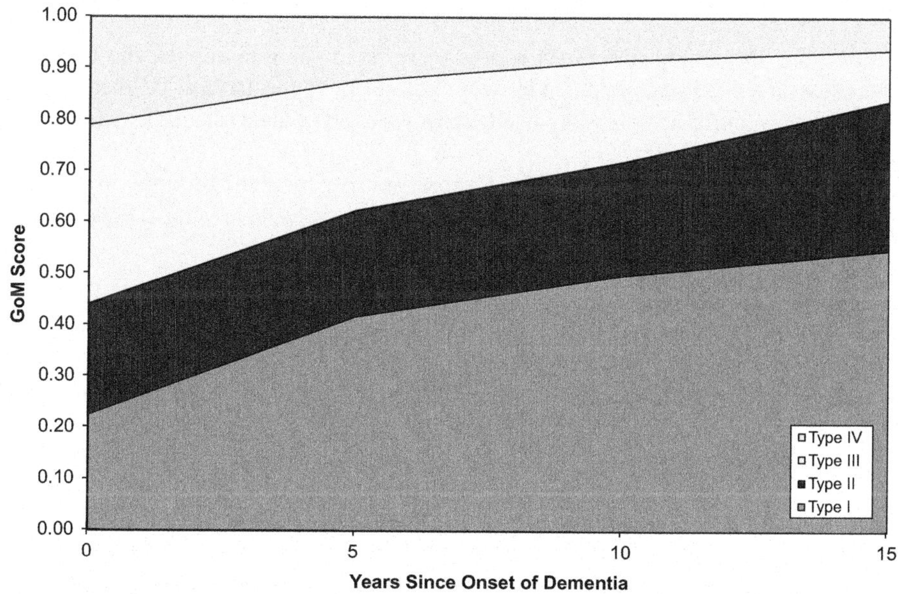

Fig. 17.2c Unadjusted GoM score distribution, female survivors

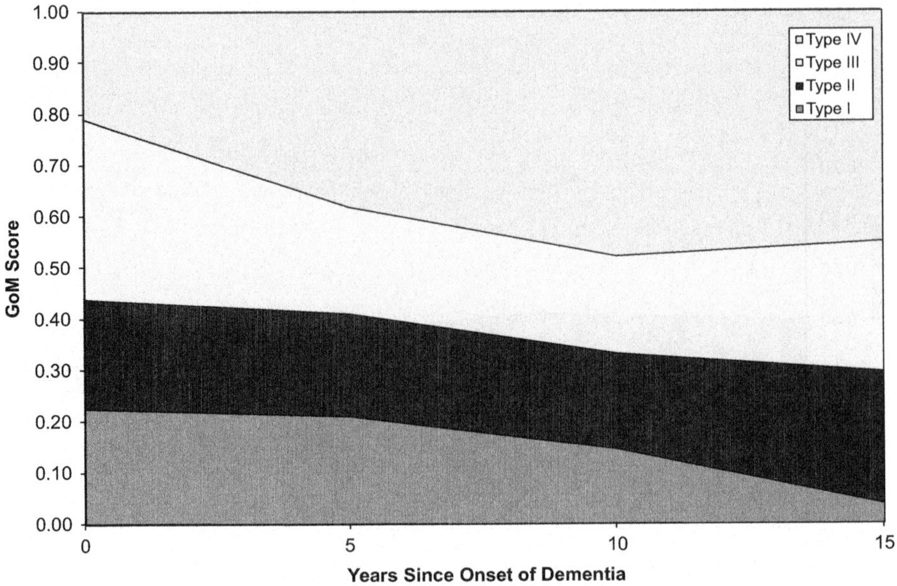

Fig. 17.2d Adjusted GoM score distribution, female survivors

- Type IV means declined a similar amount for both sexes over 15 years, from 21 % to 6 %.

Because the unadjusted GoM scores were fixed for individuals, the temporal increases in the Type I means and decreases in the Type III and IV means were solely due to mortality selection on a heterogeneous population sample (see Vaupel et al. 1979; Stallard 2007).

The time-varying (adjusted) GoM scores were generated by applying the **V**-matrices to the basic GoM scores, i.e., $\mathbf{g}'_{it} = \mathbf{g}'_i \mathbf{V}_{0,t}$. The means of the adjusted GoM scores for the four pure types exhibited the following patterns:

- Type I means increased from 22 % to 28 % for males and decreased from 23 % to 4 % for females over 15 years. The corresponding unadjusted means at 15 years duration were 74 % and 55 %, respectively.
- Type II means decreased from 19 % to 15 % for males and increased from 21 % to 26 % for females over 15 years. The corresponding unadjusted means at 15 years duration were 18 % and 29 %, respectively.
- Type III means decreased for both sexes, from 38 % to 19 % for males and from 35 % to 25 % for females over 15 years. The corresponding unadjusted means at 15 years duration were 1 % and 10 %, respectively.
- Type IV means increased for both sexes, from 21 % to 38 % for males and from 21 % to 45 % for females over 15 years. The corresponding unadjusted means at 15 years duration were 6 % for both sexes.

The application of the **V**-matrices increased the combined means of Types III–IV at 15 years duration from 7 % to 57 % for males and from 16 % to 70 % for females. Over 15 years, this produced relatively stable combined means of Types III–IV in the range 57–62 % for males and increasing combined means in the range 56–70 % for females.

17.4.3 Ancillary Analysis: Mortality

With the basic and time-varying GoM scores in hand, one can conduct a range of ancillary analyses using the GoM scores as the independent variables. To explore this type of application, we used the Medicare vital statistics data linked to the NLTCS to refine our analysis of the 5-year survival rates used in the main analysis.

The mortality data were recorded using 1-year intervals beginning with the date of onset of dementia and continuing to the corresponding anniversary date in 2004, yielding a maximum of 20 years of follow-up data for any one person. Linear interpolation was used to generate estimates of the time-varying GoM scores at the start of each 1-year follow-up interval. Conditional maximum likelihood procedures, with the g- and u-parameters fixed, were used to obtain 20 sets of λ-parameter estimates for the conditional survival probability of being alive after 1 year for a person alive at the start of each 1-year interval (i.e., equivalent to p_x in standard life table notation). The 20 sets of estimated conditional survival probabilities were chain-multiplied to produce estimates of the marginal survival function at each anniversary of onset of dementia from 1 to 20 years (i.e., equivalent to l_x in standard life table notation).

The results for 0–15 years since onset of dementia are displayed in Figs. 17.3a, 17.3b, 17.3c, and 17.3d separately by sex and by pure type, with and without adjustments for the impact of the **V**-matrices over time. The observed and predicted marginal survival functions, which are virtually identical, are displayed for comparison. For both sexes, Type I had the highest survival and Type IV the lowest. Types II and III initially declined at rates close to the marginal survival function. The adjusted survival functions for Types II and III continued their decline after 3 years while the unadjusted survival functions declined more slowly.

The survival functions for the time-varying (adjusted) pure types were obtained by application of the **V**-matrices to the 20 sets of basic **λ**-vectors for the conditional survival probabilities. In this case, the basic **λ**-vectors required an additional time index to indicate that a distinct vector, denoted as $\boldsymbol{\lambda}_{m_{jl}}(t)$, was estimated for each time t. The **V**-matrix adjustments were accomplished using $\boldsymbol{\lambda}_{m_{jlt}}(t) = \mathbf{V}_{0,t}\boldsymbol{\lambda}_{m_{jl}}(t)$, where the time subscript on $\boldsymbol{\lambda}_{m_{jlt}}(t)$ indicates adjustment by $\mathbf{V}_{0,t}$. For times $0 < t < 15$ (except $t = 5$ or 10 years), $\mathbf{V}_{0,t}$ was generated by linear interpolation from the two adjacent **V**-matrix estimates; for $t > 15$, $\mathbf{V}_{0,15}$ was used.

The 20 sets of adjusted conditional survival probabilities were chain-multiplied to produce estimates of the adjusted marginal survival function at each anniversary

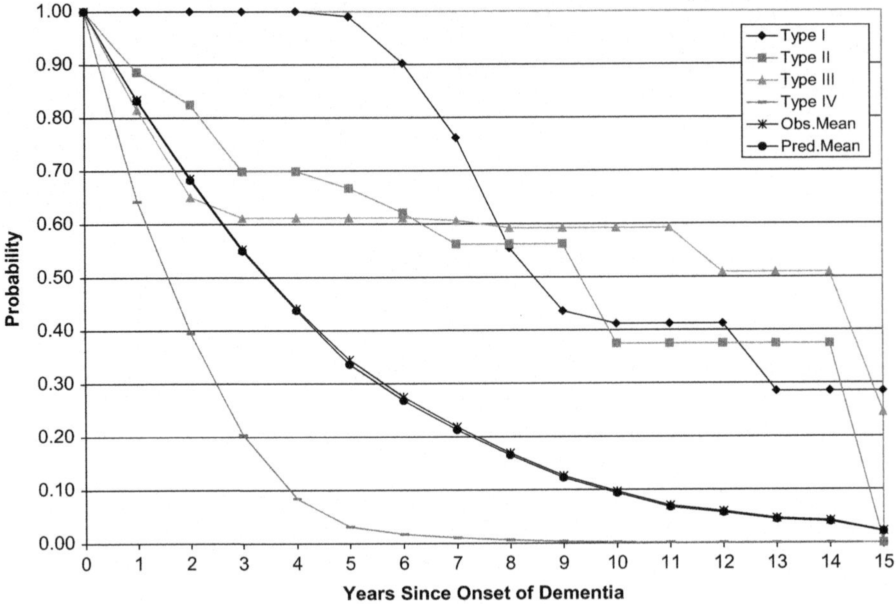

Fig. 17.3a Unadjusted survival functions, males

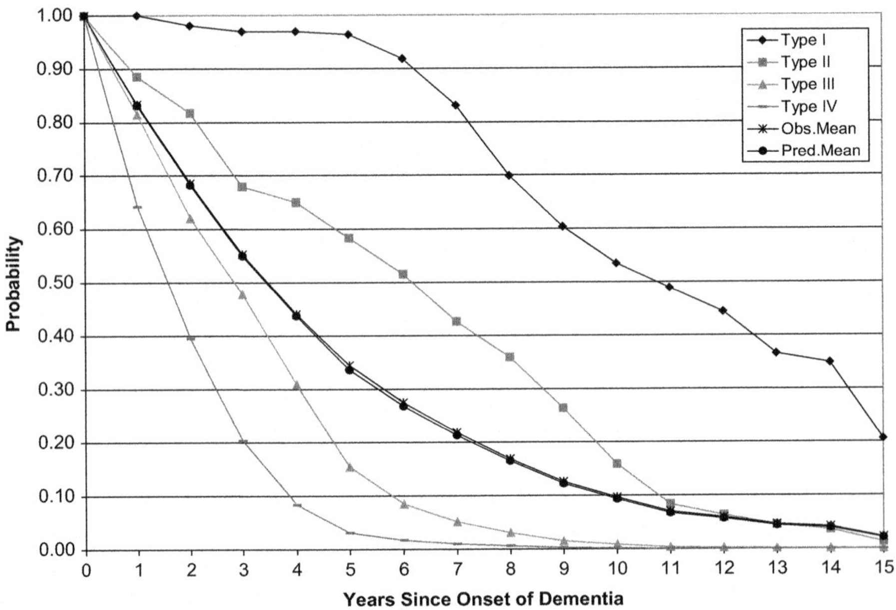

Fig. 17.3b Adjusted survival functions, males

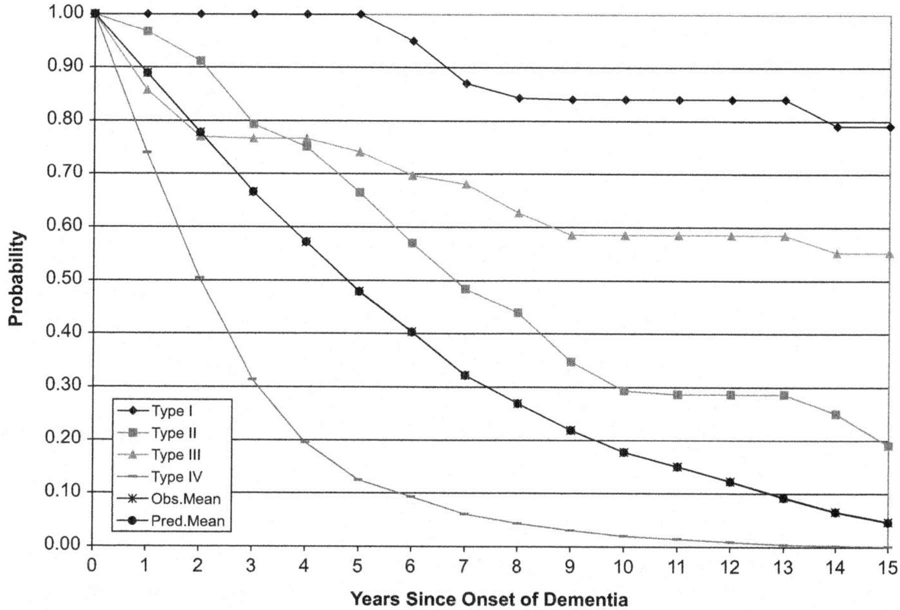

Fig. 17.3c Unadjusted survival functions, females

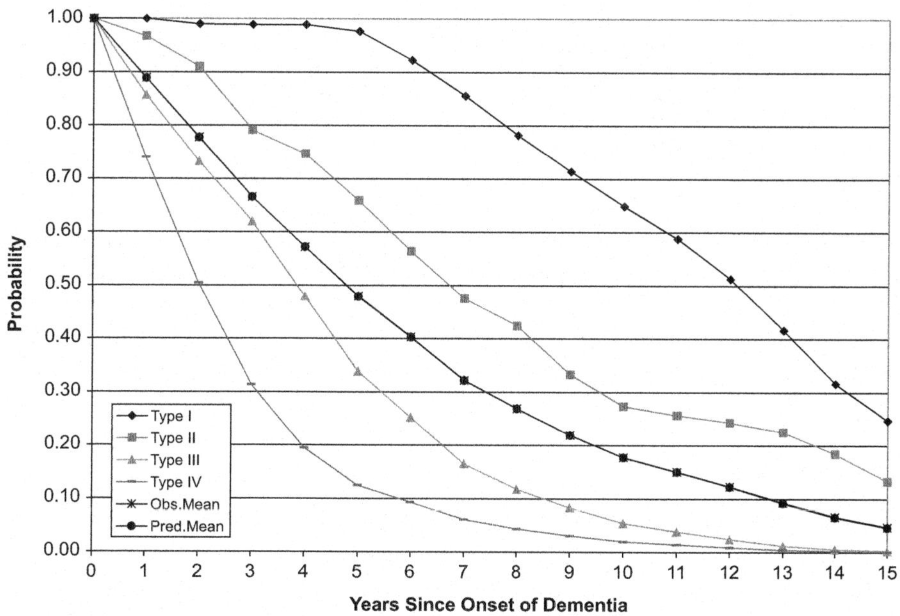

Fig. 17.3d Adjusted survival functions, females

Table 17.7 Life expectancy (in years) following onset of dementia, by sex and adjustment

Source and sex	Pure type I	II	III	IV	Observed	Predicted	Difference
Unadjusted lambdas							
Males	**11.33**	**8.46**	**10.27**	1.90	4.54	4.49	0.06
Females	**13.95**	**8.40**	**10.52**	2.66	5.85	5.84	0.01
Adjusted lambdas							
Males	**11.50**	**6.10**	3.08	1.90	4.54	4.49	0.06
Females	**11.77**	**7.99**	4.30	2.66	5.85	5.84	0.01

Note: Boldface fonts denote values greater than the observed marginal value

of onset of dementia from 1 to 20 years. The plots for Type IV and the observed and predicted marginal survival functions were unchanged. The greatest changes were for Type III, followed by Type II, and then Type I. For both sexes, the adjusted survival functions for Type III dropped below the observed and predicted marginal survival functions.

One can integrate the survival functions to compute the marginal and pure-type–specific life expectancies, assuming that the maximum survival time is 20 years; see Table 17.7. The marginal life expectancy was 4.54 years for males and 5.85 years for females. The range of life expectancies based on the adjusted lambdas was 1.90–11.50 years for males and 2.66–11.77 years for females, where the lowest values correspond to initial GoM assignment to Type IV and the highest to Type I. The ranges of life expectancies for Types II–III were substantially narrower: 3.08–6.10 years for males and 4.30–7.99 years for females. These restricted ranges are relevant to the 58 % of the sample whose GoM score on Type I was less than 0.1 (see Figs. 17.1a, 17.1b, 17.1c, and 17.1d).

The life expectancies based on the unadjusted lambdas are counterfactual because they do not account for the transitions during the 20-year follow-up interval. Alternatively, they may be interpreted as estimates of the life expectancies that would result from interventions designed to halt the progression of the dementia process. The largest benefits would accrue to persons initially like Type III, where the increase in life expectancy would be 7.2 years for males and 6.2 years for females. No benefit would accrue to persons initially like Type IV because Type IV is the endpoint of the process.

17.4.4 Ancillary Analysis: Acute and Long-Term Care

Conditional maximum likelihood procedures, with the g- and u-parameters fixed, were used to estimate λ-parameters for additional sets of covariates which were coded to reflect the use and costs of various types of acute care and LTC services, conditional on status as an institutional or community resident, on enrollment in the Medicare fee-for-service program, and on using various types of acute care or LTC services. Covariate coding for such conditional estimation was implemented by ensuring that all observations not meeting the indicated criteria were coded as "missing;" in some cases, this conditional estimation procedure required validly measured covariates for persons not meeting the indicated criteria to be recoded to "missing."

Tables 17.8 and 17.9 summarize the utilization rates and costs for males and females, respectively, for services based on data collected in the NLTCS instruments and in the linked Medicare billing records. Statistical tests of costs were performed using the Wilks chi-squared test of the non-parametric distributions of costs, using variable-sized categories covering the range of costs with cutpoints based on multiples of 10 and \$2, \$5, and \$10, with the null hypothesis being no difference in category-probabilities (λ-parameters) over the four pure types. Average costs were computed as λ-weighted averages of the means of the individual cost-categories. The differences between the observed and predicted mean costs indicated that the errors induced by this procedure were relatively small (i.e., generally 1 % or less). All costs were converted to 2004 dollars using the CPI-U.

Tables 17.8 and 17.9 show that the institutionalization rates (also reported in Tables 17.4 and 17.5) for 5–15 years duration were significantly elevated for both sexes for Type IV, whereas the rates for Types I–III were zero. Exclusion criterion E in Table 17.1 ensured that all cases of dementia in our sample were non-institutionalized at the time of onset. Hence, institutionalization was only assessed at 5, 10, and 15 years after onset of dementia.

The zero rates for Types I–III in Tables 17.8 and 17.9 do not mean that an individual with non-zero GoM scores on Types I–III had a zero probability of being institutionalized. Instead, they imply that the institutionalization probability was proportional to the individual's GoM score on Type IV. For example, Figs. 17.1a, 17.1b, 17.1c, and 17.1d show that 43 % of the male sample had Type IV GoM scores of 0.1 or higher, implying institutionalization probabilities of 7 % or higher, based on the 66.5 % value for Type IV in Table 17.8.

The estimated annual costs per institutional resident were available in the NLTCS only for 1994 and 1999 and were coded as 12 times the current monthly rates for institutional respondents. The costs did not differ significantly for either sex over the four pure types.

LTC services among community residents were modeled by estimating individual-specific probabilities that each one was currently being assisted by one or more personal helpers, and among those with such assistance, by individual-specific conditional probabilities that the personal helpers were paid. For those using one or more personal helpers, and separately for the subset using paid helpers, the intensity of care was modeled by estimating individual-specific values of the average annual hours of help per user, estimated as 52.18 times the corresponding estimate of the number of hours of help in the week prior to the NLTCS community interview (available for 1989 and later). For those using paid helpers, a second representation of the intensity of care was the individual-specific estimate of the annual cost of such care, estimated as 12 times the corresponding estimate of the

Table 17.8 Use and costs[a] of acute and long-term care, males

Item	Responses (*N*)	Pure type I	II	III	IV	Observed	Predicted	Difference	Chi-squared	d.f.	Significance	Chi-Squared Loss
Panel 1: Institutional residents												
Institutionalized at 5, 10, or 15 years	260	0.0 %	0.0 %	0.0 %	**66.5 %**	26.2 %	27.2 %	−1.0%	80.09	3		
Estimated annual costs per institutional Resident	26	$18,944	$29,732	$17,384	$36,351	$32,249	$31,873	$376	6.99	12	n.s	
Panel 2: Community residents												
Community residents with one or more helpers	1215	20.0 %	**100.0 %**	43.0 %	**99.4 %**	65.1 %	61.4 %	3.67 %	485.24	3		
Average annual hours of help per user	530	1151	1564	894	**3967**	2242	2223	19	184.58	21		
Community users with paid helpers	791	11.6 %	24.8 %	19.6 %	**40.2 %**	26.5 %	26.6 %	−0.07 %	21.75	3		
Average annual hours of paid help per user	150	519	476	391	1202	773	780	−7	19.41	21	n.s.	
Average annual cost of paid help per user	96	$9880	$3591	$2864	$10,131	$7092	$7128	-$36	28.89	24	n.s.	

Panel 3: Medicare utilization by type of service												
Acute care	1173	85.2 %	87.7 %	**91.1 %**	**92.4 %**	89.5 %	89.5 %	0.0 %	3.80	3	n.s.	
Home health care	1173	4.0 %	**26.6 %**	15.1 %	**40.2 %**	21.1 %	21.2 %	−0.1 %	79.03	3		
Hospice care	1173	0.0 %	0.0 %	**2.6 %**	**3.5 %**	1.8 %	1.8 %	0.0 %	10.18	3		
Skilled nursing facility	1173	2.6 %	**8.3 %**	6.5 %	**15.1 %**	8.2 %	8.2 %	0.0 %	19.08	3		
Total Medicare[b]	1173	85.1 %	88.3 %	**90.9 %**	**94.6 %**	90.1 %	90.1 %	0.0 %	6.64	3	n.s.	
Panel 3a: Average annual costs per user of Medicare services												
Acute care	1050	$4863	**$8750**	**$8753**	**$10,883**	$8478	$8486	−$8	82.52	48		
Home health care	248	$1727	$3059	$3510	**$7139**	$4640	$4638	$3	70.43	42		
Hospice care	21	$4343	$5447	$8849	$7275	$7389	$7455	-$66	22.35	27	n.s.	
Skilled nursing facility	96	**$8912**	**$10,122**	**$10,473**	$6311	$8715	$8641	$74	42.52	36	n.s.	
Total Medicare[b]	1057	$5227	$9303	$10,287	**$16,051**	$10,549	$10,553	-$4	106.21	48		
Panel 4: Average annual costs per enrollee for Medicare services – maximum likelihood estimates												
Acute care	1173	$4148	**$7631**	**$7933**	**$10,038**	$7589	$7601	-$12	85.97	51		0.35
Home health care	1173	$67	$706	$554	**$2553**	$981	$988	-$7	147.29	45		2.17
Hospice care	1173	$0	$48	$215	$194	$132	$132	$0	29.28	30	n.s.	3.26
Skilled nursing facility	1173	$341	**$991**	$618	**$1315**	$803	$805	-$2	56.57	39	n.s.	5.03
Total Medicare[b]	1173	$4456	$8094	$9307	**$15,095**	$9505	$9519	-$13	112.63	51		0.23

(continued)

Table 17.8 (continued)

Item	Responses (*N*)	Pure type I	II	III	IV	Observed	Predicted	Difference	Chi-squared	d.f.	Significance	Chi-Squared Loss
Panel 5: Average annual costs per enrollee for Medicare services – based on utilization probabilities and costs per user in Panels 3 and 3a												
Acute care	1173	$4145	**$7670**	**$7971**	**$10,055**	$7589	$7625	-$36	85.92	51		0.05
Home health care	1173	$68	$814	$532	**$2867**	$981	$1081	-$100	136.45	45		10.85
Hospice care[c]	1173	$0	$0	$229	$256	$132	$144	-$12	–	–	–	–
Skilled nursing facility	1173	$292	**$843**	$683	**$1273**	$803	$779	$24	50.99	39	n.s.	5.58
Total Medicare[c]	1173	$4447	$8212	$9355	**$15,182**	$9505	$9577	-$72	112.53	51		0.09

Note: Boldface fonts denote values greater than the observed marginal value
[a]All costs were converted to 2004 dollars using the CPI-U to inflate the costs for 1984, 1989, 1994, and 1999
[b]Total Medicare costs may not match the sum of the four component costs due to independent estimation of the component costs
[c]"–" indicates that the 4 pure-type model did not fit as well as the marginal model; all statistics were suppressed

Table 17.9 Use and costs[a] of acute and long-term care, females

Item	Responses (*N*)	Pure type				Observed	Predicted	Difference	Chi-squared	d. f.	Significance	Chi-squared loss
		I	II	III	IV							
Panel 1: Institutional residents												
Institutionalized at 5, 10, or 15 years	862	0.0 %	0.0 %	0.0 %	**78.9 %**	36.0 %	36.9 %	−1.0 %	354.33	3		
Estimated annual costs per institutional resident	158	$26,278	$33,777	$35,693	$44,038	$40,721	$40,720	$1	14.24	18	n.s.	
Panel 2: Community residents												
Community residents with one or more helpers	2782	11.6 %	**100.0 %**	52.9 %	**99.4 %**	68.4 %	65.0 %	3.5 %	1110.36	3		
Average annual hours of help per user	1319	386	1103	595	**4370**	2124	2069	55	606.44	21		
Community users with paid helpers	1904	10.2 %	**44.7 %**	27.1 %	**55.2 %**	38.8 %	38.8 %	−0.07 %	86.28	3		
Average annual hours of paid help per user	528	267	543	397	**3314**	1663	1676	−13	165.10	21		
Average annual cost of paid help per user	337	$3168	$6611	$4653	**$29,153**	$16,355	$16,472	-$117	107.70	36		

(continued)

Table 17.9 (continued)

Item	Responses (*N*)	Pure type I	II	III	IV	Observed	Predicted	Difference	Chi-squared	d. f.	Significance	Chi-squared loss
Panel 3: Medicare utilization by type of service												
Acute care	2835	89.8 %	**93.1 %**	89.5 %	**94.8 %**	91.8 %	91.8 %	0.0 %	10.40	3		
Home health care	2835	9.0 %	**33.5 %**	18.6 %	**32.1 %**	23.6 %	23.7 %	0.0 %	84.68	3		
Hospice care	2835	0.0 %	0.0 %	**1.8 %**	**3.9 %**	1.6 %	1.6 %	0.0 %	29.32	3		
Skilled nursing facility	2835	2.4 %	3.6 %	**13.1 %**	**16.5 %**	9.8 %	9.8 %	0.0 %	54.24	3		
Total Medicare[b]	2835	89.7 %	**93.3 %**	89.6 %	**95.4 %**	92.0 %	92.0 %	0.0 %	13.16	3		
Panel 3a: Average annual costs per user of Medicare services												
Acute care	2603	$3182	**$9814**	**$7863**	$6845	$7026	$7043	-$17	138.76	48		
Home health care	670	$2438	**$6080**	$2746	**$7285**	$5138	$5137	$1	64.60	39		
Hospice care	46	$4371	$7867	$4163	$7887	$6573	$6594	-$21	24.40	33	n.s.	
Skilled nursing facility	278	$6568	**$9029**	$7489	**$8076**	$7883	$7892	-$9	28.59	33	n.s.	
Total Medicare[b]	2609	$3419	**$12,123**	**$9759**	**$10,900**	$9285	$9316	-$31	168.12	48		
Panel 4: Average annual costs per enrollee for Medicare services – maximum likelihood estimates												
Acute care	2835	$2862	**$9064**	**$7047**	**$6494**	$6451	$6468	-$17	148.72	51		0.43
Home health care	2835	$231	**$1776**	$449	**$2304**	$1214	$1217	-$3	147.75	42		1.53
Hospice care	2835	$0	$20	$67	$292	$107	$107	$0	45.64	36	n.s.	8.08
Skilled nursing facility	2835	$145	$466	**$956**	**$1275**	$773	$773	$0	76.51	36		6.31
Total Medicare[b]	2835	$3077	**$11,185**	**$8745**	**$10,396**	$8545	$8576	-$31	180.66	51		0.63

Panel 5: Average annual costs per enrollee for Medicare services – based on utilization probabilities and costs per user in Panels 3 and 3a												
Acute care	2835	$2856	**$9133**	**$7037**	**$6487**	$6451	$6477	-$26	148.66	51		0.07
Home health care	2835	$220	**$2036**	$510	**$2340**	$1214	$1299	-$85	141.86	42		5.89
Hospice care	2835	$0	$0	$74	$306	$107	$108	-$2	13.25	36	n.s.	32.39
Skilled nursing facility	2835	$157	$325	**$979**	**$1336**	$773	$769	$4	68.99	36		7.52
Total Medicare[b]	2835	$3066	**$11,307**	**$8740**	**$10,395**	$8545	$8598	-$53	180.54	51		0.11

Note: Boldface fonts denote values greater than the observed marginal value
[a]All costs were converted to 2004 dollars using the CPI-U to inflate the costs for 1984, 1989, 1994, and 1999
[b]Total Medicare costs may not match the sum of the four component costs due to independent estimation of the component costs

monthly payment rate in the month prior to the NLTCS community interview (available for 1994 and 1999).

The λ-parameters, or λ-weighted averages of the means of the individual hour- or cost-categories, are shown in the second panel of Tables 17.8 and 17.9, with the row labels indicating increasing levels of conditioning relative to preceding rows. The Wilks chi-squared tests indicated that the pure-type differences were statistically significant for females for all measures and for males for all measures except the hours and costs of care for paid helpers. Types II and IV had helpers in 99–100 % of the cases for both sexes; in contrast, Types I and III had helpers in 20 % and 43 %, respectively, of the cases for males and 12 % and 53 %, respectively, for females. Among those who had helpers, the average annual hours of help for Type IV was about two FTE's (i.e., 3967 h for males and 4370 h for females). For Type II, males had 1564 h and females had 1103 h, on average. The averages for Type I were larger for males (1151 h vs. 386 h).

Pure-type differences in the fractions who had paid helpers, among those with one or more helpers, were statistically significant for both sexes. Type IV had the highest fractions: 40 % for males and 55 % for females. Pure-type differences in hours and costs of paid care were statistically significant for females, but not males. Moreover, the observed hours and costs of paid care for females were more than double those for males. The average number of paid hours for Type IV was 3314 h for females—2.8 times the 1202 h for males; and the average cost for Type IV was $29,153 for females—2.9 times the $10,131 cost for males.

For both sexes, the cost of paid LTC among community residents was substantially less than the cost of institutional care, even in the case of Type IV which had significantly elevated community costs for females.

Annual Medicare costs were based on tabulations of reported Medicare payments to health care providers for service periods with ending dates occurring within the month of the NLTCS interview or within one of the following 11 months. Persons enrolled in HMOs during any of these 12 months were coded as "missing" on the associated Medicare utilization and cost variables. Medicare billing records were grouped into four mutually exclusive service categories: home health care, hospice care, skilled nursing care, and a residual "acute care" category for all other medical care.

The first three categories represent services that are generally provided in non-hospital settings, which may overlap with LTC services reported by NLTCS respondents in the categories of institutional and home/community care. The acute care category includes both inpatient and outpatient services that are generally provided by medical doctors, physician assistants, and allied health professionals, with no overlap with LTC services reported by NLTCS respondents.

The Medicare utilization measures refer to fee-for-service (FFS; i.e., non-HMO) cases with billed services. Tables 17.8 and 17.9 (third panel) show that the observed average fraction of FFS cases utilizing billed services ranged from 1.6 % (female hospice) to 92 % (female acute care). Pure-type differences in the Medicare utilization rates were statistically significant for females for all five measures and for males for the three LTC categories (home health, hospice, and skilled nursing care) but not for acute care or total Medicare.

Pure-type differences in the costs *per user of Medicare* services were significant for both sexes for acute care, home health care, and total Medicare, but not for hospice and skilled nursing care. For male users, the highest total Medicare costs were for Type IV ($16,051) and the lowest for Type I ($5227); for female users, the highest total Medicare costs were for Type II ($12,123) and the lowest for Type I ($3419). For both sexes, the Type III costs for acute care were substantially higher than for Type I: for males, the Type III costs were comparable to the Type II costs; for females, the Type III costs were lower than the Type II costs but were still above the Type IV costs. The high costs for acute care for Type III were consistent with the high mortality rates for Type III; they help to distinguish Type III from Type I.

Pure-type differences in the costs *per Medicare enrollee* (Tables 17.8 and 17.9, fourth panel) were significant for females for all services except hospice care and for males for all services except hospice and skilled nursing care. The costs per Medicare enrollee were based on a marginal cost model in which a $0-category was appended to the variable-sized categories in the conditional cost model based on multiples of 10 and $2, $5, and $10. For male enrollees, the highest total Medicare costs were for Type IV ($15,095) and the lowest for Type I ($4456); for female enrollees, the highest total Medicare costs were for Type II ($11,185) and the lowest for Type I ($3077).

The fifth panel (Tables 17.8 and 17.9) presents an alternative set of estimates of the costs *per Medicare enrollee*, based on pairwise multiplication of the utilization probabilities and costs per user shown in the third panel. The entries under the heading "chi-squared loss" in the fourth panel show that the chi-squared statistics in the fourth panel were from 0.23 to 8.08 less than the sum of the chi-squared statistics from the corresponding utilization and conditional cost variables in the third panel. Similarly, the entries under the heading "chi-squared loss" in the fifth panel show that the chi-squared statistics from the fifth panel (except hospice care) were from 0.05 to 10.85 lower than those in the fourth panel Together these results indicate that the models in the fourth and fifth panels were effectively equivalent.

17.5 Discussion

This chapter had two goals, one methodological and the other substantive.

Methodologically, our goal was to present a longitudinal form of the GoM model and associated Newton-Raphson iteration procedures. The resulting model describes the natural history of dementia as a complex irreversible multidimensional process occurring within a three-dimensional state space bounded by a regular tetrahedron. Individuals can be located at any point in the state space at the time of onset of dementia. The dementia process can move them to other points in the state space over time. Longitudinal changes are governed by upper-triangular transition matrices which ensure that all changes are irreversible.

The results indicated that the description of the natural history of dementia as a multistage process with three discrete stages (Grossberg and Desai 2003), equivalent to a three-class LCA model (Lanza et al. 2007) with an ordered progression from Class I to Class II to Class III, was an oversimplification that could be resolved using the fuzzy-set conceptualization provided by GoM. The estimated GoM model identified four distinct pure types that bounded the range of variation between mild, moderate, and severe forms of dementia, and did so using a three–dimensional state space that provided a more diverse set of initial and intermediate outcomes for mild and moderate manifestations of dementia than would be otherwise possible. Moreover, the primary transitions from Types I to III, Types II to IV, and Types III to IV, were consistent with the hypothesized fuzzy-set latent membership process.

The need to generalize the discrete three-stage or three-class LCA process to the three-dimensional fuzzy-set latent membership process was supported by the finding that the optimal estimates of the GoM scores had only 17 % of cases exactly described by one pure type; 83 % of cases were best described by weighted combinations of two or more pure types, implying that the four pure types cannot be interpreted or treated as discrete stages or classes in the dementia process.

The fuzzy-set latent membership process is consistent with the use of multiple sets of rating scales for the staging of the dementia (e.g., Hughes et al. 1982; Reisberg et al. 1982; Dooneief et al. 1996), supporting the conclusions of Eisdorfer et al. (1992) and Stern et al. (1996) that simplified descriptions of dementia as having mild, moderate, and severe stages with ordered progression from one stage to the next may be highly misleading. It follows that mathematical models of this process may be seriously flawed if they assume that the stages are discrete when in fact they are abstractions from a more complex and heterogeneous continuous-state process (Green 2007; Green et al. 2011).

Substantively, our goal was to model the natural history of the loss of cognition and functioning following the onset of dementia using data from the NLTCS. The NLTCS and the linked Medicare files jointly covered a broad range of acute and long-term care services that were expected to differ according to the progression of the decline in cognitive and functional status. We expected that the natural history of dementia would be highly variable both within and between sexes with respect to cognitive and physical functioning at onset and the subsequent rates of loss of such functioning.

The results confirmed these expectations and provided numerical estimates of the different rates of decline in cognitive and functional status conditional on the initial profile of GoM scores, and of the utilization and costs of the associated acute and long-term care services. Substantial and statistically significant sex differences were found in the λ–probabilities describing the distributions of responses to the variables used in calibration of the model and in the u-parameters governing the rates of change of the GoM scores over time.

Significant sex differences involved the higher persistency of Type II for females, the association of Type II with high BMI among females, the pattern of increasing combined prevalence of Types III–IV with increasing disease duration, the dependence of acute care and paid community care on the pure types for females, and the substantially greater use of paid community care among females.

Patterns of use and costs of acute care and LTC services were as follows. For females, but not males, the use of acute care services varied by pure type. For both sexes, the costs of acute care services varied by pure type. For both sexes, the use of LTC services varied by pure type. Once nursing home and hospice services were received, however, their costs were independent of pure type. Among community residents using paid helpers for LTC services, the costs for females, but not males, varied by pure type. For both sexes, the cost of paid LTC services among community residents was substantially less than the cost of institutional care.

Estimates of life expectancy ranged from 1.9 to 11.8 years, and depended strongly on cognitive and physical functioning at onset, as summarized by the GoM scores on the four pure types.

The relatively short life expectancy of Type III was noteworthy (males, 3.1 years; females, 4.3 years), given that their initial healthy presentation at the time of onset was similar to Type I (with life expectancies of 11.5 and 11.8 years, respectively), and that they were, on average, only 3 years older at the time of onset. The primary characteristic distinguishing Type III from Type I was their relatively high acute care costs in the 12-month period following the onset interview.

Based on the Cox proportional hazard analyses (Cox 1972) of the Predictors Study in Stern et al. (1994, 1997), we expect that extrapyramidal (motor) signs and psychotic symptoms (delusions and hallucinations) could be important additional characteristics for distinguishing Type III from Type I at the onset interview. While these characteristics were not assessed in the NLTCS, they were found to be important in the Predictors Study GoM models (Stallard et al. 2010; Razlighi et al. 2014). The choice of the GoM model for the more recent and extensive analyses of the Predictors Study was motivated, in part, by the fact that the Cox model provides no mechanism for determining or describing how the time-varying covariate values change over time.

Thus, the longitudinal GoM model permits one to analyze cohort data with large numbers of time-varying covariates measured at multiple waves of follow-up. Applications of the model could be developed using data from other longitudinal studies of the general population, including persons with dementia, or from clinical data specifically focusing on dementia patients. This was done using data for AD from the Predictors Study in Stallard et al. (2010), with further data collection and development of that model now ongoing. The model could be used to better characterize the natural histories of other complex chronic diseases (e.g., cardiovascular disease or diabetes) where there are substantial differences between individuals in manifest disease symptoms, intensity, and rates of progression. Preliminary analyses based on application of this approach are being conducted at Duke University. Alternatively, the model could be used to better characterize the aging process in the general population using chronological age as the time dimension rather than time since onset of specific diseases or time since meeting specific diagnostic criteria, extending the model initially presented in Stallard (2007). Such data are becoming increasingly accessible to the research community and the longitudinal GoM model provides a potentially powerful tool for their analysis.

Acknowledgements Support for the research presented in this chapter was provided by the National Institute on Aging, through grant numbers P01-AG017937, P01-AG043352, U01-AG007198, U01-AG023712, R01-AG007370, and R01-AG046860; and by the Department of Veterans Affairs' Geriatric and Extended Care Data Analysis Center, through an IPA contract. David L. Straley provided programming support.

Appendix

Synthesis of Known Results Regarding the Consistency of the General (Cross-Sectional) Empirical GoM Model

Haberman's (1995) challenge was to identify conditions under which the conditional GoM likelihood estimator is or is not consistent; he cautioned that this would be "very difficult" to do (Haberman 1995, p. 1132). Wachter (1999) responded to this challenge by recasting the conditional GoM model in a framework based on concepts of dimensionality reduction; he commented that the theorems needed to respond directly to Haberman's consistency challenge were not "readily available" so that a direct response would take one "quickly into uncharted territory" (Wachter 1999, p. 441).

We agree with Haberman about the challenge being very difficult but we disagree with Wachter that the territory is almost completely uncharted. Indeed, our review of the statistical literature indicates that the existing theorems are highly informative. They allow unambiguous determination of consistency/inconsistency for most forms of GoM and they provide substantial insight into the issues to be resolved in such determinations. We emphasize that, for cases where the maximum likelihood estimator is inconsistent, the existing theorems (e.g., Huber 1967, Theorem 1) indicate that the resulting maximum likelihood estimates are likely to be optimal (or approximately so) in the sense that they are the closest possible to the true parameter values using the Kullback-Leibler information criterion (i.e., relative entropy, or divergence) as the distance measure (Kullback and Leibler 1951). As explained below, we conjecture that a generalization of Huber's (1967) Theorem 1 could apply to a form of GoM in which the empirical GoM-score mixing distribution is used to create an empirical marginal likelihood with the same set of λ- and g-parameters as in conditional GoM. If proven true, this would move the empirical cross-sectional GoM model into the mainstream statistical literature, thereby extending the range of applications far beyond the dimensionality-reduction applications considered by Wachter (1999).

We summarize the relevant literature, existing theorems, and implications for consistency in the form of 15 observations. Our goal is to document the findings in one accessible location and to stimulate further work on proving our conjecture:

1. Wald (1948, Theorem 2.1) provided necessary and sufficient conditions for the *existence* of a uniformly consistent estimator of a parameter (like λ; but scalar in Wald's case) in the presence of a set of vectors of incidental parameters (like the g-parameters). Tolley and Manton's (1992) proof of consistency for the marginal GoM likelihood for fixed J implies that conditional GoM meets Wald's existence conditions. Thus, marginal GoM was shown to be consistent; the consistency of conditional GoM was not addressed.
2. Tolley and Manton's (1992) marginal GoM likelihood is difficult to use in practice because it requires one to specify the form of the mixing distribution and this is generally unknown. Indeed, one major reason for performing a GoM analysis is to discover the form of this mixing distribution.
3. Kovtun et al. (2007) commented that an empirical estimator of the mixing distribution of the g-parameters can be formed directly from the estimates of the GoM scores with each individual providing a unit contribution to the histogram of the mixing distribution; in this case, they claimed that the empirical distribution converges to the true mixing distribution as J, along with N, goes to infinity. However, they did not provide a proof of convergence for this case. We refer to the marginal GoM likelihood using the empirical estimator in place of the true mixing distribution as the *empirical marginal GoM likelihood*.
4. Mak (1982, Theorem 2.1) implies that conditional and empirical marginal GoM likelihoods with fixed J and increasing N will yield estimators that converge to points in the λ-parameter space that generally differ from the true λ-parameter values; hence the associated estimators are not consistent.
5. Mak (1982, Theorem 2.1) also implies that conditional and empirical marginal GoM likelihoods with increasing J but fixed N will yield estimators that converge to points in the g-parameter space that generally differ from the true g-parameter values. Thus, the associated estimators are also not consistent for this case.
6. It follows that N and J must *both* go to infinity for consistency to be established for the conditional and empirical marginal GoM estimators; in this case, if the empirical mixing distribution converges to the true mixing distribution, then the empirical marginal GoM likelihood estimator will yield consistent estimates of the λ- and g-parameters, without prior specification of the form of the mixing distribution.
7. Substantial insight into the empirical marginal GoM model can be gained by letting J go to infinity first, and then considering the behavior of the model as N goes to infinity. Letting J go to infinity means that each observed data vector, $\mathbf{x}_i$, follows a multinomial distribution with an uncountably infinite number of cells representing all combinations of response outcomes for a countably infinite number of variables (see Feller 1971, p. 123, Theorem 1). Hence, under the general GoM model, each cell, c, will have a probability $\pi_{ic} = \lim_{J\to\infty} \prod_{j=1}^{J} \sum_{k=1}^{K} g_{ik}\lambda_{kjl_{jc}}$ for each individual i and a marginal probability $\pi_c^0 = E(\pi_{ic})$ in the population, where the expectation is taken with respect to the GoM-score

distribution. Thus, as J goes to infinity the set $\{\pi_c^0\}$ defines a multinomial distribution with a countably infinite number of λ- and g-parameters and an uncountably infinite number of cells (i.e., "points"). The presence of an uncountably infinite number of cells introduces several technical problems in defining countably additive probability measures for this distribution, but standard solutions are well-known (e.g., see Billingsley 1986). Three properties of this distribution are relevant: (1) the observed frequency distribution provides a consistent unrestricted (i.e., nonparametric) maximum likelihood estimator of $\{\pi_c^0\}$ (by a generalization of the Glivenko-Cantelli Theorem; Wellner (1981, Theorem 1)—see Gaenssler and Wellner (1981) for discussion); (2) the entropy of $\{\pi_c^0\}$ becomes infinite (because variable-specific entropies are additive over j, and J becomes infinite) (Cover and Thomas 1991, Theorem 2.6.6); and (3) the distribution $\{\pi_c^0\}$ becomes continuous (because, at the limit, no cell c carries positive probability mass; see Feller 1971, p. 137–138).

8. The conditional GoM likelihood is an "empirical estimator" in the sense that the GoM scores are directly represented via the g-parameters without consideration of a mixing distribution, and more importantly, without prior specification of the form of the mixing distribution. It can be shown that the empirical marginal GoM likelihood for fixed N and increasing J converges to a form proportional to the conditional GoM likelihood. Hence, the estimates under the conditional GoM likelihood will converge to a limit point as J goes to infinity that is the same as that of the estimates under the empirical marginal GoM likelihood: if the empirical marginal GoM likelihood estimator is consistent for infinite J, then the conditional GoM likelihood estimator will be likewise consistent. This convergence property implies that the conditional GoM likelihood estimator will provide a good approximation to the empirical marginal GoM likelihood estimator for large J.
9. Rao (1958, Assumption A_1) provided a sufficient condition for the uniform consistency of the restricted (i.e., parametric) maximum likelihood estimator for the infinite multinomial distribution as N goes to infinity: the entropy of the distribution $\{\pi_c^0\}$ must be finite. Rao (1958) emphasized that while this condition is not necessary for consistency, it is sufficient. Unfortunately, as noted in Observation 7, this condition does not hold for GoM. Nonetheless, given that the unrestricted maximum likelihood estimator is known to be consistent for the infinite multinomial distribution as N goes to infinity, it at least seems plausible that the restricted maximum likelihood estimator may also be consistent.
10. To complete our synthesis, we refer again to Mak (1982, Theorem 2.1), from which it follows that the restricted maximum likelihood estimator for GoM will converge to some point in the λ-,g-parameter space. What point is that? Mak's (1982) Theorem 2.1 does not provide an answer, but it is highly likely that Huber's (1967) Theorem 1 does, and if it does, then: it will converge to the unique point in the λ-,g-parameter space that minimizes the relative entropy

(i.e., Kullback-Leibler divergence) between the restricted and unrestricted models.

11. See McCulloch (1988) and Freedman (2006) for non-technical discussion of this result. Note that the use of relative entropies resolves the "problem" in Observations 7 and 9 that the entropy of $\{\pi_c^0\}$ becomes infinite for both the restricted and unrestricted models. If the restricted GoM model is true (i.e., is the correct model), then the proof of a generalized form of Huber's (1967) Theorem 1 will need to identify the conditions under which the parameter estimates for the restricted model converge to the true values as N and J go to infinity.
12. These must be the same values as obtained for the unrestricted model.
13. As written, Huber's (1967) assumptions for his Theorem 1 require that a sequence of maximum likelihood estimators can be formed for each N as N goes to infinity: any sequence will do. No consideration, however, was given to forming a second asymptotic sequence for J as J goes to infinity. Such consideration would clearly require some generalization of Huber's assumptions, which is what is needed to prove our conjecture. This would only work if, in fact, such sequences exist. Thus, we need to consider how at least one sequence of maximum likelihood estimators could be formed for combinations of N and J such that both N and J could go to infinity.
14. Before doing so, we first need to note that Wald (1948, Theorem 3.1) provided an additional condition for consistency which implies that the total amount of Fisher information in the empirical marginal GoM model must go to infinity as N and J go to infinity. Second, Kovtun et al. (2014, Theorem 5.4) provided additional conditions that restrict the set of admissible variables to those that yield identifiable mixture distributions with a property that they term "∞-stability." Intuitively, this restricts the set of admissible variables to some well-defined measurement domain, which should not be a serious restriction for most substantive applications. We assume that these conditions can be met, in theory, by selecting cases and variables such that the associated Hessian matrices (i.e., the "observed" Fisher information matrices) converge to block diagonal form with unbounded positive-definite diagonal blocks as N and J go to infinity. Hence, we assume that a sequence of maximum likelihood estimators that satisfy the requirements of Observation 13 can be formed using the algorithm in Sect. 17.2.8, which is justified by the convergence property in Observation 8, by letting N and J increase in fixed ratios with variables selected so that the diagonal terms of the Hessian matrices are unbounded. Convergence to block diagonal form as N and J increase follows from the structure of Eqs. (17.12) and (17.13): the only nonzero terms outside the diagonal blocks correspond to the cross-derivatives of the λ- and g-parameters and these contain only one additive term, independent of N and J. Hence, the relative sizes of these cross-derivatives will tend to zero as N and J go to infinity. Convergence to positive-definite diagonal blocks will satisfy Kuhn-Tucker Condition 5; the inverse Hessian matrices will be used in eqn. (17.36). Akaike (1973,

p. 269–270) showed the close connections between the Hessian matrix, the Fisher information matrix, and the relative entropy measures used to develop the AIC and Kullback-Leibler statistics.

15. The method described in Observation 14 allows one to construct a sequence of joint λ- and g-estimators for N and J that permit the conditions of Huber's (1967) Theorem 1 to be extended from one to two sequences. If needed, one can ensure by setting $N = J$ that the terms of the paired sequences use just a single index, say N, which would most closely match the existing form of Huber's assumptions. It remains to provide a precise specification of conditions for the asymptotic convergence of these joint sequences and to rigorously determine the changes needed in each step of Huber's proof. Given that the unrestricted maximum likelihood estimator is known to be consistent (Wellner 1981), we expect that it will be possible to specify such conditions; this expectation forms the basis of our conjecture. An essential part of Huber's proof is the assumption that the restricted maximum likelihood estimates are unique; Mak (1982, Theorem 2.1) provides conditions that justify this assumption for the GoM model and these could be incorporated into the generalized Huber theorem. Then, if the GoM model is true, the uniqueness of the limit point would ensure that the restricted and unrestricted maximum likelihood estimators would both tend to the same limit as N and J go to infinity, in which case consistency would be proven. The Kullback-Leibler divergence would converge to zero under these same conditions. For the case where the GoM model is true but one of N or J is not infinite, it would follow from the original Huber argument that the Kullback-Leibler divergence would be minimized, confirming our conjecture in its entirety.

The above synthesis clarifies Manton et al.'s (1994, p. 24) statement that the parameters obtained using the conditional GoM likelihood estimator "asymptotically maximize" the marginal GoM likelihood. Further work is needed to generalize Huber's (1967) Theorem 1 to a form directly applicable to the empirical marginal GoM estimator and to determine for practical applications how large (or small) J needs to be for the approximation in Observation 8 to apply to conditional GoM for given sizes of N and configurations of the empirical mixing distribution. Chapter 18 (Sect. 18.4.3) discusses alternative approaches based on linear latent structure (LLS) analysis to establishing conditions for consistent and asymptotically normal estimators of the λ- and g-parameters for the conditional GoM likelihood—suggesting that the consistency issue can best be completely resolved by several modes of attack. Kovtun et al. (2007) showed that J-values in the range 250–1000 were sufficient to obtain good estimates of the empirical mixing distributions for the generalized GoM scores used in the LLS model for $N = 10{,}000$. This suggests that J-values substantially below this range may have acceptable performance characteristics when considered in the context of the associated Kullback-Leibler divergences.

References

Akaike, H. (1973). Information theory and an extension of the maximum likelihood principle. In B. N. Petrov & F. Caski (Eds.), *Second international symposium on information theory* (pp. 267–281). Budapest: Akademiai Kiado.

Akaike, H. (1974). A new look at the statistical model identification. *IEEE Transactions on Automatic Control, AC–19*(6), 716–723.

Alzheimer's Association. (2016). *2016 Alzheimer's disease facts and figures*. Chicago: Alzheimer's Association.

Berkman, L., Singer, B., & Manton, K. G. (1989). Black/white differences in health status and mortality among the elderly. *Demography, 26*(4), 661–678.

Billingsley, P. (1986). *Probability and measure* (2nd ed.). New York: Wiley.

Birch, M. W. (1964). A new proof of the Pearson-Fisher theorem. *Annals of Mathematical Statistics, 35*(2), 817–824.

Bradley, R. A., & Gart, J. J. (1962). The asymptotic properties of ML estimators when sampling from associated populations. *Biometrika, 49*(1-2), 205–214.

Chanda, K. C. (1954). A note on the consistency and maxima of the roots of likelihood equations. *Biometrika, 41*(1/2), 56–61.

Cover, T. M., & Thomas, J. A. (1991). *Elements of information theory*. New York: Wiley.

Cox, D. R. (1972). Regression models and life-tables (with discussion). *Journal of the Royal Statistical Society: Series B (Methodological), 34*(2), 187–220.

Dooneief, G., Marder, K., Tang, M. X., & Stern, Y. (1996). The clinical dementia rating scale: Community-based validation of 'profound' and 'terminal' stages. *Neurology, 46*(6), 1746–1749.

Eisdorfer, C., Cohen, D., Paveza, G. J., Ashford, J. W., Luchins, D. J., Gorelick, P. B., Hirschman, R. S., Freels, S. A., Levy, P. S., Semla, T. P., & Shaw, H. A. (1992). An empirical evaluation of the global deterioration scale for staging Alzheimer's disease. *American Journal of Psychiatry, 149*(2), 190–194.

Erosheva, E. A. 2002. *Grade of Membership and Latent Structure models with application to disability survey data*. Ph.D. dissertation thesis, Department of Statistics Carnegie Mellon University, Pittsburgh, PA. http://www.stat.cmu.edu/~fienberg/NLTCS_Models/Erosheva-thesis-2002.pdf

Feller, W. (1971). *An introduction to probability theory and its applications* (2nd ed., Vol. II). New York: Wiley.

Fillenbaum, G. G., & Woodbury, M. A. (1998). Typology of Alzheimer's disease: Findings from CERAD data. *Aging and Mental Health, 2*(2), 105–127.

Folstein, M. F., Folstein, S. E., & McHugh, P. R. (1975). "Mini-mental state": A practical method for grading the cognitive state of patients for the clinician. *Journal of Psychiatric Research, 12*(3), 189–198.

Freedman, D. A. (2006). On the so-called "Huber sandwich estimator" and "robust standard errors". *The American Statistician, 60*(4), 299–302.

Freedman, V. A., Martin, L. G., & Schoeni, R. F. (2002). Recent trends in disability and functioning among older adults in the United States: A systematic review. *Journal of the American Medical Association, 288*(24), 3137–3146.

Gaenssler, P., & Wellner, J. A. (1981). Glivenko–Cantelli theorems. In S. Kotz, N. L. Johnson, & C. B. Read (Eds.), *Encyclopedia of statistical sciences* (Vol. 3). New York: Wiley.

Green, C. (2007). Modelling disease progression in Alzheimer's disease: A review of modelling methods used for cost-effectiveness analysis. *PharmacoEconomics, 25*(9), 735–750.

Green, C., Shearer, J., Ritchie, C. W., & Zajicek, J. P. (2011). Model-based economic evaluation in Alzheimer's disease: A review of the methods available to model Alzheimer's disease progression. *Value in Health, 14*(5), 621–630.

Grossberg, G. T., & Desai, A. K. (2003). Management of Alzheimer's disease. *Journal of Gerontology: Medical Sciences, 58A*(4), M331–M353.

Haberman, S. J. (1995). Book review of "Statistical Applications Using Fuzzy Sets". *Journal of the American Statistical Association, 90*(431), 1131–1133.

Huber, P. J. (1967). The behavior of maximum likelihood estimates under nonstandard conditions. In L. M. Le Cam & J. Neyman (Eds.), *Proceedings of the fifth Berkeley symposium on mathematical statistics and probability* (Vol. 1, pp. 221–233). Berkeley: University of California Press.

Hughes, C. P., Berg, L., Danziger, W. L., Coben, L. A., & Martin, R. L. (1982). A new clinical scale for the staging of dementia. *British Journal of Psychiatry, 140*(6), 566–572.

Kass, R. E., & Raftery, A. E. (1995). Bayes factors. *Journal of the American Statistical Association, 90*(430), 773–795.

Katz, S., & Akpom, C. A. (1976). A measure of primary sociobiological functions. *International Journal of Health Services, 6*(3), 493–508.

Kinosian, B., Stallard, E., Lee, J., Woodbury, M. A., Zbrozek, A., & Glick, H. A. (2000). Predicting 10-year care requirements for older people with suspected Alzheimer's disease. *Journal of the American Geriatrics Society, 48*(6), 631–638.

Kinosian, B., Stallard, E., Manton, K. G., Straley, D. L., Zbrozek, A., & Glick, H. A. (2004). The expected outcomes and costs of U.S. patients with incident suspected Alzheimer's Disease (AD) over 15 years. In *Abstract of poster session at the 9th international conference on Alzheimer's disease and related disorders*. Alzheimer's Association Conference, Philadelphia, July 17–22.

Kovtun, M., Akushevich, I., Manton, K. G., & Tolley, H. D. (2007). Linear latent structure analysis: Mixture distribution models with linear constraints. *Statistical Methodology, 4*(1), 90–110.

Kovtun, M., Akushevich, I., & Yashin, A. I. (2014). On identifiability of mixtures of independent distribution laws. *ESAIM: Probability and Statistics, PS 18*, 207–232.

Kramer, M. (1980). The rising pandemic of mental disorders and associated chronic diseases and disabilities. *Acta Psychiatrica Scandinavica, 62*(Suppl. 285), 382–397.

Kuhn, H. W., & Tucker, A. W. (1951). Nonlinear programming. In J. Neyman (Ed.), *Proceedings of the second Berkeley symposium on mathematical statistics and probability* (pp. 481–492). Berkeley: University of California Press.

Kullback, S., & Leibler, R. A. (1951). On information and sufficiency. *The Annals of Mathematical Statistics, 22*(1), 79–86.

Lanza, S. T., Collins, L. M., Lemmon, D. R., & Schafer, J. L. (2007). PROC LCA: A SAS procedure for latent class analysis. *Structural Equation Modeling, 14*(4), 671–694.

Lawton, M. P., & Brody, E. P. (1969). Assessment of older people: Self-maintaining and instrumental activities of daily living. *The Gerontologist, 9*(3), 179–186.

Lee, J., Kinosian, B., Stallard, E., Woodbury, M., Berzon, R., Zbrosek, A., & Glick, H. (1998). A comparison of the Mini-Mental State Exam and the Short Portable Mental Status Questionnaire in Alzheimer's disease (abstract). *Journal of the American Geriatrics Society, 46*(9), S97.

Mak, T. K. (1982). Estimation in the presence of incidental parameters. *Canadian Journal of Statistics, 10*(2), 121–132.

Manton, K. G., & Gu, X. (2001). Changes in the prevalence of chronic disability in the United States black and nonblack population above age 65 from 1982 to 1999. *Proceedings of the National Academy of Sciences, 98*(11), 6354–6359.

Manton, K. G., Stallard, E., & Woodbury, M. A. (1991). A multivariate event history model based upon fuzzy states: Estimation from longitudinal surveys with informative nonresponse. *Journal of Official Statistics, 7*(3), 261–293.

Manton, K. G., Stallard, E., & Singer, B. (1992). Projecting the future size and health status of the US elderly population. *International Journal of Forecasting, 8*(3), 433–458.

Manton, K. G., Woodbury, M. A., & Tolley, H. D. (1994). *Statistical applications using fuzzy sets*. New York: Wiley.

Manton, K. G., Corder, L. S., & Stallard, E. (1997). Chronic disability trends in elderly United States populations: 1982–1994. *Proceedings of the National Academy of Sciences, 94*(6), 2593–2598.

McCulloch, R. E. (1988). Information and the likelihood function in exponential families. *The American Statistician, 42*(1), 73–75.

Nagi, S. Z. (1976). An epidemiology of disability among adults on the United States. *Milbank Memorial Fund Quarterly Health and Society, 54*(4), 439–467.

Orchard, R., & Woodbury, M. A. (1971). A missing information principle: Theory and applications. In L. M. Le Cam, J. Neyman, & E. L. Scott (Eds.), *Proceedings of the sixth Berkeley symposium on mathematical statistics and probability* (Vol. 1, pp. 697–715). Berkeley: University of California Press.

Pfeiffer, E. (1975). A short portable mental status questionnaire for the assessment of organic brain deficit in elderly patients. *Journal of the American Geriatrics Society, 23*(10), 433–441.

Portrait, F., Lindeboom, M., & Deeg, D. (2001). Life expectancies in specific health states: Results from a joint model of health status and mortality of older persons. *Demography, 38*(4), 525–536.

Pressley, J. C., Trott, C., Tang, M., Durkin, M., & Stern, Y. (2003). Dementia in community-dwelling elderly patients: A comparison of survey data, medicare claims, cognitive screening, reported symptoms, and activity limitations. *Journal of Clinical Epidemiology, 56*(9), 896–905.

Raftery, A. E. (1995). Bayesian model selection in social research. *Sociological Methodology, 25*, 111–163.

Rao, C. R. (1958). Maximum likelihood estimation for the multinomial distribution with infinite number of cells. *Sankhyā: The Indian Journal of Statistics, 20*(3/4), 211–218.

Razlighi, Q. R., Stallard, E., Brandt, J., Blacker, D., Albert, M., Scarmeas, N., Kinosian, B., Yashin, A. I., & Stern, Y. (2014). A new algorithm for predicting time to disease endpoints in Alzheimer's disease patients. *Journal of Alzheimer's Disease, 38*(3), 661–668.

Reisberg, B., Ferris, S. H., de Leon, M. J., & Crook, T. (1982). The global deterioration scale for assessment of primary degenerative dementia. *American Journal of Psychiatry, 139*(9), 1136–1139.

Rubin, D. B. (1976). Inference and missing data. *Biometrika, 63*(3), 581–592.

Schwarz, G. (1978). Estimating the dimension of a model. *Annals of Statistics, 6*(2), 461–464.

Seplaki, C. L., Goldman, N., Weinstein, M., & Lin, Y. (2006). Measurement of cumulative physiological dysregulation in an older population. *Demography, 43*(1), 165–183.

Stallard, E. (2007). Trajectories of morbidity, disability, and mortality among the U.S. elderly population: Evidence from the 1984–1999 NLTCS. *North American Actuarial Journal, 11*(3), 16–53.

Stallard, E., Kinosian, B., Zbrozek, A. S., Yashin, A. I., Glick, H. A., & Stern, Y. (2010). Estimation and validation of a multi-attribute model of Alzheimer's disease progression. *Medical Decision Making, 30*(6), 625–638.

Stern, Y., Albert, M., Brandt, J., Jacobs, D. M., Tang, M. X., Marder, K., Bell, K., Sano, M., Devanand, D. P., Bylsma, F., & Lafleche, G. (1994). Utility of extrapyramidal signs and psychosis as predictors of cognitive and functional decline, nursing home admission, and death in Alzheimer's disease: Prospective analyses from the predictors study. *Neurology, 44*(12), 2300–2307.

Stern, Y., Liu, X., Albert, M., Brandt, J., Jacobs, D. M., Del Castillo-Castenada, C., Marder, K., Bell, K., Sano, M., Bylsma, F., Lafleche, G., & Tsai, W. Y. (1996). Application of a growth curve approach to modeling the progression of Alzheimer's disease. *Journal of Gerontology: Medical Sciences, 51A*(4), M179–M184.

Stern, Y., Tang, M. X., Albert, M. S., Brandt, J., Jacobs, D. M., Bell, K., Marder, K., Sano, M., Devanand, D., Albert, S. M., Bylsma, F., & Tsai, W. Y. (1997). Predicting time to nursing home care and death in individuals with Alzheimer's disease. *JAMA, 277*(10), 806–812.

Tarone, R. E., & Gruenhage, G. (1975). A note on the uniqueness of roots of the likelihood equations for vector-valued parameters. *Journal of the American Statistical Association, 70*(352), 903–904.

Taylor, D. H., Fillenbaum, G. G., & Ezell, M. E. (2002). The accuracy of Medicare claims data in identifying Alzheimer's disease. *Journal of Clinical Epidemiology, 55*(9), 929–937.

Taylor, D. H., Sloan, F. A., & Doraiswamy, P. M. (2004). Marked increase in Alzheimer's disease identified in Medicare claims records between 1991 and 1999. *Journal of Gerontology: Medical Sciences, 59A*(7), M762–M766.

Tolley, H. D., & Manton, K. G. (1992). Large sample properties of estimates of a discrete Grade of Membership model. *Annals of the Institute of Statistical Mathematics, 44*(1), 85–95.

Vaupel, J. W., Manton, K. G., & Stallard, E. (1979). The impact of heterogeneity in individual frailty on the dynamics of mortality. *Demography, 16*(3), 439–454.

Wachter, K. W. (1999). Grade of Membership models in low dimensions. *Statistical Papers, 40*(4), 439–457.

Wald, A. (1948). Estimation of a parameter when the number of unknown parameters increases indefinitely with the number of observations. *The Annals of Mathematical Statistics, 19*(2), 220–227.

Wellner, J. A. (1981). A Glivenko-Cantelli theorem for empirical measures of independent but non-identically distributed random variables. *Stochastic Processes and Their Applications, 11*(3), 309–312.

Wieland, D., Kinosian, B., Stallard, E., & Boland, R. (2013). Does Medicaid pay more to a program of all-inclusive care for the elderly (PACE) than for fee-for-service long-term care? *Journal of Gerontology: Medical Sciences, 68*(1), M47–M55.

Wilks, S. S. (1938). The large-sample distribution of the likelihood ratio for testing composite hypotheses. *Annals of Mathematical Statistics, 9*(1), 60–62.

Woodbury, M. A., & Clive, J. (1974). Clinical pure types as a fuzzy partition. *Journal of Cybernetics, 4*(3), 111–121.

Woodbury, M. A., Clive, J., & Garson, A. (1978). Mathematical typology: A Grade of Membership technique for obtaining disease definition. *Computers and Biomedical Research, 11*(3), 277–298.

Woodbury, M. A., Corder, L. S., & Manton, K. G. 1993. Change over time: Observational state, missing data, and repeated measures in the Grade of Membership model. In *Proceedings of the survey research methods section, American Statistical Association (1993)* (Vol. II, pp. 888–891). Alexandria: American Statistical Association.

Chapter 18
Linear Latent Structure Analysis: Modeling High-Dimensional Survey Data

Igor Akushevich, Mikhail Kovtun, Julia Kravchenko, and Anatoliy I. Yashin

18.1 Introduction

Survey data typically are in the form of sample-based collections of measurements made with discrete outcomes for individuals. Common properties of such datasets are high dimensionality and highly correlated measured variables. A class of methods for dealing with such properties is known as latent structure analysis. The typical assumption in such methods is that the observed structure of multiple categorical variables is generated by a small number of latent (i.e., unobserved) variables. The task of latent structure analysis is to find these latent variables, estimate parameters of their distribution, and describe their properties using a sample of high-dimensional categorical variables. Generally speaking, it is necessary to find the properties of a population associated with the latent variables and properties of the sampled individuals, based on those multiple categorical measurements. It appears that both goals may be achieved simultaneously. To increase the precision of population and individual estimates, one has to increase both the sample size (i.e., the number of individuals) and the number of measurements (i.e., questions asked for each individual).

One of the best known of methods of latent structure analysis is the latent class model (LCM), which can be characterized as a statistical method for finding discrete subtypes of related cases (latent classes) from multivariate categorical data. Other models of this type (known as latent variable models), such as item response theory and Rasch models, differ with respect to the assumptions made about the latent variable(s) (reviewed by Clogg (1995) and Collins and Lanza (2010)). One method for identifying the latent structure in large categorical data sets with a simultaneous evaluation of individual scores in a state space is Grade of Membership (GoM) analysis, initially developed by Woodbury and Clive (1974). Manton et al. (1994) provided a detailed exposition of different versions of this approach and reviewed its properties. Additional characterization of a longitudinal form of the model was provided in Chap. 17 (Stallard and Sloan 2016).

A.I. Yashin et al., *Biodemography of Aging*, The Springer Series on Demographic Methods and Population Analysis 40, DOI 10.1007/978-94-017-7587-8_18

Recently Linear Latent Structure (LLS) analysis has been developed to model high-dimensional categorical data (Kovtun et al. 2006, 2007; Akushevich et al. 2009). The LLS model assumption is that the support (i.e., the set of possible values or, formally, the smallest closed set that has probability 1) for the latent variable occupies a polyhedron of low dimensionality. The LLS model was formulated using mixing distribution theory. The specification that the latent variable occupies a polyhedron is another way of saying that the support is linear, which differs from the specifications of other latent structure models, e.g., the LCM is characterized by a mixing distribution concentrated at several isolated support points. Similar to other latent structure analyses, the goal of LLS analysis is to derive simultaneously the properties of a population and individuals, using discrete measurements. The estimation of the LLS model, however, does not use maximization of the likelihood for parameter estimation. Instead, it uses an estimator in which the LLS parameter estimates are solutions of a quasilinear system of equations.

18.2 Linear Latent Structure Analysis

18.2.1 Structure of Datasets and Population Characteristics

The typical dataset analyzed by methods of latent structure analysis can be represented by the $I \times J$ matrix constituted by categorical outcomes X_j^i of J measurements on I individuals, where $i=1,\ldots,I$ and $j=1,\ldots,J$ index the individuals and measurements, respectively. Each row in the matrix corresponds to an individual and contains an individual response pattern, i.e., a sequence of J numbers with the jth number ranging from 1 to the number of responses L_j for that variable. In most cases L_j ranges from 2 to 5–10, and rarely exceeds several dozen. Thus, the results of a survey are represented by I measurements of random variables $X_1,\ldots, X_J$, with the set of outcomes of the jth measurement being $\{1, \ldots, L_j\}$. The joint distribution of random variables $X_1,\ldots,X_J$ can be described by the elementary probabilities,

$$p_\ell = \Pr(X_1 = \ell_1 \text{ and } \cdots \text{ and} X_J = \ell_J), \tag{18.1}$$

where $\ell=(\ell_1, \ldots, \ell_J)$ is an individual response pattern and $\ell_j \in \{1, \ldots, L_j\}$. To represent marginal probabilities, we allow some components of ℓ to be 0's. For example, for three binary variables,

$$p_{(2,0,1)} = \Pr(X_1 = 2 \text{ and } X_3 = 1) = p_{(2,1,1)} + p_{(2,2,1)}.$$

Values of these probabilities p_ℓ (and only these) are directly estimable from the observations. If I_ℓ is the number of individuals with pattern ℓ, consistent estimates for p_ℓ are given by the frequency $f_\ell = I_\ell / I$.

18.2.2 LLS Task: Statistical, Geometrical, and Mixing Distribution Points of View

The analytical problem in LLS analysis is to evaluate the dimension of a hidden space, identify its location in a space of larger dimension, and to evaluate individuals' hidden characteristics (coordinates in the latent subspace) from the data. The LLS analysis is based on two assumptions. The first is the assumption of "local independence", which is common for all methods of latent structure analysis. The second is specific for LLS analysis. It assumes the existence of a low-dimensional linear subspace associated with the latent structure. We present LLS in terms of the theory of mixing distributions, and then discuss its specific assumptions from statistical and geometrical points of view.

Population characteristics are completely described by the joint distribution of random variables $X_1, \ldots, X_J$ represented by probabilities (18.1). Among all possible joint distributions, one can distinguish independent distributions, i.e. distributions satisfying

$$p_\ell = \Pr(X_1 = \ell_1 \text{ and } \cdots \text{ and} X_J = \ell_J) = \prod\nolimits_j \Pr(X_j = \ell_j). \tag{18.2}$$

The description of an independent distribution law requires only knowing $\Pr(X_j{=}l)$ which is denoted below as β_{jl}. Vectors of probabilities $\beta = (\beta_{11}, \ldots, \beta_{JL_J})$ belong to a vector space $R^{|L|}$, where $|L| = \sum\nolimits_j L_j$. Indexes of the vector components run over all possible pairs of jl, i.e., corresponding to probabilities of the first outcome to the first variable, of the second outcome to the first variable, and so on. Requirements for β_{jl} to be probabilities restricts their domain in the vector space by

$$\sum\nolimits_{l=1}^{L_j} \beta_{jl} = 1 \quad \text{and} \quad \beta_{jl} \geq 0. \tag{18.3}$$

This domain represents the direct product of J unit simplexes, each of dimensions L_j.

Since variables $X_1, \ldots, X_J$ are not independent in general, the observed distribution $\{p_\ell\}$ cannot be described by the product of independent distributions, but it can be exactly described as a mixture of independent distributions. This means that each set of independent probabilities contributes to the observed distribution with a weight function. This weight function is called a mixing distribution. It is defined in the space of independent distributions, i.e., for each vector of probabilities β satisfying (18.3). Let $F(\beta)$ be the cumulative distribution function of the mixing distribution. The probabilities p_ℓ can be represented as:

$$p_\ell = \int dF(\beta) \prod_{j=1}^{J} \beta_{j\ell_j}. \tag{18.4}$$

Thus, latent structure analysis searches for a representation of the observed distribution as a mixture of independent distributions.

Any distribution $\{p_\ell\}$ can be represented as a mixture, so the representation (18.4) does not restrict the family of distributions and further specifications are required. They are formulated as restrictions on the support of the mixing distribution or, equivalently, on a set of mixed independent distributions. The LLS specific assumption is that this set is restricted to be a K-dimensional linear subspace of the space of independent distributions, i.e., the mixing distribution is supported by the linear subspace spanned by K basis vectors $\lambda^1,\ldots,\lambda^K$. Below, this LLS assumption is considered from the point of view of pure statistical analysis and the geometry of the task.

Individual characteristics are described by individual probabilities $\beta^i_{jl} = \Pr\left(X^i_j = l\right)$ of specific outcomes ($i = 1,\ldots,I$ runs over individuals).

The LLS assumption of the existence of a low-dimensional linear subspace supporting the mixing distribution is essentially equivalent to the assumption that there exists a K-dimensional random vector G such that for every j a regression of Y_{jl} (random variable Y_{jl} equaling 1 if $X_j = l$ and 0 otherwise) on G is linear. The regression equation relates the expectation of Y_{jl}, which is β_{jl}, to the random vector G. If a specific value of G is associated with individual i (the so-called LLS scores, g_{ik}), then the regression takes the form,

$$\beta^i_{jl} = \sum_{k=1}^{K} g_{ik}\lambda^k_{jl}. \tag{18.5}$$

The sense of the regression coefficients λ^k_{jl} and model restrictions is clarified by analyzing the geometry of the LLS task.

Vectors of individual probabilities $\beta^i = \left\{\beta^i_{jl}\right\}$, of individual responses $Y^i = \left\{Y^i_{jl}\right\}$ and the regression coefficients $\lambda^k = \left\{\lambda^k_{jl}\right\}$ lie in the permitted domain (18.3) of the space of independent distributions. From a geometric point of view, LLS searches a K-dimensional subspace (represented by λ^k_{jl}) in this space, which is the "closest" to the set of I points representing individual outcomes Y^i_{jl}. This linear subspace is defined by its basis $\lambda^1,\ldots,\lambda^K$, so to find the K-dimensional subspace means finding a basis, λ^k_{jl}, ($k = 1,\ldots,K$). Every basis $\lambda^1,\ldots,\lambda^K$ defines a family of regression coefficients and vice versa. The complete set of restrictions in the LLS (which allows us to consider β^i_{jl} and λ^k_{jl} to be probabilities), is

$$\sum_{l=1}^{L_J} \lambda^k_{jl} = 1, \quad \lambda^k_{jl} \geq 0, \quad \sum_{k=1}^{K} g_{ik} = 1 \quad \text{and} \quad \sum_{k} g_{ik}\lambda^k_{jl} \geq 0. \tag{18.6}$$

The LLS scores, or g_{ik}'s, characterizing an individual i are then estimated as the expectation of vector G, conditional on the respondent's answers. Basis vectors of the subspace can be interpreted as probabilities and can define "pure types" (Manton et al. 1994). In this sense, the model decomposition (18.5) has the interpretation of a decomposition over pure types or over "ideal persons" whose individual probabilities are basis vectors of the subspace. Note that (18.6) does not contain the restriction $g_{ik} \geq 0$, $k = 1,\ldots,K$, which is a fundamental restriction of the GoM model; hence, LLS is a generalization of GoM.

Summarizing, one can say that the LLS model approximates the observed distribution of $X_1,\ldots,X_J$ by a mixture of independent distributions with a mixing distribution supported by a K-dimensional subspace of the space of independent distributions. To specify such a model distribution, it is sufficient to define the following LLS parameters:

1. A basis $\lambda^1,\ldots,\lambda^K$ of the space that supports the mixing distribution.
2. Conditional moments $\mathbf{E}(G|X = \ell)$.

This set of model parameters is not the only set possible. We chose it because of a number of useful properties listed below.

Property 1 The mixing distribution can be estimated in the style of an empirical distribution, i.e., when the estimator is a distribution concentrated at points $\mathbf{E}(G|X = \ell)$ with weights f_ℓ.

Property 2 The conditional expectations $\mathbf{E}(G\,|X = \ell)$ provide knowledge about individuals. These conditional expectations can be considered as coordinates in a phase space, to which all individuals belong. The ability to discover the phase space and determine individual positions in it is a valuable feature of LLS analysis.

Property 3 When the number of measurement, J, tends to infinity, the individual conditional expectations $g_i = \mathbf{E}\left(G|X = \ell^{(i)}\right)$, where $\ell^{(i)}$ is the vector of responses of individual i, converge to the true value of the latent variable for this individual, and estimates of the mixing distribution converge to the true one, provided some regularity conditions are satisfied (Kovtun et al. 2011).

18.2.3 Moment Matrix and the Main System of Equations

Parameter estimation is based on two properties (Kovtun et al. 2006, 2007) formulated in terms of the conditional and unconditional moments of the mixing distribution. The first is that columns of the moment matrix belong exactly to the desired linear space. The second is that they obey the main system of equations.

18.2.3.1 Unconditional Moments and the Moment Matrix

The first set of values in which we are interested consists of the unconditional moments of the mixing distribution $F(\beta)$:

$$M_\ell = \int dF(\beta) \prod\nolimits_{j:\ell_j \neq 0} \beta_{j\ell_j} = p_\ell. \tag{18.7}$$

Note an important fact regarding the above equation. The value on the left-hand-side, M_ℓ, is a moment of a *mixing distribution*, while the value on the right-hand-side, p_ℓ, comes from the *joint distribution of* $X_1, \ldots, X_J$; the equality of these values is a direct corollary of the definition of a mixture. The existence of this connection between two distinct distributions is crucial for LLS analysis.

The first corollary of Eq. (18.7) is that some moments are directly estimable from data and, therefore, the frequencies f_ℓ of response patterns ℓ observed in a sample are consistent and efficient estimators for the moments M_ℓ.

Recall that we allow some components of response pattern ℓ to be 0. In this case, the p_ℓ are marginal probabilities. In the definition of the M_ℓ, the multipliers, corresponding to 0 components of ℓ, are excluded from the product. Thus, the order of moment M_ℓ is equal to the number of non-zero components in ℓ.

All moments defined in (18.7) are estimable by frequencies; however, this definition does not cover all moments of a certain order. For example, moments of the second order with β_{jl1} and β_{jl2} (i.e., with the same j) are not estimable. This arises because the data do not include double answers to the same question. It follows that (i) all moments of first order are estimable, (ii) the proportion of estimable moments decreases with the increase of order, and (iii) no moments of order $J+1$ and higher are estimable.

The moment matrix is constructed from moments of order up to J using the following formal rules:

1. Rows of the moment matrix are indexed by response patterns containing exactly one non-zero component or, equivalently, by pair indexes jl. Thus, the moment matrix contains $|L|$ rows, and their columns can be considered as vectors in $R^{|L|}$.
2. Columns of the moment matrix are indexed by all possible response patterns, including a response pattern containing all 0's. The first column is indexed by the response pattern $(0,\ldots,0)$; the next $|L|$ columns are indexed by response patterns containing one non-zero component, and so on.
3. The element at the intersection of row ℓ' and column ℓ'' is $M_{\ell'+\ell''}$, if ℓ'' has 0 at the position of the only non-zero component of ℓ' (in this case, $\ell'+\ell''$ is a meaningful response pattern; otherwise, a question mark is placed on the position of the

intersection of row ℓ' and column ℓ''). For example, the element of the moment matrix in row (1,0,0) and column (0,2,2) is $M_{1,2,2}$, and the element in row (1,0,0) and column (1,0,2) is inestimable and denoted by a question mark.

Equation 18.8 gives an example of a portion of the moment matrix for the case of $J=3$ dichotomous variables, i.e., $L_1=L_2=L_3=2$.

$$\begin{pmatrix} M_{(100)} & ? & ? & M_{(110)} & M_{(120)} & M_{(101)} & M_{(102)} & ? & \cdots \\ M_{(200)} & ? & ? & M_{(210)} & M_{(220)} & M_{(201)} & M_{(202)} & ? & \cdots \\ M_{(010)} & M_{(110)} & M_{(210)} & ? & ? & M_{(011)} & M_{(012)} & ? & \cdots \\ M_{(020)} & M_{(120)} & M_{(220)} & ? & ? & M_{(021)} & M_{(022)} & ? & \cdots \\ M_{(001)} & M_{(101)} & M_{(201)} & M_{(011)} & M_{(021)} & ? & ? & M_{(111)} & \cdots \\ M_{(002)} & M_{(102)} & M_{(202)} & M_{(012)} & M_{(022)} & ? & ? & M_{(112)} & \cdots \end{pmatrix} \tag{18.8}$$

In this example, places for inestimable moments are filled by question marks. The first column of the moment matrix contains moments of the first order, when only one specific outcome of one specific variable is taken into account. There are no inestimable moments in the first column. Elements of this column can be denoted as components of vectors in $R^{|L|}$, i.e., as M_{jl}. The next six ($|L|$ in general) columns correspond to second-order moments. Blocks of diagonal elements are not estimable. Second-order moments can be also denoted via the pair jl of indexes as $M_{jl;j'l'}$. The last column shown corresponds to third order moments. The notations M_{jl} and $M_{jl;j'l'}$ are used below for specific columns of the moment matrix.

The part of the moment matrix consisting of second-order moments (which is an $|L| \times |L|$ square matrix) together with the column of first-order moments is of special interest. A well-known fact is that if a distribution in an n-dimensional Euclidean space is carried by a k-dimensional linear manifold, then the rank of the covariance matrix is equal to k, and the position of the manifold can be derived from the covariance matrix. This fact is the cornerstone of Principal Component Analysis (PCA). Our method is based on similar ideas, adapted to having an incomplete set of second-order moments. For a small J (as in the example), there is a relatively large fraction of non-estimable components in the second-order part of the moment matrix. For increasing J, this fraction rapidly decreases.

For a moment matrix M, let its completion $\overline{M}$ be a matrix obtained from M by replacing question marks with arbitrary numbers. The main fact with respect to the moment matrix is that the moment matrix always has a completion in which all columns belong to the supporting plane Λ. Thus, if the estimable part of the moment matrix has sufficient rank (which is the case in non-degenerate situations), a basis in Λ may be obtained from this matrix. As we have a consistent estimator of the moment matrix in the form of a frequency matrix, the supporting plane may be consistently estimated.

18.2.3.2 Unconditional Moments and Main System of Equations

Another set of values of interest are the conditional moments $\mathbf{E}(G_k|X=\ell)$, which express knowledge of the state of individuals based on measurements. They are not directly estimable from observations. The goal of LLS analysis is to obtain estimates for these conditional moments. Explicit expressions for those of the lowest order are obtained using Bayes theorem (Kovtun et al. 2007):

$$\mathbf{E}\left(G_n\middle|X=\ell\right)=\int dF(g)g_n\frac{\prod_{j:\ell_j\neq 0}\sum_k g_k\lambda_{jl}^k}{M_\ell\left(\mu_\beta\right)}. \tag{18.9}$$

Analogously, higher-order conditional moments, including products of components of G, can be constructed. As can be seen explicitly from (18.9), the relation of conditional and unconditional moments in LLS analysis can be described as

$$\sum_k \lambda_{jl}^k\mathbf{E}\left(G_k\middle|X=\ell\right)=\frac{M_{\ell+l_j}}{M_\ell}, \tag{18.10}$$

where the vector ℓ contains 0 in position j, and $\ell+l_j$ contains l in this position. Similar equations can be written for conditional moments of higher orders. We refer to the system of equations relating conditional and unconditional moments as the main system equations. Kovtun et al. (2007) proved the following properties of solutions of the main system of equations: (i) any basis λ_{jl}^k of Λ together with the conditional moments $\mathbf{E}(G_k|X=\ell)$ calculated on this basis yields a solution of the main system of equations; and (ii) under regularity conditions, every solution of the main system of equations gives a basis of Λ and the conditional moments calculated on this basis. Note, that Eq. (18.10) is linear with respect to the conditional moments.

The properties of the moment matrix and solutions of the main system of equations suggest an efficient algorithm for calculating LLS estimates. First, a basis of the supporting plane can be obtained from the moment matrix, and second, the conditional moments can be found by solving a linear system of equations.

18.2.3.3 Two Illustrative Examples

Before going into the details of the algorithm and addressing the practical tasks of data analysis, we consider two simple illustrative examples. For both of them, assume $K=2$, three dichotomous variables ($J=3$), and the basis vectors are $\lambda^1=(1,0;1,0;1,0)$ and $\lambda^2=(1/2,1/2;0,1;0,1)$. Then the independent distributions being mixed are defined by the vectors:

$$\beta=g_1\lambda^1+g_2\lambda^2=g_1\lambda^1+(1-g_1)\lambda^2,\quad 0\leq g_1\leq 1. \tag{18.11}$$

Thus, a mixing distribution can be represented by a one dimensional p.d.f. $\rho(g_1)$. For the first task, we assume that the mixing distribution is uniform $(\rho(g_1) = \vartheta(g_1)\vartheta(1 - g_1))$. In the second case, we assume the mixing distribution is concentrated at two points with $g_1 = 0.1$ and $g_1 = 0.4$ $(\rho(g_1) = \frac{1}{2}(\delta(g_1 - \frac{1}{10}) + \delta(g_1 - \frac{2}{5})))$. Unconditional moments are calculated using (18.7). Moment matrices for both cases are:

$$\begin{pmatrix} \frac{3}{4} & \frac{7}{12} & \frac{1}{6} & \frac{1}{2} & \frac{1}{4} & \frac{5}{12} & \frac{1}{3} \\ \frac{1}{4} & \frac{1}{6} & \frac{1}{12} & \frac{1}{8} & \frac{1}{8} & \frac{1}{12} & \frac{1}{6} \\ \frac{5}{8} & \frac{1}{2} & \frac{1}{8} & \frac{7}{16} & \frac{3}{16} & \frac{3}{8} & \frac{1}{4} \\ \frac{3}{8} & \frac{1}{4} & \frac{1}{8} & \frac{3}{16} & \frac{3}{16} & \frac{1}{8} & \frac{1}{4} \\ \frac{1}{2} & \frac{5}{12} & \frac{1}{12} & \frac{3}{8} & \frac{1}{8} & \frac{1}{3} & \frac{1}{6} \\ \frac{1}{2} & \frac{1}{3} & \frac{1}{6} & \frac{1}{4} & \frac{1}{4} & \frac{1}{6} & \frac{1}{3} \end{pmatrix}$$

and (18.12)

$$\begin{pmatrix} \frac{5}{8} & \frac{317}{800} & \frac{183}{800} & \frac{451}{1600} & \frac{549}{1600} & \frac{67}{400} & \frac{183}{800} \\ \frac{3}{8} & \frac{183}{800} & \frac{117}{800} & \frac{249}{160} & \frac{351}{1600} & \frac{33}{400} & \frac{117}{400} \\ \frac{7}{16} & \frac{451}{1600} & \frac{249}{160} & \frac{653}{3200} & \frac{747}{3200} & \frac{101}{800} & \frac{249}{800} \\ \frac{9}{16} & \frac{549}{1600} & \frac{351}{1600} & \frac{747}{3200} & \frac{1053}{3200} & \frac{99}{800} & \frac{351}{800} \\ \frac{1}{4} & \frac{67}{400} & \frac{33}{400} & \frac{101}{800} & \frac{99}{800} & \frac{17}{200} & \frac{33}{200} \\ \frac{3}{4} & \frac{183}{400} & \frac{117}{400} & \frac{249}{800} & \frac{351}{800} & \frac{33}{200} & \frac{117}{200} \end{pmatrix}$$

Since these matrices were constructed from mixing distributions known a priori, diagonal blocks in the sub-matrix of the second order are calculable (marked by the italic font in (18.12)). As can be seen, the rank of both matrices is 2. Conditional moments are calculated for an outcome pattern. Choose $\ell = (001)$ and $\ell + l_1(101)$. Using (18.9), we have

$$\mathbf{E}\big(G_1 \big| X = (001)\big) = 23 \text{ and } \mathbf{E}\big(G_2 \big| X = (001)\big) = 13 \tag{18.13}$$

for the first example, and

$$\mathbf{E}\big(G_1 \big| X = (001)\big) = 1750 \text{ and } \mathbf{E}\big(G_2 \big| X = (001)\big) = 3350 \tag{18.14}$$

for the second. Using corresponding elements of M_ℓ in (18.12), we can see that the l.h.s. and r.h.s of Eq. (18.10) are equal to 5/6 for first example and 67/100 for the second:

$$1 \cdot \frac{2}{3} + \frac{1}{2} \cdot \frac{1}{3} = \frac{5/12}{1/2} \quad \text{and} \quad 1 \cdot \frac{17}{50} + \frac{1}{2} \cdot \frac{33}{50} = \frac{67/400}{1/4}. \tag{18.15}$$

External indexes in this example are $j = 1$ and $l = 1$.

18.3 Computational Algorithm for Estimating LLS Model

Parameter estimates in LLS models are based on the properties of the moment matrix and the main system of equations. These properties allow us to reduce the problem of estimating the model parameters to a sequence of linear algebra problems. The algorithm based on linear algebra methods assures a low computational complexity.

Data to be analyzed are represented by a set of measurements X_j^i (See Sect. 18.2.1).

We need to find the linear space and the individual LLS scores. Estimation of the model requires four steps: (i) estimating the rank of the frequency matrix, (ii) finding the supporting plane, (iii) choosing a basis in the plane that was found, and (iv) calculating the individual conditional expectations and estimating mixing distribution. The second and fourth steps are the essence of the LLS parameter estimation problem. The first step is separable, because sometimes the desired dimensionality of the LLS model may be provided by a researcher, and this step may be skipped. The third step requires using prior information about the processes studied, so it is also examined separately.

18.3.1 Moment Matrix Calculation

An important initial step that deserves special attention is calculation of the moment matrix. The elements of the moment matrix given by M_ℓ are approximated by observable frequencies defined as $f_\ell = I_\ell / I$, where I_ℓ is the number of individuals with outcome pattern ℓ, and I is the total number of individuals having certain (not missing) outcomes for nonzero elements in ℓ. Columns of a different order have different normalizations, e.g., the sum of first-order moments corresponding to variable j is one (e.g., $M_{(010)} + M_{(020)} = 1$), while sums of columns for this j of the second-order sub-matrix are equal to the corresponding first-order moments (e.g., $M_{(110)} + M_{(120)} = M_{(100)}$). General conditions of summations of the second order moments written in terms of the notation defined after Eq. (18.8) are:

$$\sum_{l'=1}^{L_{j'}} M_{jl;j'l'} = M_{jl}. \qquad (18.16)$$

Because of missing data, the property of normalization can be violated. This property, with or without the renormalization setting the sums equal to one, is required for the analysis. The renormalization could provide the property in the presence of missing data; this approximation can be true, however, only if the missing data are random.

In addition, a matrix containing standard errors (or confidence intervals) of estimates of frequencies is calculated for each element of the frequency matrix. Standard errors for a binomial distribution, i.e., $\sigma_\ell = \sqrt{f_\ell(1 - f_\ell)/I_\ell}$, require a generalization for patterns with small I_ℓ as discussed in Brown et al. (2001).

18.3.2 Computational Rank of the Frequency Matrix

The frequency matrix can be represented as a sum of the moment matrix with rank K and a matrix with a stochastic component. To define the dimensionality of the LLS problem, we have to estimate the rank of the frequency matrix eliminating the stochastic component. Specifically, we take the greatest minor of the frequency matrix that does not contain question marks. Then we calculate the singular value decomposition (SVD) and take K equal to the number of singular values that are greater than a maximum of the total standard deviation estimated as the quadratic sum of standard errors of frequencies involved in the minor.

The choice of a minor does not essentially influence the computational rank of the frequency matrix. Indeed, the geometrically specific choice of a minor (e.g. an n-dimensional minor of maximal size in the lower left corner of the moment matrix) corresponds to the projection of part of the vectors onto an n-dimensional linear subspace. If the real rank of the moment matrix is much less than n, it is clear that the rank of the projection will not change.

18.3.3 Finding the Supporting Plane

All columns of the moment matrix belong to the supporting plane, and as the frequency matrix is an approximation of the moment matrix, a natural way to search for the supporting plane is to search for a plane that minimizes the sum of distances from it to the columns of the frequency matrix. In our case, however, this approach is complicated by three obstacles: (a) the frequency matrix is incomplete; (b) the statistical inaccuracy of the approximation of moments M_ℓ by frequencies f_ℓ

varies considerably over elements of the frequency matrix; and (c) the basis we seek should exactly satisfy the conditions $\sum_{l=1}^{L_J} \lambda_{jl}^{k} = 1$ for every k and j. These obstacles are overcome by heuristic methods: (a) An iterative procedure for completion of the frequency matrix is used: after a basis of the supporting plane is obtained, it is used to recalculate the completion of the frequency matrix. A new frequency matrix is used for adjusting the basis calculation, etc. (b) Only the first- and second-order moments are examined, so statistical errors of different columns in this matrix are compatible. (c) Rotation of each simplex (corresponding to each variable) to the hyperplane to eliminate one degree of freedom. Rotation, but not a simple projection, is required to provide the same distances between points in a simplex. Items (a) and (c) require explicit consideration.

18.3.3.1 Completion of the Moment Matrix

We consider the second-order moment matrix where, for every $\bar{j}$, there are undefined elements corresponding to repeated responses to the same variable. The intent of the completion procedure is to approximate these elements, assuming that the supporting subspace Λ is found. Since only the completed frequency matrix is used for finding the subspace Λ, and since the completion procedure uses a basis in the subspace Λ, it can be done by an iteration algorithm. For one iteration step, we need to find a symmetric matrix $B_{\bar{j}}$ of $L_{\bar{j}} \times L_{\bar{j}}$-dimension with positive elements such that the sum of elements in each column (or row) equals the corresponding moment of the first order, i.e., $\sum_l B_{\bar{j},ll'} = M_{\bar{j}l'}$. Since we know the first- and second-order frequencies (f_{jl} and $f_{jl,j'l'}; j \neq j'$), which only approximate exact moments (M_{jl} and $M_{jl,j'l'}$), special efforts are required to satisfy the properties of $B_{\bar{j}}$. Columns of the second-order sub-matrix corresponding to variable $\bar{j}$ are presented using the known frequencies $f_{jl,\bar{j}l}$; $j \neq \bar{j}$ and the inestimable elements $B_{\bar{j},l\bar{l}}$,

$$\begin{pmatrix} f_{11;\bar{j}1} & \cdots & f_{11;\bar{j}L_{\bar{j}}} \\ \cdots & \cdots & \cdots \\ f_{1L_1;\bar{j}1} & \cdots & f_{1L_1;\bar{j}L_{\bar{j}}} \\ \cdots & \cdots & \cdots \\ B_{\bar{j},11} & \cdots & B_{\bar{j},1L_{\bar{j}}} \\ \cdots & \cdots & \cdots \\ B_{\bar{j},L_{\bar{j}}1} & \cdots & B_{\bar{j};L_{\bar{j}}L_{\bar{j}}} \\ \cdots & \cdots & \cdots \\ f_{11;\bar{j}1} & \cdots & f_{J1;\bar{j}L_{\bar{j}}} \\ \cdots & \cdots & \cdots \\ f_{JL_J;\bar{j}1} & \cdots & f_{JL_J;\bar{j}L_{\bar{j}}} \end{pmatrix}. \tag{18.17}$$

The completion procedure is based on the fact that the rank of the moment matrix is K, which is much smaller than the dimension of the moment matrix, $|L|$. Therefore, only K columns are linearly independent. Each column of the moment matrix, being a vector in a K-dimensional vector space, can be expanded over the basis vectors $\lambda^1,\ldots,\lambda^K$ available after finding the subspace Λ. The known elements $f_{jl;\bar{j}\bar{l}}$ ($\bar{l} = 1, \ldots, L_{\bar{j}}$ and $j \neq \bar{j}$) of the columns of the moment matrix corresponding to variable $\bar{j}$ are expanded:

$$f_{jl;\bar{j}\bar{l}} = \sum\nolimits_k C_k^{\bar{j}\bar{l}} \lambda_{jl}^k \quad (j \neq \bar{j}). \tag{18.18}$$

If the coefficients $C_k^{\bar{j}\bar{l}}$ are found, the matrix $B_{\bar{j}}$ can be constructed as $B_{\bar{j},l'\bar{l}} = \sum_k C_k^{\bar{j}\bar{l}} \lambda_{\bar{j}l'}^k$. The number of known components of the vector $f_{jl;\bar{j}\bar{l}}$ is greater than the number of basis vectors, so the coefficients $C_k^{\bar{j}\bar{l}}$ can be calculated by ordinary least squares with restrictions: $C_k^{\bar{j}\bar{l}} \geq 0$, $\sum\nolimits_k C_k^{\bar{j}\bar{l}} = 1$, and $\sum\nolimits_k \left(C_k^{\bar{j}\bar{l}} \lambda_{\bar{j}\bar{l}'}^k - C_k^{\bar{j}\bar{l}'} \lambda_{\bar{j}\bar{l}}^k \right) = 0$. The functional to be minimized is:

$$\sum_{jl:j\neq\bar{j}} \left(f_{jl;\bar{j}\bar{l}} - \sum_k C_k^{\bar{j}\bar{l}} \lambda_{jl}^k \right)^2. \tag{18.19}$$

18.3.3.2 Removing Restrictions

The restrictions $\sum_{l=1}^{L_J} \lambda_{jl}^k = 1$ are removed by reducing the number of rows by J (one for every group of indexes $j1,\ldots,jL_j$). Specifically, we use a linear map from $R^{|L|}$ to $R^{|L|-J}$ represented by a block-diagonal matrix A with J blocks of size $L_j \times (L_j - 1)$:

$$A_j = \begin{pmatrix} -\dfrac{\sqrt{L_j} - 1}{L_j - 1} & 1 & 0 & \ldots & 0 \\ \multicolumn{5}{c}{\cdots\cdots\cdots\cdots\cdots\cdots} \\ -\dfrac{\sqrt{L_j} - 1}{L_j - 1} & 0 & 0 & \ldots & 1 \end{pmatrix}. \tag{18.20}$$

Geometrically, such a map provides an isometric rotation $\left(\bar{\lambda}^k = A\lambda^k\right)$ to the hyperplane with zero first coordinate, i.e., every block A_j defines a rotation of a unit simplex in an L_j-dimensional space around a hypersurface opposite to the first vertex; the angle of this rotation is such that the first vertex moves to the point where the first coordinate equals 0. Explicitly, this rotation is $\overline{\lambda}^k_{jl-1} = A_j \lambda^k_{jl}$ in matrix form or $\overline{\lambda}^k_{jl-1} = \lambda^k_{jl} - \frac{\sqrt{L_j - 1}}{L_j - 1} \lambda^k_{j1}$ for $l = 2, \ldots, L_j$. New vectors $\overline{\lambda}^k$ do not possess any ties. It is easy to verify that such a transformation conserves distances between points in a simplex. The reverse transformation is:

$$\lambda^k_{j1} = \frac{1 - \sum_{l=2}^{L_j} \overline{\lambda}^k_{jl-1}}{\sqrt{L_j}}, \lambda^k_{jl} = \overline{\lambda}^k_{jl-1} + \frac{\sqrt{L_j} - 1}{L_j - 1} \lambda^k_{jl}. \tag{18.21}$$

18.3.3.3 Algorithm for Identifying the Subspace

The initial completion of the moment matrix is constructed in an arbitrary way, e.g., by the unitary diagonal matrix or by completing frequencies as $f_{ij} = f_i f_j$. The next step is the rotation of each simplex (corresponding to each variable as described above) to the hyperplane to eliminate one degree of freedom. This produces n points $c^1, \ldots, c^n$ (images of the columns of the frequency matrix) in $m = (|L| - J)$-dimensional space. The problem is to find an affine plane that minimally deviates from these points in the space of individual probabilities. First, we find the center of gravity of this system

$$c^0 = \frac{1}{n} \sum_i c^i, \tag{18.22}$$

and then consider a new set of points $\overline{c}^i = c^i - c^0$ that corresponds to shifting the point of origin. Then we need to find a K-dimensional linear subspace in R^m that minimally deviates from this set of points. The solution of this problem is well-known: one has to consider an $m \times m$ matrix X with components $X_{rs} = \sum_i \overline{c}^i_r \overline{c}^i_s$; this matrix is symmetric and positive definite and thus its normalized eigenvectors are composed of an orthonormal basis in R^m. Let $\gamma_1 \geq \gamma_2 \geq \cdots \geq \gamma_m > 0$ be the eigenvalues of the matrix X, and let $Z^1, \ldots, Z^m$ be the corresponding eigenvectors. The plane of dimensionality K that minimizes the sum of squared distances from points $\overline{c}^1, \ldots, \overline{c}^n$ is spanned by $z^1, \ldots, z^m$, and the sum of squared distances is $\mathrm{tr} X - \sum_{k=1}^k \gamma_k$. Vectors $c^0, c^0 + z^1, \ldots, c^0 + z^{k-1}$ give us an affine basis of the required affine plane. Finally, we apply inverses of the transformation (18.21) to $c^0, c^0 + z^1, \ldots, c^0 + z^{k-1}$ to obtain the basis $\lambda^1, \ldots, \lambda^k$ of the subspace Λ.

18.3.4 Choice of a Basis

The basis cannot be defined uniquely, and any convex combination of basis vectors satisfying the LLS restrictions can be considered as an alternative. A choice may be made using prior information about the process of interest. The appeal of prior information at this stage is reasonable, because of the evident fact that the same dataset can be used for analyzing different (say, disability or CVD) substantive issues.

How this information is used and how the specific choice of the basis is defined is up to the researcher. Here we describe two possible schemes used in our analyses.

A researcher specifies the characteristics of "ideal" individuals based on his/her experience in the research domain. Then he/she can construct vectors of probabilities β_{jl_j} for such ideal individuals or take these individuals from the sample under consideration. The vectors of probabilities for these individuals are taken as the basis vectors. If the probability vectors are constructed by hand, they could extend beyond the polyhedron P_g, so they should be projected to P_g. The individual coordinates in this basis would represent the "proximity" of the individual to the "ideal" ones.

In another scheme, the basis is obtained using assignment of LLS scores (calculated on some arbitrary basis) to K clusters, and then the basis vectors $\lambda^1,\ldots,\lambda^k$ are calculated as the means of the probabilities β^i_{jl} over each cluster.

A researcher can develop his/her own scheme of basis selection. For example, he/she can simply use vectors already known from previous studies or construct a basis purely mathematically, e.g., from the condition of maximal linear independence of the vectors, or choose it from the set of supportive polyhedron vertexes.

18.3.5 Calculation of Individual Conditional Expectations

When a basis of the supporting plane is found, the conditional expectations can be calculated from the main system of Eq. (18.10), which is a linear system after substituting the basis. The system, however, relates conditional expectations E $(G_k|X = \ell)$ for a pattern ℓ with at least one 0th outcome. Thus, the exact system of Eq. (18.10) can be written for all patterns ℓ except patterns where all outcomes are known. For complete patterns, we can calculate J conditional expectations, subsequently excluding one of the J variables (i.e., obtaining patterns $\ell^{[j]}$, where $\ell^{[j]}$ denotes the vector ℓ with the jth coordinate set equal to 0), solving the exact system of equations for the selected patterns, and defining the LLS score for the complete pattern as a mean over J solutions for the conditional expectations for the $\ell^{[j]}$ patterns. This approach can be formalized by considering a system of J equations:

$$\sum\nolimits_k \lambda^k_{jl} \cdot g_{\ell k} \approx \frac{f_\ell}{f_{\ell^{[j]}}}. \tag{18.23}$$

This is a sparse overdetermined system that is solved by minimizing the functional

$$\sum_j \left(\sum_k \lambda_{jl}^k \cdot g_{lk} - \frac{f_\ell}{f_\ell^{[j]}} \right)^2 \tag{18.24}$$

using least squares with the restrictions $\sum_k g_{\ell k} = 1$ and $\sum_k \lambda_{jl}^k \cdot g_{\ell k} \geq 0$. The estimation procedure is implemented using SAS Proc NLP (SAS, Cary NC).

18.3.6 *Mixing Distribution*

The mixing distribution for a given set of data is approximated by an empirical distribution, where an individual provides a unit contribution to the histogram of the distribution. The support of this distribution is a set of *I* points. The probabilities of the joint distribution (18.4) are estimated as the sum over sample individuals or as the sum over possible outcome patterns:

$$p_\ell^* = \sum_i \prod_{j:\ell_j \neq 0} \beta_{j\ell_j}^i = \sum_{\ell'} f_{\ell'} \prod_{j:\ell_j \neq 0} \sum_k g_{\ell' k} \lambda_{j\ell_j}^k. \tag{18.25}$$

18.3.7 *Properties of LLS Estimator*

Kovtun et al. (2007) proved identifiability and consistency of the LLS model. The LLS model is identifiable if and only if the moment matrix has a completion with the rank equal to the maximal rank of its completed minors. This property holds for almost all (with respect to Lebesgue measure) mixing distributions; thus, LLS models are identifiable almost surely. The parameters of the LLS model are the exact solutions of the main system of equations, whose coefficients are true moments of the mixing distribution. The solutions of this system depend continuously on its coefficients; thus, consistency of the LLS estimates obtained by the above algorithm is a direct corollary of the known statistical fact that the frequencies are consistent and efficient estimators of the true moments.

Stronger results were obtained for infinite families of random variables in Kovtun et al. (2014) who described a class of mixtures with an identifiable mixing measure. This class is interesting from a practical point of view as well, as its structure clarifies the principles of selecting a "good" finite family of random variables to be used in applied research.

One domain of application of LLS analysis is the analysis of surveys, where individuals from a sample are asked multiple questions in order to obtain a description of a relatively simple (but directly unobservable) underlying phenomenon. In this context, the mixing distribution can be thought of as a description of a latent variable that characterizes the underlying phenomenon. The dimensionality of the mixing distribution corresponds to the "complexity" of the underlying phenomenon.

Another useful notion is the stability of a questionnaire. The questionnaire is stable (or, k-stable) if the dimensionality of the problem does not change after removing any k variables. Stability is a characteristic of how well a questionnaire is "balanced." A small (in comparison to the number of questions in a survey) level of stability means that a questionnaire is poorly balanced: many questions are devoted to discover one "side" of the underlying phenomenon, while only a few of them are devoted to discover another "side." From this perspective, the stability of the LLS model can be considered as a mathematical measure of the "quality" of a questionnaire. On the other hand, application of LLS analysis (as of any statistical method) is an attempt to infer something from a number of imprecise bits of data. One has to avoid inferences that are supported by a single bit of (or very few bits of) data. Thus, stability characterizes the reliability of an inference. The above arguments suggest that it is very natural to restrict consideration to stable cases.

The notion of k-stability was formally introduced in Kovtun et al. (2014) and its properties were rigorously studied. We say that a distribution of rank K is k-stable if, after removing an arbitrary k random variables from consideration, its visible rank is still K. Similarly, a supporting subspace is k-stable if its dimensionality does not decrease after removing an arbitrary k random variables. In the other words, k-stability essentially means that any dimension of a supporting subspace is confirmed by most random variables—not just one variable or a few of them. The point that we want to stress is the importance of k-stability for the reliability of LLS inferences. To give the reader a sense of numbers, let us assume that we have an observed distribution P of 100 binary random variables, which is stable, the rank of the mixing distribution is $K = 5$, and the system of variables is 90-stable (i.e., $k = 90$). In this situation (Theorem 4.12 of Kovtun et al. (2014)), there exists a five-dimensional model for P. But one may ask the question: Granted that a five-dimensional model exists does there exist a six-dimensional model that is much-much better than the five-dimensional model in a sense that was not taken into account so far? Theorem 4.18 in Kovtun et al. (2014) gives quite a strong answer to this question: There are no stable mixing measures of rank 6, 7, …, 90. One may find a mixing measure of rank 91. However, it is hard to believe that a 91-dimensional model would be better, in any sense, than a five-dimensional one.

18.3.8 Clustering

Clustering LLS scores is not necessary for finding the subspace or calculation of LLS scores; however, it is helpful in selecting a basis and cross-checking in simulation studies. Two types of clustering procedures are implemented in the algorithm. The first is the *k*-means algorithm for the situation where the number of clusters is fixed a priori. The second is a hierarchical procedure that sequentially joins clusters with minimal distance between them. Several distance definitions are possible: distance between centers of mass in clusters or between closest and the most outlying cluster members. Numerical comparisons show that the most reliable results are obtained for the center-of-mass scheme.

18.3.9 Missing Data

Missing data often create difficulties in statistical analyses. Missing data are generated by the absence of responses of an individual to specific questions. The properties of the LLS model make this type of missing data relatively easy to handle. Two main sources of missing data (e.g., responses in sample survey data) could be considered: first, when the failure to answer the question is random; and second, when the failure to answer the question correlates with answers to other questions. In the first case (missing data are random), a solution can be based on the fact that the input to the LLS algorithm consists of frequencies of partial response patterns (like the frequency of giving response C to the second question and response A to the fifth question). With missing data, such frequencies can be calculated by relating the number of individuals with a particular response pattern to the number of individuals who gave answers to the questions covered by the response pattern (rather than to the total number of individuals). The only drawback of this method is the decreased precision of the frequency estimators. As LLS scores are expectations of latent variables conditional on the arbitrary part of the response pattern for an individual, the available part of the response pattern can be used to estimate the value of the latent variable. In the second case (missing data are not random), the absence of an answer can be considered an additional alternative answer to a multi-choice question; in this case, the standard LLS analysis can be applied.

18.4 Applications

18.4.1 Simulation Studies

Three types of simulation experiments were performed to test the predictive power of the LLS model and its ability to detect and to quantitatively reconstruct a hidden latent structure. Specifically, the simulations focused on analyzing the quality of reconstruction of: (i) the linear subspace; (ii) the LLS mixing distribution; and (iii) the clustering properties. The results demonstrated an acceptable quality of reconstruction. Details of the design of these studies and results were described in Akushevich et al. (2009).

18.4.2 LLS and Latent Class Models

The geometric approach, which considers independent distributions as points in a finite-dimensional linear space and mixing distributions as measures in this space, allows us to clarify the relationships among various branches of latent structure analysis. Here we consider the relation between the LLS model and latent class models (LCM).

In geometric language, latent classes are points in the space of independent distributions. If an LCM with classes $c_1,\ldots,c_m$ exists for a particular dataset, then an LLS model also exists, and its supporting subspace is the linear subspace spanned by the vectors $c_1,\ldots,c_m$. Thus, the dimensionality of the LLS model never exceeds the number of classes in LCM. These numbers are equal if and only if the LCM classes are points in a general position (n points are said to be in a general position, if they do not belong to any linear manifold of dimensionality smaller than $n-1$).

If the LCM classes are not in a general position, however, the dimensionality of the LLS model may be significantly smaller. For example, it is possible to construct a mixing distribution such that (a) it is supported by a line (i.e., the dimensionality of the LLS model is 2); (b) there exists an LCM with J (number of variables) classes; and (c) there is no LCM with a smaller number of classes. If, however, the mixing distribution is supported by an infinite set (as in example 1 above), a latent class model does not exist at all, while an LLS model performs well. On the other hand, LLS can be used to evaluate the applicability of the LCM: if the mixing distribution in the LLS model has pronounced modality, then a LCM is more likely to exist (with the number of classes equal to number of modes). When both LCM and LLS models are applicable, the LLS model may still be the model of choice, due to its lower computational complexity.

18.4.3 LLS and Grade of Membership (GoM) Models

The parameters of the GoM model are conventionally estimated by maximizing the conditional likelihood function:

$$\prod_{\ell}\left(\prod_{j}\sum_{k} g_{\ell k}\lambda_{j\ell_j}^{k}\right)^{f_\ell}. \tag{18.26}$$

Proof of the consistency of the maximum likelihood estimator (MLE) has not been done for this form of the GoM model. Tolley and Manton (1992) presented a proof of consistency, but only for the marginal likelihood obtained by integrating (18.26) over the distribution of the $g_{\ell k}$'s, subject to the additional constraint that $g_{\ell k} \geq 0$, $k = 1,\ldots,K$. Nevertheless, there are arguments in favor of the position that a solution of the conditional GoM likelihood should provide consistent estimates. Roughly speaking, the maximum of (18.26) converges to the true values when *both* the size of the sample, N, and the number of measurements, J, tend to infinity. The idea of the proof is to show that when both N and J tend to infinity, then at the point where the maximum of (18.26) is achieved: (i) the $\lambda^1,\ldots,\lambda^K$ converge to a basis $\widetilde{\Lambda} = \{\widetilde{\lambda}^k\}$ of the support of the measure μ_β, and (ii) the g_ℓ converge to conditional expectations $\mathrm{E}(G|X = \ell)$, calculated with respect to the basis $\widetilde{\Lambda}$.

The most important question is how to define the properties that an infinite system of measurements should satisfy. The following useful property follows from Mak (1982, Theorem 2.1), for fixed N: "For sufficiently large J, at the point of the maximum of (18.26), $g_{\ell'}$ is very close to $g_{\ell''}$ for every choice of ℓ', ℓ'' that differ only in one component." The property was rigorously specified for LLS using the notion of stability defined in Kovtun et al. (2014). Now rewrite (18.26) as

$$\prod_\ell \left(\sum_k g_{\ell k} \lambda^k_{1\ell_1}\right)^{f_\ell} \cdot \ldots \cdot \prod_\ell \left(\sum_k g_{\ell k} \lambda^k_{J\ell_J}\right)^{f_\ell}. \tag{18.27}$$

Then take the jth factor of (18.27) and rewrite it as

$$\prod_{\ell' \in L^{[j]}} \left(\left(\sum_k g_{\ell'+1_j,k} \lambda^k_{j1}\right)^{f_{\ell'+1_j}} \cdot \ldots \cdot \left(\sum_k g_{\ell'+(L_j)_j,k} \lambda^k_{jL_j}\right)^{f_{\ell'+(L_j)_j}} \right). \tag{18.28}$$

Recall that the vector ℓ contains 0 in position j, and $\ell + l_j$ contains l in this position. Due to the above property, we can use the equality $g_{l'+l'_j,k} = g_{l'+l''_j,k}$ valid for every $l'_j, l''_j \in [1,\ldots,L_j]$ for $J \to \infty$. From this, we obtain the identity:

$$\sum_{l=1}^{L_j} \sum_k g_{\ell'+l_j,k} \lambda^k_{jl} = \sum_k g_{\ell',k} \sum_l \lambda^k_{jl} = \sum_k g_{\ell',k} \cdot 1 = 1. \tag{18.29}$$

Thus, in (18.28) we have a product of positive factors whose sum tends to a constant for sufficiently large J. Such a product reaches a maximum when its factors are proportional to their powers, yielding the following system of equations:

$$\left\{ \sum\nolimits_k g_{\ell'+l_j,k} \lambda^k_{jl} = \frac{f_{\ell'+l_j}}{f_{\ell'}}, l_j \in [1,\ldots,L_j]. \right. \tag{18.30}$$

It follows that the sets of $g_{\ell k}$ and λ_{jl}^{k} that produce the maximum of (18.26) also satisfy the system of Eq. (18.10) and consequently, by Theorem 5.1 of Kovtun et al. (2007) that, when the size of the sample, N, and the number of measurements, J, tend to infinity, the conditional MLEs of the g- and λ-parameters are consistent.

18.4.4 Application to the NLTCS Data

The National Long Term Care Survey is a longitudinal survey designed to study the changes over time in the health and functional status of older Americans (aged 65+). The analytic dataset used for the present application is described in Akushevich et al. (2013c).

The first 10 singular values of the frequency matrix of the analytic dataset ($\sigma_E = 0.292$) are:

σ_1	σ_2	σ_3	σ_4	σ_5	σ_6	σ_7	σ_8	σ_9	σ_{10}
39.112	3.217	1.464	0.652	0.363	0.310	0.243	0.220	0.198	0.148

When the dimensionality of the LLS-problem is fixed, we can complete the moment matrix using the algorithm described in Sect. 3.3. The sub-matrix corresponding to the first four dichotomous variables is:

$$\begin{pmatrix} 0.094 & 0.513 & 0.051 & 0.0328 & 0.011 & 0.258 & 0.012 & 0.518 & 0.014 \\ 0.906 & 0.487 & 0.949 & 0.672 & 0.989 & 0.742 & 0.988 & 0.482 & 0.986 \\ 0.264 & 0.918 & 0.196 & 0.633 & 0.128 & 0.688 & 0.051 & 0.846 & 0.153 \\ 0.736 & 0.082 & 0.804 & 0.367 & 0.872 & 0.312 & 0.949 & 0.154 & 0.847 \\ 0.335 & 0.916 & 0.275 & 0.872 & 0.142 & 0.664 & 0.164 & 0.888 & 0.230 \\ 0.665 & 0.084 & 0.725 & 0.128 & 0.858 & 0.336 & 0.836 & 0.112 & 0.770 \\ 0.160 & 0.879 & 0.085 & 0.514 & 0.034 & 0.424 & 0.027 & 0.640 & 0.069 \\ 0.840 & 0.121 & 0.915 & 0.486 & 0.966 & 0.576 & 0.973 & 0.360 & 0.931 \end{pmatrix}$$

On the basis of cluster analysis, we chose $K = 3$ clusters corresponding to (i) individuals with minor chronic diseases without disability ($k = 1$, ii) individuals with medium to severe chronic diseases, severely disabled ($k = 2$), and (iii) individuals with medium chronic diseases and minor to medium disability ($k = 3$). For the $K = 4$ case, an additional cluster $k = 4$ intermediate between ($k = 1$) and ($k = 3$) was added. An extended set of variables ($J = 230$) allowed us to identify two additional groups out of group i with similar sets and severity of chronic diseases: (a) physically and socially very active individuals without disabilities, psychologically healthy, and (b) moderately physically and socially active individuals with minor disabilities and minor to moderaSte psychological disorders.

Fig. 18.1 Polyhedrons defined by LLS constraints for $K=3$ and their filling by LLS scores for NLTCS individuals

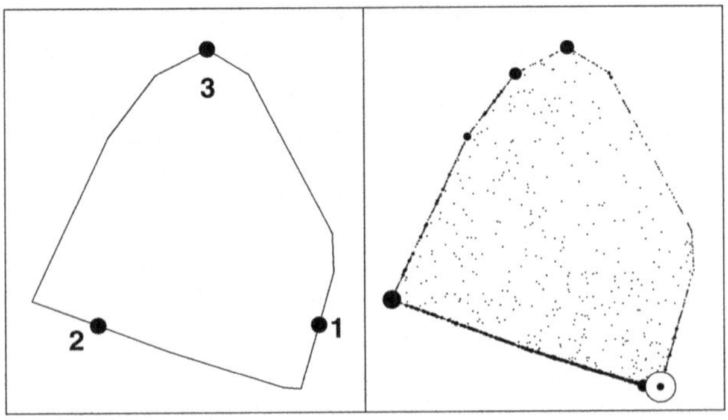

Polyhedrons defined by the LLS constraints for $K=3$ (a) and $K=4$ (c,e) and their filling by the LLS scores of NLTCS individuals can be illustrated (see Fig. 18.1 for $K=3$). The plot on the left shows the 2D-polyhedron for $K=3$. The case of $K=4$ was considered by Akushevich et al. (2009). The polyhedron was defined by the LLS restrictions. In this case, the LLS scores were restricted by 130 inequalities $\left(\sum_k g_{ik}\lambda_{jl}^k \geq 0\right)$ and one equality $\left(\sum_k g_{ik} = 1\right)$. Basis vectors produced unit simplexes labeled by numbers. Plots on the right demonstrate how the polyhedrons were filled by the population. For the filling, we assigned all individuals to 1,000 clusters. Each point in a plot represents one cluster. The area of each point is proportional to the number of individuals assigned to the corresponding cluster. An exception is the point marked by open circles with a closed point inside. About half of the total population was assigned to this cluster.

An extended set of variables ($J=230$) allowed us to identify two additional groups of individuals: (i) individuals with high physical and social activities without disabilities and psychologically healthy, and (ii) individuals with moderate physical and social activities with minor disabilities, and minor to moderate psychological disorders.

Mortality was modeled via Cox regression with vectors of predictors chosen as g_2, g_3 for $K=3$, and g_2, g_3, g_4 for $K=4$, i.e., $\mu_{(3)} = \mu_{0(3)}\exp(b_2g_2 + b_3g_3)$ and $\mu_{(4)} = \mu_{0(4)}\exp(b_2g_2 + b_3g_3 + b_4g_4)$. For both models the healthy component (g_1) was not used as a mortality predictor. This allowed us to ignore the restriction $\left(\sum_k g_k = 1\right)$ and to consider the remaining components as predictors. The estimates are $b_2=0.36\pm0.06$, $b_3=1.71\pm0.06$ for $K=3$, and $b_2=0.28\pm0.07$, $b_3=1.26\pm0.07$, and $b_4=0.01\pm0.03$ for $K=4$. All estimates for b_2 and b_3 were statistically significant with p-values less than 0.0001. Estimate for b_4 did not show a statistically significant effect. This is the expected result because of the original meaning of the fourth group of individuals as a group with intermediate state in health and disability status and because the existence of fourth component was less motivated by the above dimensionality analysis.

18.5 Discussion

LLS is a model describing high-dimensional categorical data assuming the existence of a latent structure represented by K-dimensional random vectors g_i. These vectors are interpreted as explanatory variables which can shed light on mutual correlations observed in measured categorical variables. The vectors play the role of a random variable mixing independent distributions such that the observed joint distribution is maximally close to the data. Mathematically, LLS analysis considers the observed joint distribution of categorical variables as a mixture of individual joint distributions, which are assumed to be independent. This is the standard "local independence" assumption of latent structure analysis. The specific LLS assumption is that the mixing distribution is supported by a K-dimensional subspace Λ of the space of all independent distributions or, equivalently, of the space of individual probabilities. The mixing distribution can be considered as a distribution of random vectors G, which take values in Λ. The vectors of g_i (LLS scores) are the hidden states of the individuals in which we are interested. They can be estimated as conditional expectations of G, $\mathbf{E}\big(G|X_1 = \ell_1, \ldots, X_J = \ell_J\big)$, where $\ell = (\ell_1, \ldots, \ell_J)$ is a response pattern. The support of this random vector is a K-dimensional space, the dimension of which is defined by the dataset. A distinctive feature of LLS is the linearity of the support as compared to that of other latent structure models. For example, LCM is characterized by a mixing distribution concentrated at several isolated points. Note that the specification of the space of mixed distributions as a linear space leads to fruitful developments, resulting in a new method as well as in a better understanding of existing methods.

An important distinction is the existence of an algorithm capable of estimating an LLS model for large numbers of variables and individuals. When the basis $\lambda_1, \ldots, \lambda_K$ of the linear subspace supporting the mixing distribution is known, conditional expectations g_ℓ can be calculated by solving a linear system of equations. A basis $\lambda^1, \ldots, \lambda^K$, in turn, can be identified by applying Principal Component Analysis to the moment matrix. As the choice of a basis is not unique, one has to apply substantive knowledge derived from the applied domain to make the most appropriate choice. This algorithm, being a sequence of linear algebra methods, does not use maximum likelihood methods. This is an advantage of the method, because individual information is represented via nuisance parameters, which creates difficulties in the marginal maximum likelihood approach. The LLS algorithm for parameter estimation is based on two theorems which are due to the assumption of linearity of the support of the mixing distribution. The first theorem identifies properties of the moment matrix. The second theorem presents the main system of equations. Just the existence of the system of equations that describes parameters of the model is a significant advantage of the LLS method, as it allows us to avoid using maximum likelihood estimators, which may not be consistent in the presence of nuisance parameters (see Sect. 4.3 of this chapter and Sect. 2.9 of Chap. 17 for further discussion).

As compared to other methods of latent structure analysis, LLS reduces the problem of estimating the LLS model parameters to a sequence of linear algebra problems. This assures a low computational complexity and an ability to handle large scale data that involve thousands of variables. The overall computational scheme and its components were discussed in detail in this chapter, and simulation experiments demonstrating the excellent performance of the algorithm in reconstructing model parameters were described. The technique is useful for the analysis of high-dimensional categorical data (e.g., demographic surveys or gene expression data) where the detection, evaluation, and interpretation of an underlying latent structure are required.

The estimators of the parameters may be used for construction of second-level models (for example, when the application domain justifies an assumption about the parametric structure of the mixing distribution). For such estimators, it is possible to prove consistency, to formulate conditions for identifiability, and to formulate a high-performance algorithm for the analysis of datasets involving thousands of categorical variables.

LLS can be used to analyze data where a high-dimensional measurement vector represents a hidden structure affecting the results of measurements. The most natural area for applying LLS analysis is (high-dimensional) survey data that represent sample-based collections of measurements made with discrete outcomes for individuals. Moreover, such data recently have appeared in numerous genetic studies of biological systems, where the expression of thousands of genes in cells taken from different organs and tissues of humans or laboratory animals is measured. Such measurements are performed to find appropriate regularities in biological functioning of organs and tissues of respective organisms and to detect changes in hidden biological structure due to disease, exposure to external factors, aging-related changes, etc. These analyses will help us to better understand mechanisms of genetic regulation, by identifying genes playing key roles in organizing response to changes in internal or external milieu.

In summary, Linear Latent Structure (LLS) analysis assumes that the mutual correlations observed in survey variables reflect a hidden property of subjects that can be described by a low-dimensional random vector. The statistical properties of the LLS model, the algorithm for parameter estimation and its implementation, simulation studies, and application of the LLS model to the National Long Term Care Survey (NLTCS) data were described in this chapter. The simulation studies demonstrated the high quality of reconstruction of the major model components and demonstrated its potential to analyze survey datasets with 1,000 or more questions. Step-by-step analysis of a demographic survey was presented as an empirical example: applying the LLS model to the 1994 and 1999 NLTCS datasets (5,000+ individuals) with responses to over 200 questions on behavioral factors, functional status, and comorbidities resulted in an identified population structure with a basis representing pure-type individuals, e.g., healthy, highly disabled, having chronic diseases, etc. The components of the vectors of the individual LLS scores were used to make predictions of individual lifespans.

Acknowledgements The research reported in this chapter was supported by the National Institute on Aging grants R01AG027019, R01AG030612, R01AG030198, R01AG032319, R01AG046860, and P01AG043352.

References

Akushevich, I., Kovtun, M., Manton, K., & Yashin, A. (2009). Linear latent structure analysis and modelling of multiple categorical variables. *Computational and Mathematical Methods in Medicine, 10*(3), 203–218.

Akushevich, I., Kravchenko, J., Ukraintseva, S., Arbeev, K., & Yashin, A. I. (2013). Circulatory diseases in the U.S. elderly in the linked national long-term care survey-Medicare database: Population-based analysis of incidence, comorbidity, and disability. *Research on Aging, 35*(4), 437–458.

Brown, L. D., Cai, T. T., & DasGupta, A. (2001). Interval estimation for a binomial proportion. *Statistical Science, 16*, 101–117.

Clogg, C. C. (1995). Latent class models. In *Handbook of statistical modeling for the social and behavioral sciences* (pp. 311–359). New York: Springer.

Collins, L. M., & Lanza, S. T. (2010). Latent class analysis with covariates. In *Latent class and latent transition analysis: With applications in the social, behavioral, and health sciences* (pp. 149–177). Hoboken: Wiley

Kovtun, M., Akushevich, I., Manton, K. G., & Tolley, H. D. (2006). Grade of membership analysis: One possible approach to foundations. In *Focus on probability theory* (pp. 1–26). New York: Nova Science Publishers.

Kovtun, M., Akushevich, I., Manton, K. G., & Tolley, H. D. (2007). Linear latent structure analysis: Mixture distribution models with linear constraints. *Statistical Methodology, 4*(1), 90–110.

Kovtun, M., Akushevich, I., & Yashin, A. I. (2011). Linear Latent Structure Analysis (LLS). In *JSM proceedings, section on nonparametric statistics*. Alexandria: American Statistical Association.

Kovtun, M., Akushevich, I., & Yashin, A. (2014). On identifiability of mixtures of independent distribution laws. *ESAIM: Probability and Statistics, 18*, 207–232.

Mak, T. K. (1982). Estimation in the presence of incidental parameters. *Canadian Journal of Statistics, 10*(2), 121–132.

Manton, K. G., Tolley, H. D., & Woodbury, M. A. (1994). *Statistical applications using fuzzy sets*. New York: Wiley.

Stallard, E., & Sloan, F. A. (2016). Analysis of the natural history of dementia using longitudinal grade of membership models. Chapter 17. In A. I. Yashin, E. Stallard, & K. C. Land (Eds.), *Biodemography of aging: Determinants of healthy life span and longevity*. New York: Springer.

Tolley, H. D., & Manton, K. G. (1992). Large sample properties of estimates of a discrete grade of membership model. *Annals of the Institute of Statistical Mathematics, 44*(1), 85–95.

Woodbury, M. A., & Clive, J. (1974). Clinical pure types as a fuzzy partition. *Journal of Cybernetics, 4*(3), 111–121.

Chapter 19
Conclusions Regarding Statistical Modeling of Aging, Health, and Longevity

Alexander M. Kulminski, Igor Akushevich, Kenneth C. Land, and Anatoliy I. Yashin

The analyses conducted in **Part I** did not exhaust all factors affecting age patterns of age-related changes in health and mortality. They actually provided a strong rationale for conducting more detailed analyses which require advanced methods of mathematical and statistical modeling. Development and implementation of such state-of-the-art methods is driven by two major factors. The first reflects systemic effects of various behavioral, physiological, and environmental processes on human aging and the related phenotypes. The second is that not all such processes can be readily measured and quantified in studies of human health, aging, and lifespan. In this regard, longitudinal data play a pivotal role in discovering different aspects of knowledge related to aging, health, and lifespan. A variety of statistical methods can be used to analyze longitudinal data.

Chapter 11 summarizes and discusses approaches to statistical analyses of longitudinal data on aging which are relevant to the major topic of this book, *Biodemography of Aging*, and relates this discussion to the subsequent chapters. This chapter also discusses the most essential concepts in biodemography and how they are related to longitudinal and cross-sectional data on health, aging, and lifespan.

The current situation in the study of aging, health, and lifespan is characterized by rapid accumulation of data in the relevant research areas. A better understanding of processes and mechanisms linking human aging with changes in health and survival requires integrative methods capable of taking into account relevant knowledge accumulated in the field when extracting useful information from the new data. Chapter 12 describes an approach to statistical analyses of longitudinal data based on the use of stochastic process models of human aging, health, and longevity. An important advantage of this approach compared to standard statistical methods of analyzing longitudinal data is the possibility of incorporating state-of-the-art advances in aging research into the model structure for use in statistical estimation procedures. To describe changes due to aging, the model incorporates several specific concepts characterizing resistance to stresses, adaptive capacity, and "optimal" (normal) physiological states. These concepts are incorporated in the

A.I. Yashin et al., *Biodemography of Aging*, The Springer Series on Demographic Methods and Population Analysis 40, DOI 10.1007/978-94-017-7587-8_19

model structure through the model coefficients. To capture the effects of exposure to persistent external stresses, variables in the model describing the effects of allostatic adaptation and allostatic load were also introduced. These variables facilitate the description of linkages between age-related changes in endophenotypes and morbidity or mortality risks. The model was tested in simulation experiments and applied to the analyses of the FHS data. The results of these analyses indicate that the proposed approach allows for addressing research questions important for a better understanding of the mechanisms of age-related changes that could not be addressed before. The approach provides researchers with a convenient conceptual framework for studying dynamic aspects of aging, and with an appropriate tool for analyses and systematization of information about aging and its connection with health and longevity.

The specific concepts in aging research discussed in Chap. 12 may not necessarily characterize all aspects of age-related changes in the human body. What can be done with other, unobserved information that may not yet be in the research agenda of today's science? Such information can still be included in the analyses by taking greater advantage of the power of mathematical modeling. Further insights can be gained because processes affecting age-related changes may result in population changes such as, for example, hidden or unobserved heterogeneity. Unobserved heterogeneity can arise because there may be some relevant risk factors affecting the outcomes of interest that are either unknown now or just not measured in the data. Frailty models introduce the concept of unobserved heterogeneity in survival analysis for time-to-event data. It should be noted again that the concept of demographic frailty is, in general, not the same as frailty syndrome discussed in geriatric or gerontological settings and in Chap. 7 of this book.

Chapter 13 focuses on latent class stochastic process models dealing with unobserved heterogeneity in time-to-event and longitudinal data. This discussion excludes methods focusing on analyses of longitudinal data alone where events are generally treated as a biasing factor to be adjusted for and approaches that do not include time-to-event information (e.g., onset of a disease) but include, for example, binary indicators such as prevalence of a disease. Special attention is paid to a specific class of such models that accommodates hidden heterogeneity in the population due to the presence of latent subpopulations with distinct longitudinal patterns with different relations to the risk of an event. The chapter also described the latent class stochastic process model which takes into account such hidden heterogeneity and also allows for indirect estimation of hidden components of aging that are manifested in individual age trajectories of endophenotypes measured in participants of a longitudinal study. Such hidden components of aging and their impact on the risk of events can be evaluated for latent subpopulations. This can help to unravel hidden effects from data with such a latent population structure that otherwise can remain masked if the original form of the stochastic process model which ignores this structure is applied to the data.

Despite the challenges of practical implementation of the latent class stochastic process model, it is a useful tool for researchers. This approach is also helpful for sensitivity analyses in applications of the original stochastic process model. We recommend starting the analyses with the original stochastic process model and estimating the parameters ignoring the possible hidden heterogeneity in the population. Then the latent class stochastic process model can be applied to test the hypothesis regarding the presence of hidden heterogeneity in the data, with appropriate adjustment to the conclusions if such latent structure is detected. Such an approach can be implemented not only with the original model described in Chap. 11 but also with its extensions, for example, with the genetic stochastic process model described in Chap. 14.

The field of biodemography integrates biological knowledge and methods with traditional demographic approaches. Recent revolutionary advances in genotyping of the human genome open an opportunity to greatly accelerate the progress in understanding the biology of various human health traits and lifespan. However, those phenotypes that are of primary interest from the viewpoint of improving human healthspan occur in late life. No single human study has the ability to follow large cohorts of humans from their birth to death. This implies that any study, including those of genetic origin, will face challenges including those of demographic origin as discussed in Chap. 9. Integration of genetics and demography leads to "genetic biodemography" which will continue to grow in the coming years because many studies that collected data on endophenotypes will also include genetic information.

To appreciate the rich potential of such data, special attention should be paid to the analytic approaches available to work with this diverse information. Chapter 14 discusses how genetic biodemography can advance the analyses of genetic effects on age-related phenotypes given data on health, aging, and lifespan collected in longitudinal human studies. The chapter specifically presents a longitudinal genetic-demographic model which provides a method for enhancing genetic analyses of time-to-event outcomes from longitudinal data combining several sources of information including: (i) follow-up data on the outcomes of interest (e.g., mortality) for genotyped individuals, (ii) information on the age structure of the population at the time of biospecimens collection, and (iii) follow-up data on respective events for non-genotyped participants. Such joint analysis of genotyped and non-genotyped individuals can result in substantial improvements in the power and accuracy of estimates compared to analyses of the genotyped subsample alone if the proportion of non-genotyped participants is large.

Chapter 14 also presents a genetic stochastic process model which adds a new dimension to genetic biodemographic analyses combining information on longitudinal measurements of endophenotypes with follow-up data and genetic information. Such joint analyses of different sources of information collected in both genotyped and non-genotyped individuals allow for more efficient use of the research potential of longitudinal data which otherwise would remain underused if only genotyped individuals or only subsets of available information (e.g., only follow-up data on genotyped individuals) were involved in the analyses.

In both of these models, the benefits of combining data on genotyped and non-genotyped individuals derive from the presence of common parameters describing the characteristics of the model for genotyped and non-genotyped subsamples of the data. Further generalizations of these methods were also considered.

Mortality rates are important characteristics of lifespan distributions because they integrate the influences of many external and internal factors affecting individuals in the population during their life course. The life-course accompanying processes may be manifested through several broad classes of phenotypes characterizing the ontogenetic program, individual aging processes or senescence, responses on exogenous and endogenous exposures, changes in health status, as well as effects of compensatory adaptation to damages and changes induced by all these factors. Various parametric models of human mortality rates are used in the analyses of survival data in demographic and epidemiological applications and in experimental studies of aging and longevity using laboratory animals as discussed in Chap. 11.

There is a long tradition of using patterns (shapes or parameters) of mortality rates resulting from the effects of exposures to different conditions or other interventions to characterizing aging-related processes in humans. In Chap. 15, the authors argue that this tradition has to be used with great care. This is because such characteristics are difficult to interpret in terms of properties of external and internal processes affecting the chances of death. An important question then is what kind of mortality model has to be developed to make their parameters biologically interpretable? Chapter 15 describes an approach to mortality modeling, which allows for representing mortality rates in terms of parameters of physiological changes and declining health status developing in aging human organisms. In contrast to traditional demographic and actuarial models dealing with mortality data, this model is appropriate for the analysis of longitudinal data on aging, health, and lifespan. The chapter uses a diffusion-type continuous-time stochastic process to describe the evolution of physiological states over the life course, and a finite-state continuous-time process to describe changes in health status during this period. The equations for the corresponding mortality models, and approximate changes in the physiological state by a conditional Gaussian process, given health state, were presented and discussed.

Chapter 16 describes a method of statistical modeling for joint analyses of longitudinal data on aging, health, and longevity collected using different observational plans. The method is based on the mathematical model of Chap. 15. The need for joint analyses of several longitudinal data sets arises when the number of study subjects in each dataset is not large enough to guarantee either high quality statistical estimates of dynamic characteristics in multidimensional models or of tests of statistical hypotheses related to fundamental research questions on causes and mechanisms of aging and disease development. In such cases, combining data represents a promising alternative for comprehensive analyses of mechanisms of aging-related changes, health decline, and life span. It often happens that the sets of longitudinal data available for analyses are collected using different observational plans, e.g., measurements of some important variables or health outcomes that were omitted in one dataset were measured in another dataset. It turns out that these analyses can be performed within the framework of a single comprehensive model of human aging, health, and mortality, as described in Chap. 15. Observational plans corresponding to each dataset play a crucial role in specifying the likelihood functions of the observed components of the data. The results of these analyses indicate that parameters of both continuous and jumping components of the model can be identified and estimated from the data.

Patterns of endophenotypes and risks of diseases and survival provide valuable information on integrated characteristics of human health, aging, and lifespan. Nevertheless, as discussed in Chap. 7, they do not exhaust all information collected in longitudinal studies of health and aging. An important methodology aimed to help in extracting information on age-related processes in humans was presented in Chap. 17. This chapter describes a new longitudinal form of the Grade of Membership (GoM) model which can model the dynamics of multiple time-varying factors. This chapter had two goals, one methodological and the other substantive. Methodologically, the goal was to present a longitudinal form of the GoM model and associated Newton-Raphson iteration procedures in a self-contained exposition of its estimation. The resulting model describes the natural history of dementia as a complex irreversible multidimensional process occurring within a three-dimensional bounded state space. Individuals can be located at any point in this bounded state space at the time of onset of dementia. The dementia process can move them to other points in the state space over time.

Substantively, Chap. 17 presented the results of modeling the natural history of the loss of cognition and functioning following the onset of dementia using data from the NLTCS. The NLTCS and the linked Medicare files jointly covered a broad range of acute and long-term care services that were expected to differ according to the progression of the decline in cognitive and functional status. The natural history of dementia was found to be highly variable both within and between sexes with respect to cognitive and physical functioning at onset and the subsequent rates of loss of such functioning.

The chapter shows that the GoM model permits one to analyze longitudinal cohort data with large numbers of time-varying covariates measured at multiple waves of follow-up. Applications of the model could be developed using data from other longitudinal studies of the general population, including persons with dementia, or from clinical data specifically focusing on dementia patients. The model could be used to better characterize the natural histories of other complex chronic diseases (e.g., cardiovascular disease or diabetes) where there are substantial differences between individuals in manifest disease symptoms, intensity, and rates of progression. Alternatively, the model could be used to better characterize the aging process in the general population using chronological age as the time dimension rather than time since onset of specific diseases or time since meeting specific diagnostic criteria.

Chapter 18 describes another model that is specifically designed to work with the large classes of data frequently collected in various surveys. Such surveys typically collect data representing a sample-based set of measurements made with discrete outcomes for individuals. A common property of such datasets is their high dimensionality with highly correlated measured variables. Methods dealing with such problems are known as latent structure analysis. The typical assumption in the latent structure analysis method is that the observed structure of various categorical variables is generated by a small number of unobserved factors. The goal of latent structure analyses is to find these latent factors, estimate parameters of their distribution, and describe their properties using a sample of high dimensional categorical variables. Generally speaking, it is necessary to find the properties of a population associated with the latent factors and the properties of individuals, based on those multiple categorical measurements. It was shown that both goals may be achieved simultaneously.

Within this context, Chap. 18 discusses existing latent structure models and describes the recently developed Linear Latent Structure (LLS) model for the analysis of high dimensional categorical data. The LLS specific assumption is that the support for the latent factors occupies a polyhedron of lower dimensionality. The LLS model was formulated using mixing distribution theory. Similar to other latent structure analyses, the goal of LLS analysis is to derive simultaneously the properties of a population and individuals, using discrete measurements. The LLS, however, does not use maximization of a likelihood for parameter estimation. Instead, it uses an estimator in which the parameter estimates are solutions of a quasilinear system of equations.

The LLS analysis assumes that the mutual correlations observed in survey variables reflect a hidden property of subjects that can be described by a low-dimensional random vector. The statistical properties of LLS analysis, the algorithm for parameter estimation and its implementation, simulation studies, and application of LLS model to the NLTCS data were discussed in the chapter. The results of the analyses were compared analytically to predictions of the Latent Class and GoM analyses. Simulation studies demonstrated the high quality of the reconstruction of the major model components and the potential of LLS to analyze survey datasets with 1000 or more variables. Step-by-step analysis of a demographic survey was presented as an example: applying the LLS model to the 1994 and 1999 NLTCS datasets (5000+ individuals) with responses to over 200 questions on behavioral factors, functional status, and comorbidities resulted in an identified population structure with a basis represented by either three or four sets of pure-type individuals, e.g., healthy, highly disabled, having chronic diseases, etc. The components of the vectors of individual LLS scores were used to make predictions of individual lifespans.

Part III
Conclusions

Chapter 20
Continuing the Search for Determinants of Healthy Life Span and Longevity

Alexander M. Kulminski, Anatoliy I. Yashin, Konstantin G. Arbeev, Svetlana V. Ukraintseva, Igor Akushevich, Kenneth C. Land, and Eric Stallard

Life expectancy in humans worldwide has been experiencing dramatic increases for the past two centuries (Oeppen and Vaupel 2002). In most countries, the extension of lifespan is associated with a transition from a long historical period of high fertility and high mortality (particularly infant mortality (Singh and Yu 1995)) to low fertility and low mortality. This demographic trend leads to rapidly growing populations of the elderly (e.g., the United Nations projects a nearly twofold increase in the proportion of the 60+ population from about 10–21 % over the next five decades (UN. 2007)) which raises serious concerns about a possible accompanying expansion of morbidity and disability, especially in developed countries (Olshansky et al. 2007; Robine 2003; Sierra et al. 2009). Because morbidity is in a causative pathway to disability (Verbrugge and Jette 1994), reducing the burden of morbidity could lead to compression of years of unhealthy life.

Analyses of various geriatric traits show that they tend to cluster in families (Cournil and Kirkwood 2001; Sebastiani et al. 2009) implying that they are heritable (Brown et al. 2003; Herskind et al. 1996; Iachine et al. 1998; Matteini et al. 2010) and, thus, that they can have genetic origins (Finch and Tanzi 1997; Martin et al. 2007; Vijg and Suh 2005). Studies of health in people with long life indicate that it is possible to avoid major diseases for long periods of human life (Barzilai et al. 2003; Evert et al. 2003; Kulminski et al. 2008b; Perls 2006; Willcox et al. 2008a, b). It is, therefore, possible that a basis for major breakthroughs in addressing the problem of extending the years of healthy life could be to pinpoint the role of genes in regulating health at older ages leading to longer lifespans. Understanding the genetic origins of healthspan would meet urgent health care needs in the U.S. and other developed countries. This would address the key research goals proposed in *Strategic Directions for Research in Aging* by the National Institute of Aging (Aging 2010), namely, to "...define the link between genes and lifespan, ...improve our understanding of healthy aging and disease and disability among older adults, and ... develop our understanding of how cardiovascular disease, cancer, and diabetes interface with the basic processes of aging

A.I. Yashin et al., *Biodemography of Aging*, The Springer Series on Demographic Methods and Population Analysis 40, DOI 10.1007/978-94-017-7587-8_20

which may soon . . .open doors for personalized approaches to preempt, prevent, or treat these diseases across the lifespan."

Technological breakthroughs and cost-effective solutions in genome-wide genotyping of large samples of humans have raised hopes for major progress in discovering new genes associated with better health at older ages and longer lifespans. Given these technological advances, recent genome-wide association studies (GWAS) have been conducted that shed light on genes and molecular pathways that could be involved in regulation of different traits.

The optimism is tempered, however, because studies using genome-wide resources face serious difficulties related to important limitations in currently prevailing GWAS strategies (Eichler et al. 2010; Gibson 2011; MacRae and Vasan 2011). A fundamental source of difficulties in the genetics of complex health traits in modern societies is the elusive role of evolution in these traits. Specifically, from the evolutionary viewpoint, complex geriatric diseases are a new phenomenon; they also occur mostly in late, i.e., post reproductive, life when the force of natural selection is not as strong as in the reproductive period.

This problem is strengthened by recent findings that genetic variants predisposing to geriatric traits discovered by GWAS may not be involved in the regulation of longevity (Beekman et al. 2010; Deelen et al. 2011). These findings question whether GWAS can help to effectively address the problem of extending healthspan (Bloss et al. 2011). On the other hand, various non-genetic studies show that long-living individuals typically experience better health compared to individuals having shorter lifespans (Barzilai et al. 2003; Evert et al. 2003; Kulminski et al. 2008b; Willcox et al. 2008a, b). Furthermore, candidate-gene studies document that the same genes can affect diseases *and* lifespan (Koropatnick et al. 2008; Kulminski et al. 2011). These studies underscore the need for gaining insights into the mechanisms of genetic influence on complex traits. An important property of these mechanisms should be their plasticity because they should accommodate the dynamic nature of the connections of genes with complex traits over age and time in varying environments (Kulminski 2013).

Currently, there is no broad consensus on directions for further progress in the field (Eichler et al. 2010; Visscher et al. 2012). One view is that further breakthroughs can be achieved by refocusing from GWAS to other genetic strategies (e.g., rare variants, epigenetics, copy number variants, microRNA), or to –omics (e.g., proteomics, metabolomics), or to other sources of genetic variation (e.g., stochasticity (Kirkwood et al. 2011)).

All such views on further progress for discovering the genetic origins of complex traits in late life should be in line with major empirical findings on health deterioration with aging accumulated in diverse disciplines. In this monograph, we summarized a diverse array of findings on the complex behavior of physiological markers and risks of health events and mortality over individuals' life courses. These findings suggest that biodemography may not only suggest strategies for genetic analyses and improve the quality of genetic estimates (Arbeev et al. 2011; Yashin et al. 2007c, 2013b, 2014) but may also guide genetic analyses of healthspan and lifespan. Indeed, biodemography tells us that neither the levels of physiological markers nor the risks of health/mortality events are constant over individuals' life courses. Accordingly, biodemographic processes are key factors that can modulate genetic predisposition to health, aging, and longevity phenotypes. For example, these processes imply that genetic factors have to be linked with phenotypes (e.g., physiological variables and other biomarkers) that experience age-related changes. Information about such changes and their connections with health and survival outcomes is available from longitudinal studies of individuals for whom genetic information is also collected. The use of genetic versions of dynamic stochastic process models of human aging, health, and longevity could facilitate analyses of such data (Arbeev et al. 2009).

Although the role of biodemographic processes in the expression of genetic effects is generally recognized (Gibson 2011; Kulminski 2013; MacRae and Vasan 2011), large-scale GWAS rarely address these problems (Graff et al. 2013). The importance of biodemographic processes is, however, appreciated in candidate gene studies (Atzmon et al. 2006; Kulminski et al. 2013b; Yashin et al. 1999).

The conventional GWAS research design assumes that part of the phenotypic variance of complex traits can be explained by pure genetic determinants. This assumption often is supported by estimates of narrow-sense heritability, wherein a trait is considered as an additive superposition of pure genetic factors and pure environmental factors. However, this approximation was developed for the specific circumstances of reproduction-related phenotypes used in breeding experiments with plants in a controlled environment (Lewontin 1974). These conditions generally are not met for complex traits in humans. Accordingly, the classical model which specifies an additive contribution of genes and environment to complex traits appears to be problematic (Rose 2006). This implies that the expectation of an unconditional (deterministic) set of fixed connections of genes with complex traits also appears to be problematic (Corella and Ordovas 2014; Gibson 2011; Kulminski 2013). As a result, the roles of other factors, such as those of biodemographic origin, in phenotypes of health, aging, and longevity become much more important than previously believed (Yashin et al. 2013a).

Conventional GWAS research strategies also are based on the premise that large samples are needed to robustly detect genetic associations with complex phenotypes (Locke et al. 2015; Willer et al. 2013). This makes sense in some situations if, for example, one hypothesizes unconditional connections of genes with such phenotypes. This hypothesis, however, has a limited theoretical and experimental basis (Gibson 2011; Kulminski 2013; MacRae and Vasan 2011). Assuming rather that the connections of genes with complex phenotypes are conditional, the traditional sample-size-centered GWAS strategy becomes inherently problematic. To improve it, the specific biodemographic characteristics of the studied populations should be addressed. Traditionally, GWAS addresses ancestry-related demographic structures, whereas those of biodemographic origin are typically not addressed. Thus, developing methods of uncovering heterogeneity in genetic susceptibility to complex traits could contribute to increases in the efficiency of genetic analyses.

When the assumption of the unconditional role of genes in part of the phenotypic variance is found to be problematic, the possibility that genes confer risks of these phenotypes in a complex manner through different mechanisms should be considered. One fundamental mechanism, which is the most studied to date, is associated with the biochemical genetic basis of a specific trait (Martin et al. 2007). Another mechanism, which is substantially less studied, is associated with the systemic decline in the functioning of an organism with age (Franco et al. 2009; Martin et al. 2007). A substantial evidentiary basis for this systemic mechanism comes from observations of changes in the expressions of various phenotypes, regardless of their specific details, with age, e.g., levels of physiological markers (Hershcopf et al. 1982; Scheen 2005; Yashin et al. 2006), bone mineral density (Sheu et al. 2011), or risks of aging-related diseases (Akushevich et al. 2012). Accordingly, this systemic mechanism can be associated with aging and, thus, explain genetic susceptibility to not just one, but perhaps to a major portion of health, aging, and longevity phenotypes (Franco et al. 2009; Martin et al. 2007; Yashin et al. 2007a, b). The discovery of genetic susceptibility to this systemic mechanism could lead to major breakthroughs in the extension of healthspan and lifespan (Finch and Tanzi 1997; Jazwinski 2002; Sierra et al. 2009).

Key properties of the systemic, aging-related biogenetic mechanism of healthspan and lifespan are its broad, inherently pleiotropic nature (Ukraintseva et al. 2004; Ukraintseva and Yashin 2003; Goh et al. 2007; Sivakumaran et al. 2011) and its sensitivity to age. Pleiotropy in genetic susceptibility to complex phenotypes is becoming increasingly recognized (Sivakumaran et al. 2011; Yashin et al. 2013c). Studies also provide examples of its complex forms, such as genetic trade-offs (Barnes et al. 2001; Crespi 2010; Frazer et al. 2009; Kulminski et al. 2008a, 2010b, 2011c; Wang et al. 2010; Yashin et al. 2009) and antagonistic pleiotropy (Williams 1957; Alexander et al. 2007; Kulminski et al. 2011; Martin 2007; Schnebel and Grossfield 1988; Summers and Crespi 2010; Williams and Day 2003). The concept of genetic trade-offs is a broader concept than antagonistic pleiotropy, because it refers to antagonistic effects of the same allele on different phenotypes, which may not necessarily include fitness traits. Studies also provide evidence of age-sensitive genetic effects, i.e., that the same alleles could confer different risks of the same traits at different ages (Bergman et al. 2007; De Benedictis et al. 1998; Ilveskoski et al. 1999; Jarvik et al. 1997; Kulminski et al. 2013a; Lasky-Su et al. 2008; Martin 2007; Yashin et al. 1999, 2001).

All of the above suggests a huge potential for biodemography (and related disciplines, e.g., gerontology, sociology) to advance the study of life course genetics, with the explicit goal of uncovering the mechanistisms of genetic susceptibility to phenotypes of health, aging, and longevity over individuals' life courses.

References

Aging, N. I. o. (2010). *Living long & well in the 21st century: Strategic directions for research on aging.*

Akushevich, I., Kravchenko, J., Ukraintseva, S., Arbeev, K., & Yashin, A. I. (2012). Age patterns of incidence of geriatric disease in the U.S. elderly population: Medicare-based analysis. *Journal of the American Geriatrics Society, 60*(2), 323–327.

Alexander, D. M., Williams, L. M., Gatt, J. M., Dobson-Stone, C., Kuan, S. A., Todd, E. G., Schofield, P. R., Cooper, N. J., & Gordon, E. (2007). The contribution of apolipoprotein E alleles on cognitive performance and dynamic neural activity over six decades. *Biological Psychology, 75*(3), 229–238.

Arbeev, K. G., Akushevich, I., Kulminski, A. M., Arbeeva, L. S., Akushevich, L., Ukraintseva, S. V., Culminskaya, I. V., & Yashin, A. I. (2009). Genetic model for longitudinal studies of aging, health, and longevity and its potential application to incomplete data. *Journal of Theoretical Biology, 258*(1), 103–111.

Arbeev, K. G., Ukraintseva, S. V., Arbeeva, L. S., Akushevich, I., Kulminski, A. M., & Yashin, A. I. (2011). Evaluation of genotype-specific survival using joint analysis of genetic and non-genetic subsamples of longitudinal data. *Biogerontology, 12*(2), 157–166.

Atzmon, G., Rincon, M., Schechter, C. B., Shuldiner, A. R., Lipton, R. B., Bergman, A., & Barzilai, N. (2006). Lipoprotein genotype and conserved pathway for exceptional longevity in humans. *PLoS Biology, 4*(4), e113.

Barnes, J. A., Dix, D. J., Collins, B. W., Luft, C., & Allen, J. W. (2001). Expression of inducible Hsp70 enhances the proliferation of MCF-7 breast cancer cells and protects against the cytotoxic effects of hyperthermia. *Cell Stress & Chaperones, 6*(4), 316–325.

Barzilai, N., Atzmon, G., Schechter, C., Schaefer, E. J., Cupples, A. L., Lipton, R., Cheng, S., & Shuldiner, A. R. (2003). Unique lipoprotein phenotype and genotype associated with exceptional longevity. *JAMA, 290*(15), 2030–2040.

Beekman, M., Nederstigt, C., Suchiman, H. E., Kremer, D., van der Breggen, R., Lakenberg, N., Alemayehu, W. G., de Craen, A. J., Westendorp, R. G., Boomsma, D. I., de Geus, E. J., Houwing-Duistermaat, J. J., Heijmans, B. T., & Slagboom, P. E. (2010). Genome-wide association study (GWAS)-identified disease risk alleles do not compromise human longevity. *Proceedings of the National Academy of Sciences of the United States of America, 107*(42), 18046–18049.

Bergman, A., Atzmon, G., Ye, K., MacCarthy, T., & Barzilai, N. (2007). Buffering mechanisms in aging: A systems approach toward uncovering the genetic component of aging. *PLoS Computational Biology, 3*(8), e170.

Bloss, C. S., Pawlikowska, L., & Schork, N. J. (2011). Contemporary human genetic strategies in aging research. *Ageing Research Reviews, 10*(2), 191–200. doi:S1568-1637(10)00057-7 [pii] 10.1016/j.arr.2010.07.005.

Brown, W. M., Beck, S. R., Lange, E. M., Davis, C. C., Kay, C. M., Langefeld, C. D., & Rich, S. S. (2003). Age-stratified heritability estimation in the Framingham Heart Study families. *BMC Genetics, 4*(Suppl 1), S32.

Corella, D., & Ordovas, J. M. (2014). Aging and cardiovascular diseases: The role of gene-diet interactions. *Ageing Research Reviews, 18*, 53–73.

Cournil, A., & Kirkwood, T. B. (2001). If you would live long, choose your parents well. *Trends in Genetics, 17*(5), 233–235.

Crespi, B. J. (2010). The origins and evolution of genetic disease risk in modern humans. *Annals of the New York Academy of Sciences, 1206*, 80–109.

De Benedictis, G., Carotenuto, L., Carrieri, G., De Luca, M., Falcone, E., Rose, G., Yashin, A. I., Bonafe, M., & Franceschi, C. (1998). Age-related changes of the 3′APOB-VNTR genotype pool in ageing cohorts. *Annals of Human Genetics, 62*(Pt 2), 115–122.

Deelen, J., Beekman, M., Uh, H. W., Helmer, Q., Kuningas, M., Christiansen, L., Kremer, D., van der Breggen, R., Suchiman, H. E., Lakenberg, N., van den Akker, E. B., Passtoors, W. M., Tiemeier, H., van Heemst, D., de Craen, A. J., Rivadeneira, F., de Geus, E. J., Perola, M., van der Ouderaa, F. J., Gunn, D. A., Boomsma, D. I., Uitterlinden, A. G., Christensen, K., van Duijn, C. M., Heijmans, B. T., Houwing-Duistermaat, J. J., Westendorp, R. G., & Slagboom, P. E. (2011). Genome-wide association study identifies a single major locus contributing to survival into old age; the APOE locus revisited. *Aging Cell, 10*(4), 686–698.

Eichler, E. E., Flint, J., Gibson, G., Kong, A., Leal, S. M., Moore, J. H., & Nadeau, J. H. (2010). Missing heritability and strategies for finding the underlying causes of complex disease. *Nature Reviews Genetics, 11*(6), 446–450.

Evert, J., Lawler, E., Bogan, H., & Perls, T. (2003). Morbidity profiles of centenarians: Survivors, delayers, and escapers. *The Journals of Gerontology. Series A, Biological Sciences and Medical Sciences, 58*(3), 232–237.

Finch, C. E., & Tanzi, R. E. (1997). Genetics of aging. *Science, 278*(5337), 407–411.

Franco, O. H., Karnik, K., Osborne, G., Ordovas, J. M., Catt, M., & van der Ouderaa, F. (2009). Changing course in ageing research: The healthy ageing phenotype. *Maturitas, 63*(1), 13–19.

Frazer, K. A., Murray, S. S., Schork, N. J., & Topol, E. J. (2009). Human genetic variation and its contribution to complex traits. *Nature Reviews Genetics, 10*(4), 241–251.

Gibson, G. (2011). Rare and common variants: Twenty arguments. *Nature Reviews Genetics, 13*(2), 135–145.

Goh, K. I., Cusick, M. E., Valle, D., Childs, B., Vidal, M., & Barabasi, A. L. (2007). The human disease network. *Proceedings of the National Academy of Sciences of the United States of America, 104*(21), 8685–8690.

Graff, M., Ngwa, J. S., Workalemahu, T., Homuth, G., Schipf, S., Teumer, A., Volzke, H., Wallaschofski, H., Abecasis, G. R., Edward, L., Francesco, C., Sanna, S., Scheet, P., Schlessinger, D., Sidore, C., Xiao, X., Wang, Z., Chanock, S. J., Jacobs, K. B., Hayes, R. B., Hu, F., Van Dam, R. M., Crout, R. J., Marazita, M. L., Shaffer, J. R., Atwood, L. D., Fox, C. S., Heard-Costa, N. L., White, C., Choh, A. C., Czerwinski, S. A., Demerath, E. W., Dyer, T. D., Towne, B., Amin, N., Oostra, B. A., Van Duijn, C. M., Zillikens, M. C., Esko, T., Nelis, M., Nikopensius, T., Metspalu, A., Strachan, D. P., Monda, K., Qi, L., North, K. E., Cupples, L. A., Gordon-Larsen, P., & Berndt, S. I. (2013). Genome-wide analysis of BMI in adolescents and young adults reveals additional insight into the effects of genetic loci over the life course. *Human Molecular Genetics, 22*(17), 3597–3607.

Hershcopf, R. J., Elahi, D., Andres, R., Baldwin, H. L., Raizes, G. S., Schocken, D. D., & Tobin, J. D. (1982). Longitudinal changes in serum cholesterol in man: An epidemiologic search for an etiology. *Journal of Chronic Diseases, 35*(2), 101–114.

Herskind, A. M., McGue, M., Holm, N. V., Sorensen, T. I. A., Harvald, B., & Vaupel, J. W. (1996). The heritability of human longevity: A population-based study of 2872 Danish twin pairs born 1870–1900. *Human Genetics, 97*(3), 319–323.

Iachine, I. A., Holm, N. V., Harris, J. R., Begun, A. Z., Iachina, M. K., Laitinen, M., Kaprio, J., & Yashin, A. I. (1998). How heritable is individual susceptibility to death? The results of an analysis of survival data on Danish, Swedish and Finnish twins. *Twin Research, 1*(4), 196–205.

Ilveskoski, E., Perola, M., Lehtimaki, T., Laippala, P., Savolainen, V., Pajarinen, J., Penttila, A., Lalu, K. H., Mannikko, A., Liesto, K. K., Koivula, T., & Karhunen, P. J. (1999). Age-dependent association of apolipoprotein E genotype with coronary and aortic atherosclerosis in middle-aged men: An autopsy study. *Circulation, 100*(6), 608–613.

Jarvik, G. P., Goode, E. L., Austin, M. A., Auwerx, J., Deeb, S., Schellenberg, G. D., & Reed, T. (1997). Evidence that the apolipoprotein E-genotype effects on lipid levels can change with age in males: A longitudinal analysis. *The American Journal of Human Genetics, 61*(1), 171–181.

Jazwinski, S. M. (2002). Biological aging research today: Potential, peeves, and problems. *Experimental Gerontology, 37*(10–11), 1141–1146.

Kirkwood, T. B., Cordell, H. J., & Finch, C. E. (2011). Speed-bumps ahead for the genetics of later-life diseases. *Trends in Genetics, 27*(10), 387–388.

Koropatnick, T. A., Kimbell, J., Chen, R., Grove, J. S., Donlon, T. A., Masaki, K. H., Rodriguez, B. L., Willcox, B. J., Yano, K., & Curb, J. D. (2008). A prospective study of high-density lipoprotein cholesterol, cholesteryl ester transfer protein gene variants, and healthy aging in very old Japanese-american men. *The Journals of Gerontology. Series A, Biological Sciences and Medical Sciences, 63*(11), 1235–1240.

Kulminski, A. M. (2013). Unraveling genetic origin of aging-related traits: Evolving concepts. *Rejuvenation Research, 16*(4), 304–312.
Kulminski, A. M., Ukraintseva, S. V., Arbeev, K. G., Manton, K. G., Oshima, J., Martin, G. M., Il'yasova, D., & Yashin, A. I. (2008a). Health-protective and adverse effects of the apolipoprotein E epsilon2 allele in older men. *Journal of the American Geriatrics Society, 56*(3), 478–483.
Kulminski, A. M., Ukraintseva, S. V., Culminskaya, I. V., Arbeev, K. G., Land, K. C., Akushevich, L., & Yashin, A. I. (2008b). Cumulative deficits and physiological indices as predictors of mortality and long life. *The Journals of Gerontology. Series A, Biological Sciences and Medical Sciences, 63*(10), 1053–1059.
Kulminski, A. M., Culminskaya, I., Ukraintseva, S. V., Arbeev, K. G., Land, K. C., & Yashin, A. I. (2010a). Beta2-adrenergic receptor gene polymorphisms as systemic determinants of healthy aging in an evolutionary context. *Mechanisms of Ageing and Development, 131*(5), 338–345.
Kulminski, A. M., Culminskaya, I. V., Ukraintseva, S. V., Arbeev, K. G., Akushevich, I., Land, K. C., & Yashin, A. I. (2010b). Polymorphisms in the ACE and ADRB2 genes and risks of aging-associated phenotypes: The case of myocardial infarction. *Rejuvenation Research, 13*(1), 13–21.
Kulminski, A. M., Culminskaya, I., Ukraintseva, S. V., Arbeev, K. G., Arbeeva, L., Wu, D., Akushevich, I., Land, K. C., & Yashin, A. I. (2011). Trade-off in the effects of the apolipoprotein E polymorphism on the ages at onset of CVD and cancer influences human lifespan. *Aging Cell, 10*(3), 533–541.
Kulminski, A. M., Culminskaya, I., Arbeev, K. G., Ukraintseva, S. V., Arbeeva, L., & Yashin, A. I. (2013a). Trade-off in the effect of the APOE gene on the ages at onset of cardiocascular disease and cancer across ages, gender, and human generations. *Rejuvenation Research, 16*(1), 28–34.
Kulminski, A. M., Culminskaya, I., Arbeev, K. G., Ukraintseva, S. V., Stallard, E., Arbeeva, L., & Yashin, A. I. (2013b). The role of lipid-related genes, aging-related processes, and environment in healthspan. *Aging Cell, 12*(2), 237–246.
Lasky-Su, J., Lyon, H. N., Emilsson, V., Heid, I. M., Molony, C., Raby, B. A., Lazarus, R., Klanderman, B., Soto-Quiros, M. E., Avila, L., Silverman, E. K., Thorleifsson, G., Thorsteinsdottir, U., Kronenberg, F., Vollmert, C., Illig, T., Fox, C. S., Levy, D., Laird, N., Ding, X., McQueen, M. B., Butler, J., Ardlie, K., Papoutsakis, C., Dedoussis, G., O'Donnell, C. J., Wichmann, H. E., Celedon, J. C., Schadt, E., Hirschhorn, J., Weiss, S. T., Stefansson, K., & Lange, C. (2008). On the replication of genetic associations: Timing can be everything! *The American Journal of Human Genetics, 82*(4), 849–858.
Lewontin, R. C. (1974). Annotation: The analysis of variance and the analysis of causes. *The American Journal of Human Genetics, 26*(3), 400–411.
Locke, A. E., Kahali, B., Berndt, S. I., Justice, A. E., Pers, T. H., Day, F. R., Powell, C., Vedantam, S., Buchkovich, M. L., Yang, J., Croteau-Chonka, D. C., Esko, T., Fall, T., Ferreira, T., Gustafsson, S., Kutalik, Z., Luan, J., Magi, R., Randall, J. C., Winkler, T. W., Wood, A. R., Workalemahu, T., Faul, J. D., Smith, J. A., Hua Zhao, J., Zhao, W., Chen, J., Fehrmann, R., Hedman, A. K., Karjalainen, J., Schmidt, E. M., Absher, D., Amin, N., Anderson, D., Beekman, M., Bolton, J. L., Bragg-Gresham, J. L., Buyske, S., Demirkan, A., Deng, G., Ehret, G. B., Feenstra, B., Feitosa, M. F., Fischer, K., Goel, A., Gong, J., Jackson, A. U., Kanoni, S., Kleber, M. E., Kristiansson, K., Lim, U., Lotay, V., Mangino, M., Mateo Leach, I., Medina-Gomez, C., Medland, S. E., Nalls, M. A., Palmer, C. D., Pasko, D., Pechlivanis, S., Peters, M. J., Prokopenko, I., Shungin, D., Stancakova, A., Strawbridge, R. J., Ju Sung, Y., Tanaka, T., Teumer, A., Trompet, S., van der Laan, S. W., van Setten, J., Van Vliet-Ostaptchouk, J. V., Wang, Z., Yengo, L., Zhang, W., Isaacs, A., Albrecht, E., Arnlov, J., Arscott, G. M., Attwood, A. P., Bandinelli, S., Barrett, A., Bas, I. N., Bellis, C., Bennett, A. J., Berne, C., Blagieva, R., Bluher, M., Bohringer, S., Bonnycastle, L. L., Bottcher, Y., Boyd, H. A., Bruinenberg, M., Caspersen, I. H., Ida Chen, Y. D., Clarke, R., Daw, E. W., de Craen, A. J., Delgado, G., Dimitriou, M., Doney, A. S., Eklund, N., Estrada, K., Eury, E., Folkersen, L., Fraser, R. M., Garcia, M. E., Geller, F., Giedraitis, V., Gigante, B., Go, A. S., Golay, A.,

Goodall, A. H., Gordon, S. D., Gorski, M., Grabe, H. J., Grallert, H., Grammer, T. B., Grassler, J., Gronberg, H., Groves, C. J., Gusto, G., Haessler, J., Hall, P., Haller, T., Hallmans, G., Hartman, C. A., Hassinen, M., Hayward, C., Heard-Costa, N. L., Helmer, Q., Hengstenberg, C., Holmen, O., Hottenga, J. J., James, A. L., Jeff, J. M., Johansson, A., Jolley, J., Juliusdottir, T., Kinnunen, L., Koenig, W., Koskenvuo, M., Kratzer, W., Laitinen, J., Lamina, C., Leander, K., Lee, N. R., Lichtner, P., Lind, L., Lindstrom, J., Sin Lo, K., Lobbens, S., Lorbeer, R., Lu, Y., Mach, F., Magnusson, P. K., Mahajan, A., McArdle, W. L., McLachlan, S., Menni, C., Merger, S., Mihailov, E., Milani, L., Moayyeri, A., Monda, K. L., Morken, M. A., Mulas, A., Muller, G., Muller-Nurasyid, M., Musk, A. W., Nagaraja, R., Nothen, M. M., Nolte, I. M., Pilz, S., Rayner, N. W., Renstrom, F., Rettig, R., Ried, J. S., Ripke, S., Robertson, N. R., Rose, L. M., Sanna, S., Scharnagl, H., Scholtens, S., Schumacher, F. R., Scott, W. R., Seufferlein, T., Shi, J., Vernon Smith, A., Smolonska, J., Stanton, A. V., Steinthorsdottir, V., Stirrups, K., Stringham, H. M., Sundstrom, J., Swertz, M. A., Swift, A. J., Syvanen, A. C., Tan, S. T., Tayo, B. O., Thorand, B., Thorleifsson, G., Tyrer, J. P., Uh, H. W., Vandenput, L., Verhulst, F. C., Vermeulen, S. H., Verweij, N., Vonk, J. M., Waite, L. L., Warren, H. R., Waterworth, D., Weedon, M. N., Wilkens, L. R., Willenborg, C., Wilsgaard, T., Wojczynski, M. K., Wong, A., Wright, A. F., Zhang, Q., Brennan, E. P., Choi, M., Dastani, Z., Drong, A. W., Eriksson, P., Franco-Cereceda, A., Gadin, J. R., Gharavi, A. G., Goddard, M. E., Handsaker, R. E., Huang, J., Karpe, F., Kathiresan, S., Keildson, S., Kiryluk, K., Kubo, M., Lee, J. Y., Liang, L., Lifton, R. P., Ma, B., McCarroll, S. A., McKnight, A. J., Min, J. L., Moffatt, M. F., Montgomery, G. W., Murabito, J. M., Nicholson, G., Nyholt, D. R., Okada, Y., Perry, J. R., Dorajoo, R., Reinmaa, E., Salem, R. M., Sandholm, N., Scott, R. A., Stolk, L., Takahashi, A., Tanaka, T., Van't Hooft, F. M., Vinkhuyzen, A. A., Westra, H. J., Zheng, W., Zondervan, K. T., Heath, A. C., Arveiler, D., Bakker, S. J., Beilby, J., Bergman, R. N., Blangero, J., Bovet, P., Campbell, H., Caulfield, M. J., Cesana, G., Chakravarti, A., Chasman, D. I., Chines, P. S., Collins, F. S., Crawford, D. C., Cupples, L. A., Cusi, D., Danesh, J., de Faire, U., den Ruijter, H. M., Dominiczak, A. F., Erbel, R., Erdmann, J., Eriksson, J. G., Farrall, M., Felix, S. B., Ferrannini, E., Ferrieres, J., Ford, I., Forouhi, N. G., Forrester, T., Franco, O. H., Gansevoort, R. T., Gejman, P. V., Gieger, C., Gottesman, O., Gudnason, V., Gyllensten, U., Hall, A. S., Harris, T. B., Hattersley, A. T., Hicks, A. A., Hindorff, L. A., Hingorani, A. D., Hofman, A., Homuth, G., Hovingh, G. K., Humphries, S. E., Hunt, S. C., Hypponen, E., Illig, T., Jacobs, K. B., Jarvelin, M. R., Jockel, K. H., Johansen, B., Jousilahti, P., Jukema, J. W., Jula, A. M., Kaprio, J., Kastelein, J. J., Keinanen-Kiukaanniemi, S. M., Kiemeney, L. A., Knekt, P., Kooner, J. S., Kooperberg, C., Kovacs, P., Kraja, A. T., Kumari, M., Kuusisto, J., Lakka, T. A., Langenberg, C., Le Marchand, L., Lehtimaki, T., Lyssenko, V., Mannisto, S., Marette, A., Matise, T. C., McKenzie, C. A., McKnight, B., Moll, F. L., Morris, A. D., Morris, A. P., Murray, J. C., Nelis, M., Ohlsson, C., Oldehinkel, A. J., Ong, K. K., Madden, P. A., Pasterkamp, G., Peden, J. F., Peters, A., Postma, D. S., Pramstaller, P. P., Price, J. F., Qi, L., Raitakari, O. T., Rankinen, T., Rao, D. C., Rice, T. K., Ridker, P. M., Rioux, J. D., Ritchie, M. D., Rudan, I., Salomaa, V., Samani, N. J., Saramies, J., Sarzynski, M. A., Schunkert, H., Schwarz, P. E., Sever, P., Shuldiner, A. R., Sinisalo, J., Stolk, R. P., Strauch, K., Tonjes, A., Tregouet, D. A., Tremblay, A., Tremoli, E., Virtamo, J., Vohl, M. C., Volker, U., Waeber, G., Willemsen, G., Witteman, J. C., Zillikens, M. C., Adair, L. S., Amouyel, P., Asselbergs, F. W., Assimes, T. L., Bochud, M., Boehm, B. O., Boerwinkle, E., Bornstein, S. R., Bottinger, E. P., Bouchard, C., Cauchi, S., Chambers, J. C., Chanock, S. J., Cooper, R. S., de Bakker, P. I., Dedoussis, G., Ferrucci, L., Franks, P. W., Froguel, P., Groop, L. C., Haiman, C. A., Hamsten, A., Hui, J., Hunter, D. J., Hveem, K., Kaplan, R. C., Kivimaki, M., Kuh, D., Laakso, M., Liu, Y., Martin, N. G., Marz, W., Melbye, M., Metspalu, A., Moebus, S., Munroe, P. B., Njolstad, I., Oostra, B. A., Palmer, C. N., Pedersen, N. L., Perola, M., Perusse, L., Peters, U., Power, C., Quertermous, T., Rauramaa, R., Rivadeneira, F., Saaristo, T. E., Saleheen, D., Sattar, N., Schadt, E. E., Schlessinger, D., Slagboom, P. E., Snieder, H., Spector, T. D., Thorsteinsdottir, U., Stumvoll, M., Tuomilehto, J., Uitterlinden, A. G., Uusitupa, M., van der Harst, P., Walker, M., Wallaschofski, H., Wareham, N. J., Watkins, H., Weir, D. R., Wichmann, H. E., Wilson,

J. F., Zanen, P., Borecki, I. B., Deloukas, P., Fox, C. S., Heid, I. M., O'Connell, J. R., Strachan, D. P., Stefansson, K., van Duijn, C. M., Abecasis, G. R., Franke, L., Frayling, T. M., McCarthy, M. I., Visscher, P. M., Scherag, A., Willer, C. J., Boehnke, M., Mohlke, K. L., Lindgren, C. M., Beckmann, J. S., Barroso, I., North, K. E., Ingelsson, E., Hirschhorn, J. N., Loosand, R. J., & Speliotes, E. K. (2015). Genetic studies of body mass index yield new insights for obesity biology. *Nature, 518*(7538), 197–206.

MacRae, C. A., & Vasan, R. S. (2011). Next-generation genome-wide association studies: Time to focus on phenotype? *Circulation. Cardiovascular Genetics, 4*(4), 334–336.

Martin, G. M. (2007). Modalities of gene action predicted by the classical evolutionary biological theory of aging. *Annals of the New York Academy of Sciences, 1100*, 14–20.

Martin, G. M., Bergman, A., & Barzilai, N. (2007). Genetic determinants of human health span and life span: Progress and new opportunities. *PLoS Genetics, 3*(7), e125.

Matteini, A. M., Fallin, M. D., Kammerer, C. M., Schupf, N., Yashin, A. I., Christensen, K., Arbeev, K. G., Barr, G., Mayeux, R., Newman, A. B., & Walston, J. D. (2010). Heritability estimates of endophenotypes of long and health life: The Long Life Family Study. *The Journals of Gerontology Series A: Biological Sciences and Medical Sciences, 65*(12), 1375–1379.

Oeppen, J., & Vaupel, J. W. (2002). Demography. Broken limits to life expectancy. *Science, 296*(5570), 1029–1031.

Olshansky, S. J., Perry, D., Miller, R. A., & Butler, R. N. (2007). Pursuing the longevity dividend: Scientific goals for an aging world. *Annals of the New York Academy of Sciences, 1114*, 11–13.

Perls, T. T. (2006). The different paths to 100. *American Journal of Clinical Nutrition, 83*(2), 484S–487S.

Robine, J.-M. (2003). *Determining health expectancies*. Chichester/Hoboken: Wiley.

Rose, S. P. R. (2006). Commentary: Heritability estimates – Long past their sell-by date. *International Journal of Epidemiology, 35*(3), 525–527.

Scheen, A. J. (2005). Diabetes mellitus in the elderly: Insulin resistance and/or impaired insulin secretion? *Diabetes & Metabolism, 31*(Spec No 2), 5S27–25S34.

Schnebel, E. M., & Grossfield, J. (1988). Antagonistic pleiotropy – An interspecific drosophila-comparison. *Evolution, 42*(2), 306–311.

Sebastiani, P., Hadley, E. C., Province, M., Christensen, K., Rossi, W., Perls, T. T., & Ash, A. S. (2009). A family longevity selection score: Ranking sibships by their longevity, size, and availability for study. *American Journal of Epidemiology, 170*(12), 1555–1562.

Sheu, Y., Cauley, J. A., Wheeler, V. W., Patrick, A. L., Bunker, C. H., Ensrud, K. E., Orwoll, E. S., & Zmuda, J. M. (2011). Age-related decline in bone density among ethnically diverse older men. *Osteoporosis International, 22*(2), 599–605.

Sierra, F., Hadley, E., Suzman, R., & Hodes, R. (2009). Prospects for life span extension. *Annual Review of Medicine, 60*, 457–469.

Singh, G. K., & Yu, S. M. (1995). Infant mortality in the United States: Trends, differentials, and projections, 1950 through 2010. *American Journal of Public Health, 85*(7), 957–964.

Sivakumaran, S., Agakov, F., Theodoratou, E., Prendergast, J. G., Zgaga, L., Manolio, T., Rudan, I., McKeigue, P., Wilson, J. F., & Campbell, H. (2011). Abundant pleiotropy in human complex diseases and traits. *The American Journal of Human Genetics, 89*(5), 607–618.

Summers, K., & Crespi, B. J. (2010). Xmrks the spot: Life history tradeoffs, sexual selection and the evolutionary ecology of oncogenesis. *Molecular Ecology, 19*(15), 3022–3024.

Ukraintseva, S. V., & Yashin, A. I. (2003). Individual aging and cancer risk: How are they related? *Demographic Research, 9*(8), 163–196.

Ukraintseva, S. V., Arbeev, K. G., Michalsky, A. I., & Yashin, A. I. (2004). Antiaging treatments have been legally prescribed for approximately thirty years. *Annals of the New York Academy of Sciences, 1019*, 64–69.

UN. (2007). The population division DoEaSAUNS. World population prospects: The 2006 revision. www.un.org Available from: URL: http://www.un.org/esa/population/publications/wpp2006/wpp2006_ageing.pdf

Verbrugge, L. M., & Jette, A. M. (1994). The disablement process. *Social Science and Medicine, 38*(1), 1–14.

Vijg, J., & Suh, Y. (2005). Genetics of longevity and aging. *Annual Review of Medicine, 56*, 193–212.

Visscher, P. M., Brown, M. A., McCarthy, M. I., & Yang, J. (2012). Five years of GWAS discovery. *The American Journal of Human Genetics, 90*(1), 7–24.

Wang, K., Baldassano, R., Zhang, H., Qu, H. Q., Imielinski, M., Kugathasan, S., Annese, V., Dubinsky, M., Rotter, J. I., Russell, R. K., Bradfield, J. P., Sleiman, P. M., Glessner, J. T., Walters, T., Hou, C., Kim, C., Frackelton, E. C., Garris, M., Doran, J., Romano, C., Catassi, C., Van Limbergen, J., Guthery, S. L., Denson, L., Piccoli, D., Silverberg, M. S., Stanley, C. A., Monos, D., Wilson, D. C., Griffiths, A., Grant, S. F., Satsangi, J., Polychronakos, C., & Hakonarson, H. (2010). Comparative genetic analysis of inflammatory bowel disease and type 1 diabetes implicates multiple loci with opposite effects. *Human Molecular Genetics, 19*(10), 2059–2067.

Willcox, B. J., Donlon, T. A., He, Q., Chen, R., Grove, J. S., Yano, K., Masaki, K. H., Willcox, D. C., Rodriguez, B., & Curb, J. D. (2008a). FOXO3A genotype is strongly associated with human longevity. *Proceedings of the National Academy of Sciences of the United States of America, 105*(37), 13987–13992.

Willcox, D. C., Willcox, B. J., Wang, N. C., He, Q., Rosenbaum, M., & Suzuki, M. (2008b). Life at the extreme limit: Phenotypic characteristics of supercentenarians in Okinawa. *The Journals of Gerontology. Series A, Biological Sciences and Medical Sciences, 63*(11), 1201–1208.

Willer, C. J., Schmidt, E. M., Sengupta, S., Peloso, G. M., Gustafsson, S., Kanoni, S., Ganna, A., Chen, J., Buchkovich, M. L., Mora, S., Beckmann, J. S., Bragg-Gresham, J. L., Chang, H. Y., Demirkan, A., Den Hertog, H. M., Do, R., Donnelly, L. A., Ehret, G. B., Esko, T., Feitosa, M. F., Ferreira, T., Fischer, K., Fontanillas, P., Fraser, R. M., Freitag, D. F., Gurdasani, D., Heikkila, K., Hypponen, E., Isaacs, A., Jackson, A. U., Johansson, A., Johnson, T., Kaakinen, M., Kettunen, J., Kleber, M. E., Li, X., Luan, J., Lyytikainen, L. P., Magnusson, P. K., Mangino, M., Mihailov, E., Montasser, M. E., Muller-Nurasyid, M., Nolte, I. M., O'Connell, J. R., Palmer, C. D., Perola, M., Petersen, A. K., Sanna, S., Saxena, R., Service, S. K., Shah, S., Shungin, D., Sidore, C., Song, C., Strawbridge, R. J., Surakka, I., Tanaka, T., Teslovich, T. M., Thorleifsson, G., Van den Herik, E. G., Voight, B. F., Volcik, K. A., Waite, L. L., Wong, A., Wu, Y., Zhang, W., Absher, D., Asiki, G., Barroso, I., Been, L. F., Bolton, J. L., Bonnycastle, L. L., Brambilla, P., Burnett, M. S., Cesana, G., Dimitriou, M., Doney, A. S., Doring, A., Elliott, P., Epstein, S. E., Eyjolfsson, G. I., Gigante, B., Goodarzi, M. O., Grallert, H., Gravito, M. L., Groves, C. J., Hallmans, G., Hartikainen, A. L., Hayward, C., Hernandez, D., Hicks, A. A., Holm, H., Hung, Y. J., Illig, T., Jones, M. R., Kaleebu, P., Kastelein, J. J., Khaw, K. T., Kim, E., Klopp, N., Komulainen, P., Kumari, M., Langenberg, C., Lehtimaki, T., Lin, S. Y., Lindstrom, J., Loos, R. J., Mach, F., McArdle, W. L., Meisinger, C., Mitchell, B. D., Muller, G., Nagaraja, R., Narisu, N., Nieminen, T. V., Nsubuga, R. N., Olafsson, I., Ong, K. K., Palotie, A., Papamarkou, T., Pomilla, C., Pouta, A., Rader, D. J., Reilly, M. P., Ridker, P. M., Rivadeneira, F., Rudan, I., Ruokonen, A., Samani, N., Scharnagl, H., Seeley, J., Silander, K., Stancakova, A., Stirrups, K., Swift, A. J., Tiret, L., Uitterlinden, A. G., van Pelt, L. J., Vedantam, S., Wainwright, N., Wijmenga, C., Wild, S. H., Willemsen, G., Wilsgaard, T., Wilson, J. F., Young, E. H., Zhao, J. H., Adair, L. S., Arveiler, D., Assimes, T. L., Bandinelli, S., Bennett, F., Bochud, M., Boehm, B. O., Boomsma, D. I., Borecki, I. B., Bornstein, S. R., Bovet, P., Burnier, M., Campbell, H., Chakravarti, A., Chambers, J. C., Chen, Y. D., Collins, F. S., Cooper, R. S., Danesh, J., Dedoussis, G., de Faire, U., Feranil, A. B., Ferrieres, J., Ferrucci, L., Freimer, N. B., Gieger, C., Groop, L. C., Gudnason, V., Gyllensten, U., Hamsten, A., Harris, T. B., Hingorani, A., Hirschhorn, J. N., Hofman, A., Hovingh, G. K., Hsiung, C. A., Humphries, S. E., Hunt, S. C., Hveem, K., Iribarren, C., Jarvelin, M. R., Jula, A., Kahonen, M., Kaprio, J., Kesaniemi, A., Kivimaki, M., Kooner, J. S., Koudstaal, P. J., Krauss, R. M., Kuh, D., Kuusisto, J., Kyvik, K. O., Laakso, M., Lakka, T. A., Lind, L., Lindgren, C. M., Martin, N. G., Marz, W., McCarthy, M. I., McKenzie, C. A., Meneton, P., Metspalu, A., Moilanen, L.,

Morris, A. D., Munroe, P. B., Njolstad, I., Pedersen, N. L., Power, C., Pramstaller, P. P., Price, J. F., Psaty, B. M., Quertermous, T., Rauramaa, R., Saleheen, D., Salomaa, V., Sanghera, D. K., Saramies, J., Schwarz, P. E., Sheu, W. H., Shuldiner, A. R., Siegbahn, A., Spector, T. D., Stefansson, K., Strachan, D. P., Tayo, B. O., Tremoli, E., Tuomilehto, J., Uusitupa, M., van Duijn, C. M., Vollenweider, P., Wallentin, L., Wareham, N. J., Whitfield, J. B., Wolffenbuttel, B. H., Ordovas, J. M., Boerwinkle, E., Palmer, C. N., Thorsteinsdottir, U., Chasman, D. I., Rotter, J. I., Franks, P. W., Ripatti, S., Cupples, L. A., Sandhu, M. S., Rich, S. S., Boehnke, M., Deloukas, P., Kathiresan, S., Mohlke, K. L., Ingelsson, E., & Abecasis, G. R. (2013). Discovery and refinement of loci associated with lipid levels. *Nature Genetics, 45*(11), 1274–1283.

Williams, G. C. (1957). Pleiotropy, natural-selection, and the evolution of senescence. *Evolution, 11*(4), 398–411.

Williams, P. D., & Day, T. (2003). Antagonistic pleiotropy, mortality source interactions, and the evolutionary theory of senescence. *Evolution, 57*(7), 1478–1488.

Yashin, A. I., De Benedictis, G., Vaupel, J. W., Tan, Q., Andreev, K. F., Iachine, I. A., Bonafe, M., DeLuca, M., Valensin, S., Carotenuto, L., & Franceschi, C. (1999). Genes, demography, and life span: The contribution of demographic data in genetic studies on aging and longevity. *American Journal of Human Genetics, 65*(4), 1178–1193.

Yashin, A. I., Ukraintseva, S. V., De Benedictis, G., Anisimov, V. N., Butov, A. A., Arbeev, K., Jdanov, D. A., Boiko, S. I., Begun, A. S., Bonafe, M., & Franceschi, C. (2001). Have the oldest old adults ever been frail in the past? A hypothesis that explains modern trends in survival. *Journals of Gerontology. Series A, Biological Sciences and Medical Sciences, 56*(10), B432–B442.

Yashin, A. I., Akushevich, I. V., Arbeev, K. G., Akushevich, L., Ukraintseva, S. V., & Kulminski, A. (2006). Insights on aging and exceptional longevity from longitudinal data: Novel findings from the Framingham Heart Study. *Age, 28*(4), 363–374.

Yashin, A. I., Arbeev, K. G., Akushevich, I., Kulminski, A., Akushevich, L., & Ukraintseva, S. V. (2007a). Stochastic model for analysis of longitudinal data on aging and mortality. *Mathematical Biosciences, 208*(2), 538–551.

Yashin, A. I., Arbeev, K. G., Kulminski, A., Akushevich, I., Akushevich, L., & Ukraintseva, S. V. (2007b). Health decline, aging and mortality: How are they related? *Biogerontology, 8*(3), 291–302.

Yashin, A. I., Arbeev, K. G., & Ukraintseva, S. V. (2007c). The accuracy of statistical estimates in genetic studies of aging can be significantly improved. *Biogerontology, 8*(3), 243–255.

Yashin, A. I., Ukraintseva, S. V., Akushevich, I. V., Arbeev, K. G., Kulminski, A., & Akushevich, L. (2009). Trade-off between cancer and aging: What role do other diseases play? Evidence from experimental and human population studies. *Mechanisms of Ageing and Development, 130*(1-2), 98–104.

Yashin, A. I., Arbeev, K. G., Wu, D., Arbeeva, L. S., Kulminski, A., Akushevich, I., Culminskaya, I., Stallard, E., Ukraintseva, S. (2013a) How lifespan associated genes modulate aging changes: Lessons from analysis of longitudinal data. *Frontiers in Genetics,* 4:article 3.

Yashin, A. I., Arbeev, K. G., Wu, D., Arbeeva, L. S., Kulminski, A. M., Akushevich, I., Culminskaya, I., Stallard, E., Ukraintseva, S. (2013b) How the quality of GWAS of human lifespan and health span can be improved. *Frontiers in Genetics,* 4:article 125.

Yashin, A. I., Wu, D., Arbeev, K. G., Kulminski, A. M., Stallard, E., & Ukraintseva, S. V. (2013c). Why does melanoma metastasize into the brain? Genes with pleiotropic effects might be the key. *Frontiers in Genetics, 4*, 75.

Yashin, A. I., Wu, D., Arbeev, K. G., Arbeeva, L. S., Akushevich, I., Kulminski, A., Culminskaya, I., Stallard, E., & Ukraintseva, S. V. (2014). Genetic structures of population cohorts change with increasing age: Implications for genetic analyses of human aging and life span. *Annals of Gerontology and Geriatric Research, 1*(4), 1020.

CPSIA information can be obtained
at www.ICGtesting.com
Printed in the USA
LVOW04*1152240816
501658LV00008B/41/P

9 789401 775854